GENERATION AND UTILIZATION OF ELECTRICAL ENERGY

S. Sivanagaraju

Associate Professor

Department of Electrical Engineering, University College of Engineering

JNTU Kakinada, Andhra Pradesh

M. Balasubba Reddy

Professor and Head

Department of Electrical Engineering, Prakasam Engineering College

Kandukur, Prakasam District, Andhra Pradesh

D. Srilatha

Associate Professor

Department of Electrical Engineering, Prakasam Engineering College

Kandukur, Prakasam District, Andhra Pradesh

Pearson

Dedicated to our parents

Acquisitions Editor: Sojan Jose
Associate Production Editor: Jennifer Sargunar
Composition: Mukesh Technologies Pvt. Ltd, Pondicherry
Printer:

ISBN: 978-81-317-333-25

10 9 8 7 6 5 4 3 2 1

Published by Dorling Kindersley (India) Pvt. Ltd, licensees of Pearson Education in South Asia.

Head Office: 15th Floor, Tower-B, World Trade Tower, Plot No. 1, Block-C, Sector 16, Noida 201 301, Uttar Pradesh, India.

Registered Office: 7th Floor, SDB2, ODC 7, 8 & 9, Survey No.01 ELCOT IT/ ITES SEZ, Sholinganallur, Chennai – 600119, Tamil Nadu, India.
Phone: 044-66540100
Website: in.pearson.com, Email: companysecretary.india@pearson.com

Digitally printed in India by Manipal Technologies Limited. in the year of 2022

Brief Contents

Contents

Preface

Electrical energy has a wide range of applications. The generation and utilization of electrical energy play a vital role in many areas of science and technology. Hence, this book has been designed to be useful not only to the students pursuing courses in electrical power utilization but also for the practicing engineers and those who are preparing for competitive examinations. The book covers the revised syllabus of the generation and utilization of electric energy taught at various universities.

This book is divided into 12 chapters. The first two chapters deal with the various sources of conventional and non-conventional power generation.

Chapter 3 elucidates the need for energy conservation methods, power factor improvement, various tariff methods, and power quality. It also deals with the concept of distributed generation and deregulation.

Chapter 4 describes the various electrical heating methods, the design of heating elements, and materials for heating elements. Chapter 5 deals with the various welding methods, types of welding electrodes, and the comparison of the welding methods. Chapters 6 and 7 discuss the fundamentals of illumination, illumination methods, and the design of lighting schemes.

Chapter 8 explains the types of electric drives, the selection of motors for drive applications, the characteristics of motors, the speed control of DC and AC motors, and the load equalization concept. The existing electric traction systems in India, the different methods of electric braking, and the types of services and their speed–time curves are described in Chapters 9 and 10. The basic principle and the various applications of electrolysis are delineated in Chapter 11 while Chapter 12 talks about the concepts of refrigeration and air-conditioning.

The subject is presented systematically and the topics are explained with suitable examples and figures. At the end of each chapter, the basic concepts are highlighted as key points. The book is replete with objective questions, theoretical questions, and unsolved problems to fulfill the reader's requirements.

Acknowledgments

There are several people we would like to thank. First, we would like to thank Dr Kancharla Ramaiah, Correspondent and Secretary of Prakasam Engineering College, Kandukur, Prakasam District, for his encouragement and support and for providing us with the facilities for completing this book.

Second, we would like to thank the entire faculty, staff, and students at Prakasam Engineering College, Kandukur, for their support, collaboration, and friendship.

We thank all our friends who have been involved, either directly or indirectly, in the successful completion of this book.

We owe our parents, family members, and relatives a special word of thanks for their moral support and encouragement.

We also express our gratitude to the editors at Pearson Education, particularly to those who have taken the initiative to publish this book. We thank Thomas Mathew Rajesh, Sojan Jose, M. E. Sethurajan, and Jennifer Sargunar for their efforts in bringing out the book in time.

S. Sivanagaraju
M. Balasubba Reddy
D. Srilatha

About the Authors

S. Sivanagaraju is Associate Professor, Department of Electrical Engineering, University College of Engineering, JNTU Kakinada. He graduated in 1998, completed his master's degree in 2000 from IIT Kharagpur and received his Ph.D. from Jawaharlal Nehru Technological University in 2004. A recipient of two national awards (the Pandit Madan Mohan Malavya Memorial Prize Award and the Best Paper Prize Award) from the Institute of Engineers (India) for the year 2003–2004; he is the referee for IEE Proceedings—Generation Transmission and Distribution and International Journal of Emerging Electric Power Systems. He has published about 70 publications in the journals of national and international repute and an equal number of conferences to his credit.

M. Balasubba Reddy is Professor and Head, Department of Electrical Engineering, Prakasam Engineering College, Andhra Pradesh. After graduating in engineering from Madras University, Chennai in 2000, he proceeded to receive his M.Tech. degree from NIT, Trichy, Tamil Nadu in 2004. He is currently working toward his Ph.D. degree at JNTU Kakinada, Andhra Pradesh. His research interest lies in power electronics, facts controllers, and power system optimization using artificial intelligence. A life member of Indian Society for Technical Education (ISTE), Professor Reddy has co-authored books on power semiconductor drives, power electronics, and HVDC transmission system.

D. Srilatha is Associate Professor, Department of Electrical Engineering, Prakasam Engineering College, Andhra Pradesh. She received her B.Tech. degree from JNTU Hyderabad in 2004. She is currently pursuing her M.Tech. degree at JNTU Kakinada. Her areas of interest lie in electrical machines, control systems, and power systems. She has co-authored a book on HVDC transmission system.

CHAPTER 1

Conventional Power Generation

OBJECTIVES

After reading this chapter, you should be able to:

- know the hydropower generation and the classification of hydro plants
- understand the working of thermal and nuclear power plants
- have an idea about the generation of power from diesel engine plant

1.1 INTRODUCTION

Energy provides the power to progress. The natural resources of a country may be large but they can only be turned into wealth if they are developed, used, and exchanged for other goods. This cannot be achieved without energy. Availability of sufficient energy and its proper use in any country can result in the development of its people rising from subsistence level to the highest standard of living. Based on the availability of natural sources of energy, different power plants are erected. The energy of water is utilized for hydropower generation. In recent decades, hydraulic energy has widely been utilized as one of the primary sources of electrical power generation. A hydroelectric power plant is used to supply electrical energy to consumer, where water resources are available. Thermal power plants use heat energy produced from the natural coal. Nuclear plant uses fuel (diesel) and they can be located where fuel is cheaper than coal. Electrical energy can be generated from other natural sources of energies such as sun and wind. Production of electrical power from these plants is very clean, ease of control, etc.

1.2 HYDROPOWER GENERATION

In hydroelectric power stations, electrical energy is generated by converting the energy stored in the water. Thus, the water stored at a higher level (devotion) is made to impinge on the blades of a hydraulic turbine through a penstock to covert the potential energy and kinetic energy of water into mechanical energy. The mechanical energy thus generated is used to drive the generator coupled to the turbines to produce electrical power.

Hydroelectric stations can be usually located only at such places where water is available in abundance, more over at a reasonable head (difference in levels) throughout the year. The required information can be obtained from the records maintained in respect of the annual rainfall, runoff, dry years, frequency of dry years, etc. over a period of 25–30 years.

As electrical energy is generated by the use of water in the hydroelectric stations and as such there is no cost of fuel, it may appear that the hydroelectric power is very cheap. However, this is not the case:

- The storage of water at a reasonable head requires the construction of a dam and involves many civil engineering works.
- The stations are normally located in non-popular mountainous areas, far away from the load centers, thereby necessitating longer transmission network, etc.

Because of the civil engineering works involved, the fixed costs increase; however, the running costs are much less as compared to those of the thermal power stations. Further the hydroelectric power stations may be developed as an integral part of multipurpose projects, such as irrigation and power, flood control and power or flood control, navigation, and power projects.

1.2.1 Hydrology

For the successful operation of any hydroelectric project, a huge quantity of water must be available throughout the year. So, it is necessary to obtain the stream-flow data, and hence to estimate the yearly possible flow. This necessitates having some basic ideas pertaining to hydrology.

Hydrology or hydrography deals with the occurrence and distribution of water over and under the surface of the earth. Water is received on the surface of the land in three ways such as rain, hail, or snow. This is generally referred to as the precipitation and is part of the hydrological or water cycle. The water cycle consists of evaporation, precipitation, transpiration, etc. Thus, the heat of the sun causes the evaporation of water from the seas, oceans, and other water surfaces. This leads to the formation of moist air, clouds, and air currents and the condensation of water vapor. As a result, there is precipitation or rainfall. A part of the precipitation is lost due to evaporation from the water area, soil evaporation, and transpiration, i.e., transpiration from the surface of the leaves and the water absorbed by the vegetation in the area. When the loss of water due to the various causes is subtracted from the precipitation, we get the stream flow. The stream flow is made up of the surface flow and the percolation through the ground. The amount of water that joins a stream is called 'runoff'.

1.2.2 Stream flow, hydrographs, and flow–duration curves stream flow

Stream-flow data play a vital role in considering any hydroelectric power station. From the data collected at the proposed site over a long period, the average flow and the output power can be estimated. From a survey of the site, the head available can be determined.

The stream flow is normally non-uniform. Thus, the minimum or low-water flow data used to estimate firm power of a hydroelectric station. The maximum flow data provide the information necessary for estimating the floods and for designing the spillway. Further, the maximum stream-flow conditions help in arriving at the capacity of the flood control reservoir, the purpose of which is to limit the discharge to a predetermined safe value.

In order to maintain the flow at a given value, a storage reservoir is needed. The capacity of the storage reservoir can be estimated from the stream-flow data.

1.2.3 Hydrographs

A hydrograph is a plot of the discharge (on the *y*-axis), against time (on the *x*-axis) in the chronological order. The discharge can be expressed in terms of the gauge height, cubic meters per second per square kilometer, the power that can be developed theoretically corresponding to a fall of 1 m or the energy recorded at the switch board (in kWh or MWh). Similarly, the time may be expressed in hours, days, or weeks.

An inspection of the hydrograph provides the following information.

(i) Rate of flow at any point in time.

(ii) Variation of flow with time.

A hydrograph is useful:

(i) To determine the power available at different times of the day or year.

(ii) To determine the volume of the flow up to a given point of time by measuring the area under the hydrograph up to that time.

A hydrograph is similar to the load curve. To study the effect of storage on flow, a hydrograph is required.

Flow–duration curve

A plot of flows (daily, weekly, or monthly) (on the y-axis) against percentage time (on the x-axis) is called the flow–duration curve.

Whereas the flows are plotted as they occur, i.e., chronologically on the hydrograph, the flows are plotted against the percentage of time over which the flow was either equal to or greater than a particular flow in the case of a flow–duration and the maximum flow for a smaller percentage of time. Thus, let us suppose that we have 'n' monthly discharge readings. In these, let n_q readings indicate a discharge equal to or greater than a particular discharge, say Q cubic meters per second. Then, the percentage of time over which the discharge was either equal to or greater than Q will be $(n_q/n) \times 100\%$.

The flow–duration curve can be converted to the load–duration curve of a hydroelectric plant provided the head at which the plant operates is known.

In case storage is available on the up-stream side, the flow–duration curve will be altered.

A flow–duration curve is useful:

(i) To determine the primary power (form the low-water flow data).

(ii) To determine the time during which flow may occur.

(iii) In designing the spillway to allow the escape of floodwater.

1.2.4 Mass curve

Rainfall is different during different times of the year, so the river flow also will be different at different times of the year. In order to have a uniform discharge, the water may have to be stored by means of reservoir. Thus, if the water supply is in excess of the requirement in one season, it will be stored in the reservoir to augment the supply of water during the deficient periods. The capacity of the reservoir can be determined by making the use of a 'mass curve'.

It is a plot of the cumulative volume of water that can be stored from the stream flow (on the y-axis) against time (on the x-axis).

The time may be in days, weeks, or months. Though, theoretically, the volume of water stored is to be expressed in cubic meters, it is usually expressed in day–second–meters. A day–second–meter is volume of water corresponding to a flow at the rate of 1 m^3/s for one day.

i.e., 1 day–second–meter $= 1 \times 24 \times 60 \times 60 = 86{,}400$ m^3.

1.2.5 Advantages and disadvantages of hydroelectric plants

Since electrical energy is obtained from the water in the hydroelectric stations, obviously the operating costs are less. However, since the hydroelectric stations are usually to be located far from load centers, there is a considerable expenditure involved in laying the transmission network. The various advantages and disadvantages are listed below.

Advantages of hydroelectric station

(i) Since there is no cost of fuel as such, the operating costs of a hydroelectric plant including auxiliaries are considerably less than those in the case of a thermal power station.

(ii) Hydroelectric stations do not require the purchase, transportation, and storage of large quantities of fuel as in the case of thermal stations.

(iii) There is no necessity of fuel- and ash-handling equipment.

(iv) There is no air pollution and other environmental problems.

(v) The cost per kWh of a hydroelectric station is not considerably affected by the load factor, as in the case of a thermal station.

(vi) The maintenance costs of a hydroelectric station are minimal.

(vii) Hydraulic turbines are robust. They run at low speeds of the order of 3,000–400 r.p.m., so there are no specialized mechanical problems as in the case of steam turbines, which run at 3,000 r.p.m.

(viii) The efficiency of a hydroelectric plant does not change with age.

(ix) Hydroelectric plants can respond more quickly to load changes than thermal plants.

(x) The plants are simple in construction and robust. They have a life period of 100–125 years.

(xi) Though large number of engineers and skilled workers are required during the construction phase, only a few of them are sufficient for operating the plant. Thus, plant-running cost is less.

(xii) The plants are quite neat and clean.

(xiii) A single unit of a very high output can be used.

(xiv) The water used for running the turbines may also be used for such purpose as irrigation, etc.

(xv) The cost of the land is low, since hydroelectric stations are situated far away from populated areas.

Disadvantages of hydroelectric plants

(i) Hydroelectric plants require huge quantities of water. As rainfall is at the mercy of nature, long dry seasons affect the delivery of power.

(ii) Since many civil engineering works are involved, it takes a long time for the erection of a hydroelectric plant.

(iii) As the sites for hydroelectric stations are usually far away from the load centers, the cost of transmission lines is high.

(iv) The capitals cost of generators is usually high.

1.2.6 Selection of site for hydroelectric plants

The following are the points to be considered for the selection of site for hydroelectric power station.

(i) Abundant quantity of water at reasonable head must be available.

(ii) It must be possible to construct an economical dam.

(iii) Transport facilities for workers and material must be made available, i.e., the site should easily be accessible.

(iv) Availability of labor at a cheaper rate.

(v) It should allow strong foundation with low cost.

(vi) Sittings reduce the reservoir capacity. So, the rate of sitting should not be high.

(vii) Structures of cultural or historical importance should not be damaged.

(viii) There should be no possibility of future sources of leakages of water.

(ix) A large catchments area must be available.

(x) During the construction period, it should be possible to divert the stream.

(xi) Sand, gravel, etc., should be available nearby.

1.2.7 Water power equation

In hydroelectric power station, the energy stored in the water is first converted into mechanical energy, which is used to drive the turbines to which the generators are coupled. Thus, the power developed at hydroelectric plant depends upon:

(i) the head, H (in m) and

(ii) the discharge, Q (in m^3/sec.).

We know that work done by 1 kg of water as it falls though a height of H m $=1$ (kg) $\times$ H (m) $= H \cdot$ kg-m, if the final velocity of water is zero.

Again, water discharge at a rate of Q m^3/sec, which corresponds to ($Q \times 1{,}000$) kg/sec, where 1,000 represents the weight of 1 m^3 of water.

So, the theoretical work done per second, as water falls at the rate of Q m^3/sec form a height of H m.

$$P = 1{,}000\, Q \cdot H \cdot \text{kg-m/sec.}$$

If η is the efficiency of the turbine-alternator set, the effective work done/sec:

$$P = 1{,}000\, Q \cdot H \,\text{kg-m/sec}$$

$$= \frac{1{,}000\, QH\eta}{75}\; H \cdot P \;(\because 1\, H \cdot P = 75 \text{ kg-m/sec})$$

$$= \frac{1{,}000\, QH\eta}{75} \times \frac{735.5}{1{,}000}\ \text{kW} = 9.81\, QH\eta \,\text{kW}.$$

Thus, the power output in kW $= 9.81 QH\eta$ kW.

Note: In the above equation, H is the effective head, i.e., the head available after loss of head in penstocks due to friction is taken into consideration.

1.2.8 Classification of hydroelectric plants

Hydroelectric plants are classified on different bases. Thus, they are classified according to:

(i) Head of water available.

(ii) Nature of load supplied.

(iii) Regulation of water flow.

(i) Classification according to head of water available

(a) Low-head plants

If the available water head is less than 30 m, the plant is called a low-head plant. The necessary head is created by construction of a dam or barrage. The power plant is situated near the dam. Regulating gates are provided to discharge the surplus of water. Kaplan turbines may be used. The only disadvantage is that the power output is reduced when the discharge increases as it causes an increase in the downstream water level, with a consequent reduction in the effective head. Structure of such plants is extensive and expensive. Generators used in these plants are of low speed and large diameter. Figure 1.1 shows a low-head installation.

(b) Medium-head plants

If the available water head is between 30 and 100 m, the plant is called a medium-head plant. In these plants, water is brought from the main reservoir through an open channel to the forebay. Water is led to the turbines from the forebay by the penstocks, which may be steel pipes. Forebay also stores the rejected water as the load on the turbine decreases. Francis turbines are normally used. Figure 1.2 shows a medium-head installation.

(c) High-head plants

If the available head is more than 100 m, the plant is called high-head plant. The civil works include a surge tank, the function of which is to meet the sudden changes in the

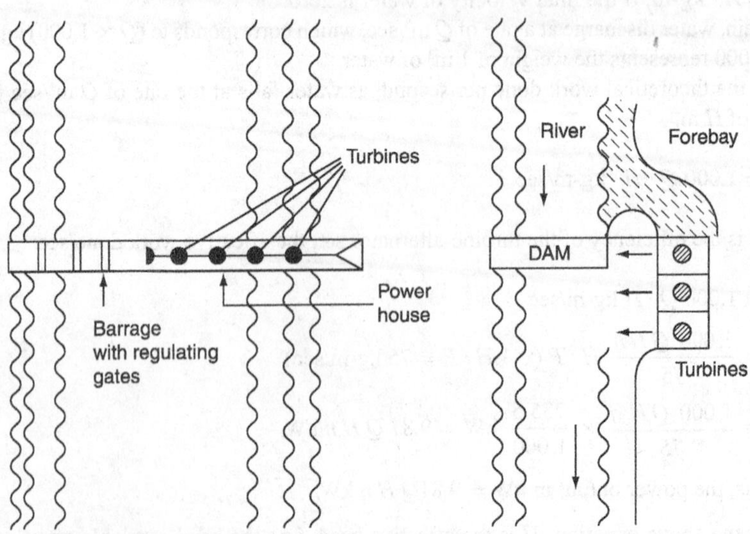

FIG. 1.1 *Low-head plant*

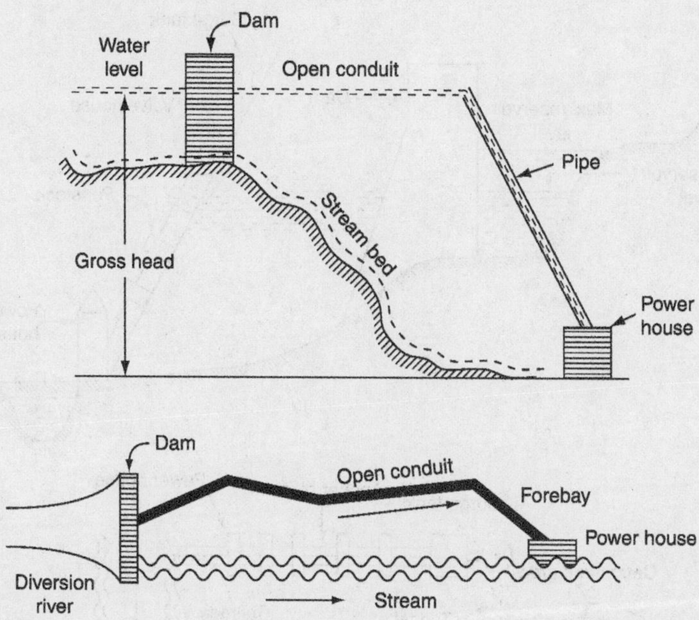

FIG. 1.2 *Medium-head plant*

requirement of water caused by the fluctuations in the system load. For heads less than 200 m, Francis turbines are used, while for higher heads, Pelton turbines are used. A pressure tunnel brings the water from the reservoir to the value house at the start of the penstocks. The generators used are of high head and small diameter. Penstocks are of large length and comparatively smaller cross-section. Figure 1.3 shows a high-head installation.

(ii) *Classification according to nature of load supplied*

Figure 1.4 shows the daily load curve of a particular system. A single plant designed to carry the entire load will have a low-load factor. So, the load is divided into two parts. They are base load and peak load. Base load is present for most of the day, while the peak load persists only for smaller period. So, the load may be supplied by two plants, one supplying the base load and the other the peak load; hence, the plants are classified as base-load plants and peak-load plants.

(a) Base-load plants

These supply the base load of the system so that the load on the plants is almost constant and hence the load factor is very high. The capacity of these plants is usually very high. Runoff river plants are without pondage or reservoir. Plants are used as base-load plants. The cost per kWh generated should be low in order that the plant be used as a base-load plant.

(b) Peak-load plants

These plants supply the peak load of the system. Reservoir plants can be used as peak-load plants. Further, runoff river plants with pondage can be operated as peak-load plants

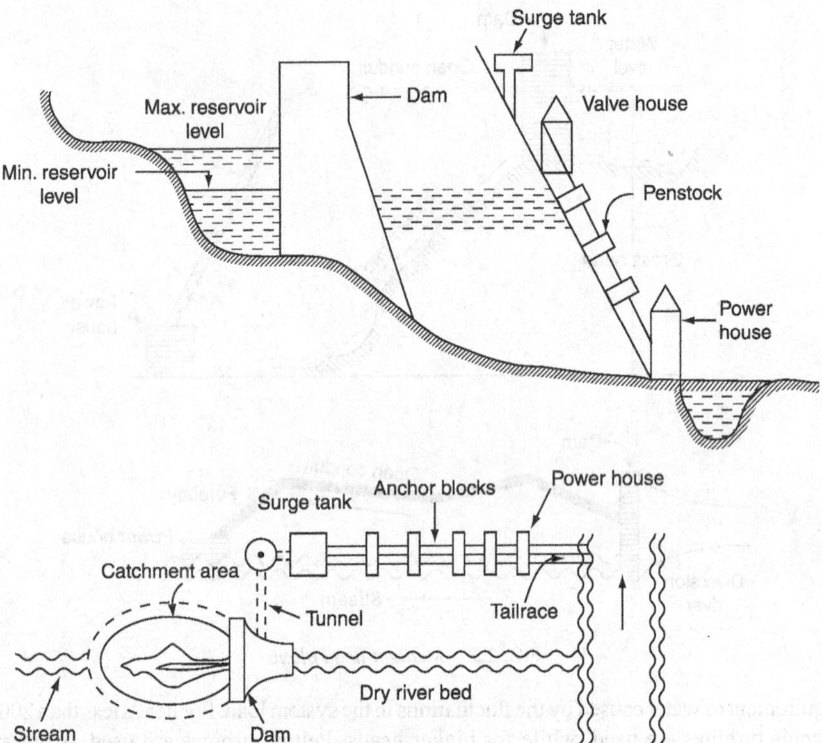

FIG. 1.3 *High-head plant*

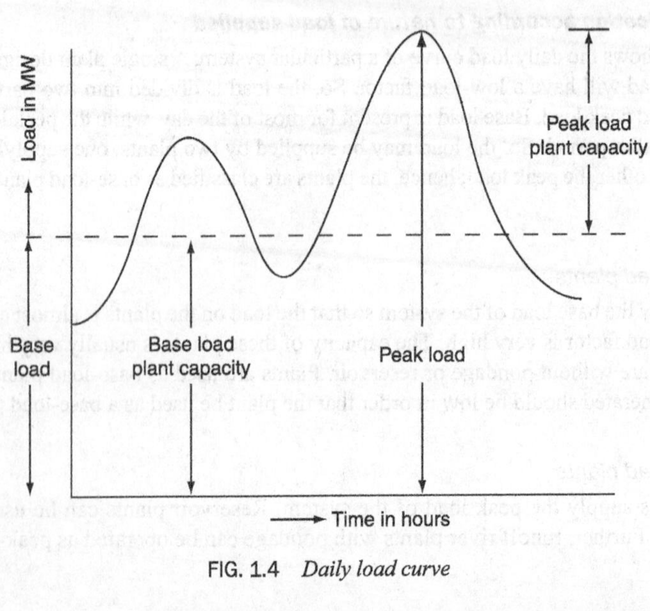

FIG. 1.4 *Daily load curve*

during the periods of lean flow. The storage of water is an essential feature of the peak-load plants. Water is stored during the off-peak period. The load factor of the peak-load plant is lower.

Pumped-storage plants also fall under the category of the peak-load plants.

Pumped storage plants

The schematic diagram of a pumped storage plant is shown in Fig. 1.5. Pumped storage plants have a small headwater pond, in addition to a tail water pond. During the peak-load period, water is drawn down from the headwater pond through the penstock to generate electric power. The water accumulated in the tail-water pond is pumped back to the headwater pond during the off-peak period. In the earlier days, the pumping was done by a separate pump. However nowadays, reversible turbine pump is used for the purpose. Thus, during the peak-load period, the turbine drives the alternator to generate electrical energy. During the off-peak period, the alternator acts as a motor deriving its power from the supply mains to drive the turbine as a pump to pump the water from the tail water pond to the head-water pond. So, the same water is used again and again to generate electrical energy. However, to take care of evaporation and seepage, some extra water is needed. The off-peak pumping helps maintain the firm capacity of the pumped storage plant.

The capacity of the reservoir should be adequate so as to enable the plant of supply the peak load for 4–11 hours.

As said earlier, during the off-peak period, the motor has to receive its power supply from the power system, which is a mixture of hydro-thermal, and nuclear power stations. The excess energy generated by steam and nuclear plant is used to drive the motor for pumping water to the headwater pond. This will result in an increase of the load factor of the steam and nuclear power stations thereby ensuring the most economic operation.

Advantages of the pumped storage plants

The following are some of the advantages of the pumped storage plants:

- Since the same water is used again, peak loads can be supplied at a cost less than that if the peak loads were to be supplied by steam or nuclear power plants.

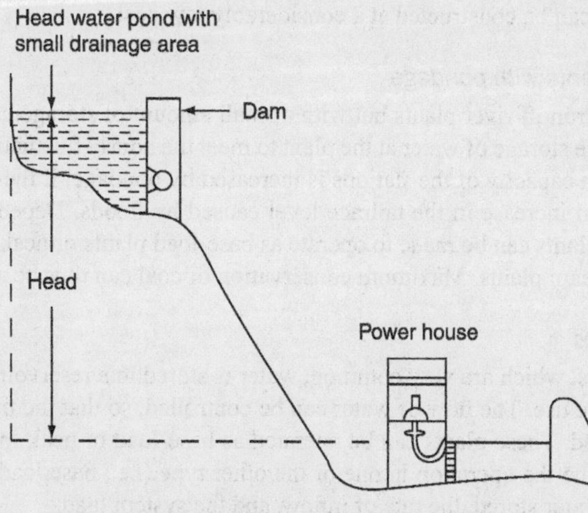

FIG. 1.5 *Schematic diagram of a pumped storage plant*

- Pumped-storage plants can pick up the load very quickly. In case of necessity, they can be started within 2 or 3 sec and can be loaded to their capacity in about 15 sec. So, they provide standby capacity on short notice.
- The excess energy generated by steam and nuclear plants during the off-peak load is utilized to drive the motors in the pumped storage plants. Consequently, the load factor of the steam and nuclear stations are improved, which contributes to their economic operation.
- The forced and maintenance outages of the base-load stations are reduced.
- The spinning reserve is reduced, since the pumped storage plants can pick up the load very quickly.
- They can be used for load frequency control.

(iii) *Classification according to regulation of water flow*

Depending upon the water flow regulation, hydroelectric plants can be classified as:

(a) Runoff river plants without pondage.

(b) Runoff river plants with pondage.

(c) Reservoir plants.

(a) Runoff river plants without pondage

The flow of water is affected by the rainfall. Thus, the flow is high when the rainfall is more and low when the rainfall is less.

In the runoff river plants without pondage, no efforts are made to regulate or control the flow of water. Water is used as it comes. Normally, in this type of plants, the generation of electrical energy is only incidental. The water may be used for such other purposes as irrigation or navigation. During high-flow periods, a substantial portion of the base load is supplied, with a consequent saving of coal which would have been otherwise required by the thermal plants. It may happen that the water is wasted during low-load periods. Further, the firm capacity of the plant is low, since the power generated during the low-flow period is low. Such plants can be constructed at a considerably low cost.

(b) Runoff river plants with pondage

These are basically runoff river plants but with a small amount of storage called pondage. Pondage refers to the storage of water at the plant to meet the hourly fluctuations of load on the station. The firm capacity of the stations is increased by pondage, if the effective head is not reduced by an increase in the tailrace level caused by floods. Depending upon the stream flow, these plants can be made to operate as base-load plants or peak-load plants in conjunction with steam plants. Maximum conservation of coal can thus be accomplished.

(c) Reservoir plants

In this type of plants, which are very common, water is stored in a reservoir behind a dam to be put to effective use. The flow of water can be controlled, so that the firm capacity of the plant is increased. These plants can be operated as base-load or peak-load plants. The factors that determine the operation in one or the other type (i.e., base load or peak load) are the amount of water stored, the rate of inflow, and the system load.

1.2.9 Function of the various components in a hydroelectric generation system

The various components in a hydroelectric generation system include:

(a) storage reservoir,
(b) dam,
(c) forebay,
(d) intake,
(e) surge tank,
(f) penstocks,
(g) spillway, and
(h) tail race.

A brief description of the various components and their functions are given below.

(a) Storage reservoir

The runoff from the rivers will be different during different seasons of the year. During rainy seasons, the runoff is high and during dry seasons it is low. To put the water to the most effective use, it becomes necessary to store the water during the rainy season when there is excess flow so that the same can be used during the periods of lean flow. This necessitates the development of a storage reservoir to help the required quantity of water to be supplied to the turbines in order that the required power can be developed by the plant.

The capacity of the storage reservoir, which can be determined from the mass curve, depends upon the difference between the maximum and the minimum runoff encountered during the high- and lean-flow periods, respectively. Low-head plants require a reservoir of a large capacity.

(b) Dam

In order to store the water and create an artificial head, a dam to be constructed. It is a highly expensive and the most important part of a hydroelectric plant. There are several types of dams, such as:

(i) masonary dams (solid gravity concrete dam, arch dam, and buttress dam),
(ii) earth dams, and
(iii) rock fill dams.

The factors that influence the type of the dam at a particular site are topography of the site, geological conditions, and subsoil conditions. The dams should be safe and economical besides having an esthetic appearance.

(c) Forebay

The water flowing from the dam is received by an enlarged body of water at the intake. It is called the forebay and it is intended to provide the temporary storage of water to meet the hour-to-hour load fluctuations on the station. The enlarged section of a canal or a pond, capable of accommodating the necessary widths of the intake, can serve the purpose of a forebay.

(d) Intake

The passage to water to the penstock, channel, or water conduit is provided by the intake. The intake structure should prevent the entry of debris and ice into the turbines. So, it is to be provided with trash racks, screens, and booms.

Intake structures are of two types: high pressure and low pressure. If the storage reservoirs are big, the high-pressure intake structures are used. In the case of ponds provided to store water to meet daily or weekly load fluctuations, the low-pressure intake structures can be used.

(e) Surge tank

The power output of a generator at a particular hydroelectric power plant is directly proportional to the discharge, i.e., $P \propto Q$ and the load on the system varies so that the load on the generator goes on fluctuation. This requires that the water intake to the turbine be regulated accordingly. Thus, when the load on the alternator is reduced, the governor closes the turbine gates. This sudden closure of the turbine gates causes an increase in the pressure in the penstock. This is referred to as water hammer. Similarly, an increased load on the alternator causes the governor to open the turbine gates to allow more water. This sudden opening of the turbine gates has a tendency to cause a vacuum in the penstocks. Both the water hammer and the negative pressure (vacuum) are detrimental to the proper functioning of the penstocks and are to be avoided. A surge tank is used to take care of these sudden changes in the water requirements and the consequent water hammer of vacuum.

In Fig. 1.6, a surge tank is shown. It acts as a relief value by allowing sufficient quantity of water to flow into or out of the surge tank. A reduction in load demand allows water to flow into the surge tank, thereby raising the water level. So, a retarding head is created and the velocity of water in the penstock is decreased. Similarly, an increased demand causes the water to flow out of the surge tank. This reduces the water level in the surge tank.

So, an accelerating head is created which increases the flow in the penstocks. This prevents the negative pressure (vacuum) to be created in the penstock. Thus, the surge tank is useful in stabilizing the velocity and pressure in the penstock, thereby reducing the water hammer and the vacuum.

The surge tank is to be located as near the powerhouse as possible.

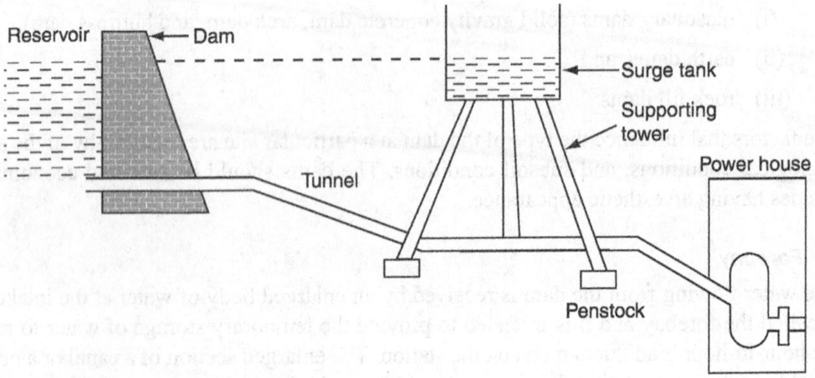

FIG. 1.6 *Surge tank*

(f) Penstock

It is a conduit system for taking water from the intake works and forebay to the turbines. These are two types and they are low- and high-pressure types. The low-pressure type consists of a canal, a flume, or a pipe line. The high-pressure type consists of steel pipe which can take the water under pressure. A penstock may be buried below the surface of the earth or it may be exposed. Penstock pipes are generally of steel for high- and medium-head plants and concrete in low-head plants. Each turbine will have its own penstock.

(g) Spillway

During floods, there will be excess water. This is to be discharged without causing any damage to the dam and allowing a predetermined head to be maintained. It will be acting as a safety valve for dam. For this purpose, a spillway which may be of the types: overflow, chute, side-channel, shaft, and siphon spillways. Alternatively, a bypass tunnel or a conduit may be used.

(h) Tailrace

The water after running the turbine is to be discharged into the river. For this purpose, a tailrace is required. Some turbines require a draft-tube while others do not. If a draft tube is used, it must be water sealed all the time. Impulse turbines can discharge the water directly into the tailrace. The tailrace should allow the free exit of water and an unimpeded passage to the jet of water leaving the turbine.

1.2.10 Location of a hydroelectric station

Generally, the hydroelectric stations are to be located at the foot of the dam and near the storage reservoir. This arrangement results in a reduction in the length of penstocks and a corresponding reduction in power loss. Further, the loss of head due to the friction in the penstocks is reduced.

The hydroelectric stations may also be located under ground. It results in safety to the installation. However, factors such as the cost of excavation, cost of tunneling, savings due to the reduction in the length of the penstock and its thickness, and the increase in the available head at the turbines are to be critically examined before arriving at the conclusion.

1.2.11 Working principle of a hydroelectric plant

The water available at a reasonable head from the river or the reservoir behind the dam is received by the intake works and the forebay, from where it is allowed to flow under pressure through the penstocks to run the turbines. In the reaction turbines, the water led to the turbine through a scroll case or scroll flame strikes the turbine vanes. It is let out through a draft tube into the tailrace without any loss of pressure. To allow the requisite quantity of water to cope up with the varying load demand, control gates are operated by a governor with the help of servo-mechanism and oil pressure system.

In the case of high-head installations, impulse turbines are used to convert the pressure head into velocity head by the nozzles at the admission of water into the turbines. The water impinging on the buckets of the runner causes the motion. After the work has been done, the water is let out into the tailrace. No draft tube is required as in the case of the reaction turbines. By varying the nozzle-opening with the help of a governor activated by a servo-mechanism, the required quantity of water can be made to impinge on the buckets of the runner of the turbine.

Reaction units are generally vertical; to arrange the draft tube etc., the power station requires many substructure and superstructure. However, in the case of impulse units, no substructure is necessary. Further, these units allow both the horizontal and vertical configurations.

The generators driven by the turbines produce the electric power. The speed of the turbine-generator set depends upon the head, specific speed of the turbine, and the power of the unit.

1.3 THERMAL POWER STATIONS

In a thermal power plant, the heat energy obtained by burning the coal in a boiler is used to raise the steam. The steam thus produced runs a steam turbine to which is coupled the alternator, which generates electrical energy. Thus, in a steam station, the boiler, the steam turbine, and the alternator constitute the main equipment. The efficient conversion of heat energy into electric energy requires a lot of auxiliary equipment. An enormous quantity of coal is required for the operation of a thermal plant. So, there must be an ample storage of coal and the coal-handling plant is required. Sometimes, the coal is used in the form of a fine powder and for this purpose, a pulverizing plant is required. There are the induced draft (I.D.) and forced draft (F.D.) fans to provide the air necessary for the combustion of coal. When coal is burnt in the boilers, large quantity of ash is produced so that there must be an ash-handling plant. Further, to deal with the flue gases, separate arrangements are required. To extract the heat from the flue gases, there will be economizers, air preheaters, etc. In addition, there will be a protection and control equipment.

1.3.1 Principle of working of a thermal power station

As discussed above, the boiler (or the steam generator), the steam turbine, and the alternator are the most important equipment in a thermal station. The coal-handling plant supplies coal to the boiler. The coal is burnt in the boiler and generates the heat that is used to convert water into steam at the required pressure and temperature, which is attained by further heating the steam in the superheater. The steam is then fed to the high-pressure turbine. The expansion of steam in the turbine produces the mechanical power at the shaft to which the alternator is coupled. The mechanical energy input to the alternator is finally converted into electric power.

As coal is burnt in the boiler, ash is formed. This is disposed off by the ash-handling plant. The air required for the combustion is taken from the atmosphere by using the F.D. or I.D. fans. Before being fed to the boiler, the air is heated in air preheater by flue gases, which are at high temperature. The flue gases are discharged to the atmosphere through the chimney after passing through the dust collector, air preheater, and economizer. The exhaust steam from the turbine is condensed by the condenser. The condensated, together with the make-up, water is passed through the economizer and then fed to the boiler and so on.

1.3.2 Factors to be considered for locating a thermal plant

The ideal location of a thermal plant is at the center of gravity (C.G.) of the load. If located at the C.G. of the load, the length of the transmission lines and the cables will be low, so that the capital cost is reduced. The C.G. of the load can easily be determined as follows:

Let X and Y be taken as the two reference lines, so that OX and OY are the axes.

Let $S_1, S_2, \ldots, S_n$ be the main loads in kVA, located at point $(x_1, y_1), (x_2, y_2), \ldots (x_n, y_n)$, respectively, on the x–y plane. Let $Q(x, y)$ be the C.G. of the loads. Then:

$$X = \frac{x_1 S_1 + x_2 S_2 + \cdots + x_n S_n}{S_1 + S_2 + \cdots + S_n} = \frac{\sum_{i=1}^{n} x_i S_i}{\sum_{i=1}^{n} S_i}$$

and $$Y = \frac{y_1 S_1 + y_2 S_2 + \cdots + y_n S_n}{S_1 + S_2 + \cdots + S_n} = \frac{\sum_{i=1}^{n} y_i S_i}{\sum_{i=1}^{n} S_i}.$$

Though the C.G. of the load is the ideal location for the thermal station; yet, it may not be possible to locate it at the C.G. of the load. Let us suppose that the C.G. of the load corresponds to a location in the heart of a city. Now, two important problems may arise: (i) the required extent of land may not be available and (ii) even if available, the cost of the load may be very high. In the latter case, the fixed costs shoot up abnormally and the cost per kWh may be very high. Besides this financial consideration, the location of a thermal station in the heart of the city may lead to atmospheric pollution due to ash and may be a source of nuisance to the public because of the noise. So, the choice of site for a thermal station takes the following points into consideration.

(i) A large extent of land is required for the erection of thermal plant. So, the cost of the land has a considerable bearing on the working of a thermal plant. So, the cost of the site should be reasonable.

(ii) The private land should be as minimum as possible.

(iii) The operation of a thermal plant requires huge quantities of water. So, it is preferable to have the site near the canal or a river.

(iv) Facilities should exist for the transport of fuel.

(v) The soil should not be too loose or too rocky.

(vi) The site should be level. There should be no excavation nearby.

(vii) The site should be far away from the residential localities so as to avoid the nuisance of smoke, noise, etc.

(viii) Future extensions of the power station should be possible.

(ix) Sufficient land must be available nearby the power station to build the residential accommodation to the operation and maintenance staff.

(x) Ash disposal should not create any problem.

(xi) To the extent possible, the thermal station should be far away from an aerodrome.

(xii) If canal or river water is used, it should not be polluted to ensure that the interests of the other users are not affected.

(xiii) The design should be in conformity with the by-laws of the land and the town planning.

(xiv) The interests of national defense must be served.

1.3.3 Schematic diagram of thermal power station

The schematic diagram of a thermal power station is shown in Fig. 1.7 and is explained briefly as follows.

In a thermal station, the fuel burnt may be a solid, a liquid, or a gaseous fuel. The solid fuels maybe bituminous coal, peat, or brown coal. Figure 1.7 depicts a coal-fired thermal power station. The coal stored in the coal storage yard is conveyed by the coal-handling plant to a high-pressure boiler.

To increase the efficiency of the boiler, the coal may be ground into a fine powder. In the boiler house, coal is burnt to convert water into high-pressure steam. The steam passes through a superheater to get superheated and then passes into the turbine to rotate the blades of the turbine. Thus, the heat energy obtained by burning the coal is converted into mechanical energy. This mechanical energy causes the rotation of the alternator, since turbine is coupled to the turbine. Thus, electrical energy is generated. There is a step-up transformer, together with its circuit breaker, etc., which enables the alternator to feed the bus-bars.

Again, the superheated steam which passes into the turbine imparts energy to the turbine rotor. In this process, the pressure decreases and the volume increases. Afterwards,

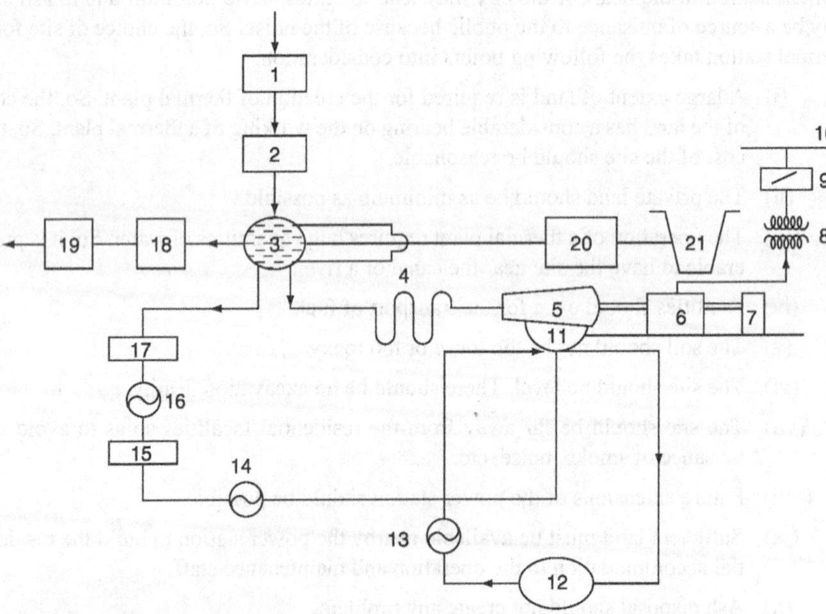

1. Coal storage
2. Coal handling
3. Boiler
4. Super heater
5. Turbine
6. Alternator
7. Exciter
8. Transformer
9. Circuit breaker
10. Bus bars
11. Condenser
12. Cooling tower
13. Cold-water circulating pump
14. Condensate extraction pump
15. Low-pressure feed-water heater
16. Boiler feed water pump
17. High-pressure feed-water pump
18. Ash-handling plant
19. Ash storage
20. Air heater
21. Chimney

FIG. 1.7 *Schematic diagram of a thermal power station*

it passes into the condenser. Cold-water circulating pump circulates water is extracted by the condensate extraction pump and is fed to the low-pressure water heater, where the low-pressure steam increases the temperature of the feed water. It is then heated in the high-pressure heater, where high-pressure steam is used for heating. The method of taking out steam from the turbines for feed-water heating is called '*bleeding*' of the turbines. This increases the overall efficiency of the boiler. The cooling towers are used to cool the water coming out of the condenser, which is rather too hot.

The ash formed after the combustion of coal is removed by the ash-handling plant and is transferred to the ash dump or ash storage, from where it is subsequently disposed.

Air is supplied to the combustion chamber of the boiler through F.D. fans and I.D. fans. The dust from the air is removed first and the air is passed through air preheater, where it is heated by the flue gases before it enters the combustion chamber. The exhaust gases after heating the incoming air are passed through dust collector and then led into the atmosphere through the chimney.

1.3.4 One-line diagram of thermal station indicating the various circuits

While the schematic diagram shown in Fig. 1.7 is helpful in understanding the general arrangement of the various components in a thermal station, Fig. 1.8 is helpful in analyzing

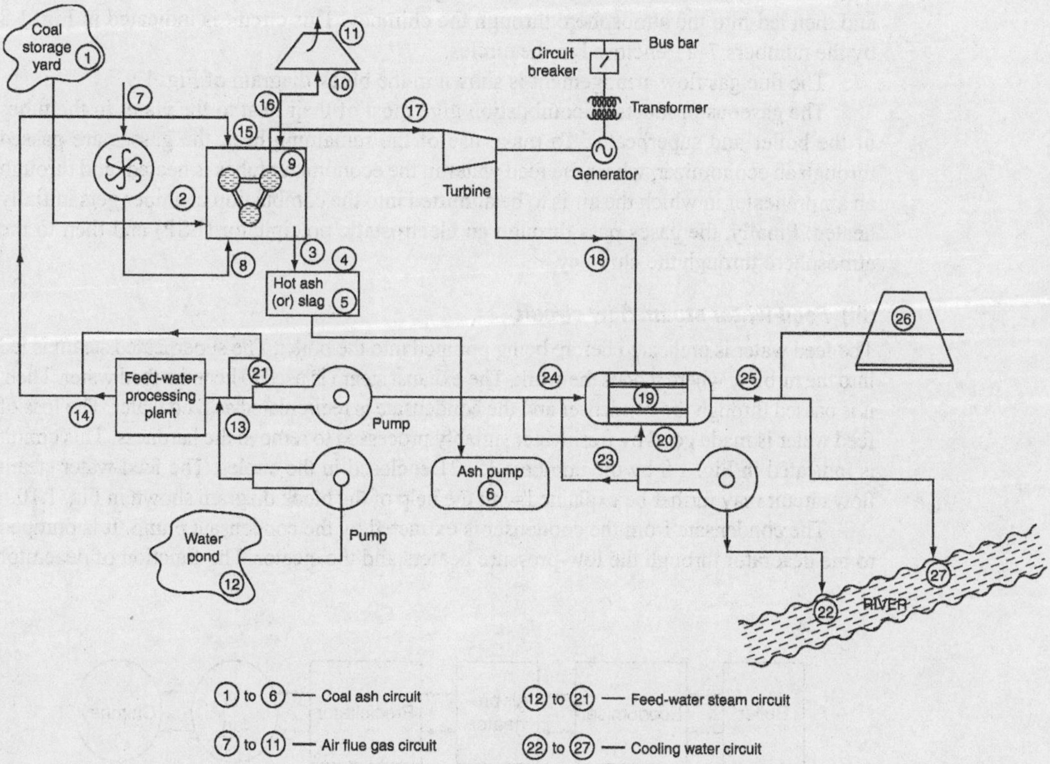

FIG. 1.8 *Flow diagram of thermal power station*

the various circuits into which a thermal station can be split up. A modern thermal station may be assumed to be consisting mainly of the following circuits:

(i) Coal and ash circuit.

(ii) Air and flue gas circuit.

(iii) Feed-water and steam-flow circuit.

(iv) Cooling water circuit.

(i) *Coal and ash circuit*

Coal from the coal storage yard is fed to the furnace of the boiler through the coal-handling equipment consisting of conveyors. The ash formed after combustion is removed and transferred to the ash dump. The coal and ash circuit is indicated in Fig 1.8 by the numbers 1–6 enclosed in the circles.

(ii) *Air and flue gas circuit*

Air is required for the combustion of the fuel. It is normally supplied to the combustion chamber of the boiler with the help of F.D. and I.D. fans in addition to the natural draft produced by the chimney. The dust from the air is removed before it is passed through the air preheater, where it is heated by the flue gases before it enters the combustion chambers. The exhaust gases after heating the incoming air are passed throughout the dust collectors and then led into the atmosphere through the chimney. This circuit is indicated in Fig. 1.8 by the numbers 7–11 enclosed in the circles.

The flue gas flow arrangement is shown in the block diagram of Fig. 1.9.

The gaseous products of combustion give most of their heat to the water in the tubes of the boiler and superheater. To make use of the remaining heat, the gasses are passed through an economizer, where the feed water in the economizer tubes is heated; and through an air preheater in which the air is to be admitted into the combustion chamber gets initially heated. Finally, the gases pass through an electrostatic precipitator (ESP) and then to the atmosphere through the chimney.

(iii) *Feed-water steam-flow circuit*

The feed water is preheated before being pumped into the boiler. The superheated steam is led into the turbine, where it does the work. The exhaust steam is used to heat the feed water. Then, it is passed through the condenser and the condensate is recirculated as feed water. The loss of feed water is made good by freshwater suitably processed to remove the hardness. This circuit is indicated in Fig. 1.8 by the numbers 12–21 enclosed in the circles. The feed-water steam flow circuit may further be explained with the help of the block diagram shown in Fig. 1.10.

The condensate from the condenser is extracted by the condensate pump. It is pumped to the deaerator through the low-pressure heaters and the ejector. The function of deaerator

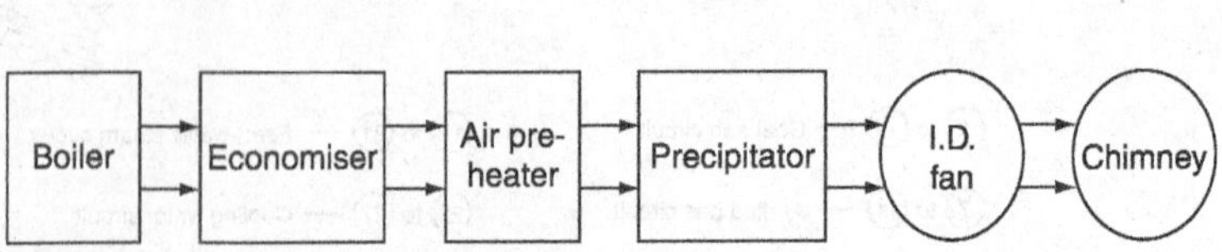

FIG. 1.9 *Flue gas flow arrangement*

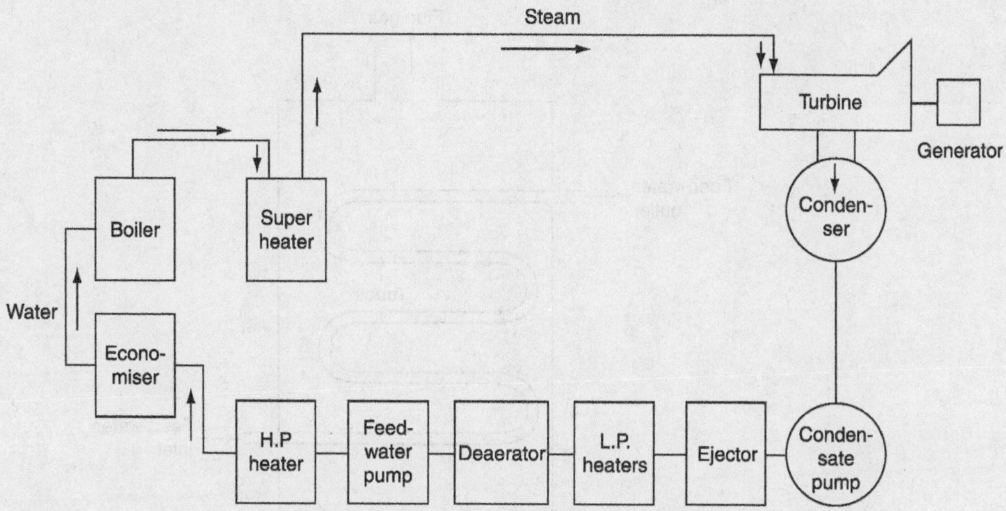

FIG. 1.10 *Block diagram of feed-water steam flow circuit*

is to reduce the dissolved oxygen in the condensate. From the deaerator, the feed water is pumped through the high-pressure heaters and the economizer to the boiler, where the steam is generated. This steam is heated in the superheater and is allowed into the turbine to do the work. After doing the work, the steam passes into the condenser and thus a regenerative cycle formed.

To make up for the loss of water owing to the leakage through steam traps, which may be of the order of 10%, demineralized water is pumped into the feed system as make-up water.

(Iv) *Cooling water circuit*

Exhaust steam in the condenser is cooled to reduce it to the condensate. A large amount of water is required for this purpose. If there is a river or a lake nearby with adequate quantity of water available throughout the year, the cooling water is pumped into the condenser from the upper side of the river. The heated water is discharged to the lower side of the river. If the quantity of cooling water is not sufficient for this open system, the heated water is cooled in the cooking towers or cooling ponds. The loss in cooling water due to evaporation is made up from the river. Such a system is called a closed system. The cooling water circuit is indicated by the circled numbers 22–27 in Fig. 1.8.

(v) *Economizer*

A huge amount of heat energy is lost in the flue gases coming out of the boiler. This loss is reduced in all modern thermal power plants by incorporating an air preheater and an economizer.

An economizer is a feed-water heater. It extracts a part of the heat carried way by the flue gases up to the chimney and uses it to heat the feed water to the boiler. An economizer is placed in the direction of flow of the flue gases from the exit of the boiler to the entry of the chimney.

By the use of an economizer, there is a considerable saving in the consumption of coal (10–25%) and an increase the boiler efficiency (10–12%). However, the incorporation of

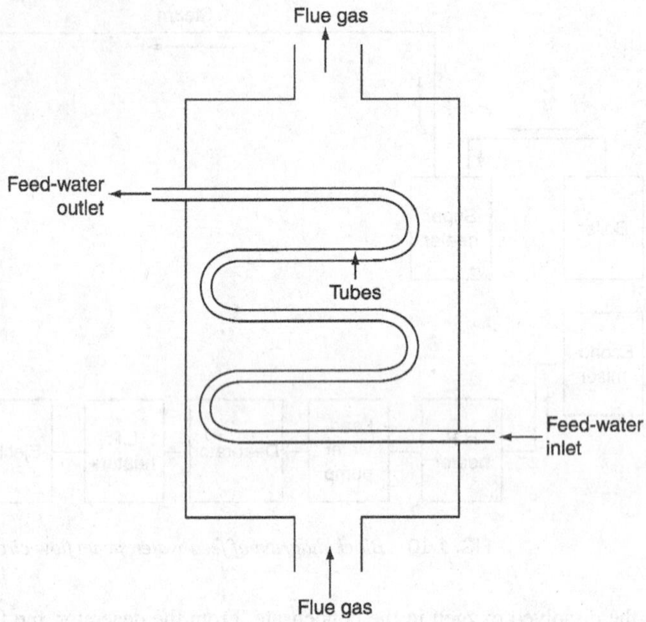

FIG. 1.11 *Schematic diagram of an economizer*

an economizer requires extra investment and increases the maintenance costs and the floor area required by the plant. The justifiable cost of an economizer depends on the increase in the boiler efficiency achieved. This in turn depends upon the flue gas temperature and the feed-water temperature.

The schematic diagram of an economizer is shown in Fig. 1.11. It consists of a large number of small diameters, thin-walled tubes placed between two headers. The feed water enters at one header, passes through the tubes, and leaves through the other header. The flue gases flow outsides the tubes. The heat extracted from the flue gases raises the temperature of the feed water.

(vi) *Feed-water heater*

The steam coming out of the turbine after doing the mechanical work is condensed in a condenser. The condensate is fed back to the boiler as feed water, after adding the make-up water. Before feeding it back to the boiler, the feed water is to be heated for the following reasons.

(i) Feed-water heating increases the boiler efficiency and thus improves the overall efficiency of the plant.

(ii) The presence of the dissolved oxygen and carbon dioxide causes the boiler corrosion. These are removed in the feed-water heater.

(iii) The thermal stresses set up by the cold water entering into the boiler drum are avoided.

(iv) Increased steam production by the boiler is achieved.

(v) The corrosion in the boiler and the condenser may cause the steam and condensate to carry some impurities. These are precipitated outside the boiler.

Feed-water heaters are of two types: contact or pen heaters and surface or closed heaters. In small thermal power plants, open type heaters are used. These heaters receive the steam from backpressure turbine or engines used for driving the auxiliaries. In large thermal plants, the heat bled from the turbines is used for feed-water heating.

In the closed feed-water heater, the steam bled from the turbines is used for heating the feed water.

(vII) *Bollers*

A boiler or a steam generator is one of the most important equipments in a thermal station. It consists of a closed vessel into which water is allowed and is heated to convert it into steam at the required pressure. The following are the requirements of a boiler.

- It should be able to produce and maintain the desired steam pressure safely.
- The boiler should have an output, capable of supplying the steam required to the turbines with 5–10% overload capacity for small durations.
- The boiler should be able to deliver the steam at the desired rate, pressure temperature, and maintaining the quality.
- As the load on the system varies, during off-peak-load hours, some of the units may be shut down. During the peak-load hours, they are restarted. So, the boilers must be able to start quickly and take load.
- Even high-ash content coals must be efficiently burnt by the boiler.
- The refractory material used must be as minimum as possible lest the efficiency should be affected adversely. Further, no joints should be exposed to the flames.
- Auxiliaries such as superheater, economizer, and air preheated may have to be provided.
- Flue gases contain a large amount of ash. About 97% of the fly ash is to be extracted so, every boiler must have an arrangement such as a mechanical ash precipitator or an electrostatic precipitator.

In general, the boiler design must be such that a maximum amount of heat produced in the process of combustion is absorbed. Heat is transferred to the boiler by conduction, convection, and radiation.

1.3.5 Types of boilers

Depending upon the contents of the tubular heating surface, the boilers are classified as fire tube boilers and water tube boilers.

(I) *Fire tube bollers*

These boilers consist of tubes through which the products of combustion and hot gases are passed. Surrounding these tubes is the water to be heated. Since water and steam are both present simultaneously in the shell of the boiler higher pressures cannot be accomplished. Pressures of the order of 17.5 kg/cm^2, with a capacity of about 9,000 kg of steam per hours, are realizable.

Depending upon wether the tubes are horizontal or vertical, whether the combustion chamber is within the boiler shell or outside the fire tube, boilers can be further subdivided into various types, as indicated in the diagram shown in Fig. 1.12.

Fire tube boilers have the advantages of simplicity, compactness, and rugged construction, besides an initial low cost. Further, they can easily meet the fluctuation in steam demand.

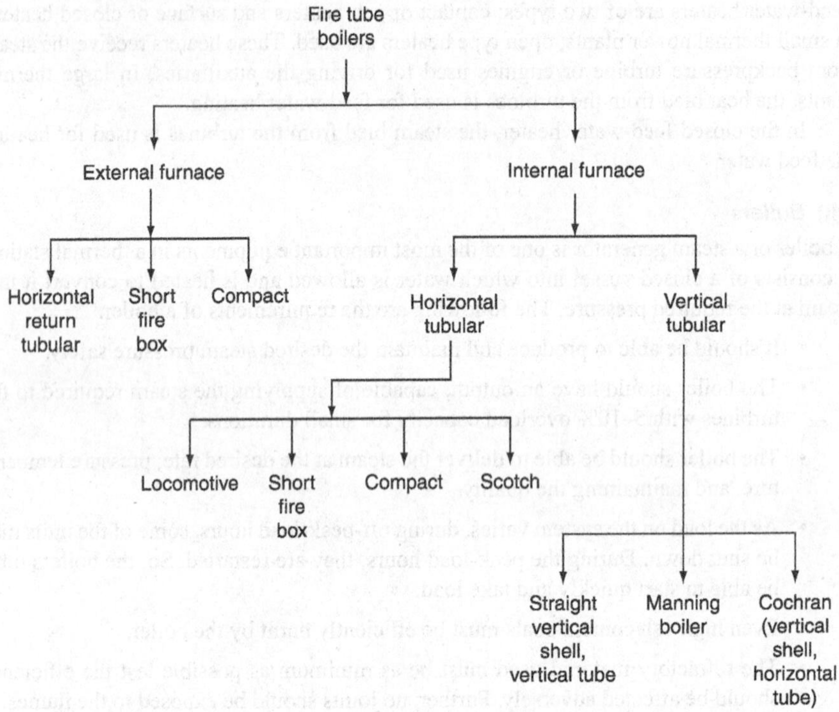

FIG. 1.12 *Fire tube boilers*

However, they have the following disadvantages.

(1) Larger time is required for steam rising. This is due to large quantity of water present in the drum.

(2) Higher pressures than 17.5 kg/cm^2 cannot be attained, since water and steam are simultaneously present in the drum.

(3) The steam is wet and the output of the boiler is not high.

Horizontal return tube boilers are used in thermal plants of low capacity and they occupy a higher floor space.

Vertical fire tube boilers occupy less floor space. They are economical for low pressures. They are available in small sizes with steam capacity of bout 15,000 kg/hour.

(II) *Water tube boilers*

A water tube boiler consists of one or more drums and tubes. Water flow inside the tubes and hot flue gases flow outside the tubes. The tubes are always external to the drum and are interconnected to common water channels and to the steam outlet. The drum stores water and steam. The drums are built in smaller diameters and hence they can withstand higher pressures. Most of the conventional water tube boilers depend upon the natural circulation of water through the tubes. However, pumps may be used to obtain forced circulation of water in modern high-pressure steam boilers.

Forced circulation of water has several advantages.

(i) The weight of the boiler is less and the foundations are cheap.

(ii) The tubes are lighter and scaling problems are not present.

(iii) Greater flexibility in the configuration of the furnaces, tubes, etc.

(iv) Uniform heating of all parts and an increase in the efficiency of the boiler.

(v) Better control of temperature and quicker response to changes in the load.

The disadvantages of forced circulation water include higher investment, increased cost of maintenance, and power consumption of the auxiliaries.

Though water tube boilers with a single drum can operate satisfactorily water tube boilers of two- or three-drum type are commonly used in the thermal stations. Due to the development of high-pressure boilers, the capacities of the boilers have increased. Thus, boilers units with capacities of 1,000 ton/hour at pressures as high as 168 kg/cm^2 (gauge) are available.

The classification of the water tube boilers are shown in the Fig. 1.13.

Depending upon whether the tubes are arranged in horizontal, vertical, or inclined, the water tube boilers are classified as horizontal, vertical, or inclined tube boilers, respectively. The number of drums may be one or more.

The advantages of the water tube boilers are given as:

(i) By increasing the number of tubes, a large heating surface can be obtained.

(ii) Greater efficiency of the boiler can be achieved since the movement of water in the tubes is high with a consequent increase in the rate of heat transfer.

(iii) Because of the large heating surface available, steam can be raised easily.

(iv) Very high pressures can be obtained.

The approximate efficiency of water tube boilers using coal as fuel and without any heat recovery can be taken as about 75–77%. With the addition of heat recovery apparatus (such as economizer, superheated, and air preheater), efficiencies of the order of 85–90% can be achieved. Use of oil as fuel may cause an increase in the efficiency to the extent of about 2–3%.

Finally, the choice of a boiler is based on the initial cost, availability labor and maintenance costs, requirement of space, and the cost of the fuel.

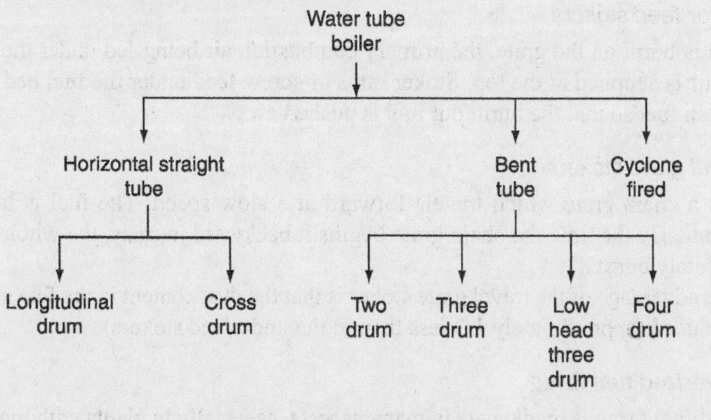

FIG. 1.13 *Water tube boiler*

1.3.6 Methods of firing boilers

There must be efficient combustion of fuel used in the boilers. This is ensured by (i) the proper quantities of the primary and secondary air needed for combustion, (ii) the necessary stoker or grate area needed for burning the coal, (iii) the designed temperature to be attained, and (iv) the non-formation of caking during the burning of the fuel.

There are several methods of firing boilers, two are important. They are (i) solid fuel firing and (ii) pulverized fuel firing.

(i) *Solid fuel firing*

The solid fuel firing of boilers may be accomplished in two ways. They are:

(a) hand firing and
(b) mechanical stoker firing.

(a) Hand firing

This is suitable for boilers of a small output. The grate consists of bar over which coal is put. Dampers are used to regulate the primary and the secondary air required for the combustion of the fuel.

(b) Mechanical stoker firing

Boilers of large output may require a lot of coal to be burnt in the furnace. In such cases, the fuel is fed to the furnaces by means of mechanical stokers. The advantages of this type of stoker firing are given below.

(i) As the coal is fed by the stokers, the labor cost is reduced.
(ii) The fuel can be fed at a uniform rate.
(iii) Fluctuations in the load demand can be met by a proper control of the combustion.
(iv) By burning more amount of coal, the boiler output can be increased.
(v) Poor grades of coal can be burnt with a proper control of the primary and secondary air.

Again, mechanical stokers are of two kinds: (a) under feed stokers and (b) travel grate stokers.

(a) Under feed stokers

The fuel is burnt on the grate, the primary combustion air being fed under the grate. Secondary air is supplied at the top. Stoker rams or screw feed under the fuel bed are used to force fresh fuel so that the burnt out fuel is pushed away.

(b) Travel grate stokers

There is a chain grate which travels forward at a slow speed. The fuel is burnt on the chain grate. By the time the chain grate begins it backward journey, the whole of the fuel is completely burnt.

The advantage of the travel grate stoker is that the dust content in the flue gases is very much reduced (approximately 1/3 less than in the underfeed stokers).

(ii) *Pulverized fuel firing*

The solid fuel firing is inadequate in many aspects, especially in plants with higher capacities (100 MW or more). The conventional methods are unable to meet the fluctuations in the load. They are not suitable for plants burning coal with high-ash content, since the ash content interferes with the combustion process. To overcome these difficulties, pulverized fuel firing is resorted.

In pulverized fuel firing, the coal is ground into a fine powder in a grinding mill. It is led into the combustion chamber with the help of hot primary air currents. (Depending upon the type of pulverized used, the primary air may vary from 10% to 100% of the total air requirement.) To complete the combustion, an additional amount of air called the secondary air is circulated in the combustion chamber.

Advantages of pulverized fuel firing

- The requirement of air for complete combustion is reduced, because of the increased surface area per unit mass of coal.
- Even low-grade coals with higher ash content can easily be burnt.
- The firing can be controlled to match the load requirements.
- There are no clinker and slagging problems.
- The system can work successfully even in combination with gas and oil fuels.
- Highly preheated air (350°C) can be used as secondary air to help rapid prorogation of flame in the combustion chamber.
- Since the pulverizing system is located outside the furnace, it can be repaired without cooling the furnace.
- Larger steam capacities of the order of 2,000 ton/hour can be realized.
- Rapid and efficient starting of the boilers from cold.
- The burning losses are lower in the stoker firing system.
- Since smokeless combustion is possible, the external heating surfaces are free from corrosion.
- Since there are no moving parts in the furnace which are subjected to high temperature, the system has a long and trouble-free life.
- The ash-handling problems are reduced to a minimum, i.e., practically there are no ash-handling problems.
- Less furnace volume.
- Because of the smaller requirement of air and thorough mixing of air and fuel, very high-combustion temperatures can be attained.
- Even fine wet coal can be used if the conveying equipment can carry it to the pulverizing mill.

Disadvantages of pulverized fuel firing

- The investment cost of the plant is increased due to the high-initial cost of the pulverization plant.
- The operating cost is more than that of a stoker-fired system.
- The high-furnace temperatures, unburnt fuel, etc. deteriorate the refractory material.
- Because of the higher combustion temperatures, the thermal losses in the flue gasses are increased.

- There is a danger of explosion hazards so, skilled operating personnel are required.
- Auxiliary power consumption is increased.
- Fly ash, i.e., ash in the form of a fine dust is produced. Costly equipment, such as electrostatic precipitators, is required for its removal.
- The extra equipment such as mills and burners are needed.
- Special equipment is required for the removal of the slag deposited on the lower rows of boiler tubes.
- Difficulty in arresting the fine particles of coal going into the flue gases.
- The storage of powdered coal requires special care and protection against fire hazards.
- The fine grinding of fuel is not possible at all loads, in a unit system.
- Special starting up equipment is required.

The advantages of using pulverized fuel outweighed the disadvantages, so that all modern power plants use pulverized coal. For pulverizing the coal, pulverizing mills are used. These are classified as contact mills, ball mills, and impact mills.

Different systems of pulverized fuel operation

There are different systems of pulverized fuel operation. They are (a) central system, (b) unit system, and (c) bin system.

(a) Central system: The coal pulverized at a central plant is distributed to all the boilers. This method has a high degree of flexibility and ease of control over the quantity of fuel and air. However, a separable space is required to house the coal preparation plant besides a separate crew of operators. Consequently, the installation and operation costs shoot up. Further, there are fire and explosion hazards. So, the unit system is preferred.

(b) Unit system: Each boiler is provided with its own pulverizing plant to prepare and pulverize the fuel. The coal is led to the pulverizing mill by automatic control. This control also adjusts the supply of coal and air in accordance with the load. So, the pulverizing mill receives the warm air from the preheater. There is no necessity of separate drying. Pulverized coal is carried to the boiler by the primary air. The secondary air added around the burner mixes with the pulverized coal and the primary air. Combustion takes place with the fuel in suspension. It is simple and cheap in installation and operation and easy in regulation.

(c) Bin system: The coal is ground at a constant rate. It is transported to the bin or the pulverized fuel store, from where it flows through the feeder to the burners. The speed of the feeder is adjusted to suit the varying load conditions.

1.3.7 Furnaces

The efficient utilization of the pulverized coal depends to a large extent on the ability of the burners to produce a uniform mixing up of air and coal, and the turbulence within the furnace. Again, the design of a furnace is based on the following factors:

(i) The amount of fuel to be burnt.

(ii) The type of the fuel to be burnt.

(iii) The type of firing.

(iv) The load on the boiler and the maximum steam output required.

(v) The operating pressure and the maximum steam output required.

(vi) The degree of heat recovery required.

In the furnaces fired by pulverized fuel, the combustion equipment has burners. The flame may be a short flame, a long flame, or a tangential. The furnace can be classified as:

(a) dry bottom furnaces,

(b) slagging furnaces, and

(c) cyclone-fired furnaces.

(a) Dry bottom furnaces: Fuels with medium or high-ash fusion temperatures are fired in these furnaces. As the fuel is burnt, about 40% of the ash contents fall into the ash pit because of the force of gravitation. On the other hand, if the ash is deposited on the tubes, it may fall due to gravity if the amount deposited is high. The deposited ash may be blown off at the time of soot blowing also.

The draw back of this furnace is that the ash particles are picked up along with the air intake through pit doors. Therefore, the ash content in the flue gases is very high, which is about six to seven times that of under-feed stokers.

(b) Slagging furnaces: These furnaces use fuels which have lower ash fusion temperatures. The particles become molten after combustion. The tubes and walls get pasted with this sticky ash, which subsequently entraps the flue ash particles escaping with the products of combustion. The sticky flue ash particles escaping with the products of combustion. The sticky layers thus formed slide down into an ash pit, where they are cooled.

(c) Cyclone-fired furnaces: It is a high-turbulence furnace used with some modern boilers. It is a wet-bottom furnace. The cyclone furnace is a horizontal cylinder of water-cooled construction: with its inner surface lined with chrome one. Primary air and partially crushed fuel are admitted tangentially to a small scroll section at the end of the cyclone. The swirling motion imparted is amplified by the secondary air admission tangential to the inner surface. There is combustion at a rapid rate and temperature of the order of 1,650°C can be attained. The heat release of the furnace may be as high as 3.5 kcal/cm^3/hour.

The ash is removed in the molten form. The combustion air pressure is of the order of 700–1,000 mm of water gauge. I.D. fans are not normally required. Even if used, there are fewer burdens on the I.D. fans.

In order that the boilers respond to quick load changes, it can have multiple cyclone installations instead of single one. Such boilers can handle 40–110% load conditions.

In a cyclone-fired furnace, the boiler can be fired with dry pulverized fly ash of the adjacent dry bottom installation units.

1.3.8 Superheaters and reheaters

Superheater is one of the auxiliary equipment used to increase the efficiency of a boiler, in addition to such others as air preheaters (economizers) feed-water heaters, etc. A superheater is used to remove the last traces of moisture from the saturated steam which is leaving the boiler tube and to raise the temperature of the steam.

Without the use of a superheater, the steam produced by a boiler has a dryness fraction of 98%, i.e., nearly saturated steam. If this steam (saturated steam) was admitted into the turbine, steam exhaust from the turbine will have low-dryness fraction. It may be practically wet steam, with the presence of moisture. The presence of moisture not only reduces the efficiency of the turbine, but also causes corrosion of its parts. To avoid this, the temperature of the steam at the point of admission into the turbine must be increased. This in turn requires that the temperature of the steam from the boiler output be raised. This is accomplished by 'superheating' the steam with the help of a 'superheater' to get 'superheated steam'. 'Superheated steam' is meant that steam which contains more heat than the saturated steam at the small pressure. It is the steam heated to temperature higher than that corresponding to its pressure. The heat contained in the combustion gases from the furnace is used for superheating.

The use of superheated steam increases the efficiency of the turbine. Superheated steam causes lesser corrosion of the turbine blades. It can be transmitted over longer distances with little heat loss.

(i) *Types of superheaters*

Superheaters may be classified into the following types:

(a) convection type,

(b) radiant type, and

(c) the combination of convection and radiant types.

The convection type of superheater utilizes the heat in the flue gases to heat the saturated steam. It is placed somewhere in the gas stream to receive most of the heat by convection. A radiant superheater is located in or near the furnace, customarily in the surface between the furnace wall tubes to absorb the heat from the luminous fuel by radiation.

With an increase in the output of the boiler, a convection superheater exhibits a rising characteristic, while a radiant superheater exhibits a falling characteristic.

To produce steam at constant high temperature, a combined superheater, i.e., a radiant superheater in series with a convection superheater, is used. Thus, the steam leaving the boiler drum passes through the convection section first and through the radiant section next. Finally, it passes to the steam heater.

In addition to the superheater, reheater is also provided in the modern boiler. The reheater superheats the expanded steam from the turbine, so that the steam remains dry through the lost stage of the turbine.

Just as a superheater, a reheater may be of the convection or radiant type or a combination of both the types is used. Modern boilers employ twin furnaces, one containing a superheater and the other a reheater.

1.3.9 Steam turbines

As discussed earlier, the mechanical energy required to drive the alternators in a thermal power station obtained by converting the heat energy of steam. For this purpose, a steam turbine is used. It works on the principle that high velocity is attained by the steam issuing from a small opening. The velocity attained during the expansion of steam depends on the difference between the initial and final heat content of the steam, which represents the amount of heat energy converted into kinetic energy. The steam turbines are of two types. They are:

(a) impulse turbines and

(b) reaction turbines.

In both the impulse and the reaction turbines, the pressure drop takes place in several stages. The number of stages in a reaction turbine is more than that in an impulse turbine of the same rating. Steam turbines of rating up to 100,000 H.P. or even more are available. They have horizontal configuration. The standard speeds are 3,000 and 1,500 r.p.m. (to drive 2-pole and 4-pole alternators, respectively for 50-Hz operation).

Speed governors are used to maintain the speed constant at all loads either centrifugal or hydraulic type governors may be used.

(i) *Impulse turbines*

In the turbines, the steam expanded in the nozzles attains a high velocity. The steam jet impinges on the blades of rotor, which may be a built-up rotor or an integral rotor. In a built-up rotor, separate forged steel discs are shrunk and keyed onto a forged shaft. A built-up rotor can be manufactured easily and it is cheap. However, there is a possibility for the discs to become loose. In an integral rotor the wheels and the shaft are formed from a single solid forging, so that the discs cannot became loose. For high and intermediate pressures, integral rotors are used.

In the impulse turbines, the steam pressure remains the same during the flow of steam over the turbine blades, since complete expansion takes place in the nozzles. The pressure is the same on the profile of the blades.

(ii) *Reaction turbines*

In a reaction turbine, the expansion of the steam takes place only partially in the nozzle. As the steam flows over the rotor blades, the further expansion takes place and the relative velocity of steam increases. Unlike in the impulse turbine, the pressure is not the same on the two sides of the moving turbine blades, which have an aerofoil section.

Though designated as a reaction turbine, in reality, it is an impulse-reaction turbine, since there is a partial expansion of steam in the nozzle which is an impulse action.

Modern reaction turbines have both moving and stationary blades. The blades are similar and arranged such that the area through which the steam leaves is less than that through which it enters. There is pressure drop in both the stationary and moving blades, the velocity of the steam leaving the blades is increased because of the restricted area at the outlet of the blades.

1.3.10 Condensers

A condenser, as the very name implies, condenses the steam exhausted from the turbine. It helps maintain a low pressure (below the atmospheric pressure) at the exhaust. This use of a condenser in a power plant improves the efficiency. Further the steam condensed by the condenser may be used as a good source of feed water to the boiler. This results in a reduction of the work on the water treatment plant. The efficient operation of the condenser requires a high vacuum to be maintained in the condenser. Any leakage of air into the condenser destroys the vacuum. However, the leakage of air cannot be completely eliminated. So, a vacuum pump is absolutely necessary to remove the air leaking into the condenser.

(i) *Types of condensers*

Basically, there are two types of condensers. They are:

(a) mixing type or jet condensers and

(b) non-mixing type or surface condensers.

(a) Mixing type condensers

The exhaust steam from the turbine and the cooling water come into direct contact. The steam condenses in the water directly. The condensate is not free from salts and other pollutants, so that it may not be reused as feed water. These condensers are rarely used in modern power plants.

(b) Non-mixing type or surface condensers

In these condensers, the steam and the cooling water do not come into contact with each other. Cooling water passes through the tubes attached to the condenser shell and steam surrounds the tubes. The condensate coming out from the condenser can be used as feed water. These condensers are used in all high-capacity modern power plants.

Figure 1.14 shows the schematic diagram of a surface condenser. It consists of a cast iron air-tight cylindrical shell closed at each end. A number of water tubes are fixed in the tube plates located between the cover head and the shell. The exhaust steam from the turbine enters at the top of the condenser. It surrounds the condenser tubes through which cooling water is circulated under force. The steam gets condensed as it comes into contact with the cold surface of the water tubes. The cooling water flows in one direction through a set of tubes located in the lower half of the condenser and returns through the other set in the upper half. The cooling water coming from the condenser is discharged into a river or pond.

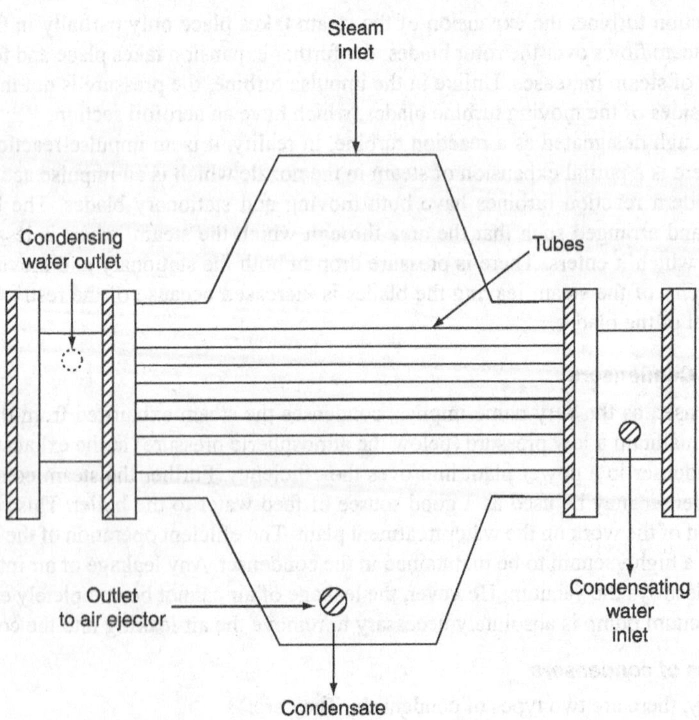

FIG. 1.14 *Schematic diagram of a surface condenser*

The condensed steam is taken out of the condenser by a separate extraction pump. Air is removed by an air pump.

The surface condensers are generally used where large quantities of poor quality cooling water are available and pure feed water to the boiler must be used very economically.

1.3.11 Cooling towers

A cooling tower is a steel or concrete hyperbolic structure. There is reservoir at the bottom for storing the cold water. Water is circulated from the basin of the cooling tower through the condenser. It absorbs latent heat from the steam to get warm. This hot water is return to the cooling tower. It is dropped from a height of about 8–10 m. The cooling tower reduces the temperature of the hot water by about 7°C–10°C, as it falls down into the basin at the bottom of the cooling tower. This water at the reduced temperature is circulated through the condenser and the cycle is repeated.

The reduction in the temperature of the water is brought about by allowing the air flows from bottom to the top. The water drops, as they falls from the top, come into contact with the air and lose heat to the air and get cooled.

(I) *Types of cooling towers*

Depending upon the method of creating air movement through the cooling towers, they can be classified as:

(i) natural draught cooling towers,

(ii) forced draught cooling towers, and

(iii) induced draught cooling towers.

(i) *Natural draught cooling towers*

In these towers, air movement is induced by a large chimney and the difference in the densities of air inside and out side the chimney. These towers have relatively better output at the lower wet bulb. Relative humidity influences buoyancy drive and chimney effect. At high-relative humidity, the performance of these towers is better

Figure 1.15 shows the details of a natural draught cooling tower. Circulating water is diverted in small channels all-round the tower and toward the center and arranged to fall in droplets. This results in a considerable evaporation and cooling. The difference in the pressure of the hot air column inside the tower and the equivalent column of cold air outside the tower predicted the necessary draught. Water from the base of the cooling tower is pumped into the condenser and the cycle is repeated.

(ii) *F.D. cooling towers*

Figure 1.16 shows the arrangement of forced draught tower. The fan is located at the bottom of tower and air is blown by the fan up through the descending water. The hot water from the condenser enters the nozzle and falls in the pond through the hurdles. The entrained water is removed by drift eliminator provided on the top.

(iii) *I.D. cooling towers*

Figure 1.17 shows the arrangement of I.D. tower. The difference between F.D. and induced drought lies in supply of air. In this case, the fan is located at the top of the cooling tower and air enters through the louvers located on the sides of the towers as shown in Fig. 1.17.

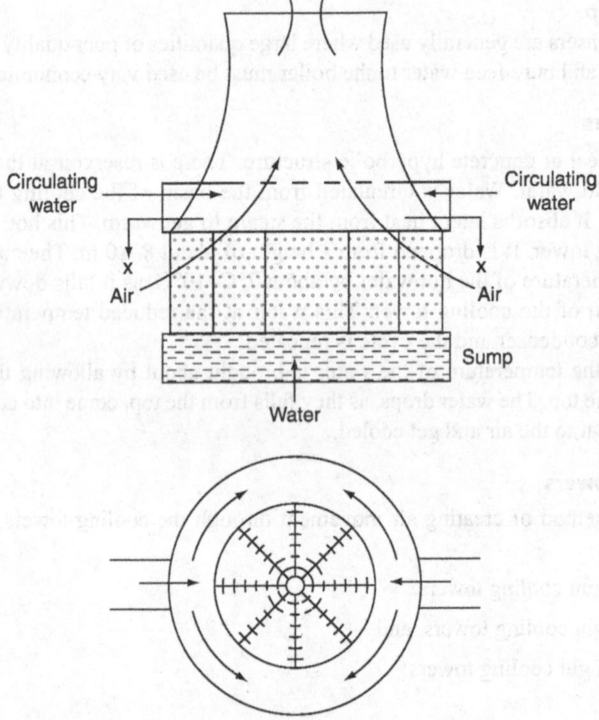

FIG. 1.15 *Natural draught cooling tower*

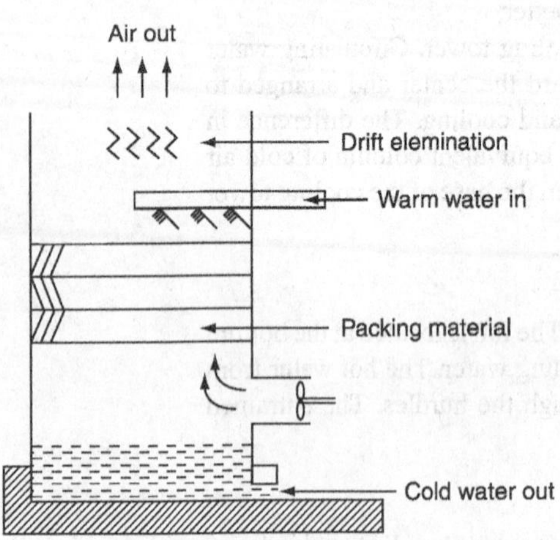

FIG. 1.16 *Forced draught cooling tower*

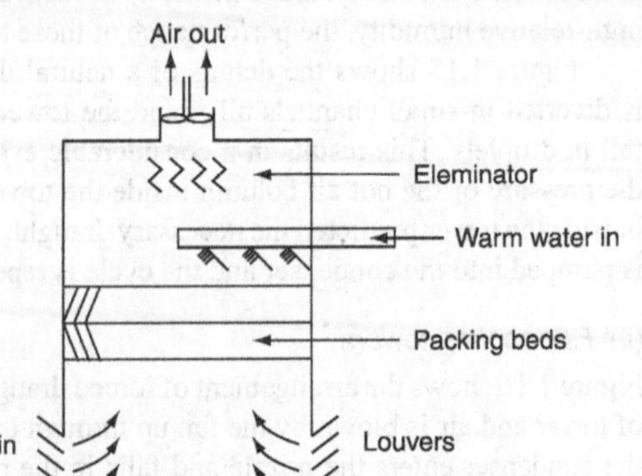

FIG. 1.17 *Induced draught cooling tower*

The fans pull the air upwards from the cooling tower and the hot air is exhausted at a considerable velocity after cooling the water on its way. These types of cooling towers are popular for very large capacity installations.

1.3.12 Chimneys

In modern power plants, the purpose of the chimney is to discharge the exhaust gases into the atmosphere at a high elevation so as to avoid the nuisance to the people living in the locality. The reasons for providing a chimney are:

- To discharge the products of combustion at a great height to avoid nuisance.
- To create more draught to pull the products of combustion.

The diameter at the base of the chimney and the connecting ducts should be adequate to allow the volume of gases to pass through without the necessity of the gases to acquire high speed. The chimney should be firmly supported and anchored to withstand high wind. The main load acting on the chimney are its own load and wind pressure. The chimney must be designed for structural stability against these factors.

Types of chimneys

The three types of chimneys mainly used are:

(i) steel chimneys,
(ii) site constructed chimneys, and
(iii) plastic chimneys.

(i) *Steel chimneys*

These are used for short exhaust stacks, where the draught is created by a fan. They are lined with brick to increase the life. They can be erected in a short time. Self-supporting steel stacks located on the roof of the power house must be enclosed carefully and sufficient structural steel bracing should be used to carry the load to the building column.

(ii) *Site constructed chimneys*

These are built of brick or concrete with mineral or steel liners. Though in the earlier days common bricks were used, nowadays, perforated radial bricks are used for best results. The performance aid the structural stability. The heat insulating properties of the dead air space formed are advantageous for getting maximum draught performance of the chimney. Since the construction process is very slow, brick chimneys are rarely used in large thermal power stations.

(iii) *Plastic chimneys*

These chimneys are built of glass with reinforced plastic. However, these chimneys did not stand well against gas temperature. These are used wherever there is a requirement for a low stress, low-temperature chimney for corrosive effluents.

1.4 NUCLEAR POWER GENERATION

In the previous units, we discussed in detail the hydroelectric and thermal power stations. Hydroelectric stations are to be backed up by the thermal stations since the operation of hydroelectric stations is very much dependent on the rainfall, etc. Again, the thermal power stations require huge quantities of coal. The coal reserves are getting fastly depleted. So, the alternative sources of energy generation are to be sought. The nuclear energy is one among them. The discovery of utilization of nuclear fuel for electric power plants has been presently taken high importance. The nuclear fuel is highly concentrated from the heat energy.

It has been found that 1 kg of atomic material (i.e., uranium) can produced as much energy as produced by burning 3,000 ton of high-grade coal. This shows that nuclear energy can be successfully employed to bridge the gap caused by inadequate coal and oil supplies.

1.4.1 Working principle of a nuclear power station

The schematic diagram of nuclear power station is shown in Fig. 1.18. A generating station in which nuclear energy is converted into electrical energy is known as nuclear power station. The main components of this station are nuclear reactor, heat exchanger or steam generator, steam or gas turbine, AC generator and exciter, and condenser.

The reactor of a nuclear power plant is similar to the furnace in a steam power plant. The heat liberated in the reactor due to the nuclear fission of the fuel is taken up by the coolant circulating in the reactor. A hot coolant leaves the reactor at top and then flows through the tubes of heat exchanger and transfers its heat to the feed water on its way. The steam produced in the heat exchanger is passed through the turbine and after the work has done by the expansion of steam in the turbine, steam leaves the turbine and flows to the condenser. The mechanical or rotating energy developed by the turbine is transferred to the generator which in turn generates the electrical energy and supplies to the bus through a step-up transformer, a circuit breaker, and an isolator. Pumps are provided to maintain the flow of coolant, condensate, and feed water.

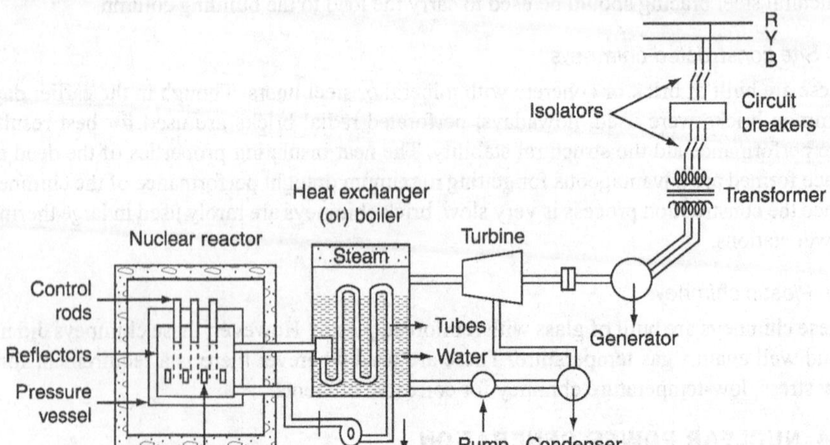

FIG. 1.18 *Schematic diagram of nuclear power station*

1.4.2 Advantages and disadvantages of nuclear power plants

Some of the advantages of nuclear power plants are:

(i) They reduce the demand for coal, gas, and oil.

(ii) Fuel required is only in kilograms; hence, there is no problem for transportation, storage, etc.

(iii) It requires less area as compared to any other plant of the same size.

(iv) The running costs are less.

(v) For large capacity, nuclear power plants are more economical,

(vi) The cost per unit decreases when the power generated is in large.

(vii) The output control is extremely flexible.

(viii) These are not affected by adverse weather conditions.

Some of the disadvantages of nuclear power plants are:

(i) High initial capital cost as compared to other types of power plants.

(ii) These plants are not suitable for varying loads, as reactors cannot be easily controlled.

(iii) It is difficult to shield the plant from radioactive radiation.

(iv) The disposal of fission products is a big problem.

(v) The maintenance cost is high.

1.4.3 Location of nuclear power station

Some of the important points to be kept in view in choosing a site for the location of a nuclear power plant are:

(i) *Proximity to the load center:* The nuclear power station should be located as near to the load center as possible in order to reduce transmission losses.

(ii) *Availability of water supply:* The cooling water requirement of the nuclear power station is more than double that of a coal plant of the same size. So, it is preferable to locate the plant near a river or a lake.

(iii) *Distance from population area:* As per as possible it should be away from thickly populated area, in view of danger of radio activity in the vicinity of the plant.

(iv) *Accessibility:* A nuclear plant requires very little fuel. Therefore, rail facilities are not required for the transport of fuel. However, transport facilities are required during the construction stages.

(v) To improve the reliability of supply to the whole area, nuclear plants may be located far removed from coal fields and hydro sites.

(vi) *Radioactive waste disposal:* The location must be suitable for short-time storage and long-term burial of the radioactive waste.

1.4.4 Energy–mass relationship: Einstein's law

According to Einstein's theory of relativity, mass and energy are interchangeable. Energy can be produced by destroying mass and mass can be produced by the expenditure of energy. Mathematically:

$$E = mc^2, \tag{1.1}$$

where E is the energy in J, m is the mass in kg, and c is the velocity of light in m/sec (= 3×10^8 m/sec).

Nuclear energy is produced by the destruction mass. Thus, if 1 kg of mass is destroyed, according to Equation (1.1), the energy produced is:

$$\begin{aligned} E &= 1 \times (3 \times 10^8)^2 \text{ kg-m}^2/\text{sec}^2 \\ &= 9 \times 10^{16} \text{ N-m/sec or J or W-sec} \\ &= (9 \times 10^{16}) \left[\frac{1}{60} \times \frac{1}{60} \times 10^{-6}\right] \text{MW hours} \\ &= 25 \times 10^{16} \text{ MWh.} \end{aligned}$$

For convenience, energy is expressed in units of 'electron-volts' in nuclear engineering.
Electron-volt: It is the energy gained by an electron in falling through a potential difference of 1 V. An electron has a negative charge of 1.602×10^{-19} C. Now, 1 eV = 1.602×10^{-19} J, so that 1 million eV (MeV) = 1.602×10^{-13} J.

$$\text{And,} \quad 1 \text{ J} = \frac{1}{1.602 \times 10^{-13}} \text{ MeV.} \tag{1.2}$$

Again the energy corresponding to one atomic mass unit (a.m.u.) is equal to 1.494×10^{-10} J, so that:

$$1 \text{ J} = \frac{1}{1.494 \times 10^{-10}} \text{ a.m.u.} \tag{1.3}$$

From Equations (1.2) and (1.3), we have:

$$1 \text{ a.m.u.} = \frac{1.494 \times 10^{-10}}{1.602 \times 10^{-13}} = 931.1 \text{ MeV.} \tag{1.4}$$

1.4.5 Mass defect and binding energy

We know that an atom consists of protons, neutrons, and electrons; each one of which possesses a finite mass. However, the weight of an atom is always less than the sum of the weights of its protons, neutrons, and electrons. The difference is known as the *mass defect.* The mass defect of a given nucleus can be calculated as:

$$\text{Mass defect} = Z \cdot m_p + (A–Z)\, m_n - \text{nuclear mass,} \tag{1.5}$$

where m_p and m_n are the masses of the proton and the neutron, respectively, Z is the atomic number, and A is the mass number.

The energy equivalent of the mass defect is called the *binding energy.* It can be calculated from the relation.

$$\text{Binding energy (MeV)} = 931 \left\{ \begin{array}{l} \text{Atomic number} \times \text{mass of proton (in a.m.u.)} + \\ \text{number of neutrons} \times \text{mass of neutron (in a.m.u.)} - \\ \text{actual mass of assembled nucleons (in a.m.u.)} \end{array} \right\}. \quad \textbf{(1.6)}$$

1 a.m.u. of mass defect equals 931 MeV of binding energy. An amount of mass equals to mass defect is converted into potential energy to hold the nucleus together.

The binding energy per nucleon varies from element to element. The binding energy (or mass deficiency) is highest at the center of the periodic table (or elements). So, if lighter elements are fused together or heavier elements are split, there will be a release of energy.

Thus, the energy release can be obtained:

(i) by combining light nuclei, the process being know as fusion.

(ii) by breaking up heavy nuclei into nuclei of intermediate size, the process being known as fission.

In nuclear power plants, the fission process is used for generation of energy. It results in an increase in the binding energy per nucleon. Kinetic energy and heat are developed by the change in the binding energy.

1.4.6 Nuclear reaction

The naturally occurring nuclear disintegration is slow and uncontrolled. However, different types of nuclear reaction can be produced by particle accelerators. When charged particles such as protons or α-particles are accelerated, they acquire sufficient energy to cause nuclear reactions when they hit a '*target nucleus*'. The particles that are commonly used to start and accelerate nuclear reactions are:

(i) proton ${}_1H^1$,

(ii) α-particle ${}_2He^4$,

(iii) dueteron ${}_1H^2$,

(iv) neutron ${}_0n^1$, and

(v) γ-rays.

The nuclear reactions depend upon the bombarded element or isotope and the kind of the bombarding particle. When the nucleus of an element is bombarded, the result any be a stable or an unstable atom, smaller nucleus of the atom of a different type. Before finally reaching a stable state, the smaller atom thus formed, if radioactive, may emit energy in the form of radiation or particles.

Though an element may be bombarded by different methods, the neutron bombardment results in many advantages. In particular, the neutrons move through matter for longer distances without being stopped, as they have no charge. Neutrons can be produced by various methods:

(a) Particle accelerators such as cyclotrons or vande graft generators speed up charged particles to bombard a target nucleus such as lithium and beryllium, which produces a neutron beam.

(b) α-Particle reactions use α-emitters such as radium to bombard a light element such as beryllium or boron:

$$_4Be^9 + {_2He^4} \rightarrow {_2C^{12}} + {_0n^1}. \tag{1.7}$$

(c) Bombardment of light element such as beryllium by γ-rays:

$$_4Be^9 + {_0\gamma^0} \rightarrow {_4Be^8} + {_0n^1}. \tag{1.8}$$

(d) Neutrons used to produce the fission reaction in nuclear reactors, produce some more high-speed neutrons.

1.4.7 Nuclear fission

We have already discussed that the energy release may be due to fusion or fission. In nuclear reactors, the fission process is used. Nuclear fission is the process in which heavy nucleus is split when it is bombarded by certain particles.

A thermal neutron, i.e., a neutron with a speed corresponding to the speed of molecules in a gas at normal temperature and pressure viz. 2.2×10^3 m/sec, bombarding a heavy atom can cause fission. Thus, if a U^{235} atom is bombarded by a neutron, the nucleus splits up to give the nuclei of some other elements. One of the possible reactions is:

$$\text{Uranium}^{235} + \text{neutron} \rightarrow \text{lanthanum}^{148} + \text{bromine}^{85} + \text{three free neutrons}. \tag{1.9}$$

The mass equation of this reaction is:

$$235.124 + 1.009 \rightarrow 147.961 + 84.938 + 3 \times 1.00897. \tag{1.10}$$

In Equation (1.10), the right-hand side indicates the sum of the masses of the fission products viz. 235.926 a.m.u. It is less than the mass indicated on the left-hand side (236.133 a.m.u.) by an amount of 0.207 a.m.u. Thus, the mass defect is 0.207 a.m.u., so that the equivalent energy is 0.207×931, i.e., 193 MeV. As an approximation, we can assume that one fission of U^{235} causes a release of 200 MeV of energy.

Now, $200 \text{ MeV} = 200 \times 1.6 \times 10^{-13} = 3.2 \times 10^{-11}$ W-sec or J.

Therefore, 1 W requires $\dfrac{1}{3.2 \times 10^{-11}} = 3.1 \times 10^{10}$ fissions per second.

Again 1 kg of U^{235} contains 25.64×10^{23} atoms. If these were fissioned, the energy released would be equivalent to that contained in 3×10^6 kg of coal with a calorific value of 6,000 k-cal/kg. Natural uranium contains only 0.7% of U^{235}. If we assume a fission efficiency of 50%, i.e., if only a half of the total atoms take part in fission, then the fission of 1 kg of natural uranium would give energy equivalent to $(3 \times 10^6) \times (0.7/100) \times 50/100 = 10{,}500$ kg of coal.

(i) *Cross-section (attenuation coefficient) for nuclear reaction*

A concept that can be applied to all possible nuclear reactions, such as fusion, photo-disintegration by γ-rays, and fission, is the 'cross-section for nuclear reaction'. It is a measure of the probability of a given nuclear reaction to occur. The cross-section may be microscopic or macroscopic depending on whether the reference is to a single nucleus or to the nuclei contained in a unit volume of material. The unit for cross-section is 'Barn' ($=10^{-24}$ cm^2).

(ii) *Canning materials*

In order to ensure that the fuel does not contaminate the coolant, the fuel element in the nuclear reactor is canned. Canning eliminates the radiation hazards also. The materials used for canning are aluminum, magnesium, beryllium, and stainless steel. The canning material is chosen based on the fuel used in the reactor.

(iii) *Coolant*

Coolant removes heat from the fuel elements and transfers it to the water. For a material to be used as a coolant, it must have the following properties.

(i) It should not absorb neutrons.

(ii) It should be non-oxidizing.

(iii) It should be non-toxic and non-corrosive.

(iv) It should have high chemical and radiation stability.

(v) It should have good heat transfer capability.

The material used as coolants are carbon dioxide, air, hydrogen, helium water, heavy water, and liquid metal: sodium or sodium potassium.

1.4.8 Nuclear chain reaction

Consider Equation (1.9) for a typical fission reaction,

Uranium^{235} + neutron → lanthanum^{148} + bromine^{85} + three free neutrons.

When the nucleus captures a neutron, an unstable compound nuclease is formed. It splits up into fragments and releases binding energy. The most important aspect, however, is that three free neutrons are ejected. If conditions are favorable, the neutrons ejected by the first fission may be captured by other nuclei to cause the second and subsequent fission reactions'. It may be recalled that bombarding a nucleus with a neutron is easier (than with a proton or α-particles).

Neutrons produced by the fission process are known as *fast neutrons.* They are ejected from the nucleus at a velocity of nearly 1.5×10^7 m/sec and thus possess a very high-kinetic energy. U^{233}, U^{235}, and Pu^{239} are the elements that can under go a fission reaction with fast neutrons.

Natural uranium contains 99.23% of U^{238} and only 0.7% of U^{235}. Unless the proportion U^{235} in the metal is increased to more than 10%, chain reaction is not possible. This is due to the fact that U^{238} atoms absorb fast neutrons to such an extent that the neutrons produced by the fission reactions are absorbed before they can reach a U^{235} nucleus to cause a further fission. This absorption effect is overcome by increasing the proportion of U^{235} in reactors known as '*fast reactors*'.

For effective use in nuclear reactors, the fast neutrons are slowed down to a speed of 2.2×10^3 m/sec. These are called slow or thermal neutrons. When slow neutrons are used, the absorption properties of U^{238} are reduced so that a chain reaction can sustain. A reactor in which natural uranium (containing 99.23% of U^{238}) is bombarded by slow neutrons is called a '*thermal reactor*'.

In thermal reactors, the fissile material is mixed with another material known as a *moderator*. The moderator provides nuclei with which the fast neutrons may collide by elastic collisions so that successive collisions will slow them down to the required speed.

(i) *Multiplication factor*

For a chain reaction to sustain, at least one neutron is to be produced in each fission reaction to initiate the next fission reaction. The possibility or otherwise of the chain reaction to sustain is indicated by a '*multiplication factor*'. It is the ratio of the neutrons in one generation and the immediately preceding generation. Thus, the multiplication factor is:

$$K = \frac{\text{no. of neutrons in the } n\text{th generation}}{\text{no. of neutrons in the } (n-1)\text{th generation}} \quad (n \geq 2). \qquad \textbf{(1.11)}$$

If $K < 1$, it implies that the number of neutrons that can initiate the fission reactions gradually decreases and therefore the process dies down.

If $K > 1$, it implies that more and more neutrons will be produced as the number of fission reaction increases and as a consequence a nuclear explosion occurs as in the case of an atomic bomb.

Therefore, the value of K must be maintained equal to unity (theoretically) in order that the chain reaction be possible. However, there will be loss of neutrons due to leakage, capture in the canning material, and control rods. To overcome this, K is maintained slightly greater than one, the approximate value being 1.04.

The most difficult problem in the control of the reactors is to maintain the value of K at the exact value required.

(ii) *Critical size*

If the core of the reactor was to be infinitely large, there would be no leakage of neutrons. The multiplication factor of a reactor having a core of infinite dimensions is referred to as K_∞. However, if the core were to be very small, there would be excessive leakage of neutrons so that the multiplication factor would be less than unity, with the obvious die-down of the chain reaction. As a result, the reactor should be of a certain minimum size in order that the chain reaction may continue. This size of the reactor is referred to as the critical size of the reactor.

1.4.9 Main parts of a nuclear rector and their function

Reactor is a part of nuclear power plant where nuclear fuel is subjected to nuclear fission and the energy released in the process is utilized to heat the coolant which may in turn generate steam.

Figure 1.19 shows the main parts of a nuclear reactor. They are (a) reactor core, (b) moderator, (c) reflector, (d) shielding, and (e) cooling system.

(a) Reactor core: This contains a number of fuel rods made of fissile material.

(b) Moderator: The neutrons speed is enormously high.

The speed of the neutrons is to be moderated or reduced to such a value as to increase the probability of the occurrence of fission. For this purpose, a material known as a moderator is used. The fast neutrons collide with the nuclei of the moderator material. In the process, the neutrons lose their energy and get slowed down. A good moderator material should have the following properties.

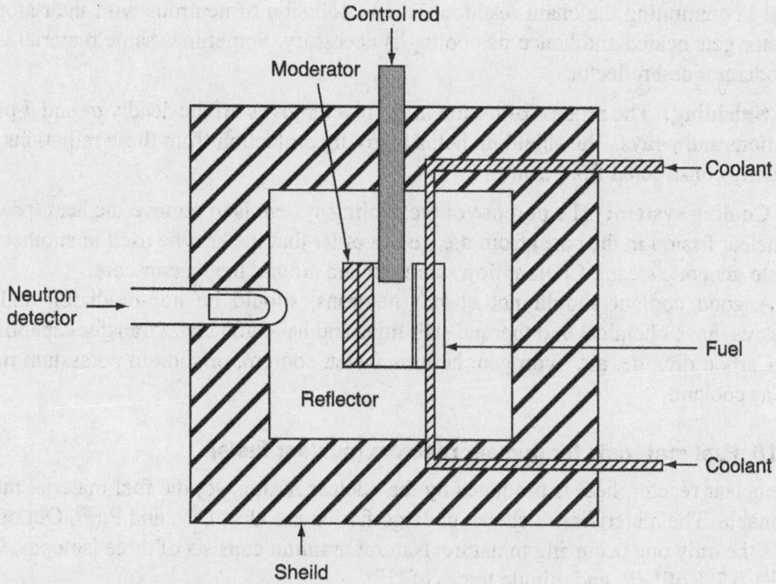

FIG. 1.19 *Nuclear reactor*

(i) It must not react with neutrons. Neutrons captured in nuclear reactions are lost to the fission process, so that the reactor becomes inefficient.

(ii) It should not be very costly.

(iii) It must be non-corrosive.

(iv) Chemical and radiation stability.

(v) High-thermal conductivity.

Elements to the top of the periodic table or compounds with small molecular weight can be used as moderator materials.

Moderator materials

Gases (having small atomic mass) are not suitable as moderator materials since their densities are low and consequently the number of collisions will be small. Helium and beryllium are costly. Boron and lithium have a high-neutron absorption tendency.

Heavy water, inspires of its high cost, is an ideal moderator material and is used in many reactors.

Carbon, which is cheap and satisfactory, is used in many reactors. It can be obtained with any degree of purity.

The moderator and the fuel may be intimately mixed to get an arrangement called '*homogenous*' arrangement. By scattering the fuel in discrete lumps throughout the moderator, a '*heterogeneous*' arrangement can be realized.

(c) Reflector: This completely surrounds the reactor core within the thermal shielding arrangement. The reflector helps in bouncing the escaping neutrons back into the core. This results in conserving the nuclear fuel, since the low-speed neutrons thus returned are useful in continuing the chain reaction. Due to collision of neutrons with their atom, the reflector gets heated and hence its cooling is necessary. Sometimes same material is used in moderator and reflector.

(d) Shielding: The process of fission in the reactor gives off the deadly α- and β-particle radiations and γ-rays. The shielding helps in giving protection from these radiations and it is usually constructed from iron.

(e) Cooling system: The purpose of the cooling system is to remove the heat (produced by nuclear fission in the core) from the core in order that the heat be used in another apparatus to generate steam. Coolant flows through and around the reactor core.

A good coolant should not absorb neutrons, should be non-oxidizing and non-corrosive, have chemical and thermal stabilities and have good heat transfer capability.

Carbon dioxide, air, hydrogen, helium, water, sodium, or sodium potassium may be used as coolant.

1.4.10 Fuel materials for nuclear reactors (nuclear fuels)

In a nuclear reactor, heat is produced by the nuclear fission, so, the fuel material must be fissionable. The materials which can undergo fission are U^{235}, U^{233}, and Pu^{239}. Out of these U^{235} is the only one occurring in nature. Natural uranium consists of three isotopes; 99.3% of U^{238}, 0.7% of U^{235}, and minute traces of U^{234}.

U^{238} and Th^{232} are not fissionable. However, they can be converted into Pu^{239} and U^{233}, respectively, indicated as:

$$_{92}U^{238} + {}_0N^1 \rightarrow {}_{92}U^{239} + \gamma$$
$$_{92}U^{239} + {}_1e^0 \rightarrow {}_{93}Np^{239}$$
$$_{92}Np^{239} + {}_{-1}e^0 \rightarrow {}_{94}Pu^{239}.$$

The above process is called *conversion.* The Pu^{239} formed can be used as fuel.

Again, $$_{90}Th^{232} + {}_0n^1 \rightarrow {}_{90}Th^{233} + \gamma$$
$$_{90}Th^{233} + {}_{-1}e^0 \rightarrow {}_{91}Pa^{233}$$
$$_{91}Pa^{233} + {}_{-1}e^0 \rightarrow {}_{92}U^{233}.$$

The above process is called *breeding.*

1.4.11 Control of nuclear reactors

The heat output of a nuclear reactor is to be controlled. This can be accomplished by controlling the neutron flux. Automatic control is employed to start, operate, and shut down a reactor.

Reactor control depends on changing the value of the multiplication factor, K for normal operation, i.e., to keep the rate of output constant, K must be maintained at unity.

To start the reactor, K is raised to a value of greater than unity, with a consequent increase in the power level. When the required power level is reached, K is reduced to unity and maintained constant at that value slightly less than unity. When the required power level is reached, K is once again made equal to unity and maintained at that value. To shut down the reactor, K is reduced to a value less than unity, so that the chain reaction dies down. For these purposes, the control rods are used.

(i) *Control rods*

Maintaining the multiplication factor at unity ensures the neutron flux, which is held at a constant value. Materials such as boron, hafnium, and cadmium having a high-absorption cross-section are inserted to absorb the neutrons. They are generally alloyed with steel and made into control rods. They can be moved in and out of the channels in the core. Generally, a large number of rods (usually more than 100) is employed to ensure even distribution of neutron flux. When the rods are fully inserted, neutron absorption will be a maximum, so that K is less than unity and the reactor is shut down.

The control rods are of three categories: shut off rods, coarse regulation rods, and find regulation rods.

Shut off rods are normally kept out. They are used for reducing the reactivity in case on an emergency.

For starting and continuous control, the regulation rods are used. After the reactor is started, it can be taken to the required power level by the coarse control rods. The coarse control rods are be charged at a dangerously high rate.

The fine control rods are used to maintain the reactor 'critical', when running under normal conditions. They can adjust the reactivity to a fine degree of accuracy.

(ii) *Control through flow of coolant*

In addition to control by using control rods, an appropriate relation between the mass flow of coolant and power is to be maintained. At constant temperature, the power output is proportional to the rate of flow of the coolant (which removes the heat from the fuel elements and transfers it to the heat exchanger). Coolant temperature recorders, coolant flow indicators, and operating switches are necessary for this purpose.

1.4.12 Classification of nuclear reactors

Nuclear rectors can be classified on several bases such as the purpose for which the rectors are used, the type of fusion, and the fuel used. A few of them are listed below.

(a) Purpose

A reactor can be used for different purposes. Thus, a reactor can be used for:

(i) *Research and development purposes*: To test new reactor designs and for research.

(ii) *Production:* To convert fertile materials into fissile materials.

(iii) *Power*: Electric power generation.

(b) Type of fission

Depending upon the kinetic energy associated with the neutrons, a reactor can be classified into:

(i) *Slow:* Kinetic energy less than 0.1 eV.

(ii) *Intermediate:* Kinetic energy between 0.1 eV and 0.1 MeV.

(iii) *Fast:* Kinetic energy equals to 1 MeV or so.

(c) Fuel used

The fuel used in the reactor may be:

(i) natural uranium,

(ii) enriched uranium, or

(iii) plutonium.

(d) State of fuel

The fuel may be a:

(i) solid or
(ii) liquid.

(e) Fuel cycle

To appreciate the classification of reactors based on fuel cycle, it is necessary to have some understanding of fissile and fertile materials and the breeding/conversion ratio '*r*'.

The materials which can undergo neutron fission are known as *fissile materials.* U^{235}, U^{233}, and Pu^{239} are fissile materials; U^{235} only occurs in nature. Pu^{239} is obtained by converting U^{238}, while U^{233} can be obtained by converting Th^{232}; U^{238} and Th^{232} are not fissionable. However, they can be converted into Pu^{239} and U^{233} which are fissile. U^{238} and Th^{232} are called *fertile materials.* Thus, a fertile material is non-fissionable material which can be converted into a fissile material.

Each fission process of U^{235} produces about 2.5 neutrons per fission. Of these, only one neutron is required to sustain the chain reaction. The excess fission neutrons can be used to activate the isotopes of fertile materials to produce new fuel atoms. In the conversion process, we define a parameter.

$$r = \frac{\text{no. of fertile atoms consumed (i.e., number of new fuel atoms formed)}}{\text{(no. of original fuel atoms consumed in the fission and radioactive capture process)}}.$$

If $r \geq 1.0$, the reactor is called a *breeder reactor* and *r* is called the *breeding ratio.*
If $r < 1.0$ and not equal to zero, i.e., $0 < r < 1.0$, the reactor is called a *converter reactor* and *r* is called the *conversion ratio.*

Now, based on the fuel cycle, the reactors are classified thus:

(i) *Burner (thermal):* Designed for producing only heat. There is no recovery of converted fertile material.

(ii) *Converter:* Converts fertile material into fissile material. The converted fissile material is not the same as the one initially fed into the reactor. r (<1.0) is the conversion ratio.

(iii) *Breeder:* Converts fertile material into fissile material. The converted fissile material is the same as that initially fed into the reactor. r (>1.0) is the breeding ratio.

(f) Arrangement of fuel and moderator

(i) *Homogeneous:* Fuel and moderator are mixed.

(ii) *Heterogeneous:* Fuel in discrete lumps in moderator.

(g) Moderator material

(i) Heavy water,

(ii) Graphite,

(iii) Ordinary water,

(iv) Beryllium water, and

(v) Organic.

(h) Arrangement of fissile and fertile material

(i) *One region:* Fissile and fertile materials are mixed.

(ii) *Two regions:* Fissile and fertile materials are separate.

(I) Cooling system

(i) *Direct:* The liquid fuel is circulated from the reactor to the heat exchanger, where steam is generated.

(ii) *Indirect:* Coolant is passed through the reactor and then through the heat exchanger for the generation of steam.

(J) Coolant used

(i) Gas,

(ii) Water,

(iii) Heavy water, and

(iv) Liquid metal.

1.4.13 Commercial types of reactors

There are several types of commercial reactors such as calderhall reactor, pressurized water reactor, boiling water reactor, gas cooled reactor, candu type reactor, sodium graphite reactor, and fast breeder reactors. We shall discuss a few types of reactor as follows:

(i) Pressurized water reactor (PWR)

Figure 1.20 shows the schematic arrangement of a pressurized water reactor. Enriched uranium oxide, clad in zinc alloy, is used as the fuel. The pressure vessel is made of steel. Water under pressure is used both as a coolant and moderator. PWR is designed to prevent

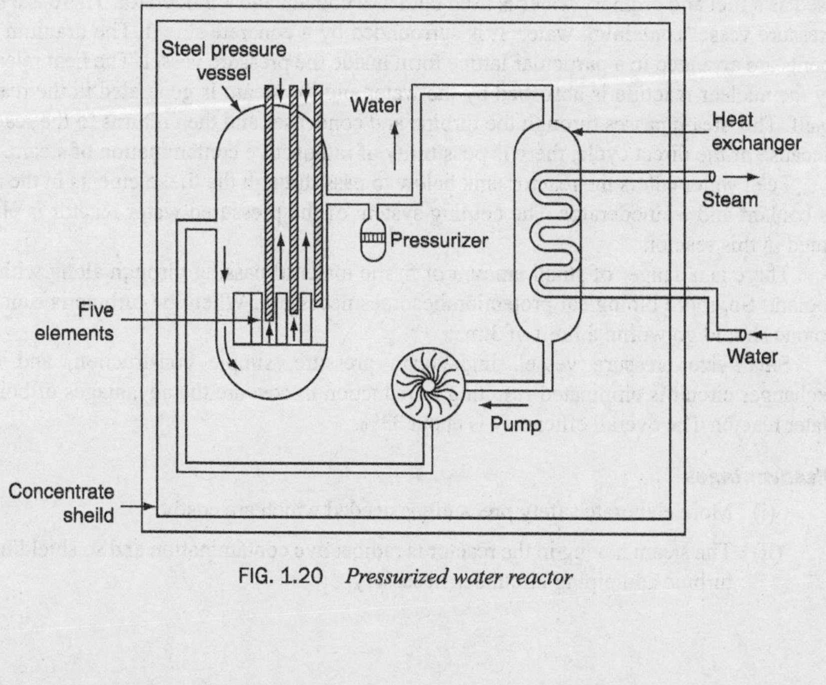

FIG. 1.20 *Pressurized water reactor*

the boiling of the water coolant in the uranium core. Water under pressure is circulated (by a pump) round the core. Water in the liquid state absorbs heat from uranium and transfers it to the boiler having a heat exchanger and a steam drum. The pressure vessel and the heat exchanger are surrounded by a concrete shield. To maintain the pressure in the water system, a pressurized tank tapped into the pipe loop is used. Water with a temperature of 190°C and pressure of 140 kg/cm^2 (2,000 p.s.i.) is passed into the reactor and is discharge from the reactor at 270°C. This water is passed into the heat exchanger, where the steam is raised. The temperature and the pressure of the steam area around are 250°C and 42 kg/sq.-cm (600 p.s.i.), respectively. This steam of poor quality is condensed in the condenser and the condensate return to the heat exchanger, thus forming a closed circuit.

The advantage of the PWR reactor are compactness, high-power density, less number of control rods is required, and water used in reactor (as coolant com moderator) is cheap.

However, it suffers from the following disadvantages.

(i) High-strength pressure vessel is required.

(ii) Severe corrosion problem.

(iii) High losses from heat exchanger.

(iv) Fuel element fabrication is expensive.

(v) Auxiliaries consume high power.

(vi) Low temperature steam is formed.

Reactors using heavy water as coolant moderator are called 'pressurized heavy water reactors' (PHWR). The atomic reactors such as Rajasthan Atomic Power Station, Madras Atomic Station, and Narora Atomic Power Project are PHWRs.

1.4.14 Boiling water reactor (BWR)

Figure 1.21 shows a boiling water reactor. In this reactor also enriched uranium oxide is used as a fuel and ordinary water is used both as a coolant and a moderator. There is a steel pressure vessel containing water. It is surrounded by a concrete shield. The uranium elements are arranged in a particular lattice form inside the pressure vessel. The heat released by the nuclear reaction is absorbed by the water and the steam is generated in the reactor itself. This steam passes through the turbine and condenser and then returns to the reactor. Because of the direct cycle, there is possibility of radioactive contamination of steam.

Feed water enters the reactor tank below to pass through the fuel elements in the core as coolant and as moderator. The cooling system of the pressured water reactor is eliminated in this reactor.

There is a danger of small amount of fissile material passing through along with the coolant. So, more biological protection becomes necessary. When the turbine is running, no one should go within a limit of 3 m.

Small-size pressure vessel, high-steam pressure, simple construction, and heat exchanger circuit is eliminated resulting in reduction in cost are the advantages of boiling water reactor. The overall efficiency is about 33%.

Disadvantages

(i) More elaborate safety precautions needed which are costly.

(ii) The steam having in the reactor is radioactive contamination and so shielding of turbine and piping circuits is necessary.

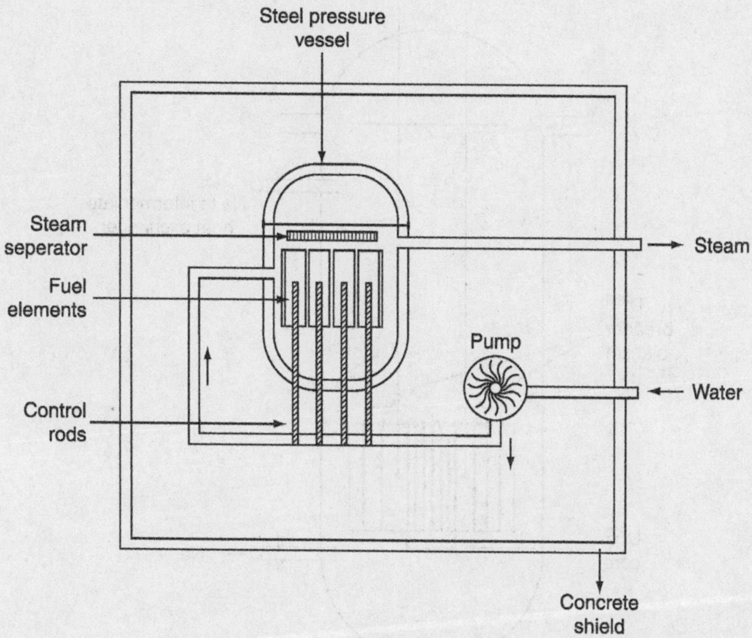

FIG. 1.21 *Boiling water reactor*

(iii) Thermal efficiency on part load operation is low due to wastage of steam.

(iv) It cannot meet a sudden increase in load.

The reactors at Tarapore Atomic Power Station are of the boiling water type.

(I) *Fast breeder reactors*

As the name implies, fast neutrons are used in these reactors. Further, there is breeding, i.e., more fissile material is kept. The fuel is enriched uranium or plutonium. There is no moderator in this reactor. The vessel is surrounded by a thick blanket of depleted uranium, which is a fertile material. By absorbing neutrons from the new fissile material, the fertile material is converted into fissile material. A reactor working under these conditions is called a breeder. Hence, it is called fast breeder reactor, which is shown in Fig. 1.22.

There are two heat exchanges in this reactor. The reactor core is cooled by a liquid metal-sodium or potassium. The secondary heat exchanger uses liquid sodium/potassium as coolant. It transfers heat to the feed water. The neutron shield between the core and the primary heat exchanger is provided by the use of boron, light water, oil, or graphite. The shielding against gamma (γ) rays is accomplished by lead, concrete with magnetite or barium added.

The core of a fast reactor needs a high enrichment (i.e., above 10% of fissile material). The core consists of 30% fuel, 50% coolant, and 20% canning and structural material by volume.

In addition to producing power, fast reactors can produce Pu^{239} and U^{238}. The possibility of the core getting overheated and destroyed has to be considered carefully. The heat transfer and control problems need special attention. The thermal efficiency is of the order of 43%.

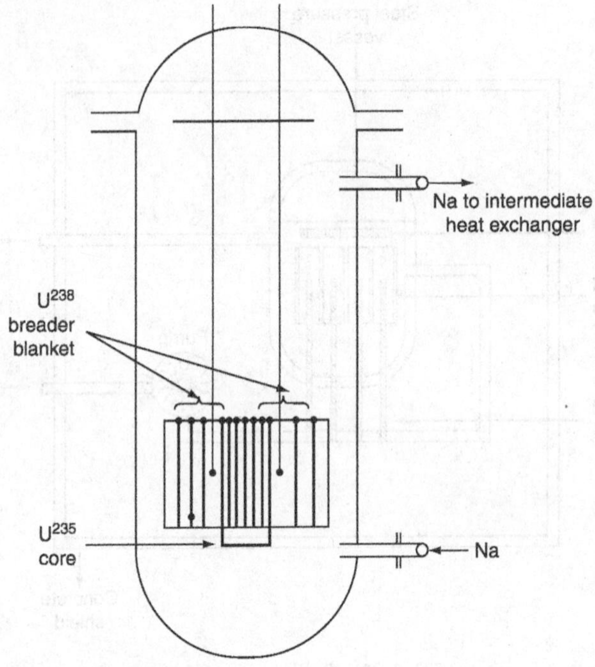

FIG. 1.22 *Fast breeder reactor*

Advantages

(i) Require small core.

(ii) Moderator is not required.

(iii) High breeding is possible.

(iv) Control is easy.

(v) Greater inherent safety.

Disadvantages

(i) Highly enriched fuel is required.

(ii) Specific power of the reactor is low.

(iii) Size and weight of reactor per unit power are higher.

(iv) Choice of fuel is restricted.

(II) *Comparison of thermal and fast breeder reactors*

Depending upon the neutron kinetic engines, the neutrons may be thermal neutrons or fast neutrons. Accordingly, the reactors may be thermal reactors or fast reactors.

In thermal reactors, moderators are used to slow down the neutrons to a speed at which they can react readily with U^{235} and cause fission. The energy of these neutrons corresponds to the ambient temperature inside the reactor. In fast reactors, the neutron energies are high and are nearly equal to those with which they emerge as fragments from the fissile materials during fission.

We had already discussed the significance of breeding. The advantage and disadvantages of the thermal reactors compared to fast breeder reactors are given below.

Advantages

(i) Greater inherent safety.

(ii) The heat generated per unit volume of ore or per unit area of fuel is less.

(iii) Easy controllability.

Disadvantages

(i) The choice of fuel is severely restricted from the viewpoint of neutron economy when uranium is used as the fuel.

(ii) The size and weight of the reactor per unit power are higher.

(iii) The fissile material consumed is more than that could automatically be replaced.

The net fuel consumption in fast reactors is much less since more fertile can be converted to fissile material.

1.4.15 Gas-cooled reactor

In this reactor, a pressurized carbon-dioxide gas is used as a coolant instead of water. It uses a lattice of graphite in the form of blocks as moderator. It is necessary to protect graphite from contact with coolant, either by canning the graphite or by piping the coolant through the reactor in metal tube. The tubes used will absorb the neutrons to some extent only thus, reducing radio activity of the system.

The CO_2 gas flows to the heat exchanger where it transfers its heat to water which gets converted into steam. The steam flows into the turbine which in turn drives the alternator to generate electricity.

Advantages

(i) Less corrosive problem.

(ii) Capability for the use of natural uranium as fuel.

(iii) Greater safety as compared to the water-cooled reactor.

(iv) Simple for processing the fuel.

Disadvantages

(i) Require large size of reactor.

(ii) Very low-power density.

(iii) For coolant circulation, more power is required.

(iv) Complicate in control.

1.4.16 Radiation

In the operation of a nuclear power plant, the disposal of the solid, liquid, and gaseous waste and effluent is an important problem to be tackled effectively, since the radiation can give rise to several hazards, both internal and external. Thus, it is essential that adequate shielding be provided to guard the operating personnel and delicate instruments, in particular, and the environment in and around nuclear plant, in general. In addition, certain safety measures (or precautions) are to be followed to overcome the ill effects of radiation.

1.4.17 Types of radiations

The important types of radiations that can cause hazards are the α-, β-, and γ-rays and the neutrons.

(i) *α-Rays:* These are the nuclei of helium atom, $_2He^4$. They carry a positive charge. They cannot penetrate the skin. However, they can cause internal hazard, if ingested.

(ii) *β-Rays:* These are electrons that travel at the speed of light since they are smaller in size; they have grater penetrating power than α-rays.

Over exposure to β-rays can cause skin burns. Malignant growth may result because of repeated over exposure to β-rays. Since their penetrating power is not high, a thin sheet of metal or a brick wall can stop them.

(iii) *γ-Rays:* These are electromagnetic radiations of a very short wave length. They have high energy and penetrating power. They can cause considerable damage, especially to organic materials.

Over exposure to γ-rays can lead to blood diseases, anemia, and some undesirable genetic effects. Larger exposure may cause death in a few hours. The effects of slow exposure may become apparent only after several years.

There is no material which can stop the γ-rays completely, though thick sheet of lead and concrete can alter them considerably.

(iv) *Neutrons:* These are produced in fission. They have a wide range of energies up to 10 MeV. They possess no charge, but they are highly penetrating.

The effects of neutrons are similar to those of the γ-rays.

1.4.18 Radiation hazards

The hazards caused by radiations can be either internal or external.

(i) Internal hazards: Food inhalation or breaking in of the skin by radioactive radiation is the causes of internal hazards. The tolerance of the body to the radiations depends on several variables.

(a) Degree of retention of the radioactive material in the body. The longer the retention, the greater is the harm caused.

(b) The fraction of the radioactive material passed to the critical tissues by the blood stream. The greater the fraction conveyed, the greater will be the harm.

(c) Radio sensitivity of the tissues: Different tissues will have different radiosensitivities. Thus, bone, lymph glands, ovaries, and testes are more vulnerable to the energetic radiation.

(d) Size of the organ involved: The smaller the organ, the greater the concentration of the radioactive material and hence the greater is the damage.

(e) Essentiality of the organ: The most essential organ damaged can cause early death.

(f) The type of radiation: The important radiations are α-, β-, and γ-rays and the neutrons.

(ii) External hazards: Before discussing the external hazards, it is necessary to have some preliminary ideas pertaining to health physics. These are presented below.

Health physics: Units of contamination and radiation.

The unit of contamination is a measure of the amount of radioactive material present in any material. The unit is curie. One curie is equal to 3.7×10^{10} nuclear disintegrations per second. Curie is rather a large unit. So, submultiples, such as millicurie and microcurie are normally used. The curie indicates the number of radioactive atoms in a material.

Rontgen is the classical unit for measuring radiation. The quantity of x or γ radiation necessary to produce one electrostatic unit of charge, be it positive or negative, in one cubic centimeter of dry air at standard temperature and pressure (S.T.P.) is termed one rontgen.

One rontgen = 86.9 ergs of energy absorbed per gram of air. Though rontgen is the basic unit, another unit called RAD is commonly used. RAD is equivalent to 100 ergs of energy absorbed per gram of irradiated material at the point of interest. It should be noted that the absorption of energy from the radiations, rather than the radiations, is of great concern. The absorption of radiations depends upon the material absorbing them.

Since the unit RAD specifies the absorbing material, it is more specific and relevant.

Another unit is 'rontgen equivalent man' or 'rontgen equivalent mammal' abbreviated as REM. REM and RAD are related through a factor R.B.E., which is the abbreviated from of 'relative biological effectiveness'. R.B.E makes allowance for the different biological damage resulting from unit energy absorption from different radiations.

REMs = R.B.E. × RADs.

External hazards due to radiations: An external hazard is caused when the body is bombarded by energetic radiation from radioactive sources. The radiations may penetrate through the skin and damage the internal tissues.

We had already noted that REM is the unit of radiation dose rate. The maximum integrated dose depends upon the age of the person. For a person of age, A years, the maximum integrated does allowed is (A-18) × 5 REM. [18 is used under the assumption that a person below the age of 18 years need not have to work in a place where there is a possibility of damage due to radiations.]

The effect of radiation on the body can be summarized as given in the Table 1.1. It may be noted that the values represented in Table 1.1 are some typical values. Though an irradiation dose of about 400–450 REM to the whole body may result in death, the individual parts may withstand comparatively higher doses of radiation.

TABLE 1.1 *Energetic radiation effect on human body*

		Dose (REM)
(i)	Amount which is detectable	25
(ii)	Radiation sickness in 50% of exposed	200
(iii)	Eventual death in 50% of exposed	450
(iv)	Certain death in 24 hours	600–1,000
(v)	Death in few hours	1,000

When a part of the body is exposed to a radiation dose of about 200 REMs, the result may be a temporary effect of radiation sickness such as shock symptoms and nausea.

1.4.19 Shielding

To guard the personnel and delicate instruments, adequate shielding is to be provided by suitable shielding materials.

The effectiveness of the material in providing the shielding depends to a large extent upon its density. Hence, the shielding purpose are given below.

(i) *Lead:* Density is 11.3 g/cm^3. It is a commonly used shielding material because of its low cost and high density.

(ii) *Concrete:* Density is 2.4 g/cm^3. It is less efficient than lead.

(iii) *Steel:* Density is 7.8 g/cm^3. Though steel has good structural properties, it is not an efficient shielding material. So, it is used as an alternating shield.

(iv) *Cadmium:* Density is 8.65 g/cm^3. Slow neutrons of nuclear reaction can be absorbed by cadmium.

It should be noted that no shielding material is effective in shielding all types of radiations. For example, materials containing hydrogen (such as water and polythene) are used to slow down fast neutrons. For absorbing thermal neutrons, born or steel may be used. On the other hand, for absorbing γ-rays, a heavy material such as lead is needed to act as a shield.

In nuclear power reactors, there will be a thermal shield of several centimeters thick steel. It is surrounded by about 3-m thick concrete shield. The water, in concrete, slows down the fast neutrons. To attenuate the γ-rays and to absorb thermal neutrons, iron, barium, or steel turnings.

1.5 GAS POWER GENERATION

In some case of incorporating intermittent (or) peak-load plants in combination with the base-load plants, the gas turbines are preferable which are the cheapest types of plants available. In large system, the size of these gas turbine plants are from 10 to 25 MW, and the largest size of plant being used is about 50 MW. The thermal efficiency of the gas turbine plant (20–25%) is less when compared to the steam power plant (25–30%). This lower thermal efficiency reduces the load factor and increases the fuel cost but which can be compensated in gas plant by lowering the fixed, operating, as well as maintenance charges. In India, gas turbine plant of size 70 MW was situated at Namrup in Assam, working as a base-load plant with natural gas fuel. And, the second one uran gas turbine power plant was placed at Maharastra.

Types of gas turbine power plants:

(i) According to the process of combustion takes place in the combustion chamber, gas turbine power plants are classified into two types:

(a) *Continuous combustion constant pressure type:* In this type, the combustion cycle working on constant pressure principle known as joule (or) Brayton cycle.

(b) *The explosion (or) constant volume type:* In this type, the combustion cycle working on constant volume principle known as Atkinson cycle.

(ii) According to the path of working substance, gas turbine plants are classified into following types.

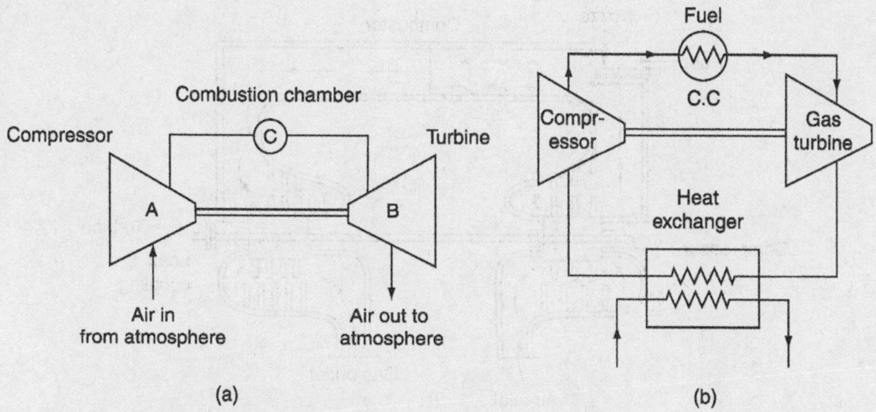

FIG. 1.23 *Schematic diagram of open and closed cycle gas turbine*

(a) *Open-cycle gas plant:* In this type of plant, working substance is air. This working fluid enters into the compression chamber from atmosphere, there compressed after compressed air is fed into the combustion chamber, to raise the temperature of it by burning the fuel. Now, the final product of combustion chamber is mixed up with excess air and pumped through the turbine, developing power, and then exhausted into the atmosphere. The above process is repeated for every cycle. Figure 1.23 (a) shows open-cycle gas turbine.

(b) *Closed-cycle gas turbine plant:* In this type of plant, working fluid, i.e., air is heated in the preheater by burning fuel external to it. Hot air is fed to the turbine to generate power. The hot air expands in the turbine and then cooled in a precooler for cooling purpose. Now cooled air is passed to the compressor and again fed back to generate power. Again, the same air circulates over the system continuously. Figure 1.23 (b) shows closed-cycle gas turbine.

1.5.1 A simple gas turbine power plant

A simple gas turbine power plant consists of units such as compressor, combustion chamber, and turbine. In addition to the above main components, plant has some other auxiliaries such as starting device, fuel system, duct system, and auxiliary lubrication system. Schematic diagram of gas turbine power plant is shown in Fig. 1.24.

A simple gas turbine plant uses air as a working fluid. When the plant units run atmospheric air admitted into the combustion chamber to raise the pressure of air to several times of atmospheric pressure.

Compressed air is fed into the combustion chamber with mixture of gasses at high pressure to increase the temperature. Now, the working fluid is passed through the turbine at which gasses expand thereby developing heavy motive forces which will drive the shaft of turbine. After the expansion, the gasses will exhausted out from the turbine to the atmosphere.

The temperature of the products in the combustion chamber is about 1,000–1,500°F. The temperature of the exhaust gas is about 900–1,100°F. In this model, compressor, turbines, generator are arranged on the same shaft. In this, some of the energy (60% of power generation) developed by the turbine is used to drive the compressor and remaining is for the generation of the electric power by the generator.

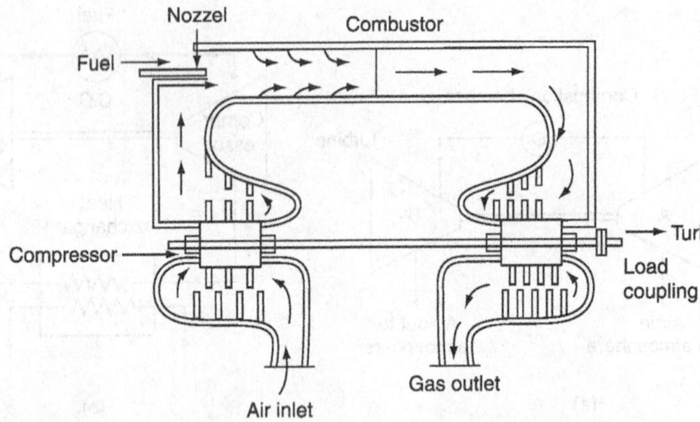

FIG. 1.24 *Schematic diagram of gas turbine power plant*

1.5.2 Applications of gas turbine plants

Gas turbine plants have the following applications.

- Gas turbine plants are used as substitute plants for the hydroelectric power plants.
- These plants are used to supply peak loads in steam, diesel (or) hydro plants, etc.
- The plants are used to supply mechanical drive for auxiliaries.
- Gas turbine plants are widely used in aircrafts and ships.
- These plants sometimes used as engines for automobile use.

1.5.3 Advantages and disadvantages of gas turbine plants

Gas turbine plants have the following advantages.

- The size and weight of the gas turbine plant are less for large capacities compared to the steam power plant.
- Water needed to run the gas turbine plant is less compared to the steam plant.
- Gas turbine plants can be put on load easily and they can be started quickly.
- The maintenance cost of the gas turbine plant is less.
- The installation of the gas turbine plant is easier because of the absence of boiler, evaporator, condensating system, etc.
- Heavy foundations and buildings are not requiring for the gas turbine power plant.

Disadvantages

In addition to the above advantages, the gas turbine plants have the following disadvantages.

- Net output from the gas power plant is low; this is because major portion of the energy is required to drive the compressor.
- Temperature of the combustion products is too high. So, even at moderate pressure, more care should be taken.

1.6 DIESEL POWER GENERATION

Diesel engine power plants are more efficient than the other types of engine plant for the same capacity. The diesel engine plants are more suitable for low- and medium-power outputs. These plants are commonly employed where fuel prices are low and water availability is limited. The capacity of the diesel power plants is about 5 MW, such plants are used as standby plants to hydro- and diesel power plants for small output. These plants do not require large amount of water for cooling. Diesel engines are widely used in railroad locomotives, road buildings, ship propulsion electric generators for feeding supply to public industrial and institutional purpose, etc. A wide application of diesel engine is mainly due to the less cost fuel than gasoline products.

1.6.1 Diesel engine power plant

Diesel engine power plant has the following systems by means of which electric power can be generated. Schematic diagram of diesel engine power plant is shown in Fig. 1.25.

(i) Air intake and exhaust system: This system consists of pipes arrangement for admitting fresh atmospheric air into the diesel engine, and also to pump out the exhaust gases to the atmosphere. Filters are needed at the air inlet to remove dust particles etc. from the incoming air. At the outlet of the system silencer is provided to reduce the noise when the exhaust gases are coming out from the engine.

In order to reduce the specific fuel consumption and to increase the engine capacity, the intake system must have to maintain minimum pressure loss.

(ii) Fuel supply system: This system consists of fuel tank to store fuel, and fuel pumps and filters to transfer and inject fuel into the diesel engine. Fuel oil is supplied by trucks, rail, cars, etc. at the plant site.

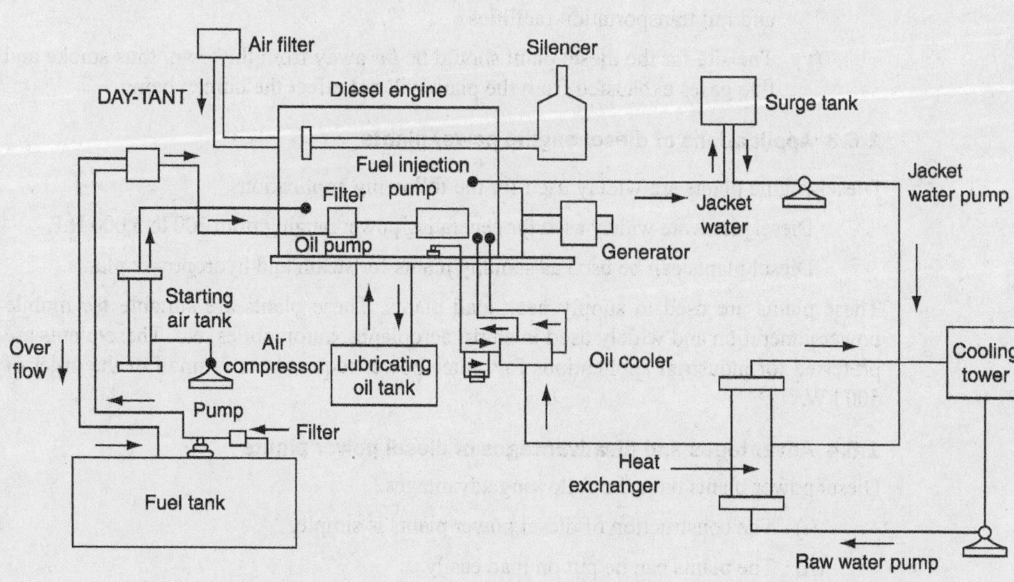

FIG. 1.25 *Diesel engine power plant*

(iii) Cooling system: This system circulates sufficient amount of water around the engine in order to maintain desired temperature. The hot water recooled in the cooling ponds again recirculated into the system.

(iv) Lubricating system: This system is necessary to reduce ware on the rubbing parts and friction. It consists of lubricating oil tank, pumps, filters, etc.

(v) Starting system: Starting system is essential for the initial starting of engine. It consists of compressor, battery, and electric motor (or) self-starter.

Diesel engine is nothing but internal combustion engine, in which fuel is ignited by injecting into system thereby compression. So that these engines are also called as compression ignition engines. This engine will convert heat energy into mechanical work. In the combustion chamber, fuel burns rapidly and gases attain very high temperature and produces extremely hot compressed gases. These gases expand and push back the piston of engine. This is nothing but the power stroke in which mechanical work is done. This work is helpful to rotate the crank shaft on which generator is mounted, which converts mechanical power into electric power.

1.6.2 Site selection for diesel power plants

While selecting a site for the diesel engine power plant, it is necessary to consider the following factors.

(i) Site for the diesel power plant should be nearer to the load center; this is to reduce the cost of transmission of power and also to reduce the power loss.

(ii) Plant should be located where plenty of water is available.

(iii) The site for the diesel plant should nearer to the source of fuel supply, to decrease the transportation charges.

(iv) The selection of the site for the plant should be in such a way that, it has road and rail transportation facilities.

(v) The site for the diesel plant should be far away from the town, thus smoke and flue gases exhausted from the plant will not affect the human being.

1.6.3 Applications of diesel engine power plants

Diesel engine plants are widely used for the following applications.

Diesel plants are widely used for generating power ranging from 100 to 5,000 H.P.

Diesel plants can be used as standby plants for steam and hydropower plants.

These plants are used to supply peak-load plants. These plants are suitable for mobile power generation and widely used in ships, aeroplanes, automobiles, etc. These plants are preferred for industrial applications for which power requirement is small of the order of 500 kW.

1.6.4 Advantages and disadvantages of diesel power plants

Diesel power plants have the following advantages.

(i) The construction of diesel power plants is simple.

(ii) The plants can be put on load easily.

(iii) The plants can be started quickly.

(iv) No need of requiring large amount of water for cooling.

(v) The size of diesel engine plant is small compared to the steam plant for the same capacity of generation.

(vi) The maintenance cost of the plants is less.

(vii) The thermal efficiency of diesel plants is high compared to the steam plants.

(viii) There is no standby loss for diesel power plants.

(ix) The plants can be easily located nearer to load center.

Disadvantages

In addition to the above advantages, the diesel plants have the following disadvantages.

(i) Diesel plants do not work satisfactorily for over loads.

(ii) The cost of diesel is high.

(iii) The plant capacity is limited.

(iv) The life of diesel plants is less compared to the steam plant.

In addition to the above disadvantages, the system of generation suffers from the operational and constructional difficulties.

KEY NOTES

- The available water head is less than 30 m, such plants are called low-head water plants.
- The available water head is between 30 and 100 m, such plants are called medium-head water plants.
- The available water head is more than 100 m, such plants are called high-head water plants.
- Hydroelectric plants can be classified as:
  (i) Runoff river plants without pondage.
  (ii) Runoff river plants with pondage.
  (iii) Reservoir plants.
- The water flowing from the dam is received by an enlarged body of water at the intake. It is called the forebay.
- An economizer is a feed-water heater. It extracts a part of the heat carried away by the flue gases up to the chimney and uses it to heat the feed water to the boiler.
- A boiler or a steam generator is one of the most important equipment in a thermal station. It consists of a closed vessel into which water is allowed and is heated to convert it into steam at the required pressure.
- Depending upon the contents of the tubular heating surface, the boilers are classified as:
  (i) fire tube boilers and
  (ii) water tube boilers.
- Superheater is one of the auxiliary equipment used to increase the efficiency of a boiler, in addition to such others as air preheaters (economizers) and feed-water heaters.
- A superheater is used to remove the last traces of moisture from the saturated steam which is leaving the boiler tube and to raise the temperature of the steam.
- Superheaters may be classified into
  (a) convection type, (b) radiant type, and
  (c) the combination of the convection and radiant types.

- In the turbines, the steam expanded in the nozzles attains a high velocity. The steam jet impinges on the blades of rotor, which may be a built-up rotor or an integral rotor.
- In a reaction turbine, the expansion of the steam takes place only partially in the nozzle.
- A condenser condenses the steam exhausted from the turbine. It helps maintain a low pressure (below the atmospheric pressure) at the exhaust. This use of a condenser in a power plant improves the efficiency.
- The condensers are of two types. They are:
  (i) mixing type or jet condensers and
  (ii) non-mixing type or surface condensers.
- Depending upon the method of creating air movement through the cooling towers, they can be classified as:
  (i) natural draught cooling towers,
  (ii) forced draught cooling towers, and
  (iii) induced draught cooling towers.
- In modern power plants, the purpose of the chimney is to discharge the exhaust gases into the atmosphere at a high elevation so as to avoid the nuisance to the people living in the locality.
- Coolant removes heat from the fuel elements and transfers it to the water.
- Multiplication factor (k) is the ratio of the neutrons in one generation and the immediately preceding generation.
- If $k < 1$, it implies that the number of neutrons that can initiate the fission reactions gradually decreases and therefore the process dies down.
- If $k > 1$, it implies that more and more neutrons will be produced as the number of fission reaction increases and as a consequence, a nuclear explosion occurs as in the case of an atomic bomb.
- The thermal efficiency of gas turbine plant is 20–25%.
- The thermal efficiency of steam power plant is 25–30%.
- In India, gas turbine plant of size 70 MW was situated at Namrup in Assam, working as a base-load plant with natural gas fuel.
- Second one Uran Gas Turbine Power Plant was placed at Maharastra.
- Continuous combustion constant pressure type combustion cycle working on constant pressure principle known as joule (or) Brayton cycle.
- The explosion or constant volume type combustion cycle working on constant volume principle known as Atkinson cycle.

SHORT QUESTIONS AND ANSWERS

(1) What are the types of gas turbine power plants based on the process of combustion?

(a) Continuous combustion constant pressure type.

(b) The explosion (or) constant volume type.

(2) What are the types of gas turbine power plants according to the path of working substance?

(a) Open-cycle gas plant.

(b) Closed-cycle gas plant.

(3) Give any two applications of gas turbine plants?

Applications of gas turbine plants are:

(i) The plants are used to supply mechanical drive for auxiliaries.

(ii) Gas turbine plants are widely used in aircrafts and ships.

(4) What is lubricating system?

This system consists of lubricating oil tank, pumps, filters, etc., which is necessary to reduce ware on the rubbing parts and friction.

(5) What is circulating system?

This system circulates sufficient amount of water around the engine in order to maintain desired temperature. The hot water recooled in the cooling ponds again recirculated into the system.

(6) Give any two applications of diesel plants?

(i) Diesel plants are widely used for generating power ranging from 100 to 5,000 H.P.

(ii) Diesel plants can be used as standby plants for steam and hydropower plants.

(7) What is the necessity of energy auditing?

The main aim of the energy auditing is to reduce the energy consumption without loosing the quantity or the quality of the product or to reduce the operating cost.

(8) What are the types of energy auditing?

The types of energy auditing are:

(i) preliminary auditing and

(ii) detailed auditing.

(9) What is preliminary auditing?

This auditing is carried out within time from 1 to 10 days. Preliminary audit survey reveals the energy cost and the wastages in the major process.

(10) What is detailed auditing?

This auditing gives the detailed analysis of well-defined projects with their priorities. Detailed auditing is carried out in the limited time of 1–10 weeks.

MULTIPLE-CHOICE QUESTIONS

(1) Hydro plant installed capacity depends on:

(a) Storage

(b) Discharge

(c) Head

(d) All

(2) For high-power output, which of the following turbine is used?

(a) Francis

(b) Kaplan

(c) Pelton

(d) Propeller

(3) For medium output about (330,000 H.P.), which of the following turbines is used?

(a) Franics

(b) Kaplan

(c) Pelton

(d) Propeller

(4) For low output (about 150,000 H.P.), which of the following turbines is used?

(a) Franics

(b) Kaplan

(c) Pelton

(d) Propeller

(5) For high-head plants, the permissible velocity of water through the penstock is:

(a) 8 m/sec

(b) 4 m/sec

(c) 2 m/sec

(d) 10 m/sec

(6) For medium-head plants, the permissible velocity water through the penstock is:

(a) 8 m/sec

(b) 4 m/sec

(c) 2 m/sec

(d) 10 m/sec

(7) For low-head plants, the permissible velocity of water through the penstock is:

(a) 8 m/sec

(b) 4 m/sec

(c) 2 m/sec

(d) 10 m/sec

(8) A Kaplan turbine is:

(a) Inward flow, impulse turbine

(b) Outward flow, reaction turbine

(c) A high-head mixed-flow turbine

(d) Low-head mixed-flow turbine

(9) A mass curve can be plotted form:

(a) Load–duration curve

(b) Chronological load curve

(c) Energy load curve

(d) Both a and b

(10) A mass curve drawn for hydroelectric power stations essentially gives:

(a) Storage requirement

(b) Number of units generated

(c) Amount of water utilized

(d) None

(11) In India, the first tidal power plant is likely to come up in:

(a) Bay of Bengal

(b) Korba

(c) Singrauli

(d) Gulf of Kutch

(12) Pelton Turbines suitable for:

(a) Low head

(b) Medium heads

(c) High heads

(d) Low and medium heads

(13) The efficiency of pumped storage plant is generally about:

(a) 95%

(b) 85%

(c) 65%

(d) 40%

(14) Which of the following have generally salient pole construction?

(a) Alternators used in thermal plants

(b) Alternators used in hydro plants

(c) Both a and b

(d) Neither a nor b

(15) The load factor for which of hydroelectric plants is high:

(a) Runoff river plants without pondage

(b) Runoff river plants with pondage

(c) Storage reservoir plants

(d) Pumped storage plants

(16) A surge tank stabilizes:

(a) Generation

(b) Sudden rise of electric power

(c) Pressure difference in water

(d) Sudden loss of electric load

(17) The purpose of trash rack is:

(a) To reduce water hammer effect

(b) To reduce cavitation effect

(c) To avoid entry of debris

(d) To avoid excess water to enter in

(18) A surge tank is provided near:

(a) Penstock

(b) Trash rack

(c) Spillway

(d) Turbine

(19) The regulating reservoir storing water temporarily when load on plant is reduced is:

(a) Forebay

(b) Surge tank

(c) Reservoir

(d) None

(20) The purpose of spillway is:

(a) To discharge water

(b) For irrigation purpose water connection

(c) It discharges excess water

(d) None

(21) Run away speed of turbine is:

(a) Full load speed

(b) Running speed

(c) Speed of turbine under no load and governor failure

(d) None

(22) The type of water turbine used for heads greater than 600 m is:

(a) Francis

(b) Pelton

(c) Kaplan

(d) Propeller

(23) In an hydroelectric power plant, a conduit system for carrying water from the intake to the turbines is:

(a) Reservoir

(b) Penstock

(c) Dam

(d) Surge tank

(24) In high-head hydroelectric power plant, the velocity of water flow in penstock is around:

(a) 2m/sec

(b) 4m/sec

(c) 7m/sec

(d) 10m/sec

(25) The first nuclear power plant was commissioned in:

(a) USA

(b) USSR

(c) India

(d) France

(26) The function of a surge tank is to:

(a) Supply water at constant pressure

(b) Relieve water hammer pressures

(c) Produce surge in the pipe line

(d) None of the above

(27) Water hammer is developed in:

(a) Surge tank

(b) Water turbine

(c) Penstock

(d) Draft tube

(28) A graphical representation of the discharge and time is known as:

(a) Load curve

(b) Load-duration curve

(c) Monograph

(d) Hydrograph

(29) The enriched uranium consists of approximately:

(a) 10% of U^{235} and 90% of U^{238}

(b) 20% of U^{235} and 80% of U^{238}

(c) 30% of U^{235} and 70% of U^{238}

(d) None of these

(30) The electrical power developed by a hydroelectric plant kW is given by the expression:

(a) $\frac{0.736}{75}\,WQH\eta$

(b) $\frac{75}{0.736}\,WQH\eta$

(c) $175 \times 0.736\,WQH\eta$

(d) $\frac{7}{75\times0.736}\,WQH\eta$

(31) The specific speed (N_s) of a turbine is given by the expression:

(a) $N_s = \frac{\sqrt{P}N}{H^{1.5}}$

(b) $N_s = \frac{\sqrt{N}P}{H^{1.5}}$

(c) $N_s = \frac{\sqrt{P}N}{H^{1.25}}$

(d) $N_s = \frac{\sqrt{P}N}{H^{2/3}}$

(32) An hydrograph indicates:

(a) The discharge at any time during the period under consideration

(b) The maximum and minimum runoff during the period

(c) The average runoff during the period

(d) All of the above

(33) The rotor used in alternators of hydroelectric stations is:

(a) Cylindrical rotor

(b) Cage rotor

(c) Salient rotor

(d) Round rotor with A.C. excitation

(34) The plot of water flow available in a stream against percentage of time used for assessment hydropower is called:

(a) Hydrograph

(b) Mss curve

(c) Flow-duration curve

(d) Load flow

(35) Thermal power station is suitable as:

(a) Base-load plant

(b) peak-load plant

(c) Base or peak load

(d) None

(36) Whenever the boiler pressure exceeds 70 kg/cm² invariably we use:

(a) Superheater

(b) Condenser

(c) Turbine

(d) Economizer

(37) The temperatures inside the water tube boiler is:

(a) 300°C

(b) 400°C

(c) 560°C

(d) 720°C

(38) Which of the following is considered as superior quality of coal?

(a) Peat

(b) Coke

(c) Bituminous coal

(d) Lignite

(39) Ash content of Indian coal is approximately:

(a) 5%

(b) 8%

(c) 10%

(d) 25%

(40) Water is supplied to boiler:

(a) At 120 kg/m^2

(b) At atmospheric pressure

(c) At slightly more than atmospheric pressure

(d) At more than the steam pressure in the boiler

(41) The efficiency of ESP is as high as:

(a) 99.6%

(b) 90%

(c) 85%

(d) 80%

(42) The addition of ESP apparatus in a thermal plant:

(a) Increases the efficiency of the plant

(b) Increases the reliability of the plant

(c) Decreases both the efficiency and the reliability

(d) Has no efficiency and reliability of plant

(43) The useful life of thermal and hydroplants are:

(a) 25 and 125 years

(b) 40 and 160 years

(c) 45 and 175 years

(d) 100 years

(44) The machines used in hydro and thermal plants run at:

(a) 300 and 3,000 r.p.m.

(b) 500 and 6,000 r.p.m.

(c) 3,000 and 300 r.p.m.

(d) 6,000 and 600 r.p.m.

(45) Thermal power station works on:

(a) Carnot cycle

(b) Otto cycle

(c) Rankine cycle

(d) Diesel cycle

(46) Which of the following coal is superior?

(a) Bituminous

(b) Peat

(c) Lignite

(d) Coke

(47) The overall efficiency of thermal power station is less than 30% this mainly due to:

(a) Power consumption of its auxiliaries is high

(b) Heat loss at various components

(c) Because of using low-quality grade coal

(d) Because of using low-efficiency turbine generator

(48) An ESP remove dust particles in:

(a) Bottom

(b) Fly Ash

(c) Coal

(d) None

(49) In Indian largest thermal power station is located at:

(a) Chandrapur

(b) Kota

(c) Neyveli

(d) Semi

(50) In a steam turbine cycle, the lowest pressure occur in:

(a) Condenser

(b) Turbine inlet

(c) Boiler

(d) Superheater

(51) In which part of thermal power plant, the steam pressure is less than that of atmosphere?

(a) Boiler

(b) Turbine

(c) Superheater

(d) Condenser

(52) The major function of the condenser is to:

(a) Remove the condensate for boiler feed water

(b) Condense steam

(c) Reduce the back pressure so that minimum heat energy can be extracted from steam

(d) Provide a closed cycle

(53) The largest size of steam turbine installed in India is:

(a) 100 MW

(b) 250 MW

(c) 500 MW

(d) 1,000 MW

(54) Pulverized coal is:

(a) Non-smoking

(b) Coal free from ash

(c) Coal broken in fine particles

(d) Coal which burns for long time

(55) The coal of the lowest calorific value is:

(a) Anthracite

(b) Bituminous coal

(c) Lignite

(d) Steam coal

(56) The coal having the highest calorific value is:

(a) Anthracite

(b) Steam coal

(c) Char cola

(d) Coke

(57) Nuclear power station is suitable as:

(a) Base-load plant

(b) peak-load plant

(c) Base or peak load

(d) None

(58) Pick out fertile material:

(a) U^{235}

(b) U^{233}

(c) P_u^{239}

(d) U^{238}

(59) Pick out fissile material:

(a) U^{238}

(b) Th^{232}

(c) U^{235}

(d) None

(60) The purpose of control rods in nuclear reactor:

(a) Slowing down fast neutrons

(b) Absorbing neutrons

(c) To reflect neutrons

(d) None

(61) The material used for control rod is:

(a) Heavy water (D_2O)

(b) Carbon

(c) Cadmium

(d) Boron

(e) Both c and d

(62) Commonly used coolant is:

(a) Hydrogen

(b) CO_2

(c) Sodium

(d) Helium

(63) The moderator used in fast breeder reactor is:

(a) Heavy water (D_2O)

(b) Carbon

(c) Boron

(d) None

(64) Half-life means:

(a) Life of nuclear reactor

(b) Life or radioactive material

(c) Time required disintegrating the nuclei to one-half of its original value

(d) None

(65) The moderator used in first breeder reactor is:

(a) Heavy water

(b) Graphite

(c) Ordinary water

(d) Any of the above

(66) Natural uranium contains:

(a) 0.7% U^{235}

(b) 50% U^{235}

(c) 99.3% U^{235}

(d) 100% U^{235}

(67) The energy released per neutron in fusion process:

(a) Greater than fission

(b) Equal to fission

(c) Less than fission

(d) No comparison possible

(68) Which of the following material is used as moderator?

(a) Graphite

(b) Boron

(c) Sodium potassium liquid

(d) Plutonium

(69) It is desirable to use reactor core as:

(a) Cubical or cylinder

(b) Cubical or spherical

(c) Cylindrical or spherical

(d) Spherical

(70) If natural uranium is used as the fuel, the moderator to be used is:

(a) Heavy water

(b) Graphite

(c) Ordinary water

(d) None of the above

(71) The pressurized water reactors employ:

(a) Light water and natural uranium

(b) Heavy water and enriched uranium

(c) Light water and enriched uranium

(d) None of the above

(72) Which of the following has maximum number of auxiliaries?

(a) Hydro

(b) Thermal

(c) Nuclear

(d) None

(73) The largest size of hydroelectric generating unit and thermal generating unit in India are:

(a) 500 MW and 500 MW

(b) 310 MW and 310 MW

(c) 165 MW and 500 MW

(d) 210 MW and 165 MW

(74) Large size steam plants and nuclear plants are suitable for:

(a) Base loads

(b) Intermittent loads

(c) Peak loads

(d) Both a and c

(75) A Nuclear power plant should preferably be located:

(a) Near a coal field

(b) Near a reservoir

(c) Away from coal field and reservoir dam

(d) Near load center

(76) A moderator material should have:

(a) Low atomic mass

(b) Large atomic mass

(c) Moderate atomic mass

(d) Both a and b

(77) A breeder reactor is one which:

(a) Produces heat without any recovery of fertile material

(b) Converts fertile material into fissile material

(c) Produces heat and consumers fissile material

(d) Produces heat and converts fertile material into fissile material

(78) Which of the following plants consume more power for their auxiliaries to operates?

(a) Thermal

(b) Hydro

(c) Nuclear

(d) Diesel

(79) The energy equivalent of mass defect is called:

(a) Recombining energy

(b) Mass energy

(c) Fission energy

(d) Binding energy

(80) One a.m.u. is approximately equal to:

(a) 1.66×10^{-24} kg

(b) 1.66×10^{-27} kg

(c) 1.6×10^{-16} kg

(d) 1.6×10^{-13} kg

(81) Heavy water is:

(a) H_2O

(b) HO_2

(c) D_2O

(d) B_2O

(82) The function of reflector in a nuclear reactor is to:

(a) Bounce back most of the neutrons that escape from the fuel core

(b) Reduce the speed of the neutrons

(c) Stop the chain reaction

(d) All of the above

(83) Reflectors of a nuclear are made of:

(a) Cast iron

(b) Beryllium

(c) Steel

(d) Boron

(84) The essential requirement(s) of the power plants to be operated as base-load plants is/are:

(a) Low-operating cost

(b) The capability of operating continuously for long periods

(c) Requirement of few operating personnel and economical repair

(d) All of the above

(85) Interconnected systems have the advantage of:

(a) Reduced reserve plant capacity, capital cost per kW, and economy in operation

(b) Improved load factor, diversity factor and operation efficiency, and increased reliability of supply

(c) All of the above

(d) None of the above

(86) An interconnected system has the following power plants:

(a) Nuclear

(b) Steam

(c) Diesel

(d) Gas turbine

(e) Hydro with storage

(f) Runoff the river

(g) Pumped storage

(87) Energy produced by fission reaction of uranium having mass of atom *m* and velocity of *j* of light c is:

(a) mc

(b) $1/2\ (mc^2)$

(c) mc^2

(d) (m^2c)

REVIEW QUESTIONS

(1) What is hydrology? Give the significance of hydrograph regarding to hydro plant.

(2) Give the advantages and disadvantages of hydroelectric plants.

(3) Enumerate the factors affecting the site selection of hydroelectric plants.

(4) Explain in detail the classification of hydroelectric plants based on water head level.

(5) Give the classification of hydroelectric plants based on load level.

(6) Discuss the classification of hydroelectric plants based on regulation of water flow.

(7) Explain in detail various components and function of a hydroelectric generation system.

(8) Draw a neat sketch and explain the function of thermal power plant.

(9) Explain the following:

(i) Feed-water heater and

(ii) Boiler.

(10) What are the various types of steam turbines and give their uses.

(11) What is condenser and explain its function.

(12) Explain in detailed the working principle of a nuclear power station.

ANSWERS

1. d	23. b	45. c	67. a
2. a	24. c	46. a	68. a
3. c	25. a	47. b	69. a
4. b	26. b	48. b	70. b
5. a	27. c	49. a	71. c
6. b	28. d	50. a	72. b
7. c	29. a	51. d	73. c
8. d	30. a	52. c	74. a
9. b	31. c	53. c	75. c
10. a	32. d	54. c	76. a
11. d	33. a	55. c	77. d
12. c	34. b	56. a	78. a
13. c	35. c	57. c	79. c
14. b	36. d	58. d	80. b
15. d	37. c	59. c	81. c
16. c	38. c	60. d	82. a
17. c	39. d	61. e	83. b
18. d	40. d	62. c	84. d
19. a	41. a	63. d	85. c
20. c	42. c	64. c	86. b
21. c	43. a	65. d	87. b
22. b	44. a	66. a	

Non-conventional Power Generation

OBJECTIVES

After reading this chapter, you should be able to:

- know various non-conventional energy sources
- generate electric power by utilizing non-conventional energy

2.1 INTRODUCTION

A plenty of energy is needed for industrial growth and agricultural production. The world's fossil fuels or the conventional sources of energy such as coal, oil uranium, petroleum, and natural gas are not adequate for future increasing energy demands and may be depleted and exhausted in few hundred years until we exploit other sources of energy. Consequently, non-conventional and renewable sources have to be developed by the scientists for future energy requirements.

2.2 GENERATION OF ELECTRICAL POWER BY NON-CONVENTIONAL METHODS

The various non-conventional energy sources are:

1. solar energy,
2. wind energy,
3. tidal energy,
4. geothermal energy,
5. magneto-hydrodynamics (MHD) generator,
6. thermionic converter,
7. energy from biogas and biomass,
8. ocean thermal energy conversions,
9. hydrogen energy,
10. fuel cells, and
11. thermo-electric power.

The percentage use of various sources for the total energy consumption in the world is given in Table 2.1.

TABLE 2.1 *Energy consumption in the world*

Coal	32.5	92%
Oil	38.3	
Gas	19	
Uranium	0.13	
Water	2	
Wood	6.6	8%
Dung	1.2	
Wastage	0.3	

Referring to Table 2.1, the world's energy supplied from commercial or conventional energy such as coal, oil, gas, uranium, and water up to 92%. In many developing countries, non-conventional energy such as wood, animal dung, and agricultural wastage would serve 8% of total energy used in the world.

Main advantages of non-conventional energy sources are:

- Atmospheric pollution is less.
- These sources are available in large scale at free of cost.
- These sources are well suited for decentralized use.
- Maintenance is less.

2.3 SOLAR ENERGY

Solar energy is very large and inexhaustible source of energy. It comes from the sun to the earth. This energy is cheap and free from pollution. The earth receives nearly 4,000 trillions kWh of energy from the sun. Normally, solar power at the atmosphere around the sun is 10^{17} W but solar power at the atmosphere around the earth is 10^{6} W. Now, total power world requires for all needs of civilization is only 10^{13} W, i.e., sun gives nearly 1,000 times more than energy what we actually need. If we use only 5% of this energy, it is sufficient for the worldwide energy requirement.

Solar radiation, which is not absorbed or scattered by the atmosphere, reaches the ground directly from the sun is known as direct radiation. The radiation received after scattering is called diffuse radiation. The diffuse radiation comes to earth from all parts of the sky. The total solar radiation received at any point on the earth's surface is the sum of total direct radiation and diffuse radiation. Figure 2.1 shows the solar energy storage.

2.3.1 Solar energy collector

Solar energy collectors are used to collect and absorb the solar energy radiated from the sun. The solar energy collectors are essential devices for the system of converting the solar energy into the desired form such as heat or electricity.

Generally used solar energy collectors are of two types. They are:

(i) Non-concentrating or flat plate type solar collector.

(ii) Concentrating or focusing type solar collector.

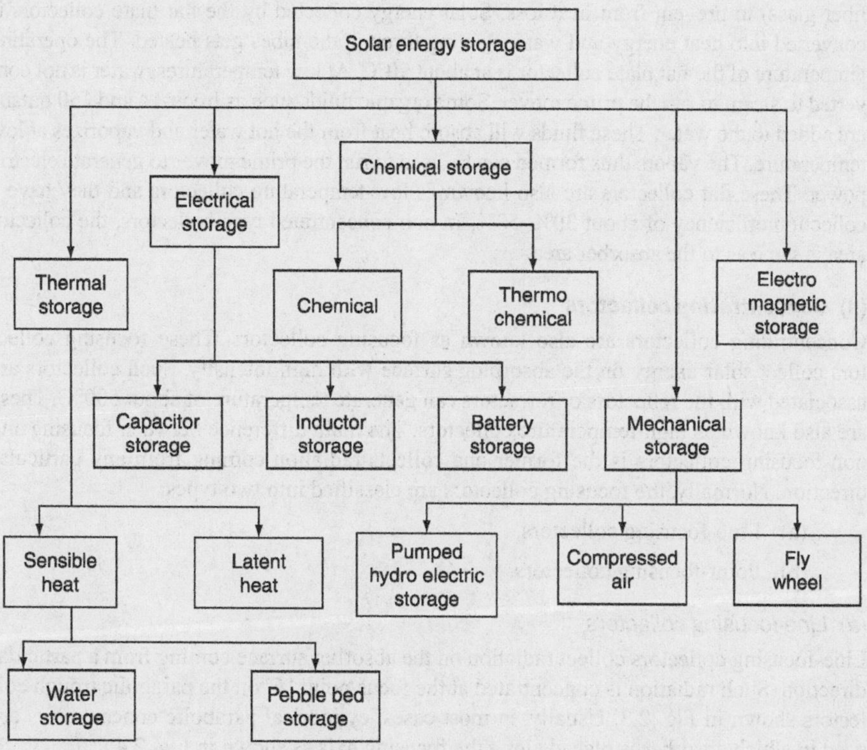

FIG. 2.1 *Solar energy storage*

(I) *Non-concentrating or flat type solar collectors*

Non-concentrating or flat type solar collectors are solar energy collectors which may collect and absorb both direct and scattered solar radiation. These collectors are made in the form of rectangular panels with an area of about 1.7–2.9 sq. m. Construction of such flat plate collector is quite simple and is shown in Fig. 2.2.

The absorbing surface of the solar flat plate collector is made up of copper, aluminum, or steel coated with carbon, which absorbs solar energy. The solar collectors are associated with the water-circulating tubes; these tubes are coated with insulating materials (such as

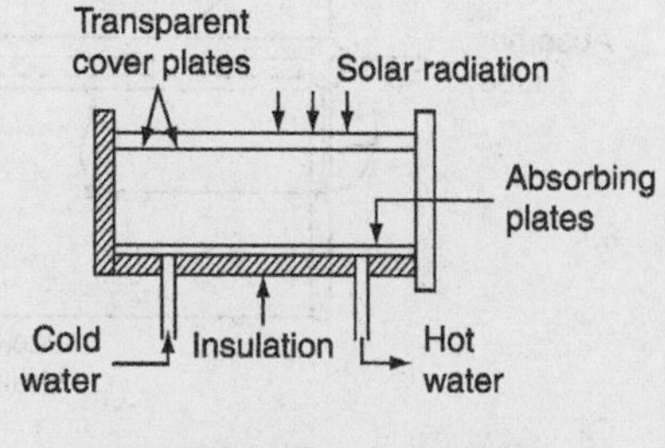

FIG. 2.2 *Flat plate collector*

fiber glass) to prevent from heat loss. Solar energy collected by the flat plate collectors is converted into heat energy and water flowing through the tubes gets heated. The operating temperature of the flat plate collector is at about 90°C. At low temperatures, water is not converted to steam to run the prime mover. Some organic fluids such as freon-14 and 150 butane are added to the water. These fluids will absorb heat from the hot water and vaporizes at low temperature. The vapors thus formed can be used to run the prime mover to generate electric power. These flat collectors are also known as low-temperature collectors and they have a collection efficiency of about 30%–50%. In non-concentrated type collectors, the collector area is same as to the absorber area.

(ii) *Concentrating collectors*

Concentrating collectors are also known as focusing collectors. These focusing collectors collect solar energy on the absorbing surface with high intensity. Such collectors are associated with the reflectors or refractors can generate temperature of about 500°C. These are also known as high-temperature collectors. The main difference between focusing and non-focusing collectors is the former one collects radiation coming from any particular direction. Normally, the focusing collectors are classified into two types:

(a) Line-focusing collectors.

(b) Point-focusing collectors.

(a) Line-focusing collectors

Line-focusing collectors collect radiation on the absorber surface coming from a particular direction. Such radiation is concentrated at the focus point '*F*' on the parabolic trough collectors shown in Fig. 2.3. Usually, in most cases, cylindrical parabolic concentrators are used in which absorber is placed along the focusing axis as shown in Fig. 2.4.

The length of the reflector unit is about 3–5 m and width is about 1.5–2.4 m. Parabolic reflector is usually made up of polished aluminum, silvered glass, etc.

(b) Point-focusing collectors

A point-focusing collector is in the form of a paraboloidal shape. A paraboidal dish reflector concentrates solar radiation at a focus point shown in Fig. 2.5. A paraboloidal dish is made

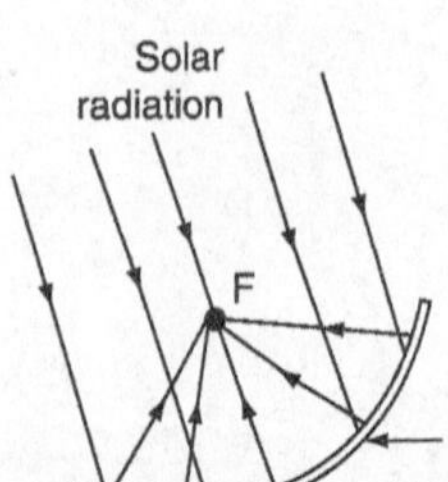

FIG. 2.3 *Parabolic trough collector*

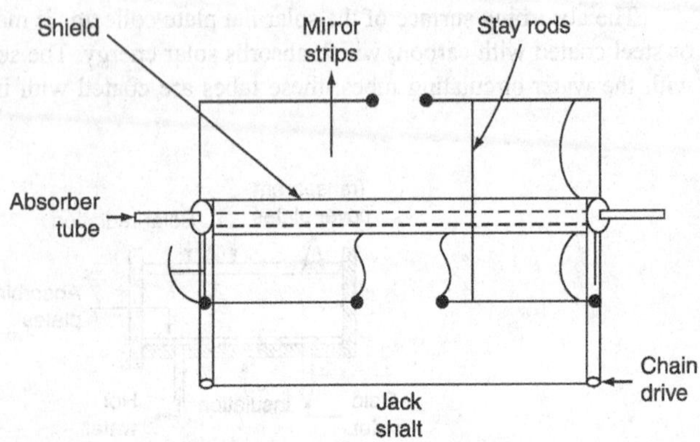

FIG. 2.4 *Cylindrical parabolic concentrator*

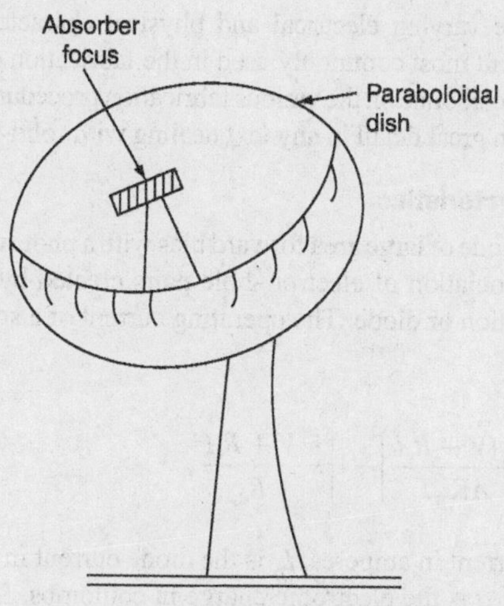

FIG. 2.5 *Point focusing collector*

with 200-curved mirror segments and each of them is known as heliostat. The dish diameter is about 6.5 m. The absorber is a cavity and is made up of zirconium–copper alloy and is coated with black chrome, which is located at the focus point.

In these collectors, the heat transferred into and out of the absorber cavity through pipes bonded to the interior dish structure. The dish can be moved in any direction thereby focusing the sun rays on the absorber properly.

2.4 POINT-FOCUSING COLLECTOR

Concentrating collectors have many advantages over flat collectors

- The structure of reflecting surface is less.
- Collecting system cost is less.
- The generating temperature of the concentrating collectors is higher than the flat collectors.
- The absorber area of the concentrator system is smaller than the flat plate system. Thus, the intensity of sun radiation will be more.
- The concentrated collectors have more efficiency.

Disadvantages

- The initial cost of concentrated collectors is high.
- Flux distribution over the absorber area is non-uniform but whereas the flat collectors flux distribution is uniform.
- Reflector system to track the sun is costlier.

2.4.1 Photovoltaic cells or solar cells

The solar cell is the basic unit of the photovoltaic generator. The solar cell is the device that transforms the sun's rays or photons directly into electricity. There are various models of solar cells made with different technologies available in the market today.

These models have varying electrical and physical characteristics depending on the manufacturer. The element most commonly used in the fabrication of solar cells is silicon. In this research, we will not elaborate on the various fabrication procedure processes or techniques. This subject is covered in great detail in any text dealing with solid-state electronics.

2.4.2 Solar cell characteristics

A solar cell is simply a diode of large area forward bias with a photovoltage. The photovoltage is created from the dissociation of electron–hole pairs created by incident photons within built-in field of the junction or diode. The operating current of a solar cell is given by:

$$I = I_{ph} - I_D$$

$$= I_{ph} - I_o\left[\exp\left(\frac{q(V+R_sI)}{AK_BT}\right)-1\right] - \frac{V+R_sI}{R_{sh}}, \qquad (2.1)$$

where I_{ph} is the photocurrent in amperes, I_D is the diode current in amperes, I_o is the saturation current in amperes, q is the electronic charge in coulombs, K_B is the Boltzmann constant in joules per kelvin, T is the junction temperature in kelvin, R_s is the series resistance in ohms, R_{sh} is the shunt resistance in ohms, and A is the ideality factor.

Under the darkness, the solar cell is not an active device. It functions primarily as a diode. Externally, the solar cell is an energy receiver that produces neither a current nor a voltage. Under this condition, if the solar cell is connected to an external supply, theory shows that the voltage and current are related by the diode equation given by:

$$I_D = I_o\left[\exp\left(\frac{q(V+R_sI)}{AK_BT}\right)-1\right]. \qquad (2.2)$$

Since the ultimate photovoltaic generator will be composed of N cells in series and M cells in parallel, the I–V characteristics of the whole generator can be derived by scaling the I–V characteristics of one cell with a factor of N in voltage and M in current. This approach is correct only when the cells are identical.

Electrical characteristics of solar cells

The graph of current as a function of voltage ($I = f(V)$) for a solar cell passes through three significant as illustrated in Fig. 2.6.

(a) Short-circuit current

The short-circuit current, I_{sc}, occurs on a point of the curve, where the voltage is zero. At this point, the power output of the solar cell is zero.

(b) Open-circuit voltage

The open-circuit voltage, V_{oc}, occurs on a point of the curve, where the current is zero. At this point, the power output of the solar cell is zero.

(c) Operation at maximum power

The maximum power output occurs at point A on the curve. The point A is usually referred to as the 'knee' of the V–I curve. The electrical characteristics of the solar cells are based on their V–I curves. The V–I curve is based on the cell being under the standard conditions of

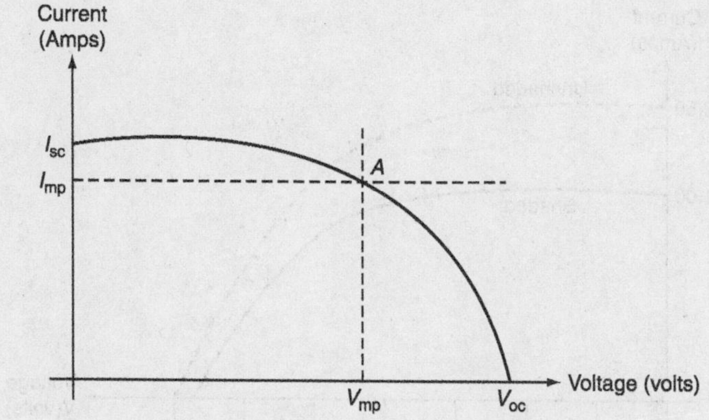

FIG. 2.6 *Solar cell V–I characteristics*

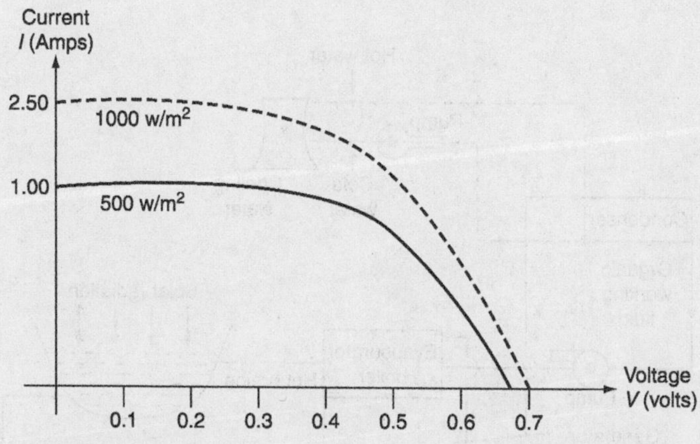

FIG. 2.7 *Solar cell V–I characteristics at one sun and one half suns*

sunlight and cell temperature, and assumes there is no shading in the cell. Standard sunlight conditions on a clear day are assumed to be 1,000 W of solar energy per square meter (1,000 W-m^{-2} or 1 kW-m^{-2}). This condition is sometimes called 'one sun' or 'peak sun' when the cell is operating in conditions less than one sun, the current output of the cell is reduced as shown in Fig. 2.7. Since PV cells are electrical semiconductors, partial shading may cause the cell to heat up. Under this condition, the cells act as an inefficient conductor rather than an electrical generator. Partial shading may run shaded cells and also affect the power output of the cell. Figure 2.8 shows the *V–I* characteristics of shaded and unshaded cell.

2.4.3 Solar power generation

Solar power generation plant is shown in Fig. 2.9. Solar power generation plant employs different power cycles depending upon the temperature of working fluid, as low-, medium-, and high-temperature cycles. A low-temperature cycle uses flat plat collectors to collect solar energy. So, the maximum temperature of the fluid is limited to 100°C. Medium-temperature

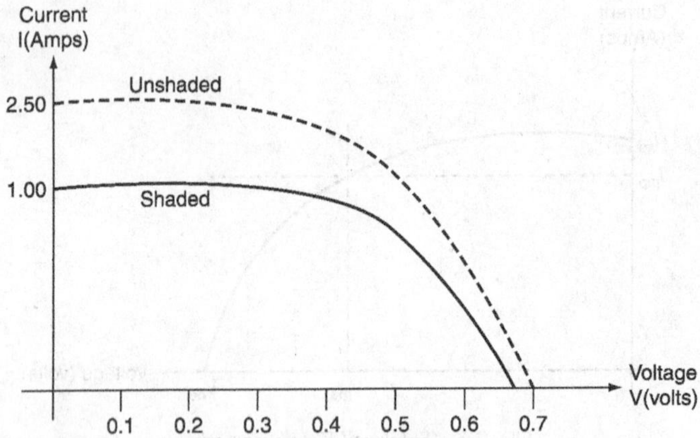

FIG. 2.8 *V–I characteristics of a shaded and unshaded solar cell*

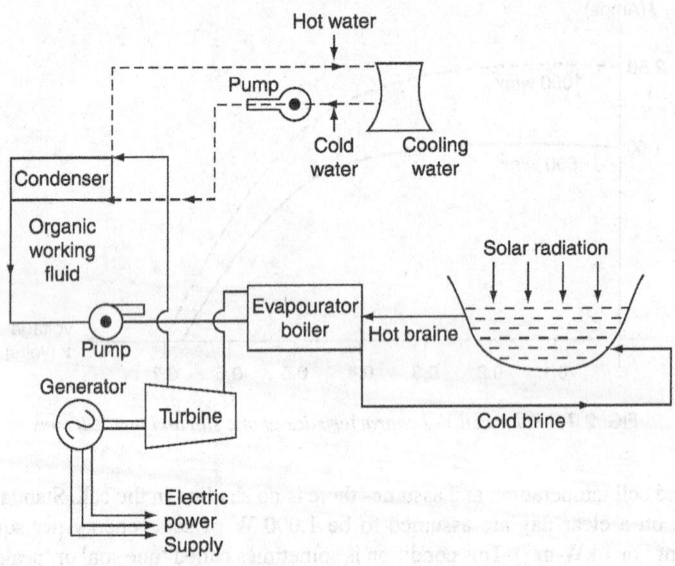

FIG. 2.9 *Solar power generation plant*

and high-temperature cycles use the concentrating collectors to collect solar energy, so the maximum temperature of the fluid is limited from 150°C to 300°C for the medium-temperature cycles and above 300°C for the high-temperature cycles.

Thermodynamic cycles preferred for low and medium temperature are the rankine cycles; for high temperatures, Brayton and Stirling cycles are also used.

In a solar power generation plant, solar energy is collected by the solar pond and flat plate collectors. The solar energy collected by the flat plate collector is utilized to raise the temperature of fluid. The fluid from pond may be directly used for various cycles such as rankine or Brayton or passed through the heat exchanger; there organic fluids are heated and converted into vapor or steam. The vapor or steam is fed to the turbine blades used to rotate the shaft of electric generator coupled the turbine.

The vapor from the turbine is fed to the condenser, where cold water from the cooling tower condenses the vapor into liquid and is again fed back to the boiler, where fluid is reheated to convert it into steam then pumped to the turbine, and the cycle is repeated.

2.4.4 Advantages and disadvantages of solar power

Some of the advantages of converting solar energy into electric power are:

- Solar power conversion system has no moving parts.
- Absence of pollution.
- Highly reliable.
- Less maintenance cost.
- The average life of photovoltaic cells is high.
- The efficiency of conversion system is high because of the absence of moving parts.
- Solar energy is available at free of cost, thus there is no consumption of fuel.
- The power-handling capability of system is very large.

The main disadvantage of the solar power generation system is high initial cost; this is mainly due to the absence of the sun light during night time, so that additional equipment such as batteries are used to store the energy.

2.4.5 Applications of solar energy

Solar energy has wide applications such as:

- Water pumping for drinking water supply.
- Irrigation purpose in rural areas.
- Street lighting.
- Battery charging and weather monitoring.
- Railway signaling equipment.

2.5 WIND ENERGY

Wind results from air in motion. Air in motion arises from the pressure gradient. The wind is basically caused by solar energy radiating the earth. The useful work done for the conversion of kinetic energy of the wind into mechanical energy can be utilized to generate the electricity. Most of the machines for converting wind energy into mechanical energy consist of number of sails, vanes, or blades radiating air from the hub or the central axis. When wind blows against the vanes or the blades, they rotate about the axis and the rotational motion can be used to perform the useful work.

Wind energy conversion devices are known as wind turbines, because they convert wind stream into energy of rotation because the wind turbine produces rotational motion. Wind energy is readily converted into electrical energy by connecting the turbine to an electric generator.

2.5.1 Basic principle of wind energy conversion

Wind possesses energy by virtue of its motion. Any energy conversion device can extract this and convert it into useful work depending on:

(i) the wind speed,

(ii) the cross-section of wind swept by the rotor, and

(iii) the overall efficiency of the rotor and generator efficiency.

The power in the extracted wind can be found out by kinetics concept. The amount of air passing in unit time through an area '*A*' with velocity '*V*' is $A \times V$.

And, mass is given by:

$$M = pAV,$$

where '*p*' is the density of the air; kinetic energy of the particle is given by:

$$E = \frac{1}{2}MV^2 \quad \textbf{(2.3)}$$

$$= \frac{1}{2}pAV^3. \quad \textbf{(2.4)}$$

Equation (2.4) gives maximum wind energy available and is proportional to the cube of the wind speed. Hence, it is observed that small increase in wind speed can have noticeable effect on the power in the wind.

Since power available is proportional to density, it may vary 10–15%, because of pressure and temperature change. It is also shown that by doubling the velocity, the power available increases by eightfold. As power available is directly proportional to the cross-sectional area, it decides the diameter of the vanes for the required power. Since the area is normally circular of diameter '*D*' in horizontal axis aero turbines, then:

$$A = \frac{\pi}{4}D^2 \quad \text{sq. mtrs.}$$

Available wind power,

$$P_a = \frac{1}{2}p \times \frac{\pi}{4}D^2 \times V^3 = \frac{1}{8}pD^2v^3 \text{ watts.}$$

Strictly noted that it is not possible to convert all the wind energy into any other form of energy because the load would reduce the wind speed to zero.

2.5.2 Basic components of wind energy conversion plant

The block diagram representation of the wind energy conversion system is shown in Fig. 2.10.

The main components of the wind energy conversion system are:

(i) *Aero turbine:* Aero turbines convert wind energy into rotary mechanical energy. This block requires pitch and yaw, i.e., direction of wind flow control for proper operation.

(ii) *Mechanical interface (coupling & gearing):* A suitable mechanical gear should be provided to transmit mechanical energy into electric generator.

(iii) *Electric generator:* Generator that converts mechanical energy from the aero turbine into electrical energy and is connected to the load and or power grid.

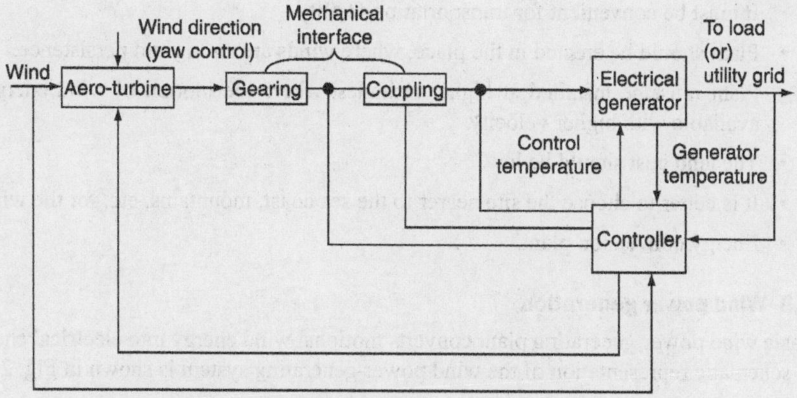

FIG. 2.10 *Block diagram of wind energy conversion system*

(iv) *Controller:* Controller that senses wind speed, wind direction, and shaft speeds. The output power from the generator and temperature is sensed by the controller and if necessary controller will send appropriate signal to the wind energy input to protect the system from abnormal conditions.

2.5.3 Types of wind mills

A wind mill is machine, which plays major role in wind energy conversion. Wind turbine that converts the kinetic energy of the wind motion to the mechanical energy transferred to an electric generator through the shaft. Electric generator converts mechanical energy into electrical energy.

Normally, based upon the axis of rotation of turbine, wind mills are classified into two types. They are:

(i) Horizontal axis wind mill.

(ii) Vertical axis wind mill.

In horizontal axis, wind mill uses motional wind energy for the rotation of shaft, in which the axis of rotation of the shaft is along horizontal axis and the aero turbine plane is vertically facing to the wind. In vertical axis wind mill type, the axis of rotation of the shaft is along the vertical axis and the aero turbine plane is horizontally facing the wind.

Horizontal axis type wind mills are further classified into various types such as single-bladed, double-bladed, multibladed, and bicycle multibladed type.

Vertical type wind mills are further classified as savonius or 's' type rotor mill and darrieus type rotor mill. Vertical axis type wind mills are having simple structure and easier to design compared to horizontal axis type wind mills.

2.5.4 Site selection for wind energy conversion plant

Various factors on site selection that are need to be considered while erecting a wind energy conversion plant is:

- Site for the wind plant should be nearer to the consumers of the generated electrical energy.

- It must be convenient for transportation facility.
- Plant should be erected in the place, where winds are strong and persistence.
- Plant must be installed at higher attitudes, where the motion of wind energy is available with higher velocity.
- The land cost should be low.
- It is better to choose the site nearer to the sea coast, mountains, etc. for the wind.
- Energy conversion plant.

2.5.5 Wind power generation

A basic wind power-generating plant converts motional wind energy into electrical energy. The schematic representation of the wind power-generating system is shown in Fig. 2.11.

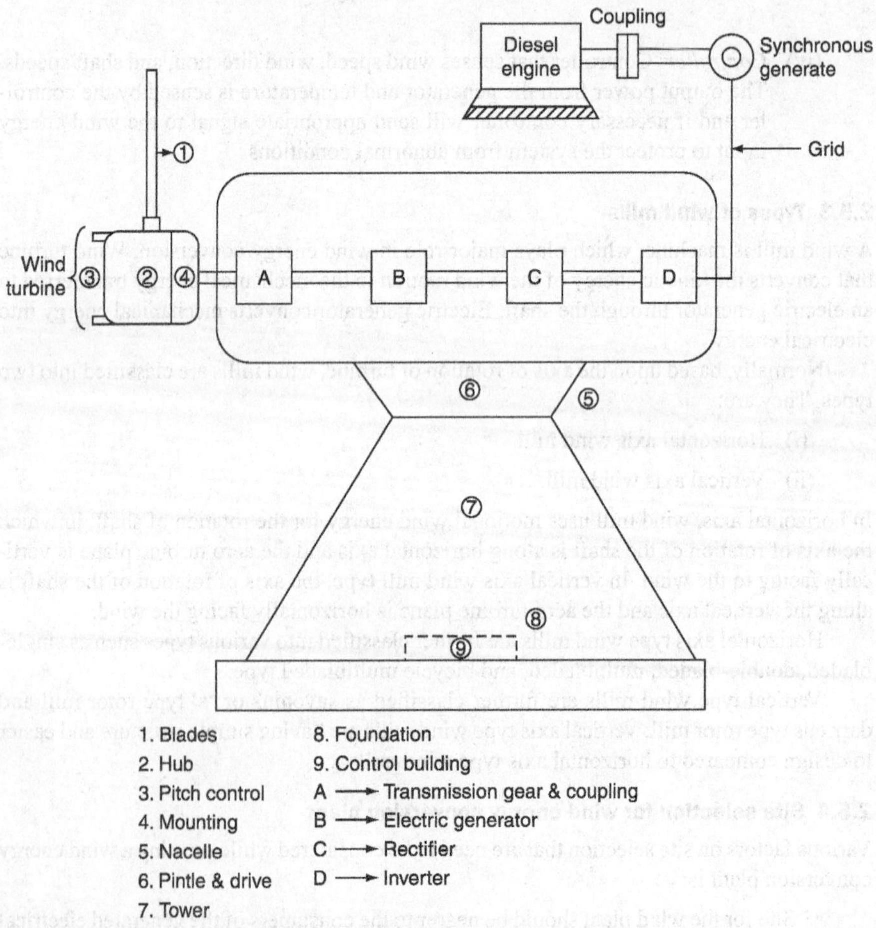

FIG. 2.11 *Wind power generating plant*

In wind energy-generation system, wind turbine converts kinetic energy of wind motion into mechanical energy with the help of blades. The direction of wind flow control, i.e., pitches and yaw control is required for the proper operation. A suitable mechanical transmission gear is provided to transmit the mechanical energy from the wind turbine to electrical generator.

An electric generator converts mechanical energy into electrical energy and is fed to the rectifier thereby converting fixed AC to variable DC supply. Further DC is fed to an inverter, which converts DC into variable AC supply, transmitted to grid system for utility purpose.

A diesel engine is used to drive a synchronous machine when there is no wind energy as input to the aero turbine.

2.5.6 Advantages and disadvantages of wind power

Advantages

- Wind is renewable source of energy.
- There is no need of using fuel for wind energy conversion system.
- There is no need of transportation facility.
- It is pollution free.
- The maintenance cost of wind energy conversion system is less for low power generation.

Disadvantages

- The availability of wind energy is fluctuating in nature.
- The auxiliary storage devices such as battery must be provided for wind energy conversion system because of the fluctuation of the wind in nature.
- Wind energy conversion systems are noisy.
- More space should be needed for wind power generation.
- The structure of wind power conversion system is complex and the weight of system is also high due to the construction of high towers.

2.5.7 Applications of wind energy

- Wind machines can generate low power for space heating and cooling of homes.
- The electric energy generated from the wind stations can be adoptable for domestic appliances.
- Low power wind energy conversion systems have been used for corrosion protection of buried metal pipelines.
- Wind power turbines up to 50 kW can be used for irrigation pumps, navigational signals, remote communications, etc.

2.6 TIDAL POWER

The periodic and continuous raise and fall of water on the surface of the sea is known as tide. These tides are caused mainly due to the gravitational force of the moon and sun on the water of the oceans. Mainly 70% of the gravitational force to produce a tide exists between the moon and the surface of the seawater and only 30% of the force to produce a tide is due to the force of the sun on the water of the ocean. Thus, the moon plays a major role in the formation of tide on the ocean surface.

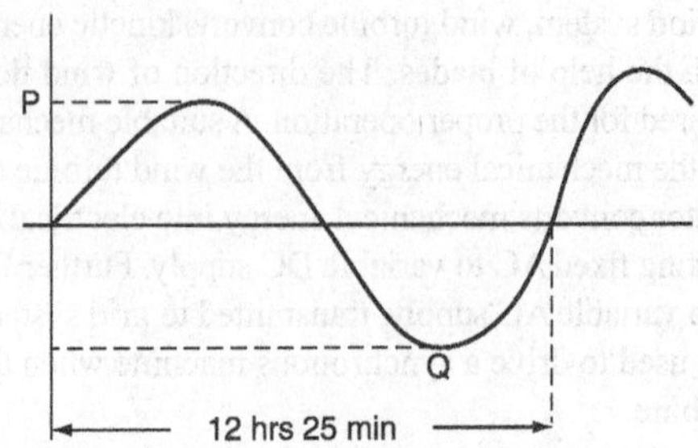

FIG. 2.12 *Nature of tide*

On the surface of the ocean, most of the water is pulled away from the solid earth surface or toward the moon and at the same time, the earth is moving away from the water in the opposite direction, so that high tides occur at these two areas and low tides will occur at the center of these two water and earth.

Normally, over the surface of seawater, two high tides and two low tides will be produced within a span of 24 hr 50 min, such types of tides are known as 'semidiurnal tides'. Usually, these tides are sinusoidal in nature as shown in Fig. 2.12.

In Fig. 2.12, *P* and *Q* indicate high- and low-tide points, respectively. The difference between these two points is known as tide range.

2.6.1 Components of tidal power plant

The tidal power plant has the following main parts:

(i) *Dam or dyke*: A dam or dyke is nothing but a barrier that exists between sea level and a basin or between a basin and the other in case of a multiple basin.

(ii) *Power house*: The tidal power plant equipment such as turbine, electric generator, and other auxiliary devices are placed in the power house.

(iii) *Sluice-ways*: It is nothing but a gate-controlled way either to fill the basin during high-tide period or it will keep empty during low tide.

2.6.2 Site selection of tidal power plant

Various factors that are needed to be considered for the location of tidal power plant are:

- The location of the plant must be nearer to the ocean.
- Site selection for the plant should be in such a way that the tidal range of ocean is large.
- The geographic features of the plant must be encloses of large areas with short dams.
- The sluice gates of dam should allow water to or from the basins.

2.6.3 Tidal power generation

The electrical energy generated by the generator in a tidal power plant mainly depends upon the raising and falling level of water above the surface of the sea. A simple single-basin tidal power plant is shown in Fig. 2.13. In this arrangement, both the basin and the sea are separated by a dam or dyke at which power house, houses turbine, or generator to generate electric energy.

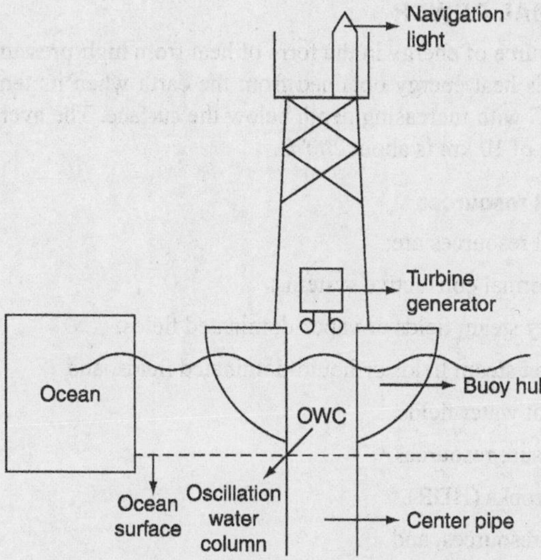

FIG. 2.13 *Simple single basin tidal power plant*

During flood-tide period, sluice gates get opened and water is allowed into the basin on the other side of the dam, through the turbine. Then, it will rotate and is coupled to the generator thereby generating electrical energy.

The turbine causes to generate electric power only during the high-tide period and begins to drop. During the low-tide period, water head of the sea will gradually fall down and is not sufficient to generate electric power to meet the no load losses.

2.6.4 Advantages and disadvantages of tidal power

Advantages

(i) Tidal power is free from pollution.

(ii) Tidal power generation is not affected by the rain.

(iii) The land cost of the tidal plant is less because such plants are located at seashore.

(iv) The plant does not require large space.

Disadvantages

(i) The power out of the plant will fluctuate continuously, because it depends on tidal range.

(ii) The construction of a tidal plant in a sea is complex.

(iii) The transmission cost of tidal power is costlier because such plants are located far away from the load center.

(iv) The initial cost of the plant is high.

(v) plant equipment will be subjected to corrosion due to seawater.

(vi) The efficiency of plant will be affected due to the variable tide range.

2.7 GEOTHERMAL POWER

It is a renewable source of energy in the form of heat from high-pressurized steam coming from the earth. This heat energy obtained from the earth when its temperature increases rapidly up to 180°C with increasing depth below the surface. The average temperature of the earth at a depth of 10 km is about 200°C.

2.7.1 Geothermal resources

Various geothermal resources are:

1. hydrothermal convective systems:
   - (i) dry steam fields or vapor-dominated fields,
   - (ii) wet steam fields or liquid-dominated fields, and
   - (iii) hot water fields.
2. geopressure resources,
3. hot dry rocks (HDR),
4. magma resources, and
5. volcanoes.

Nowadays, hydrothermal convective systems and hard rocks are being used as geothermal resources for energy generation.

2.7.2 Geothermal power generation

Normally, hydrothermal convective systems are widely used as geothermal resources to generate electric energy.

Hydrothermal convective systems are broadly classified into the following three categories. By utilizing these resources, geothermal power will be generated. Some methods of generating geothermal power are explained below.

(i) *Dry steam or vapor-dominated fields*

Figure 2.14 shows the dry steam geothermal power generation. In vapor-dominated systems, geothermal zone has well, which delivers steam with little or without water at a temperature of 150–250°C. These fields are the most attractive geothermal resources. The dry steam supplied by well is delivered to the steam turbine, which drives an electric generator, generates electric energy.

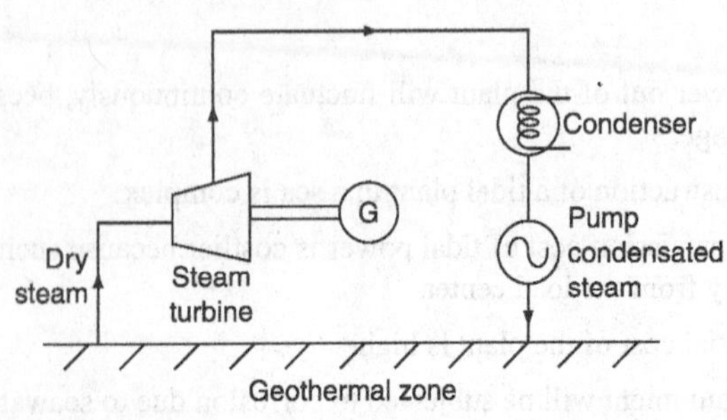

FIG. 2.14 *Dry steam geothermal power generation*

In this scheme of generation, the main difference between system and conventional steam turbine is only geothermal steam that is supplied from nuclear or fossil fuels at low temperature and pressure.

(ii) Wet steam or liquid-dominated fields

Figure 2.15 shows the wet steam geothermal power generation. In this method of power generation, liquid-dominated fields are the geothermal resources. These wet steam fields are available 20 times more than the dry steam fields. In the wet steam reservoir, the water temperature is above the boiling point of 100°C but it is under pressure, thus water does not boil and remains in the liquid state.

When the water comes to the surface of the earth, the pressure decreases and then the liquid subjected to rapid heat and flashes into a mixture of water and steam. Now, the mixture of water and steam supplied by well is delivered to the flash separator to separate steam and hot water then steam is fed to steam turbine to drive generator thereby generating electricity. Steam from the turbine is pumped into the condenser, then condensate steam and hot water from flash chamber are fed to the reservoir.

2.7.3 Advantages and disadvantages of geothermal power

Advantages

(i) Geothermal energy is quite cheaper.

(ii) Less pollution.

(iii) Geothermal energy can be utilized for various purposes from a single resource.

(iv) Geothermal resources delivers net energy compared to other resources.

(v) It is versatile.

Disadvantages

(i) The efficiency of the power generation is less about 1.5% compared to the other systems of generation.

(ii) Noisy operation.

(iii) Large area is required for the geothermal power generation.

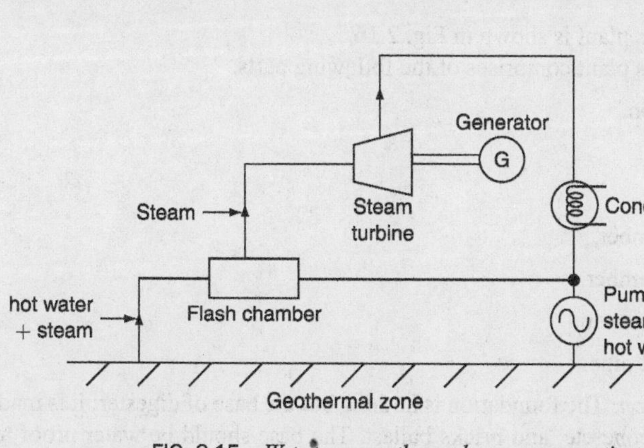

FIG. 2.15 *Wet steam geothermal power generation*

2.7.4 Applications of geothermal energy

Main applications of geothermal energy are:

(i) Electric power generation.

(ii) The heating of buildings.

(iii) Industrial heating purpose such as drying timber, wool washing, and crop drying.

2.8 BIOMASS AND BIOGAS

Biomass is the natural source of energy such as animal waste, wood, agricultural residues, dung, vegetable waste, and plant waste. Biogas is produced by decomposing the biomass. The conversion of biomass into biogas takes place through the process of digestion, pyrolysis, or hydro-gasification.

Energy from the biomass is obtained from the following ways.

(i) The biomass such as wood, dung, and agricultural residues is burnt directly to obtain energy.

(ii) The biomass is converted to fuels such as ethanol and methanol, which can be used as liquid fuels in engines.

(iii) The biomass is subjected to fermentation process to obtain a gaseous fuel called biogas.

2.8.1 Biogas generation

Biogas is produced from the decomposition of the biomass. It is a mixture of 55–65% of methane, 30–40% carbon dioxide, and some impurities such as H_2, H_2S, and nitrogen.

Biogas can be produced from the biomass through various processes such as digestion, pyrolysis, or gasification.

Digestion is the process of decomposition of the biomass in the absence oxygen and in the presence of anaerobic organisms at ambient temperature of 35–70°C. The device or container used for digestion process is known as digester.

Biomass gasification is the process of converting a solid or liquid into a gaseous without leaving a carbon residue.

Equipment used for gasifying biomass such as agricultural waste and wood waste is known as gasifier.

A simple biogas plant is shown in Fig. 2.16.

A simple biogas plant comprises of the following parts.

(i) foundation,

(ii) digester,

(iii) dome,

(iv) inlet chamber,

(v) outlet chamber,

(vi) mixing fans, and

(vii) gas outlet pipe.

(i) *Foundation:* The foundation is nothing but the base of digester; it is made up of cement, concrete, and bricks ballast. The base should be water proof to avoid the water leakage.

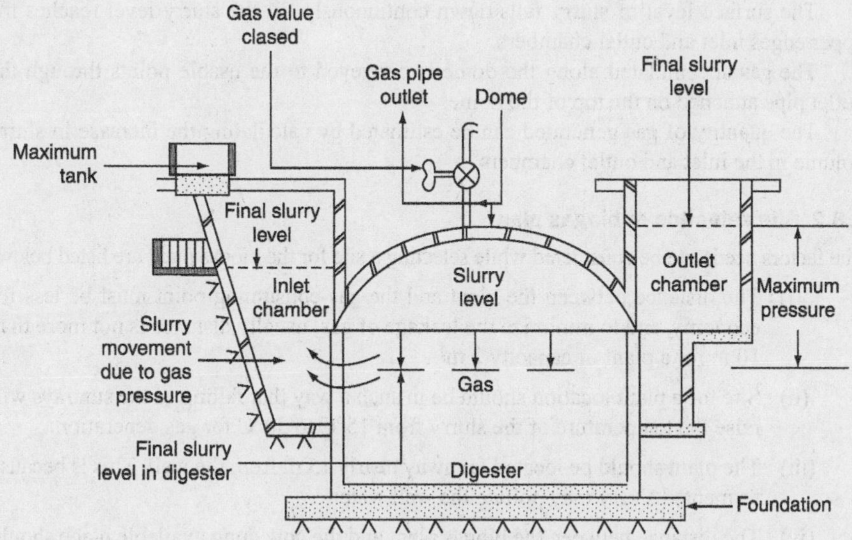

FIG. 2.16 *Biogas plant*

(ii) *Digester:* It is a container made up of bricks, sand, and cement. In this digester, fermentation of biomass such as dung, animal waste takes place, thus it is also known as fermentation tank.

(iii) *Dome:* It is the roof of the digester; after the decomposition of biomass, gas gets collected in the space of the dome over the slurry (mixture of water, dung, animal waste, etc.) in the digester.

(iv) *Inlet chamber:* An inlet chamber is made with bricks, cement, and sand. It is of bell mouth shape. It is the opening valve to admit slurry into the digester.

(v) *Outlet chamber:* It is the part of the plant of rectangular cross-section through which the final slurry moves out of the digester, after the digestion process.

(vi) *Mixing tank:* It is a tank placed on the top of the inlet chamber in which dung and water are mixed properly to make slurry and then admitted into the digester through the inlet chamber.

(vii) *Gas outlet pipe:* It is an outlet pipe fitted on the top of the dome of the digester to take away the gas for the utility purpose. A valve is provided to control the flow of gas to usage.

Plant operation

Initially, slurry is prepared by mixing the cow dung and water properly in the ratio of 1:1 and then the digester is completed filled with the slurry up to the dome level.

The fermentation of slurry takes place in the digester; gas will be generated due to the fermentation process and is accumulated along the dome.

The gas accumulated along dome exerts pressure on the slurry and displaces into the inlet and outlet chambers.

The surface level of slurry falls down continuously till the slurry level reaches the upper edges inlet and outlet chambers.

The gas accumulated along the dome is conveyed to the usable points through the outlet pipe attached on the top of the dome.

The quantity of gas generated can be estimated by calculating the increase in slurry volume in the inlet and outlet chambers.

2.8.2 Site selection of biogas plant

The factors needed to be considered while selecting a site for the biogas plant are listed below:

(i) The distance between the plant and the gas-consuming point must be less for economy, and to minimize the leakage of gas, usually distance is not more than 10 m for a plant of capacity 2 m^3.

(ii) Site for a plant location should be in such a way that falling of the sun rays will raise the temperature of the slurry from 15°C to 30°C for gas generation.

(iii) The plant should be located far away nearly 15 m from the well. This is because fermented slurry may pollute the well water.

(iv) The distance between the biogas plant and the cow dung available place should be less to minimize the transportation cost.

(v) The plant should be located in underground so that slurry can be filled and removed easily.

2.8.3 Advantages and disadvantages of biogas

Advantages

(i) The initial cost of the biogas plant is low.

(ii) The byproducts of the biogas plant can be used again for biogas generation.

(iii) It is pollution free.

(iv) Biogas can be conveyed to consumer point through GI pipes.

(v) Biogas can be easily stored in any container and can be transported to the consumers.

Disadvantages

(i) The land required for the biogas plant is relatively large so land cost is high.

(ii) Various nutrients must be added to the slurry for developing the bacteria.

(iii) The cost of producing energy is high.

(iv) Sometimes, the addition of fertilizer will reduce the gas production.

2.9 MHD GENERATIONS

MHD generation is one of the methods of generating electrical energy, which is highly efficient and low pollution one. In advanced countries, MHD generators are widely used but in developing countries, those are still under construction.

MHD generators are devices, which convert heat energy of a fuel into electric energy. The principle of the MHD generator is electromagnetic induction, 'when an electric conductor is passed through a magnetic field some voltage is induced'. This principle is the same as the conventional generator, the only difference being that a solid electrical conductor is used.

2.9.1 MHD generation

An MHD generator is a simple device to convert heat energy into electrical energy. Generally used methods of generating MHD power are:

(i) open-cycle generations and

(ii) closed-cycle generations.

In open-cycle generation of MHD power, the working fluid after the generation of electrical energy is released to the atmosphere but in closed-cycle generation, the working fluid is continuously recirculated.

(i) *Open-cycle MHD generation*

Figure 2.17 shows the schematic arrangement of open-cycle MHD generator.

An open-cycle MHD generator generates electric power. In this generator, fuel is admitted into the combustion. Initially, atmospheric air is fed to preheater then hot air is passed through the combustion chamber, which helps to burn the fuel. Hot gases in combustion chamber are mixed with ionized alkali metals such as cerium and potassium to increase the electrical conductivity of the hot gas. Thus, seeded material potassium ionizes by the hot combustion gas. The operating temperature of the combustion chamber is 2,300°C to 2,700°C. The hot gases from the combustion chamber are then passed to the magnetic field created by the permanent magnets. Thus, the MHD generator is able produce direct current and then is converted to AC power with the use an inverter.

The hot gases passed away from the generator are then heated again in air preheater to increase the temperature. These hot gases are converted to steam in a steam generator and then passed to the steam turbine to drive a synchronous generator thereby generating electrical energy. Remaining steam from the steam generator passed to the atmosphere through stacks.

(ii) *Closed-cycle MHD generation*

Figure 2.18 shows the schematic arrangement of closed-cycle MHD generation.

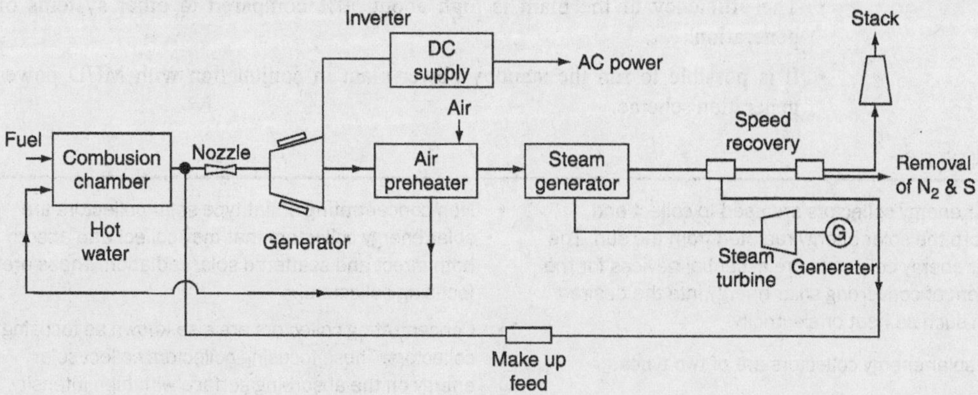

FIG. 2.17 *Open-cycle MHD generator*

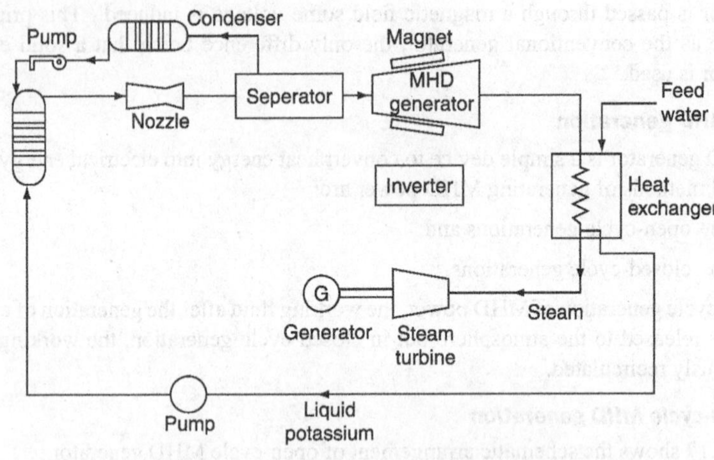

FIG. 2.18 *Closed-cycle MHD generation*

In this scheme of MHD power generation, liquid potassium is used as working fluid. Fluid from the breaker is passed through the nozzle to increase the speed of fluid. The working fluid is then passed through the MHD generator thereby generating energy fluid coming out from the MHD generator is passed through the heat exchanger converting into steam to run the steam turbine as well as generator to generate electric power, and remaining is pumped back to the reactor.

2.9.2 Advantages and disadvantages of MHD power generation

Various advantages of MHD power generating system are:

- Large amount of electric power generation is possible.
- It is highly reliable, as the system is having no moving parts.
- Closed-cycle system of MHD power generation is pollution free.
- The size of the power plant is small.
- The efficiency of the plant is high about 50% compared to other systems of generation.
- It is possible to run the standby power plant in conjunction with MHD power generation scheme.

KEY NOTES

- Solar energy collectors are used to collect and absorb the solar energy radiated from the sun. The solar energy collectors are essential devices for the system of converting solar energy into the desired form such as heat or electricity.
- The solar energy collectors are of two types.
- Non-concentrating or flat plate type solar collector.
- Concentrating or focusing type solar collector.
- Non-concentrating or flat type solar collectors are solar energy collectors that may collect and absorb both direct and scattered solar radiation. These are focusing collectors.
- Concentrating collectors are also known as focusing collectors. These focusing collectors collect solar energy on the absorbing surface with high intensity.
- Focusing collectors are classified into two types. They are:

(i) Line-focusing collectors.

(ii) Point-focusing collectors.

- The main difference between the focusing and non-focusing collectors is former one collects radiation coming from any particular direction.
- Solar cell is the device that transforms the sun's rays or photons directly into electricity. The element is most commonly used in the fabrication of solar cells is silicon.
- A basic wind power generating plant converts motional wind energy into electrical energy.
- Geothermal power is a renewable source of energy in the form of heat from high-pressurized steam coming from the earth.
- Biogas is produced from the decomposition of biomass. It is a mixture of 55–65% of methane, 30–40% carbon dioxide, and some impurities such as H_2, H_2S, and nitrogen.
- Biogas can be produced from the biomass through various processes such as digestion, paralyses, or gasification.
- Biomass gasification is the process of converting a solid or liquid into a gas without leaving a carbon residue.
- MHD power generation methods are:

(i) Open-cycle generations.

(ii) Closed-cycle generations.

SHORT QUESTIONS AND ANSWERS

(1) List out some of the non-conventional energy sources.

Non-conventional energy sources are:

- solar energy,
- wind energy,
- tidal energy, and
- geothermal energy.

(2) What are solar energy collectors?

Solar energy collectors are collecting plates used to collect and absorb the solar energy radiated from the sun.

(3) What are the generally used solar energy collectors?

The Generally used solar energy collectors are of two types.

(i) Non-concentrating or flat plate type solar collectors.

(ii) Concentrating or focusing type solar collectors.

(4) What is non-concentrating or flat type solar collector?

Non-concentrating or flat type solar collector is solar energy collectors that may collect and absorb both direct and scattered solar radiation.

(5) What are concentrating collectors or focusing collectors?

Concentrating collectors are also known as focusing collectors. These focusing collectors collect solar energy on the absorbing surface with high intensity.

(6) What are the types of focusing collectors?

Focusing collectors are classified into two types.

(a) Line-focusing collectors.

(b) Point-focusing collectors.

(7) What are the advantages of focusing collectors?

(i) The structure of reflecting surface is less.

(ii) The cost of collecting system is less.

(8) What are the main components of wind energy conversion system?

(i) Aero turbine.

(ii) Mechanical interface (coupling and gearing).

(iii) Electric generator.

(9) What is the function of wind turbine?

Wind turbine converts kinetic energy of wind s motion to mechanically energy transferred to an electric generator through the shaft.

(10) What are the types of wind turbines based on rotation of shaft?

Based upon the axis of rotation of turbine, wind mills are classified into two types.

(i) Horizontal axis wind mill.

(ii) Vertical axis wind mill.

(11) What is meant by geothermal power?

It is a renewable source of energy in the form of heat from high pressurized steam coming from the earth. This heat energy obtained from the earth when its temperature increases rapidly up to 180°C with increasing depth below the surface.

(12) What is meant by biomass?

Biomass is the natural source of energy such as animal waste, wood and agricultural residues, dung, vegetable waste, and plant waste.

(13) What is meant by biogas?

The gas produced by decomposing biomass. The conversion of biomass into biogas takes place through the process of digestion, paralysis, or hydro-gasification.

MULTIPLE-CHOICE QUESTIONS

(1) The energy obtained directly from the sun is called:

(a) Nuclear energy.
(b) Solar energy.
(c) Thermal energy.
(d) Hydroenergy.

(2) Which of the following is unconventional source of electrical power?

(a) Coal.
(b) Diesel.
(c) Geothermal.
(d) Nuclear.

(3) The main daily solar radiation at many places in India is about:

(a) 100 kwh m^{-2}.
(b) 20 kwh m^{-2}.
(c) 5 kwh m^{-2}.
(d) 1 kwh m^{-2}.

(4) The ocean thermal energy is larger than:

(a) Wave energy.
(b) Tidal energy.
(c) Wave and tidal energies.
(d) None.

(5) Ocean thermal energy is:

(a) Low-quality heat.
(b) High-quality heat.
(c) Median quality heat.
(d) None.

(6) The instrument used to measure the solar radiation is:

(a) Thermometer.
(b) Thermocouple.
(c) Monometer.
(d) Pyrheliometer.

(7) Winds are caused due to:

(a) The absorption of solar energy by the earth and the atmosphere.
(b) The rotation of the earth about its axis and around the sun.
(c) Both (a) and (b).
(d) None.

(8) Much of wind energy utilization is closed to the ground level within:

(a) 1 m.
(b) 5 m.
(c) 50 m.
(d) 500 m.

(9) From wind energy viewpoint, wind measurements were conducted since:

(a) 1904.
(b) 1968.
(c) 1986.
(d) 1995.

(10) The main cost component in the wind farm project is:

(a) The post.
(b) Generator.

(c) Exciter.

(d) Wind turbine.

(11) The approximate life time of wind turbine is:

(a) 1 year.

(b) 2 years.

(c) 20 years.

(d) 50 years.

(12) The ocean power plants are existing at:

(a) Kodaikanal.

(b) Kothagudem.

(c) Ramagundam.

(d) None.

(13) The source of power for satellite is:

(a) Wind energy.

(b) Thermionic converter.

(c) Solar cells.

(d) Microwave energy reflector.

(14) The method of generating power from seawater is more advantageous is:

(a) Ocean currents.

(b) Tidal power.

(c) Wave power.

(d) None.

(15) In fuel cell, the electrical energy is generated from:

(a) Mechanical energy.

(b) Heat.

(c) Sound.

(d) Chemical.

(16) Tidal power plant is being installed in India in:

(a) Tarapur.

(b) Vijayawada.

(c) Vijjeshwaram.

(d) Gujarat.

(17) The conductor used in MHD generator is:

(a) Gold.

(b) Silver.

(c) Copper.

(d) Gas.

(18) Wind energy is:

(a) Generated as a supplement to other power.

(b) Developing power proportional to wind power.

(c) Clean, free, and domestically produced.

(d) All the above.

(19) The power constant in a wind mill depends on:

(a) Wind speed.

(b) The shape of rotor blades.

(c) The type of rotor blades.

(d) All the above.

(20) The secondary source of energy are:

(a) Coal, oil, and uranium.

(b) Wind, tide, and the sun.

(c) Hydrogen, oxygen, and water.

(d) None.

(21) In a fuel cell, the electrical energy is obtained from:

(a) Chemical energy.

(b) Mechanical energy.

(c) Electrical energy.

(d) Heat energy.

(22) The sun gives:

(a) Heat.

(b) Light.

(c) Both (a) and (b).

(d) None.

(23) Wind energy is first converted into:

(a) Electrical energy.

(b) Mechanical energy.

(c) Chemical energy.

(d) None.

(24) The current developed by MHD generator is:

(a) AC.

(b) DC.

(c) Either AC or DC.

(d) None.

REVIEW QUESTIONS

(1) List out the various non-conventional energy sources and their availability.

(2) What are solar energy collectors and also explain the use of them.

(3) Write short notes on photovoltaic cells.

(4) Give the applications of solar energy.

(5) Discuss the function of basic components of wind energy conversion plant.

(6) With the help of neat sketch explain the function of wind power generation system.

(7) Give the advantages and disadvantages of wind power generation system.

(8) Discuss in detail about the components of tidal power plant.

(9) What is the significance of geothermal power and list out the resources.

(10) Write short notes on biogas and biomass.

(11) Draw a neat sketch and explain the function of biogas plant.

(12) Write short notes on MHD power generation.

ANSWERS

1. b	7. c	13. c	19. a
2. c	8. c	14. b	20. b
3. c	9. c	15. d	21. a
4. c	10. d	16. d	22. c
5. a	11. c	17. d	23. b
6. d	12. d	18. d	24. b

Conservation

OBJECTIVES

After reading this chapter, you should be able to:

- know various types of load curves
- know various tariff plans of electrical energy
- understand the need for energy conservation methods
- know the causes of low pf and pf improvement

3.1 INTRODUCTION

A power system consists of several generating stations (where electrical energy is generated) and several consumers (for them the electrical energy is generated). The objective of any power system is to generate electrical energy in sufficient quantities at the best-suited locations and to transmit it to the various load centers and then distribute to the various consumers maintaining the quality and reliability at an economic price. Quality implies that the frequency be maintained constant at the specified value (50 Hz in our country; though 60 Hz systems are also prevailing in some countries) and that the voltage be maintained constant at the specified value. Further, the interruptions to the supply of energy should be as minimum as possible.

One important characteristic of electric energy is that it should be used as it is generated stated otherwise, the energy generated must be sufficient to meet the requirements of the consumers at all times. Because of the diversified nature of activities of the consumers (e.g., domestic, industrial, agricultural, etc.), the load on the system varies from instant to instant. However, the generating station must be in a 'state of readiness' to supply the load without any intimation from the consumer. This 'variable load problem' is to be tackled effectively ever since the inception of a power system. This necessitates a thorough understanding of the nature of the load to be supplied, which can be readily obtained from the load curve, the load duration curve, etc., which form the contents of this chapter.

3.2 LOAD CURVE

A load curve is a plot of the load demand (on the y-axis) versus the time (on the x-axis) in the chronological order.

Out of the load connected, a consumer uses different fractions of the total load at various times of the day as per his or her requirements. Since a power system is to supply load to all such consumers, the load to be supplied varies continuously with time and does not remain constant. If the load is measured (in units of power) at regular intervals of time, say, once in an hour (or half-an-hour) and recorded, we can draw a curve known as the load curve.

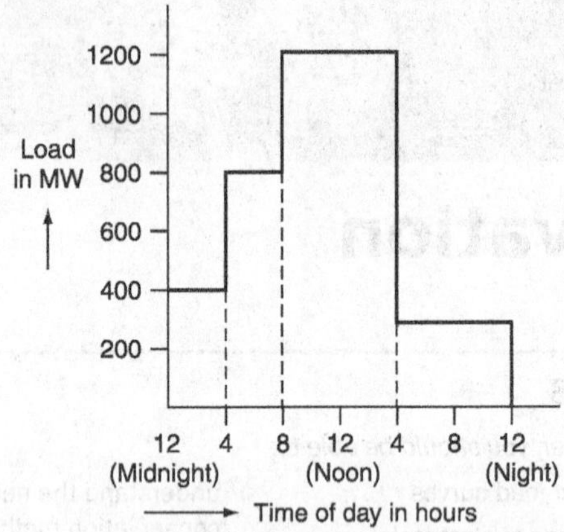

FIG. 3.1 *Daily load curve*

A period of 24 hr only is considered. The resulting load curve is called a 'daily load curve' as shown in Fig. 3.1. However, to predict the annual requirements of energy, the occurrence of load at different hours and days in a year and in the power supply economics, the 'annual load curves', are used.

An annual load curve is nothing but a plot of the load demand of the consumer against time in hours of the year (1 year = 8,760 hr).

Significance: From the daily load curve, shown in Fig. 3.1, the following information can be obtained.

- Observe the variation of load on the power system during different hours of the day.
- The area under this curve gives the number of units generated in a day.
- The highest point in that curve indicates the maximum demand on the power station on that day.
- The area of this curve divided by 24 hr gives the average load on the power station in the day.
- It helps for selecting the rating and number of generating units required.

3.2.1 Load duration curve

The load duration curve is a plot of the load demands (in units of power) arranged in a descending order of magnitude (on the *y*-axis) and the time in hours (on the *x*-axis). The load duration curve can be drawn as shown in Fig. 3.2.

3.2.2 Definition of terms and factors

Several terms are used in connection with the power supply to an area, whether it is for the first time (as is the case when the area is being electrified for the first time) or subsequently (due to the load growth). These terms are explained below.

(i) *Connected load*

A consumer, for example a domestic consumer, may have several appliances rated at different wattages. The sum of these ratings is his or her connected load.

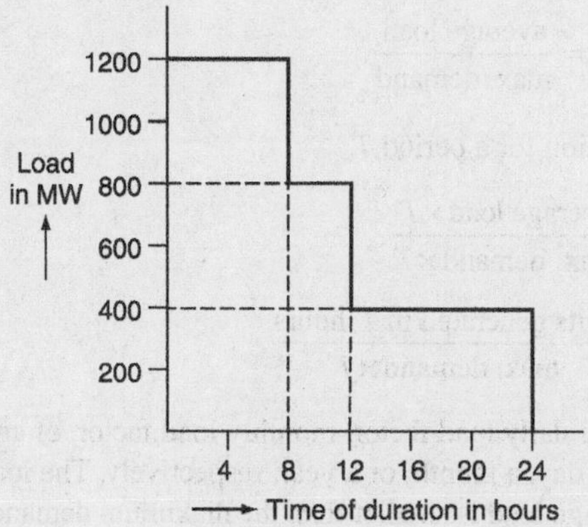

FIG. 3.2 *Load duration curve*

The connected load is the sum of the ratings (w, kW, or MW) of the apparatus installed on a consumer's premises.

(ii) *Maximum demand*

It is the maximum load used by a consumer at any time. It can be less than or equal to the connected load. If all the devices connected in the consumer house run to their fullest extent simultaneously, then the maximum demand will be equal to the connected load. But, generally, the actual maximum demand will be less than the connected load since all the appliances never use at full load at a time.

The maximum demand is usually measured in units of kW or MW by a maximum demand indicator. (Usually, in the case of HT consumers, the MD is measured in terms of kVA or MVA.)

(iii) *Demand factor*

The ratio of the maximum demand and the connected load is called the 'demand factor'.
Note: (a) The maximum demand and the connected load are to be expressed in the same units (W, kW, or MW).

(iv) *Average load*

If the number of kWh supplied by a station in one day is divided by 24 hr, then the value so obtained is known as daily average load.

$$\text{Daily average load} = \frac{\text{kWh in one day}}{24}$$

$$\text{Monthly average load} = \frac{\text{kWh in one day}}{30 \times 24}$$

$$\text{Yearly average load} = \frac{\text{kWh in one day}}{365 \times 24}.$$

(v) *Load factor*

The ratio of the average demand and maximum demand is called the load factor.

$$\text{Load factor (LF)} = \frac{\text{average load}}{\text{max. demand}}.$$

If the plant is in operation for a period T,

$$\text{Load factor} = \frac{\text{average load} \times T}{\text{max. demand} \times T}$$

$$= \frac{\text{units generated in } T \text{ hours}}{\text{max. demand} \times T}.$$

The load factor may be daily load factor, monthly load factor, or annual load factor; if the period considered in a day, a month, or a year, respectively. The load factor is always less than one because average load is smaller than the maximum demand. It plays an important role in determining the overall cost per unit generated. Higher the load factor of the power station, lesser will be the cost per unit generated.

(vi) *Diversity factors*

The diversity factor is the ratio of the sum of the maximum demands of a group of consumers and the simultaneous maximum demand of the group of consumers.

$$\text{Diversity factor} = \frac{\text{sum of individual max. demand}}{\text{max. demand on system}}.$$

A power system supplies load to various types of consumers, whose maximum demands generally do not occur at the same time. Therefore, the maximum demand on the power system is always less than the sum of individual maximum demands of the consumers.

A high diversity factor implied that with a smaller maximum demand on the station, it is possible to cater to the needs of several consumers with varying maximum demands occurring at different hours of the day. The lesser the maximum demand, the lesser will be the capital investment on the generators. This helps reduce the overall cost of the unit (kWh) generated.

Thus, a higher diversity factor and a higher load factor are the desirable characteristics of the load on a power station. The load factor can be improved by encouraging of the consumers to use power during off-peak hours with certain incentives such as offering a reduction in the cost of energy consumed during off-peak hours.

(vii) *Plant capacity*

It is the capacity or power for which a plant or station is designed. It should be slightly more than maximum demand. It is equal to the sum of the ratings of all the generators in a power station.

(viii) *Plant capacity factor*

It is the ratio of the average demand on the station and the maximum installed capacity of the station.

Or, capacity factor = load factor × utilization factor.

Reserve capacity = plant capacity – maximum demand.

(ix) *Utilization factor (or plant use factor)*

It is the ratio of kWh generated to the product of the plant capacity and the number of hours for which the plant was in operation.

$$\text{Plant use factor} = \frac{\text{station output in kWh}}{\text{plant capacity} \times \text{hours of use}}.$$

Example 3.1: A generating station has a maximum demand of 35 MW and has connected load of 60 MW. The annual generation of units is 24 × 10^7 kWh. Calculate the load factor and the demand factor.

Solution:

No. of units generated annually = 24 × 10^7 kWh.

No. of hours in a year (assuming a 365 day in year) = 365 × 24
= 8,760 hr.

$$\therefore \text{ Average load on the station} = \frac{24 \times 10^7}{8,760} = 27,397.26 \text{ kW} = 27.39726 \text{ MW}.$$

$$\therefore \text{ Load factor} = \frac{27.39726(\text{MW})}{35(\text{MW})} = 0.7828 \text{ or } 78.28\%.$$

$$\text{Demand factor} = \frac{35(\text{MW})}{60(\text{MW})} = 0.583 \text{ or } 58.3\%.$$

Example 3.2: A generating station supplies four feeders with the maximum demands (in MW) of 16 MW, 10 MW, 12 MW, and 7 MW. The overall maximum demand on the station is 20 MW and the annual load factor is 45%. Calculate the diversity factor and the number of units generated annually.

Solution:

Sum of maximum demands = 16 + 10 + 12 + 7 = 45 MW.

Simultaneous maximum demand = 20 MW.

$$\therefore \text{ Diversity factor} = \frac{45}{20} = 2.25.$$

Average demand = (maximum demand) × (load factor)
= (20) × (0.45) = 9 MW.

∴ No. of units generated annually = 9 × 8,760 MWh
= 78,840 MWh.

Alternatively,

$$\text{Annual load factor} = \frac{\text{no. of units generated annually}}{(\text{max. demand}) \times 8,760}.$$

$$\text{i.e., } 0.45 = \frac{\text{no. of units generated annually}}{20 \times 8,760}.$$

So that the number of units generated annually = 0.45 × 20 × 8,760 MWh
= 78,840 MWh.

Example 3.3: The yearly load duration curve of a power plant is a straight line. The maximum load is 30 MW and minimum load is 20 MW. The capacity of the plant is 35 MW. Calculate the plant capacity factor, load factor, and utilization factor.

Solution:
No. of units generated per year = area OACD = area OBCD + area BAC

$$= 20 \times 8{,}760 + \frac{1}{2}(30 - 20) \times 8{,}760$$

$$= 8{,}760 \left(20 + \frac{1}{2} \times 10\right)$$

$$\doteq 8{,}760 \times 25 = 219{,}000 \text{ MWh.}$$

$$\therefore \text{ Average annual load } = \frac{8{,}760 \times 25}{8{,}760} = 25\text{MW.}$$

$$\therefore \text{ Load factor } = \frac{25}{30} = 0.833.$$

$$\text{Plant capacity factor } = \frac{\text{average annual load}}{\text{rated plant capacity}} = \frac{25}{35} = 0.714.$$

$$\text{Utilization factor } = \frac{\text{maximum demand}}{\text{rated capacity}} = \frac{30}{35} = 0.857.$$

Alternatively,

Capacity factor = 0.714.

$$\text{Utilization factor } = \frac{\text{capacity factor}}{\text{load factor}} = \frac{0.714}{0.833} = 0.857.$$

Problem 3.4: A central station supplied energy to two substations A and B; four feeders take off from each of the substations as shown in Fig. 3.3. The maximum demands are as given below.

Central station: 10 MW
Substation A: 6 MW
Substation B: 8 MW
Feeders on substation A: 1.50, 2.0, 5.0, 3.0 MW
Feeders on substation B: 2.0, 4.0, 5.0, 1.0 MW

Calculate the diversity factors among (a) substations, (b) feeders on substations A, and (c) feeders on substation B.

Solution:
The sum of the maximum demands on:

Substations A and B = 6 + 8 = 14 MW.
Maximum demand on the central station = 10 MW.

$$\therefore \text{ Diversity factor between substations } = \frac{14}{10} = 1.4.$$

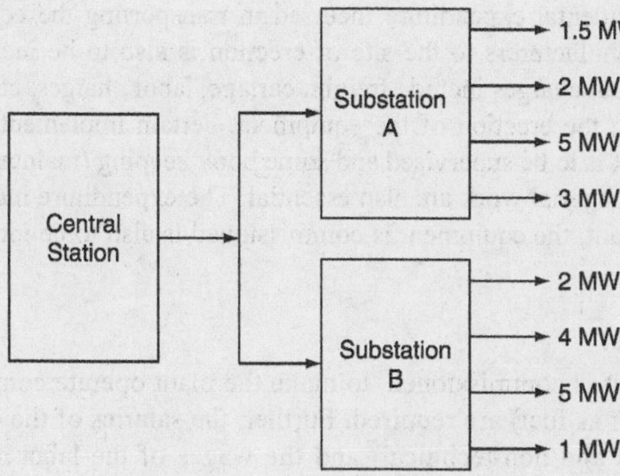

FIG. 3.3 *Load distribution*

The sum of the maximum demands on the feeders of substations A = 1.5 + 2 + 5 + 3 = 11.5 MW.

$\therefore$ Diversity factor between feeders of substation A $= \dfrac{11.5}{6} = 1.917$

Similarly, diversity factor between the feeders of substation

$$B = \frac{2+4+5+1}{5} = \frac{12}{8} = 1.5.$$

3.3 COST OF ELECTRICAL ENERGY

To run any business, a certain investment (capital) is required. The return (revenue) is realized by selling the end product. For the business to prosper, the revenue returns must be more than the expenditure incurred in running the business and producing the end product. The business of 'electrical energy' is no exception to this basic rule.

In any business, the vendor has to provide the customer with a 'price list' or 'rate schedule'. The rate schedule pertaining to electrical energy is called a 'Tariff'. The formulation of a good 'tariff' involves a thorough study of the various economic aspects of the electric supply.

3.3.1 Cost of generation station

For taking up any project, a certain amount of capital is required. The total capital required can be subdivided into two major heads:

(i) fixed capital and

(ii) running capital.

(i) *Fixed capital*

The operation of a power system requires that a considerable amount to be spent on the purchase of certain assets such as land, plant, and equipment, to start with. The fixed capital of an electrical installation may be grouped under the following needs.

(i) The capital cost of generating equipment.

(ii) The capital cost of transmission system.

(iii) The capital cost of the distribution system, both HT and LT.

In addition, the incidental expenditure incurred in transporting the equipment from the premises of the manufacturers to the site of erection is also to be included in the fixed capital. The incidental charges include freight, cartage, labor charges, etc.

Further, during the erection of the equipment, certain implements may have to be purchased; the work is to be supervised and some book keeping (maintaining the accounts and stores) and managerial work are also essential. The expenditure incurred on all these items up to the instant, the equipment is commissioned is also to be included in the fixed capital.

(ii) *Running capital*

After the equipment is commissioned, to make the plant operate continuously, several raw materials (such as fuel) are required. Further, the salaries of the operating personnel (both technical and non-technical) and the wages of the laborers are to be paid. The capital required for the continuous operation of the project is called the '*running cost*'.

3.3.2 Annual cost

The total expenditure to be incurred annually is called the '*annual cost*'. The economy of the project is judged from this cost (but not from the total investment). The annual cost comprises of:

(i) The running cost or operating cost

(ii) The fixed charges (not fixed capital).

(i) *Running charges*

In general, the annual running charges include:

- *The cost of fuel:* The higher the number of units (kWh) generated, the higher is the fuel consumption.
- *The maintenance and repair charges of the equipment in all the three sections:* generation, transmission, and distribution.
- The wages of the operational staff.
- The salaries of the supervising staff.
- The cost of water (in thermal stations, the feed water is to be treated before it is boiled to raise the steam) and the cost of the lubricating oil, etc.

(ii) *Fixed charges*

The various items given below are included in the fixed charges.

(a) Huge investment running into, maybe, several hundreds of crores of rupees, is required in any power project. Normally, a part of this amount is borrowed from the public, offering them a certain rate of interest. This interest has to be paid annually, whether or not the plant is in an operative state.

(b) Certain taxes are to be paid annually.

(c) Insurance charges.

(d) The salaries of management and clerical staff.

(e) The annual instatement to build up the 'depreciation reserve'.

3.3.3 Factors influencing the formulation of tariff

An electric supply company earns its revenue by selling the kWh of electrical energy generated. However, the capital investment and a major part of the operating expenses are decided by the capacity of the plant. So, a consumer has to pay for both the expenses on the capital investment and the operating costs. Therefore, care must be taken in designing the '*tariff*' or '*rate schedule*'. Several factors influence the formulation or design of the tariff are given below.

- The rates must be uniform throughout the area supplied by the company.
- The company is not allowed to build up a reserve for the slack period by charging higher rates.
- Some risks are inevitable in any business. However, the company should not aim to complete protection for its investment.
- The company should save all the customers without any discrimination.
- The larger the utility system, the greater is the loss of units it can afford without affecting the tariff.
- Bulk consumers may be allowed to have certain plant installed at their cost to serve them.
- The consumers likely to create higher maximum demand on the station are to be charged at higher rates.
- Incentives are to be offered to consumers having desirable loads, i.e., with higher load factors.
- Price structure should have a component to take care of the fixed charges and another component to take care of the running charges.
- The rates should be fixed keeping the future investment in view. Sufficient revenue must be realized to cover future plans.
- The lower the pf, the higher will be the kVA for the same kW demand (and hence for the same kWh consumption). So, a certain reference value, say, 0.85 lagging pf, may be fixed. For the loads with pf less than the reference value, higher rates may be charged and for the load with pf greater than the reference value, lower rates may be charged by offering an incentive, say, a reduction in the bill amount by a small percentage. This may encourage the consumer to improve the pf of the load, by installing pf improvement devices, for example, static condensers.
- The consumers using power during the off-peak load hours, say, from 10 PM to 5 AM may be offered concessional rates. This will improve the revenue returns without any increase in the capacity of the plant.
- Tariffs, once fixed, may have to be in force for a considerable period. Frequent changes may lead to criticism from consumers.
- Rate schedules should be as simple as possible so that the consumers can understand them easily.

3.3.4 Factors to be considered in fixing up the tariff

The salient points to be considered in fixing up a tariff.

(i) *Annual cost of production:* This aspect should be given the almost importance. The future of the project depends on this, since it has a direct bearing on the revenue returns.

(ii) Electricity may be used for different purposes, for example, lighting, heating, etc., in the case of a domestic consumer. A higher rate per kWh may not bother a consumer using electricity only for lighting while it may cause some concern to a consumer using electricity for heating purposes. So, same sort of a grid must be there among the consumers using electricity for different purposes: For example, a farmer (agriculturist) may have to be charged at a low rate.

(iii) The standard of living and the ability of the consumer to pay the electricity bill must be given due weightage. As an example, let us consider two consumers: one using electricity for domestic purposes and the other for running an industry. The domestic consumer can afford to pay a higher rate since his or her consumption of energy is less. This is because increased energy consumption spreads the fixed charges over the greater number of units, hence reducing the overall generating cost.

(iv) The tariff should be simple in calculation to understand by all the consumers.

(v) Encourage the consumes to create an extended use of power.

(vi) For low power factor, the consumers are penalized.

3.3.5 Types of tariffs

There are several types of tariffs and they are:

(i) Simple tariff, (ii) flat-rate tariff, (iii) block-rate tariff, (iv) two-part tariff, (v) maximum demand tariff, and (vi) power factor tariff.

(i) *Simple type of tariff*

This type of tariff is based on the idea that the cost per unit equals the total amount spent in producing the energy divided by the number of units supplied to the consumers. (Usually, the number of units sold to the consumers is less than the number of units generated. The difference is due to the transmission and distribution losses. However, the revenue returns come only from the units supplied to the consumer.)

Thus,

$$\text{Cost/unit} = \frac{\text{annual running charges} + \text{annual fixed charges}}{\text{the total number of units supplied to the consumers}}.$$

The only merit of this tariff is simple in calculations to understand by all the consumers.

Demerits

- The calculated cost per unit will be higher.
- No distinction is made between bulk and small (or domestic) consumers.

This tariff can be more suitable by having the following modifications.

- A discount may be given to a consumer depending upon the quantity of energy consumed by him or her.
- Even domestic consumers can be charged at two different rates, higher rates being collected for the kWh used for lighting and fan loads than those used for heating purposes.
- Consumers using energy during off-peak load hours may be allowed a discount by offering them a tariff called 'off-peak tariff'.
- A provision must be made to suit metered load factor and power factor.

(II) *Flat-rate tariff*

In the flat-rate tariff, the consumers are categorized depending upon the main purpose for which electrical energy is used. Thus, the consumers may be categorized as domestic, industrial, commercial, agricultural, etc. consumers; each type of consumer is charged at different rates. If a particular consumer uses electrical energy for two different purposes, say, for lighting and fans and for industrial purposes, two meters are installed in his or her premises to measure the energy used for each of the purposes and the bill is accordingly prepared. In addition, the supplier may collect the meter rent. Sometimes, discount on the bill amount may be allowed for prompt payment.

In arriving at the rates payable by different types of consumers, the load factor and the diversity factor of each type of consumers are taken into consideration.

The advantage of this tariff is that it can easily be understood by the consumers.

The demerits of this tariff are:

- Separate meter is required for different types of supply.
- Difficulty dishonesty in assessing the load factor and the diversity factor of each type of consumer factors correctly.

(III) *Block-rate tariff*

The block-rate tariff is based on the fact that the greater the number of units of electrical energy generated, the lower will be the cost of generation per unit. So, a consumer having a large demand of number of units will be charged at lower rates at higher slabs of energy consumption (different blocks or slabs will be specified in the tariff).

Thus, the block rate tariff may be of the form given below.

First 50 units .... Rs. 1/unit.
Next 100 units .... Rs. 0.9 unit.
Next 200 units .... Rs. 0.75/unit.
Over and above 350 units .... Rs. 0.6/unit.

The merit of this tariff is that if the consumer consumes more energy, he or she gets an incentive. This leads to increase the load factor of the power system, so reduces the cost of generation. The drawbacks of this tariff are that it lacks a measure for the demand of the consumer and it is suitable only for residential and small commercial consumers.

(iv) *Two-part tariff*

In the two-part tariff, the fixed and the running charges are separated. Thus, let us suppose a consumer has a connected load of 1 MW. He or she has the right to use 1 MW of power at any time without any intimation what so ever to the supply authorities. So, it may be assumed that 1 MW of the installed capacity is earmarked for this particular consumer. Hence, he or she has to pay his or her share of the annual fixed charges. This forms the first component of the two-part tariff. Again, depending upon the amount of electrical energy consumed, he or she has to pay more or less to meet the running charges. Thus, the general form of the two-part tariff may be expressed as:

$$\text{Total energy charges} = \text{Rs. } (a \times \text{kW} + b \times \text{kWh}),$$

Where a is the charge per kW of connected load and b is the charge per kWh of energy consumed.

The advantages of this tariff are that it can easily be understood by the consumers and depends upon the maximum demand it recovers the fixed charges. However, the consumer may be at a disadvantage, sometimes. Irrespective of consumption, he or she has to pay fixed charges. For example, due to some unforeseen circumstance in any industry such as workers strike or a lockout, there is no industrial activity for one month, even though he or she has to pay fixed charges. Similar may be the case with a consumer who may be out of station for a considerable period.

This type of tariff is suitable for medium industrial consumers.

(v) *Maximum demand tariff*

The drawback of the two-part tariff is overcome in the maximum demand tariff. In this tariff, the maximum demand of the consumer is actually measured by installing a 'maximum demand meter', in the consumer's premises. Thus, the fixed charges component will be proportional to the (actual) maximum demand rather than to the connected load. The general form may be as follow:

$$\text{Total energy charges} = \text{Rs. } (a \times \text{kW} + b \times \text{kWh}),$$

where a is the charge per kW of maximum demand and b is the charge per kWh of energy consumed

This type of tariff may be detrimental to the interests of the supply authorities. Thus, the industrial consumer need not pay even a single paisa to the supply company for the period his or her industry was closed (due to strike, etc.).

This type of tariff is suitable for big industrial consumers.

So, a realistic tariff aims at protecting the interests of both the supplier and the consumer.

(vi) *Power-factor tariffs*

We know that the efficiency of the plant and equipment is affected by the power factor. The maximum utility of the plant is obtained when it operates at the most economical power factor (not equal to upf). So, some tariffs are designed that take the pf into consideration. They are discussed as follows:

(a) kVA maximum demand tariff

In this tariff, the maximum demand of consumers is measured in kVA rather than in kW and the charges is collected based on the kVA demand. This, a given kW of load (kW = kAV cos ϕ) gives rise to a higher kVA demand if the power factor is poor; and to a lower kVA demand if the power factor is high. This encourages the consumer to improve the pf of his or her load by the installation of power factor improvement devices. Improved pf operation of the individual consumer leads to a reduction in the kVA demand on the generating station.

(b) kWh and kVArh tariff

We know that kVAr = kVA sin ϕ. So far a given kVA, the smaller the value of kVAr, the smaller is the value of sin ϕ and hence, the higher is the value of cos ϕ, i.e., pf. So, in this type of tariff, the consumer is charged for both the kWh and the kVArh separately. If the kVArh is low, the consumer will pay less, else, more. This encourages the consumer to improve the pf of the load.

(c) Average (or sliding scale) power factor tariffs

In these tariffs, an average value of power factor, say, 0.8 or 0.9 = 85 lagging, is assumed as the reference should the pf fall below the reference value a surcharge at a specified rate is levied for every 0.01 fall in the pf below the reference pf. Similarly, an incentive is given for an improvement in the power factor above the reference value. This encourages the consumer to improve the power factor of his or her load.

(vii) *Three-part tariff*

In this type of tariff, the total charge to be made from the consumer is split into three parts: (i) fixed charges, (ii) semifixed charges, and (iii) running charges.

The general form may be:

Total energy charges = Rs. $(a + b \times \text{kW} + c \times \text{kWh})$,

where a is the fixed charge made during billing period, b is the charge per kW of connected demand, and c is the charge per kWh of energy consumed.

When considering the fixed charges in addition to two-part tariff, it becomes a three-part tariff.

Example 3.4: The average motor load of a consumer is 250 kW at a pf 0.85 lag. The consumer is charged electricity at the tariff of 50 Rs./kVA of maximum demand plus 10 paise per unit consumed. Determine the consumer's annual bill for a load factor of 70%.

Solution:

Consumer motor load = 250 kW.

Maximum demand in kVA at a pf of $0.85 = \dfrac{250}{0.85} = 294.1.$

Units consumed/year = Max. demand × LF × hours in a year
= 250 × 0.7 × 8,760
= 1,533,000 kWh.

Annual bill = Max. demand charges + energy charges
= Rs. $50 \times 294.1 + 0.1 \times 15.33 \times 10^5$.

Example 3.5: An industry daily load is 250 kW for first 2 hr, 100 kW for next 8 hr, 150 kW for next 6 hr, and 5 kW for the remaining time. Calculate the electricity expenditure per year, if the tariff in force is Rs. 1,200/kW of maximum demand per annum plus Rs. 2.0/kWh.

Solution:

Daily energy consumption = 250 × 2 + 100 × 8 + 150 × 6 + 5 × 8
= 2,240 kWh.

Annual energy consumption = 2,240 × 365 = 817,600 kWh.

Maximum demand = 250 kW.

Demand charges per annum = Rs. 1,200 × 250
= Rs. 3,00,000.00.

Energy charges per annum = Rs. 2.0 × 817,600
= Rs. 16,35,200.00.

Total electricity expenditure per annum = Rs. (3, 00,000 + 1,635,200)
= Rs. 19,35,200.00.

Example 3.6: An industrial consumer having a maximum demand of 120 kW, maintain a load factor of 65%. The tariff rates are Rs. 950/kVA of maximum demand per annum plus Rs. 2/kWh of energy consumed. If the average pf is 0.707 lagging, determine the following:

(i) the total energy consumed per annum,

(ii) the annual electricity bill, and

(iii) the overall cost per kWh consumed.

Solution:

Maximum demand = 120 kW.

Load factor = 65%.

Average power factor = 0.707.

$$\text{Maximum demand in kVA} = \frac{\text{max. demand in kW}}{\text{average power factor}} = \frac{120}{0.707} = 169.7.$$

(i) Total energy consumed per annum = max. demand in kW × load factor × 8,760
= 120 × 0.65 × 8,760
= 6,83,280 kWh.

Annual demand charges = Rs. 950 × 169.7 = Rs. 1,61,215.

Annual energy charges = Rs. 2.0 × 6,83,280 = Rs. 13,66,560.

(ii) Annual electricity charges = Rs. (161,215 + 1,366,560) = Rs. 1,527,775.

(iii) Overall cost per kWh supplied $= \text{Rs.}\ \frac{1,527,775}{683,280} = \text{Rs. } 2.23.$

Example 3.7: An industry has a maximum load of 250 kW at 0.707 pf lag, with an annual consumption of 30,000 units. The tariff is Rs. 60/kVA of maximum demand plus 15 paise per unit. Calculate the following:

(i) the flat rate of energy consumption and

(ii) the annual saving if the pf is raised to units.

Solution:

Maximum load = 250 kW.

Power factor = 0.707 lag.

Annual consumption = 30,000 kWh.

Maximum demand in kVA at a pf of 0.707 $= \frac{250}{0.707} = 353.606.$

(i) Annual bill = demand charges + energy charges
= Rs. (60 × 353.6 + 0.15 × 30,000)
= Rs. 25,716.

$$\text{Flat rate/unit} = \frac{25,716}{30,000} = \text{Rs. } 0.8572.$$

(ii) When pf is raised to unity, the maximum demand in kVA $= \frac{250}{1} = 250.$

Annual bill = Rs. 60 × 250 + 0.15 × 30,000 = Rs. 19,500.
Annual saving = Rs. (25,716 – 19,500) = 6,216.

Example 3.8: Annual consumption of consumer energy is 60,000 kWh, the charge is Rs. 110/kW of maximum demand plus 6 paise per kWh.

(i) Determine the annual bill and the overall costs per kWh if the load factor is 50%.

(ii) What is the overall cost per kWh if consumption was reduced by 30% with the same load factor?

Solution:

Energy consumption = 60,000 kWh.
Let us take load factor as 100%.
Energy consumed/annum = MD × LF × 8,760
60,000 = MD × 1 × 8,760
MD = 6.85 kW.

(i) At 50% load factor:

Energy consumed/annum = 6.85 × 0.5 × 8,760 = 30,003 kWh.

Annual bill = 110 × 6.85 + 0.06 × 30,003 = Rs. 2,553.7.

$$\text{Cost/kWh} = \frac{2,553.7}{30,003} = 8.5 \text{ paise.}$$

(ii) Energy consumption = 0.7 × 60,000 = 42,000 kWh.

$$\text{MD} = \frac{42,000}{8,760} = 4.794 \text{ kW.}$$

Annual bill = 110 × 4.794 + 0.06 × 42,000 = Rs. 3,047.34.

$$\text{Cost/kWh} = \frac{3,047.34}{42,000} = 7.25 \text{ paise.}$$

Example 3.9: An industry has a connected load of 200 kW. The maximum demand is 160 kW. On an average each machine works for 70% time. If the tariff is Rs. 1,400 + Rs. 140/kW of maximum per year + Re. 0.2/kWh, calculate the yearly expenditure on electricity.

Solution:

Energy consumption in one year = 160 × 0.7 × (365 × 24) = 981,120 kWh.
Total electricity bill = Rs. (1,400 + 140 × 160 + 0.2 × 981,120) = Rs. 220,024.

Example 3.10: A power station has got maximum demand of 40 MW with annual load factor is 60%. Determine the cost per kWh generated from the following data.

Capital cost = Rs. 80 × 10^5, annual cost of fuel and oil = Rs. 8 × 10^5, taxes, wages, and salaries = Rs. 5 × 10^5, and the rate of interest and depreciation is 12%.

Solution:

Maximum demand = 40 MW.

Capital cost = Rs. 80 × 10^5.

The rate of interest and depreciation = 12%.

$$\text{Annual fixed charges} = \text{Rs. } 80\times10^5\times\frac{12}{100} = \text{Rs. } 9.6\times10^5.$$

Total running charges = Rs. (8 × 10^5 + 5 × 10^5)
= Rs. 13 × 10^5.

Total annual cost = Rs. (9.6 × 10^5 + 13 × 10^5)
= Rs. 22.6 × 10^5.

$$\text{Annual load factor} = \frac{\text{no. of units delivered}}{\text{maximum demand}\times 8{,}760}.$$

No. of units delivered = 0.6 × 40,000 × 8,760 = 210.24 × 10^6.

$$\text{Cost per unit} = \frac{\text{total annual cost}}{\text{no. of units delivered}} = \frac{22.6\times10^5}{210.24\times10^6}\times100 = 1.075 \text{ paise.}$$

Example 3.11: Determine the consumer's annual bill for a load factor of 75% for the following tariff.

Rs. 60/kVA of his or her maximum demand + 4 paise/unit consumed. The customer has an average motor load of 225 kW at power factor of 0.7 lagging.

Solution:

Average motor load = 225 kW.

Load factor = 75%.

$$\text{Load factor} = 0.75 = \frac{\text{no. of units consumed in a year}}{\text{maximum demand}\times 8760}.$$

No. of units consumed in a year = 225 × 8,760 × 0.75 = 14.78 × 10^5 kWh.

$$\text{The annual cost of energy consumed} = \text{Rs. } \frac{14.78\times10^5\times4}{100} = \text{Rs. } 5.913\times10^4.$$

$$\text{Maximum demand in kVA} = \frac{\text{kW}}{\text{pf}} = \frac{225}{0.7} = 321.43 \text{ kVA.}$$

The annual cost of maximum demand = Rs. 321.4 × 60 = Rs. 19,285.7.

Total annual bill = Rs. (59, 130 + 19,285.7) = Rs. 78,415.7.

Example 3.12: Installed capacities of generating station is 25 MW and generated 200 × 10^6 units/annum. Calculate the cost per unit generated, if the annual fixed charges are Rs. 150/kW installed and running charges are 5 paise/kWh.

Solution:

The installed capacities of the generating station = 25 MW.

No. of units generated per annum = 200 × 10^6 kWh.

Annual fixed charges = 150 × plant capacity = Rs. 150 × 25,000 = Rs. 37.5 × 10^5.
Annual running charges = Rs. 0.05 × 200 × 10^6 = Rs. 100 × 10^5.
Total annual charges = Rs. (37.5 × 10^5 + 100 × 10^5) = Rs. 137.5 × 10^5.

$$\text{Cost per unit} = \text{Rs.}\frac{137.5 \times 10^5}{200 \times 10^6} = \text{Rs.}0.687 = 6.8 \text{ paise.}$$

Example 3.13: The data of a power station as follows:
Installed capacity = 200 MW.
Capital cost = Rs. 350 × 10^6.
Rate of interest and depreciation = 20%.
Annual cost of fuel oil, salaries, and taxation = Rs. 40 × 10^6.
Load factor = 0.5.
Determine the cost of generation and the cost of saving per kWh if the annual load factor is raised to 0.6.

Solution:
Assuming maximum demand equal to the capacity of the power plant.

$$\text{Load factor} = \frac{\text{average load}}{\text{maximum demand}}$$

$$0.5 = \frac{\text{average load}}{200}.$$

∴ Average load = 0.5 × 200 = 100 MW.
Energy generated per annum = 100 × 10^3 × (365 × 24) = 876 × 10^6 kWh.
Fixed cost = interest and depreciation on capital cost

$$= \frac{20}{100} \times 350 \times 10^6 = \text{Rs.}70 \times 10^6.$$

Running (operating cost) = cost of fuel oil, salaries, and taxation
= Rs. 40 × 10^6.
Total annual cost = Fixed cost + operating cost
= Rs. (70 + 40) × 10^6
= Rs. 110 × 10^6.

$$\therefore \text{ Cost per kWh} = \frac{110 \times 10^6}{876 \times 10^6} \times 100 = 12.55 \text{ paise.}$$

When the load factor is raised to 0.6:
Average load = load factor × maximum demand
= 0.6 × 200 = 120 MW.
Energy produced per annum = 110 × 10^3 × 365 × 24
= 963.6 × 10^6 kWh.
Total annual cost will not change.

$$\therefore \text{ Cost per kWh} = \frac{110 \times 10^6}{963.6 \times 10^6} \times 100$$

= 11.41 paise.
∴ Saving cost per kWh = 12.55 – 11.41 = 1.14 paise.

Example 3.14: A customer takes a constant load of 200 kW at a pf of 0.85 lagging for 12 hr/day and 365 days/annum. Calculate the annual payment under each of the following tariffs.

(i) Rs. 1.3/kWh + Rs. 1,000/kVA/annum.

(ii) Rs. 1.25/kWh + Rs. 1,000/kW/annum + 30 paise kVArh.

Solution:

Maximum demand = 200 kW.
Power factor = 0.85 lag.
Annual energy consumption = load in kW × working hours per day × working day per annum

$= 200 \times 12 \times 365$

$= 8{,}76{,}000$ kWh.

$$\text{Maximum demand in kVA} = \frac{\text{kW}}{\text{pf}} = \frac{200}{0.85} = 235.294 = 235.3.$$

kVArh consumed per annum = kWh × tan ($\cos^{-1}$ pf)

$= 876{,}000 \times \tan(\cos^{-1} 0.85)$

$= 876{,}000 \times \tan 31.78$

$= 876{,}000 \times 0.619$

$= 542{,}896.$

Annual payment under tariff (i) = Rs. (1.3 × 876,000 + 1,000 × 235.3)

= Rs. 13,74,100.00.

Annual payment under tariff (ii) = Rs. (1.25 × 876,000 + 1,000 × 200 + 0.30 × 542,896)

= Rs. 1,457,868.8.

Example 3.15: Calculate the number of units to be consumed so that the annual bill based on the two-part tariff is same for the following data:

Maximum demand = 15 kW.

Two-part tariff Rs. 1,000/annum/kW of maximum demand plus Rs. 1.6 per unit consumed. Flat rate tariff Rs. 2.5 per unit.

Solution:

Maximum demand = 15 kW.

Let the consumption be x units so that the annual bill based on the two-part tariff and the flat-rate tariff is the same.

Annual bill under the two-part tariff = Rs. $1{,}000 \times 15 + 1.6 \times x$

$= 1.6x + 15{,}000.$

Annual bill under the flat-rate tariff = Rs. $2.5 \times x$.

Given annual bill will be same if $(1.6x + 15{,}000) = 2.5 \times x$

$$x = \frac{15{,}000}{0.9} = 16{,}666.7$$

Rs. 1,68,005.

3.4 NEED FOR ELECTRICAL ENERGY CONSERVATION—METHODS

Generally, the investment for energy conservation is to be judged exactly like any other forms of capital investment.

Energy conservation may be classified into three categories. They are:

(i) short-term energy conservation,

(ii) medium-term energy conservation, and

(iii) long-term energy conservation.

The short-term energy conservation measures changes in operating equipment with little or no capital expenditure. Medium conservation refers to low-cost modifications and improvements to existing equipments. In long-terms schemes capital costs are very high, which seeks for frequent implementation of new techniques, and technologies.

Short-term energy conservation method

The items belonging to short-term energy conservation schemes are given below.

(i) *Heat exchangers:* Heat exchangers through which heat is transformed from product streams to feed streams. In this scheme of frequent monitoring, optimal number of cycles can be determined and also frequent cleaning ensures heat recovery improvement.

(ii) *Good house-keeping:* Whenever plenty of natural light is available then artificial light is to be avoided for better energy conservation.

(iii) *Electrical power:* It is better to use the off-peak electricity for energy conservation.

(iv) *Furnace efficiency:* Minimum amount of air is to be maintained for combustion and always it is necessary to monitor the oxygen levels in flue gasses continuously. Oil burners should be cleaned frequently.

Medium-term energy conservation methods

The medium-term energy conservation refers to low-cost modification and improvements to existing equipment. In this case, pay-back period is often less than two years. Some of the examples for medium-term energy management are given below.

(i) Leakage of air in a room is to be prevented by improving the insulation levels which can be done by estimating the optimum thickness of insulation.

(ii) Effective temperature control is to be achieved.

(iii) Air compressors are to be replaced whenever essential.

(iv) Power factor of electrical systems is suitably adjusted to have the considered savings in energy.

Long-term energy conservation methods

In long-term investment, saving on capital may not be good compared to the medium term. But some of the modifications are suggested to assure the economic viability.

Heater modifications:

- Heating tubes and air preheaters may be installed to extract more heat from furnace.
- Additional insulation is to be provided for heated storage tanks.
- Heat recovery can be improved by providing additional heat exchangers in the processing areas.

3.4.1 Energy efficient equipment

Electric motors are the most extensively used power-consuming equipment in an industry. In India, 72–75% of total electricity used in industrial sectors is consumed by electric motors. In United States, the country that produces nearly one-third of world's electricity.

For many years, the trend was toward small and lighter motors in order to reduce cost. Many users have been chosen to purchase the lowest first cost motor without considering the power factor and efficiency. Since 1975, as the cost of electrical energy increases, manufacturers have been tried to improve the efficiency of the motor to have the significant saving in electrical energy.

Need of efficient motor:

As we know that the increase in the cost of electric power began in 1972, so that it was highly expensive to use inefficient motor. From the past data, the cost of increase in electric power is 11.5% per year from 1972 to 1979. From that year onwards still cost of electric power increases at an average annual rate of 6% per year. Thus, the manufacturers of electric motors were for methods to improve efficiency of motors.

What is efficiency?

The efficiency of electric motor is a measure of the ability of an electric motor to convert electrical energy to mechanical energy.

$$\text{Efficiency} = \frac{\text{mechanical output}}{\text{electrical input}} \times 100.$$

The electrical energy input to the motors is supplied to its terminals and the horse power of the mechanical energy is taken out of the motor from the rotating shaft.

Where,

Mechanical energy output = Electrical energy input – motor losses.

Or,

Electrical energy input = mechanical energy output + motor losses.

Here, the efficiency of motor can be improved by reducing the amount of electrical power consumed by the motor.

Losses in motors:

Various losses occurring in a motor are:

(i) Magnetizing loss, core loss, or iron loss is dependent on voltage but independent on load on the motor.

(ii) Winding loss or copper loss is proportional to the square of current and dependents on the load.

(iii) Friction and windage loss.

(iv) Stray loss.

Iron losses are further classified into two types namely hysteresis and eddy current loss.

Hysteresis loss which occurs in a magnetic material subjected to continuous reversal of applied field. An empirical formula for hysteresis loss is given by:

$$W_h = k_h B_m^n f v \text{ W},$$

where k_h is hysteresis loss coefficient, depends on type of magnetic material. B_m is the maximum flux density, f is the frequency of supply, v is the volume of core, and n is the Steimentz constant typically 1.5–2.5.

Eddy current losses occur because the magnetic core material itself consists of materials that conduct electricity. As the voltages are induced in the material by alternating magnetic fields causes to circulate current through core called eddy currents. Expression for eddy current loss is given by:

$$W_e = k_e B_m^2 f^2 t^2 v \, W,$$

where k_e is the eddy current loss coefficient.

The other losses friction and windage loss caused by the bearings and motor fan, and stray loss occurred due to electronic harmonics and stray currents. The friction losses in the bearings are a function of bearing size, speed, type of bearing, load, and lubrication used. This loss is relatively fixed for a given design and since it is a small percentage of total motor losses. Reduction in these losses will not affect the efficiency of the motor.

Stray load losses are residual losses in the motor that are difficult to determine by direct measurement or calculation. These losses are load related and are generally assumed to vary as the square of the output torque.

Criteria for selection of motors:

Selection of electric motors are based on the following factors.

(i) The initial cost of motor is in significant compared to the running cost.

(ii) Besides cost, the efficiency of motor is important.

(iii) The speed of motor should be very high. These speed motors are having high efficiency.

(iv) The power factor of motor should be high.

3.4.2 Energy management

The energy management system involving the energy area of business and it is more than conservation. Energy like any other resources such as men, material, plant, and capital is subjected to management process considered as edgiest resource which can be controlled. Thus by definition, energy management is the most cost effected and efficient use of energy. This energy management takes into account for planning, communication, salesmenship, and marketing.

Energy management includes energy productivity and energy awareness. Energy awareness is essential in motivating employees to save energy. The energy management encodes the aspects of load management efficient end use, fuel conservation, heat recovery efficient process, and equipments.

Steps for energy management program

There are three basic steps that are necessary for energy management program.

(i) *Initiation face:* In this face, the resource to be managed must be realized and it is necessary to decide energy management is indeed applied. It is nothing but the commitment of management by energy management program. Later, energy management coordinator is assigned then energy management committee is created to bring new ideas and coordinate plants, etc.

(ii) *Audit and analysis phase:* The execution of the program involves the commitment on the part of the management involves the following steps.

- Analyzing the historical patrons of fuel and energy use.
- Making walk-through survey.

- Making data sheets and equipment specifications.
- Conducting energy audit.
- Calculation of annual energy use.
- Comparing the above analysis with historical results.
- Simulation of evaluating energy management options.

(iii) *Implementation phase:* This is the controlling phase. This phase involves the following steps.

- Establishing energy efficient goals for individual plants.
- Determining capital investment required.
- Making measurements and reporting procedures.
- Providing periodic review and evaluation of overall energy management program.

3.4.3 Energy auditing

Energy auditing can be defined as an audit that serves the purpose of identifying the usage of energy or the conservation opportunities for any plant or a while performing function.

The energy audit is an attempt to determine the use of energy attributed to each of the major components of energy use. This auditing gives a brief overview of understanding present system operation, and future modifications and making new decisions.

The main objective of energy auditing are:

- To know the energetic behavior of plant.
- To identify the excess of energy consumed.
- To understand the purpose of alternate solutions.
- To determine ideal energy consumer profiles.

Usually, the energy auditings are classified into three categories.

1. Walk through.
2. Mini audit.
3. Maxi audit.

The walk-through type of auditing identifies the preliminary energy savings and it is the least costly method. This periodic inspection is made to determine maintenance and operation saving opportunities to know more detailed analysis.

Mini audit requires test to know the energy uses and losses in order to determine economics for change.

Maxi audit goes one step further than the mini audit. This audit gives energy usage for each function. It requires some analysis to determine energy use patterns on annual basis.

3.5 POWER FACTOR IMPROVEMENT

The electrical power or energy is generated at generating station and transmitted through transmission lines and then distributed to the consumer. The quantity of power transmitted and distributed is based on the power factor of load and the parameters of lines.

Almost all the power system loads are of inductive type and have low lagging pf, which is undesirable. The power factor of the power system close to unity is desirable for economical and better distribution of electrical energy.

3.5.1 Causes of low power factor

The induction motors work at low lagging pf at light loads and improved pf with increased loads.

i. The transformers have lagging power factor because they draw magnetizing current.
ii. The miscellaneous equipment such as arc lamps, electric discharge lamps, and welding equipment operate at low power factor.
iii. The industrial heating furnaces operate at a low lagging power factor.
iv. The variation of load on the power system also causes low pf.

3.5.2 Effects or disadvantages of low power factor

$$\text{The load current, } I_L = \frac{P}{\sqrt{3}V_L \cos\phi},$$

where P is the real power (watts).

From the above expression for a given load, it is clear that if the pf is low, the load current will be higher. The larger the load current due to low pf results in the following effects.

(i) Effect on transmission lines: For the fixed active power to be transmitted over the line, the lower the pf, the higher will be the load current to be carried by the line. Since the maximum permissible current density of the line conductor is fixed, the cross-sectional area of the conductor is to be increased in order to carry larger current. This results in an increased volume of the conductor material which in turn increases the capital cost of transmission lines.

Further, increase in the current causes increase in the line losses with a reduction in the efficiency of the line. Also the line voltage regulation is poor.

(ii) Effect on transformers: A reduction in the pf causes a reduction in the kW capacity of a transformer.

(iii) Effect on switch gear and bus bar: The lower the pf at which a given power is to be supplied, the larger is the cross-sectional area of the bus bar and the larger is the contact surface of the switch gear.

(iv) Effect on generators: With a lower pf, the kW capacity of a generator is reduced. The power supplied by the exciter is increased. The generator copper losses are increased, which results in low efficiency of the generator.

(v) Effect on prime movers: When the pf is decreased, the alternator develops more reactive kVA, i.e., the reactive power generated is more. This requires a certain amount of power to be supplied by the prime mover. So, a part of the prime mover capacity is idle and it represents a dead investment. The efficiency of the prime mover is reduced.

(vi) Effect on existing power systems: For the same active power, the operation of an existing power system at a lower pf necessitates the overloading of the equipment during full load.

3.5.3 Advantages of power factor improvement

From the earlier discussions, it is observed that if the power station works at low power factor, the capital cost of generation, transmission, and distribution systems is increased. Higher capital charges means higher annual fixed charges, which will increase the cost-per-unit supply to the consumer. So, the maintenance of high pf (closed to unity) is always advantageous for both consumers and suppliers.

Following are the advantages of power factor improvement.

- The kW capacity of the prime movers is better utilized due to decreased reactive power.
- This increases the kilowatt capacity of the alternators, transformers, and the lines.
- The efficiency of the system is increased.
- The cost per unit decreases.
- Improves the voltage regulation of the lines.
- Reduction in power losses in the system due to reduction in load current.
- The cost of generators, transformers, and transmission lines per kW of load supplied decreases.
- Reduction in kVA demand charges for large consumers.

3.5.4 Methods of improving power factor

In case of inductive loads, the pf is lagging. This lagging pf can be compensated by using some devices that are called compensators. These are:

(i) static capacitors,

(ii) synchronous condensers, and

(iii) phase advancers.

(i) *Static capacitor*

Static capacitors are connected across the mains at the load end as shown in Fig. 3.4(b). This supplies a reactive component of the current to reduce the out-of-phase component of current required by an inductive load, i.e., it modifies the characteristics of an inductive load by drawing a leading current that counteracts or opposes the lagging component of the inductive load current at the point of installation. So, the reactive volt ampere transmitted over the line is reduced, thereby the voltage across the load is maintained within the specified limits.

By the application of the shunt capacitor to a line, the magnitude of source current can be reduced, the pf can be improved, and consequently the voltage drop between the sending and receiving ends is also reduced as shown in Fig. 3.5(b). However, it is important to note that it does not affect the current or pf beyond their point of installation.

The voltage drop of the line without shunt capacitors is given by:

$$V_d = I_r R \cos\phi + I_r X_L \sin\phi. \tag{3.1}$$

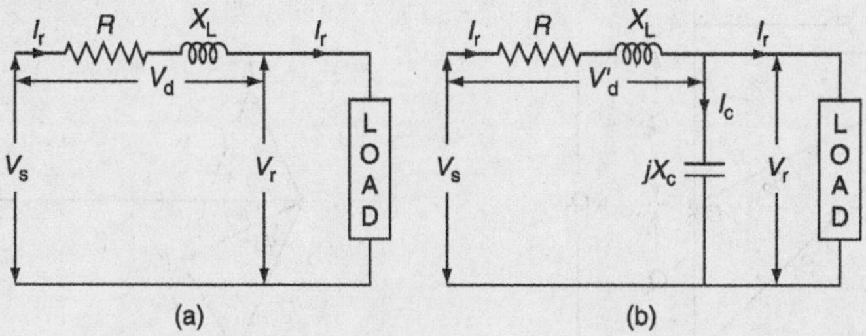

FIG. 3.4 *Single-line diagram without and with shunt capacitive compensation*

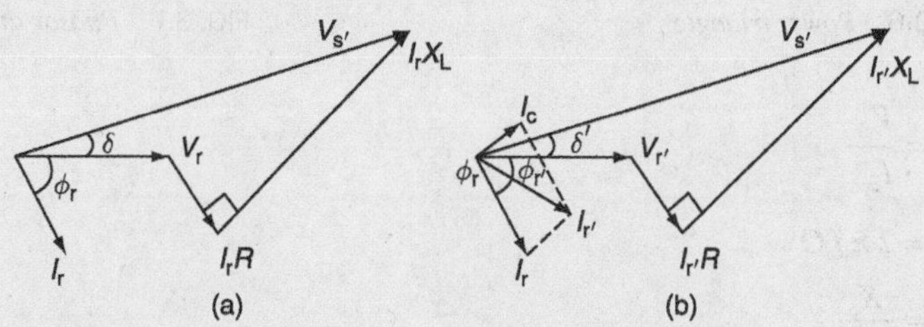

FIG. 3.5 *Phasor diagram of Fig. 3.4 (a) and (b), respectively*

With shunt capacitor, $V'_d = I_r R \cos\phi + (I_r - I_C) X_L \sin\phi$, **(3.2)**

where I_c is the reactive component of current leading the supply voltage by 90°.

The voltage rise due to the location of the capacitor is the difference between the voltage drops determined by using Equations (3.1) and (3.2) and is given as:

Voltage rise $= I_c X_L$

and improved pf $= \dfrac{P}{\sqrt{P^2 + (Q_L - Q_C)^2}}$.

Calculation of shunt capacitor rating

The power factor correction can be determined from the power triangle shown in Fig. 3.6. From Fig. 3.6, the cosine of the angle $\angle OAB$ is the original pf ($\cos\phi_1$), whereas the cosine of the angle $\angle OAC$ is the improved pf ($\cos\phi_2$). It may be observed that the active power (OA) does not change with pf improvement. However, the lagging kVAr of the load is reduced by the pf correction equipment, thus improving the pf to $\cos\phi_2$.

Leading kVAr (Q_c) supplied by pf correction equipment as:

$$\begin{aligned} BC &= AB - AC \\ &= Q_1 - Q_2 \\ &= OA(\tan\phi_1 - \tan\phi_2) \\ &= P(\tan\phi_1 - \tan\phi_2). \end{aligned} \quad \textbf{(3.3)}$$

Knowing the leading kVAr (Q_c) supplied by the pf correction equipment, the capacitor current can be calculated.

$$I_c = \frac{Q_c}{\sqrt{3}V_{L-L}}$$

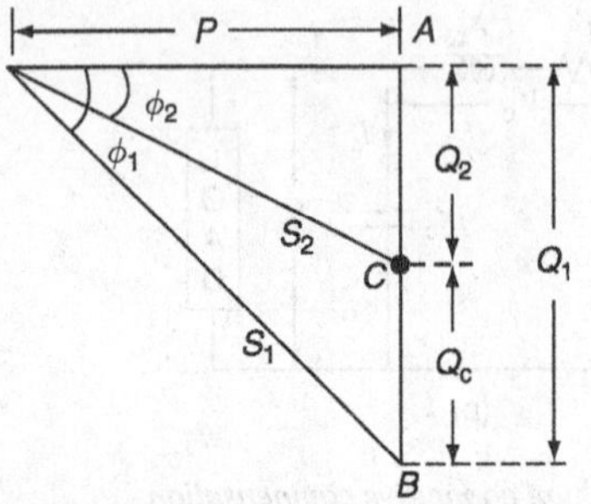

FIG. 3.6 *Power triangle*

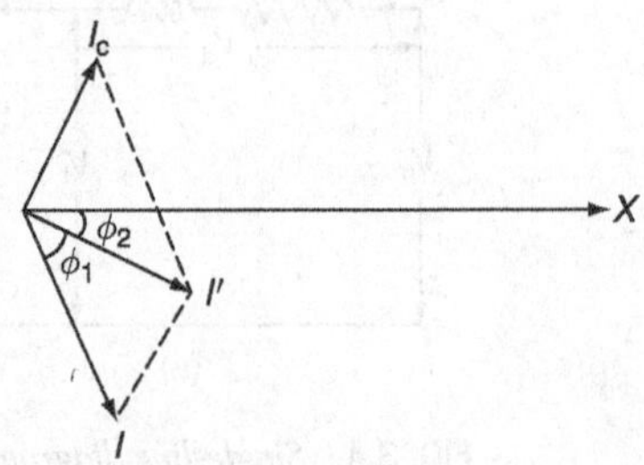

FIG. 3.7 *Phasor diagram*

$$X_c = \frac{V_{ph}}{I_{cph}}$$

$$= 2\pi f C$$

$$\therefore C = \frac{X_c}{2\pi f}.$$

Alternative method

Consider a single-phase load that is taken as lagging current I. When the capacitor is connected across the load, the current taken by the capacitor (I_c) leads the supply voltage (V) by 90°. The resultant current I' is the vector sum of I and I_c and its angle of lag is ϕ_2, which is less than ϕ_1, i.e., cos ϕ_1 is less than cos ϕ_2 as shown in Fig. 3.7. For three-phase loads, the capacitors can be connected in star or delta.

If the pf of the load is cos ϕ_1 and is improved to cos ϕ_2 then the value of capacitor '*C*' can be calculated as follows.

The active component of load current, $I_a = I \cos \phi_1$.

The reactive component of load current, $I_{r1} = \frac{I_a}{\cos\phi_1} \times \sin \phi_1 = I_a \tan \phi_1$.

At fixed load, the reactive component with increased pf, $I_{r2} = I_a \tan \phi_2$.

Current taken by the capacitor, $I_c = I_a (\tan \phi_1 - \tan \phi_2)$.

Applied voltage, $V = I_C X_C = I_C \frac{1}{2\pi f C}$ volts.

Value of capacitor, $C = \frac{I_C}{2\pi f V}$ farads.

If the power factor of the load is to be improved up to unity, the phasor diagram is shown in Fig. 3.8.

Then, $I_c = I \sin \phi_1$.

And the capacitance, $C = \frac{I_C}{2\pi f V}$ farads.

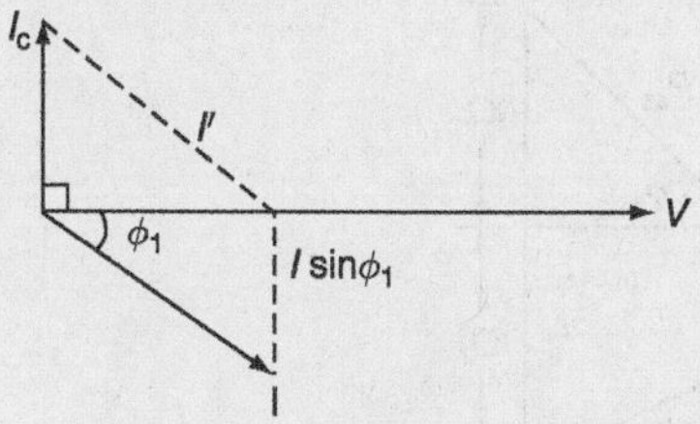

FIG. 3.8 *Phasor diagram for unity power factor*

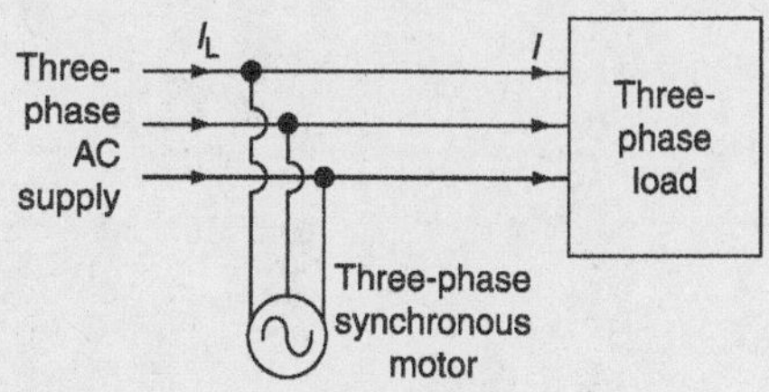

FIG. 3.9 *Synchronous machine is connected in parallel with the supply*

Advantages of shunt capacitor

- Losses are low.
- It requires less maintenance, because there are no rotating parts.
- Easy installation.

Disadvantages of shunt capacitor

- Less service life.
- Easily damaged due to excess of voltage.
- Its repair is uneconomical.
- Difficult to control because of removing or adding the capacitors in the circuit for different power factors.

(ii) *Synchronous condenser*

The synchronous condenser is a synchronous motor running without a mechanical load. A synchronous motor takes a leading current when overexcited and, therefore, behaves as a capacitor. It is connected in parallel to the supply or load as shown in Fig. 3.9. It generates leading current to neutralize the lagging component of the load current, results improve the power factor.

Let P_1 be the active power of load.
Cos ϕ_1 is the pf of load without synchronous condenser.
Cos ϕ_2 is the pf of load after using synchronous condenser.
P_s is the active power taken by the synchronous condenser from the supply.

The rating and the pf at which the synchronous condenser is operating can be calculated as follows.

Reactive power of load, $Q_1 = P_1 \tan \phi_1$.
When synchronous condenser is connected in the circuit:
Total load (active power) $P = P_1 + P_s$.
Total reactive power, $Q_2 = P \tan \phi_2$.
Reactive power supplied by the synchronous condenser, $Q = Q_1 - Q_2$.
The phasor diagram is shown in Fig. 10.8.
From Fig. 3.10, the rated kVA of synchronous condenser, $P_{as} = \sqrt{P_s^2 + Q_r^2}$.

And, power factor of synchronous condenser, $\cos\phi_2 = \dfrac{P_s}{P_{as}}$.

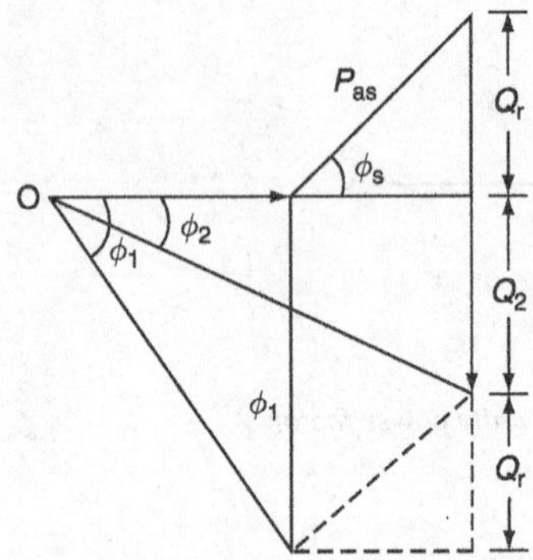

FIG. 3.10 *Phasor diagram*

Advantages

- A synchronous condenser has an inherently sinusoidal wave form and the voltage does not exist.
- It can supply as well as absorb kVAr.
- The pf can be varied in smoothly.
- It allows the overloading for short periods.
- The high inertia of the synchronous condenser reduces the effect of sudden changes in the system load and improves the stability of the system.
- It reduces the switching surges due to sudden connection or disconnection of lines in the system.

Disadvantages

- Power loss is more.
- For small rating it is uneconomical.
- It is not possible to add or take away the units and to alter the rating of the synchronous condenser.
- The cost of maintenance is high.
- It produces noise.

(iii) *Phase advancers*

There are special commutator machines, which are used to improve the power factor of the induction motor. When the supply is given to the stator of an induction motor, it takes a lagging current. So, the induction motor has low lagging power factor. For compensating this lagging current, a phase advancer (mounted on same shaft) is used. It supplies mmf to the rotor circuit at slip frequency.

Advantages

- The lagging kVAr drawn by the motor is reduced by compensating the stator lagging current at slip frequency.

1. Where the use of synchronous motors is not suitable, phase advancer can be used. Generally, these compensators are not recommended for the economical operation of motors of low rating below 200 HP.

Example 3.16: A single-phase motor connected to a 230-V and 50-Hz supply takes 30 A at a pf of 0.7 lag. A capacitor is shunted across the motor terminals to improve the pf to 0.9 lag. Determine the capacitance of the capacitor to be shunted across the motor terminals.

Solution:

Motor current, $I_{m1} = 30$ A.

The active component of motor current (I_{m1}) at pf 0.7 lag, $I_{a1} = I_{m1} \cos \phi_1$

$= 30 \times 0.7$

$= 21$ A.

The reactive component of motor current (I_{m1}) at pf 0.7 lag, $I_{r1} = I_{m1} \sin \phi_1$

$= 30 \times 0.714$

$= 21.42$ A.

The active component of current at improved pf of 0.9 lag is same as the active component of current at a pf of 0.7 lag as shown in Fig. 3.11.

The active component of motor current at a pf of 0.9, $I_{a2} = 21$ A.

$$\text{The reactive component of motor current at a pf of 0.9, } I_{r2} = \frac{21}{0.9} \times 0.436 = 10.173 \text{ A.}$$

The reactive component of motor current to be neutralized, $I_c = I_{r1} - I_{r2}$

$= 21.42 - 10.173$

$= 11.247$ A.

$$\text{It is also equal to } (I_C) = \frac{V}{X_C}$$

$$= V \times 2\pi f C$$

$$11.247 = 230 \times 2\pi \times 50 \times C.$$

$\therefore$ The capacity of capacitance connected across the motor terminals, $C = 155.65$ µF.

Example 3.17: A single-phase, 400-V, and 50-Hz motor takes a supply current of 50 A at a power factor of 0.8 lag. The motor pf has been improved to unity by confectioning a condenser in parallel. Calculate the capacity of the condenser required.

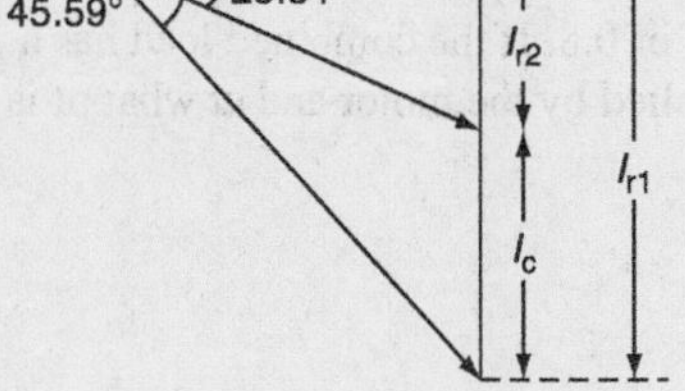

FIG. 3.11 *Phasor diagram*

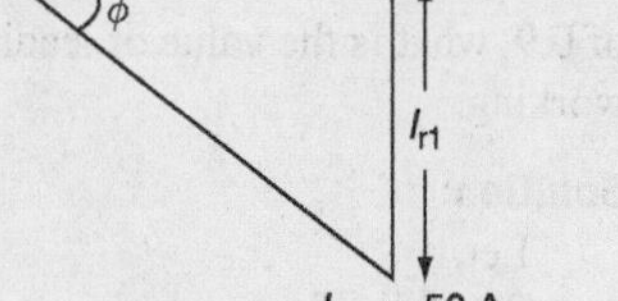

FIG. 3.12 *Phasor diagram*

Solution:

The motor current at a pf 0.8 lag, $I_{m1} = 50$A.

The active component of motor current (I_{m1}) at pf 0.8 lag, $I_{a1} = I_{m1} \cos \phi_1$

$= 50 \times 0.8$

$= 40$ A.

The reactive component of motor current (I_{m1}) at pf 0.8 lag, $I_{r1} = I_{m1} \sin \phi_1$

$= 50 \times 0.6$

$= 30$ A.

The active component of motor current at improved pf of unity is same as the active component of motor current at a pf of 0.8 lag, for a fixed load as shown in Fig. 3.12.

∴ The active component of motor current at a pf of 0.9 $I_{a2} = 40$ A.

The reactive component of motor current at a pf of unity, $I_{r2} = \dfrac{40}{1} \times 0 = 0.0$ A.

The reactive component of motor current to be neutralized, $I_c = I_{r1} - I_{r2}$

$= 30 - 0 = 30$ A.

It is also equal to $(I_C) = \dfrac{V}{X_C}$

$= V \times 2\pi f C$

$30 = 400 \times 2\pi \times 50 \times C.$

∴ The capacity of capacitance connected across the motor terminals, $C = 238.73$ μF.

Example 3.18: A 440-V and 3-ϕ distribution feeder having a load of 100 kW at lagging power factor and the load current is 200 A. If it is desired to improve the pf. Determine:

(i) the uncorrected pf and reactive load and

(ii) the new corrected pf after installing a shunt capacitor of 75 kVAr.

Solution:

(i) Uncorrected pf $= \cos\phi = \dfrac{P}{\sqrt{3}V_L I_L} = \dfrac{100\times10^3}{\sqrt{3}\times 440\times 200} = 0.656$ lagging

$Q_L = P \tan \phi = 115.055$ kVAr

$Q_c = 75$ kVAr.

(ii) Corrected pf $= \dfrac{P}{\sqrt{(P^2 + (Q - Q_c)^2)}} = \dfrac{100}{\sqrt{(100)^2 + (115.055 - 75)^2}}$

$= 0.928$ lagging.

Example 3.19: A synchronous motor having a power consumption of 50 kW is connected in parallel with a load of 200 kW having a lagging pf of 0.8. If the combined load has a pf of 0.9, what is the value of leading reactive kVA supplied by the motor and at what pf is it working.

Solution:

Let,

Pf angle of motor $= \phi_1$.

Pf angle of load $= \phi_2 = \cos^{-1}(0.8) = 36.87°$.

Combined pf angle (both motor and load), $\phi = \cos^{-1}(0.9) = 25.84°$.

Tan ϕ_2 = tan 36°87′ = 0.75; tan ϕ = tan 25°84′ = 0.4842.
Combined power $P = 200 + 50 = 250$ kW.
Total kVAr of combined system = P tan ϕ_t = 250 × 0.4842 = 121.05.
Load kVAr = 200 × tan ϕ_2 = 200 × 0.75 = 150.
:. The leading kVAr supplied by synchronous motor =150 – 121.05 = 28.95.
Pf angle at which the motor is working, $\phi_1 = \tan^{-1} 28.95/50 = 30.07°$.
Pf at which the motor is working = cos ϕ_1 = 0.865 (lead).

Example 3.20: A 3-ϕ and 5-KW induction motor has a pf of 0.85 lagging. A bank of capacitor is connected in delta across the supply terminal and pf raised to 0.95 lagging. Determine the kVAr rating of the capacitor in each phase.

Solution:

The active power of the induction motor, $P = 5$ kW.
When the pf is changed from 0.85 lag to 0.95 lag, by connecting a condenser bank.
The leading kVAr taken by the condenser bank = P (tan ϕ_2 – tan ϕ_1)
= 5(0.6197 – 0.3287) = 1.455.
∴ The rating of capacitor connected in each phase = 1.455/3 = 0.485 kVAr.

Example 3.21: A 3-phase, 500-HP, 50-Hz, and 11-kV star-connected induction motor has a full load efficiency of 85% at lagging pf of 0.75 and is connected to a feeder. If it is desired to correct the pf of load to 0.9 lagging. Determine:

(a) the size of the capacitor back in kVAr and

(b) the capacitance of each unit if the capacitors are connected in Δ as well as in Y.

Solution:

Induction motor output = 500 HP.
Efficiency η = 85%, where η = output/input.
Input of the induction motor, P = output/η = 500/0.85 = 588.235 HP
= 588.235 × 746 = 438.82 kW.

Initial pf, (cos ϕ_1) = 0.75 ⇒ tan ϕ_1 = 0.88.
Corrected pf (cos ϕ_2) = 0.9 ⇒ tan ϕ_2 = 0.48.
Leading kVAr taken by the capacitor bank, $Q_c = P$ (tan ϕ_1 – tan ϕ_2)
= 438.82 (0.88 – 0.48) = 175.53 kVAr.

Line current drawn, $I_L = \dfrac{Q_c}{\sqrt{3}V_{L-L}} = \dfrac{175.53}{\sqrt{3}\times 11} = 9.213$ A.

Case I: Delta connection

Charging current per phase $I_c = \dfrac{I_L}{\sqrt{3}} = 5.319$ A.

The reactance of capacitor bank per phase $X_c = \dfrac{V_{L-L}}{I_C} = \dfrac{11\times 10^3}{5.319} = 2.068$ KΩ

$$X_c = \frac{1}{2\pi f_c} \Rightarrow c = \frac{1}{2\pi f X_c}.$$

The capacitance of capacitor bank $C = \dfrac{1}{2\pi \times 50 \times 2.068 \times 10^3} = 1.539$ μF.

Case II: Star connection

$I_L = I_c = 9.213$ A.

The reactance of capacitor bank per phase $X_c = \frac{V_{L-N}}{I_c} = \frac{11 \times 10^3}{\sqrt{3} \times 9.213} = 0.689\ K\Omega$.

The capacitance of capacitor bank, $C = \frac{1}{2\pi f X_C} = 4.619\ \mu F$.

3.5.5 Most economical power factor when the kW demand is constant

For improving the pf of load at consumer end, the consumer must provide equipment for improving pf. So, there is a capital investment on correction equipment. At the same time, there is a savings due to reduced demand in kVA. Therefore, the net annual savings is equal to the difference between the annual saving in maximum demand charges and the annual expenditure incurred on pf correction equipment.

The value of a power factor at which the net annual saving is maximum is known as the most economical power factor.

Consider a load of P kW that is taken by the consumer at a power factor of cos ϕ_1 and is represented in Fig. 3.13.

Then, the reactive component of load P is, $Q = P \tan \phi_1$.

And, kVA demand of load $P, S = \frac{P}{\cos \phi_1}$.

Let charge at the rate of Rs. X per kVA maximum demand per annum.

Suppose the consumer improves the power factor to cos ϕ_2 by installing pf correction equipment. Let the expenditure incurred on the pf correction equipment be Rs. Y per kVAr per annum. The power triangle at the original pf cos ϕ_1 is OAB and for the improved pf cos ϕ_2, it is OAC (are shown in Fig. 10.9.).

kVA maximum demand at cos ϕ_1 $\text{kVA}_1 = \frac{P}{\cos\phi_1}$.

kVA maximum demand at cos ϕ_2, $\text{kVA}_2 = \frac{P}{\cos \phi_2}$.

Because of the improvement in the pf, the kVA maximum demand is reduced from kVA_1 to kVA_2 (since the real power remains unchanged).

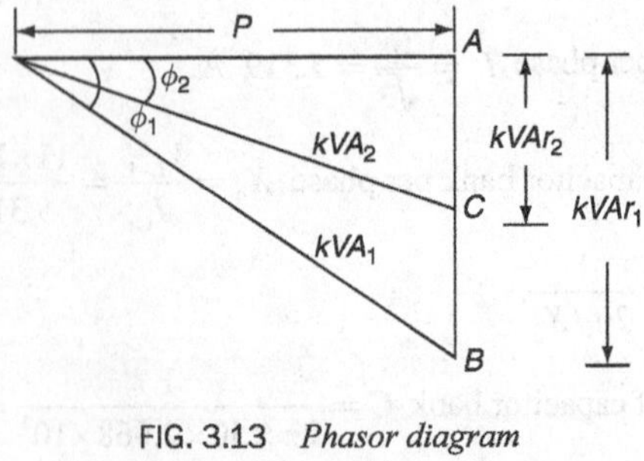

FIG. 3.13 *Phasor diagram*

Annual saving in maximum demand = Rs. $X(\text{kVA}_1 - \text{kVA}_2)$

$$= \text{Rs. } X\left(\frac{P}{\cos\phi_1} - \frac{P}{\cos\phi_2}\right)$$

$$= \text{Rs. } XP\left(\frac{1}{\cos\phi_1} - \frac{1}{\cos\phi_2}\right).$$

Reactive power at $\cos\phi_1$, $\text{kVAr}_1 = P\tan\phi_1$.
Reactive power at $\cos\phi_2$, $\text{kVAr}_2 = P\tan\phi_2$.
Leading kVAr supplied by pf correction equipment $= P(\tan\phi_1 - \tan\phi_2)$.
So, the annual charges toward phase advancing plant = Rs. $YP(\tan\phi_1 - \tan\phi_2)$.
Net annual saving toward phase advancing circuit:

$$S = XP\left(\frac{1}{\cos\phi_1} - \frac{1}{\cos\phi_2}\right) - YP\left(\tan\phi_1 - \tan\phi_2\right).$$

In this expression, only ϕ_2 is variable while all other quantities are fixed. For maximum net annual saving:

$$\frac{d}{d\phi_2}(S) = 0$$

$$\frac{d}{d\phi_2}\left[XP(\sec\phi_1 - \sec\phi_2) - YP(\tan\phi_1 - \tan\phi_2)\right] = 0$$

$$\frac{d}{d\phi_2}(XP\sec\phi_1) - \frac{d}{d\phi_2}(XP\sec\phi_2) - \frac{d}{d\phi_2}(YP\tan\phi_1) + \frac{d}{d\phi_2}(YP\tan\phi_2) = 0$$

$$0 - XP\sec\phi_2\tan\phi_2 - 0 + YP\sec^2\phi_2 = 0$$

$$\text{or}\quad -X\tan\phi_2 + Y\sec\phi_2 = 0$$

$$\tan\phi_2 = \frac{Y}{X}\sec\phi_2$$

$$\sin\phi_2 = \frac{Y}{X}.$$

Most economical power factor,

$$\cos\phi_2 = \sqrt{1 - \sin^2\phi_2}$$

$$= \sqrt{1 - \left(\frac{Y}{X}\right)^2}$$

$$\text{or}\quad \cos\left(\sin^{-1}\frac{Y}{X}\right).$$

3.5.6 Most economical power factor when the kVA maximum demand is constant

This contingency arises in the case of power supply agencies. They try to improve the pf so that the kVA maximum demand on the station is reduced. Since the cost of the plant is proportional to the kVA installed, an improvement in the pf reduces the cost of the plant. Further, the revenue returns are the function of active power supplied.

The phasor diagram is shown in Fig. 3.14. The kVA output is constant and equal to S. By the addition of Q kVAr leading, the pf is improved from cos ϕ_1 to cos ϕ_2. Consequently, the real power is increased from P_1 to P_2.

Let the annual interest and depreciation charges for a capacitor = X Rs. per kVAr.

Let the net return per kW of installed capacity per year = Rs. Y.

From phasor diagram, leading kVAr supplied by the pf improvement equipment is:

$$Q = S(\sin\phi_1 - \sin\phi_2).$$

∴ Annual charge on capacitor installation =Rs. $XS(\sin\phi_1 - \sin\phi_2)$.

Annual increase in revenue return because of increase in the real power

$= \text{Rs. } Y(P_2 - P_1)$

$= Y(S\cos\phi_2 - S\cos\phi_1)$

$= YS(\cos\phi_2 - \cos\phi_1)$.

Net saving = $YS(\cos\phi_2 - \cos\phi_1) - XS(\sin\phi_1 - \sin\phi_2)$.

In this expression, only ϕ_2 is variable while all other quantities are fixed.

For maximum net annual saving:

$$\frac{d}{d\phi_2}(\text{Net saving}) = 0$$

$$\frac{d}{d\phi_2}\left[YS(\cos\phi_2 - \cos\phi_1) - XS(\sin\phi_1 - \sin\phi_2)\right]$$

or $\quad YS(-\sin\phi_2 - 0) - XS(0 - \cos\phi_2) = 0$

or $\quad YS(-\sin\phi_2) = XS(-\cos\phi_2)$

or $\quad \tan\phi_2 = \dfrac{X}{Y}$.

Most economical power factor, when the kVA maximum demand is a constant

$$\cos\phi_2 = \cos\left(\tan^{-1}\frac{X}{Y}\right).$$

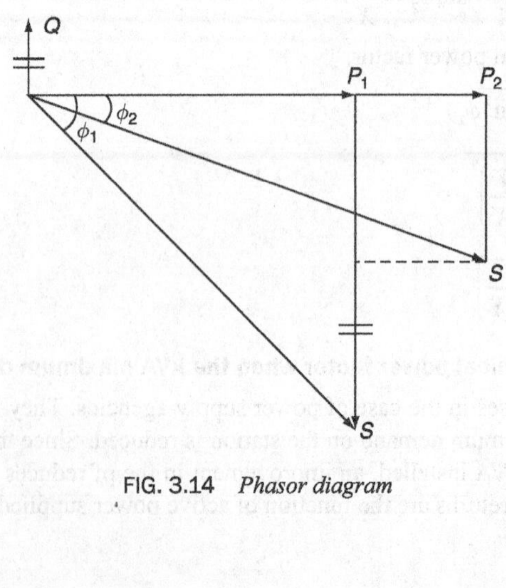

FIG. 3.14 *Phasor diagram*

It may be noted that the most economical power factor (cos ϕ_2) depends upon the relative costs of supply and the pf correction equipment but is independent of the original pf cos ϕ_1.

The following are the results that we can observe after the power factor improvement.

i. The circuit current I^1 after pf correction is less than the original circuit current I.

ii. The active or watt-ful component of current remains the same before and after the pf correction because only the lagging reactive component is reduced by the capacitor.

$I \cos \phi_1 = I^1 \cos \phi_2$.

iii. The lagging reactive component of current is reduced after pf improvement and is equal to the difference between lagging reactive component of load and capacitor current.

$I^1 \sin \phi_2 = I \sin \phi_1 - I_C$.

iv. As $I \cos \phi_1 = I^1 \cos \phi_2$,

$VI \cos \phi_1 = VI^1 \cos \phi_2$.

Therefore, active power (kW) remains unchanged due to power factor improvement.

v. As $I^1 \sin \phi_2 = I \sin \phi_1 - I_C$,

$VI^1 \sin \phi_2 = VI \sin \phi_1 - VI_C$.

Therefore,

Net kVAr after pf correction = lagging kVAr before pf correction – leading kVAr of equipment.

Example 3.22: A consumer is charged at the rate of Rs. 75 per annum per kVA of maximum demand plus a flat rate per kWh. The phase-advancing plant can be purchased at a rate of Rs. 70/kVA. The rate of interest and the depreciation on the capital is 12.5%. Find the most economical pf to which it can be improved.

Solution:

Annual charges toward the interest and depreciation of the phase advancing equipment

$$= \text{Rs.}\, 70 \times \frac{12.5}{100} = \text{Rs.}\, 8.75.$$

Annual charge per kVA = Rs. 75.00.

Let ϕ_2 be the angle corresponding to the most economical pf:

$$\sin \phi_2 = \frac{8.75}{75} = 0.1166.$$

The most economical pf to which it can be improved is:

$\cos \phi_2 = \cos (\sin^{-1} 0.1166) = 0.9932$ lagging.

3.6 CONCEPT OF DISTRIBUTED GENERATION

Distributed or dispersed generator may be defined as a generating resource, other than central generating station, that is placed close to load being served, usually at customer site. It may be connected to the supply side or demand side of meter. It can be renewable sources based microhydro, wind turbines, photovoltaic, or fossil fuel based fuel-cells, reciprocating engines, or microturbines. In term of size, DG may range from a few kilowatts to over 100 MW. Employing DG in a distribution network has several advantages and a few disadvantages.

Advantages

- Improving of the voltage profile.
- Reduction of line losses.
- Transmission and distribution capacity (congestion) relief.
- Overall improvement in system efficiency.
- Potential for increased security.

Disadvantages

- Reverse power flow as a result of connecting DG in the system causing the malfunction of protection circuits, as they are configured at present.
- Harmonics injection into the system by asynchronous DG source that uses inverters for interconnection.
- Increased fault currents depending on the location of DG units.

3.7 DEREGULATION

The deregulation of electric sector is nothing but its privatization. While the two words are different literally, 'deregulation' often starts with the sale of state-owned utilities to the private sector. This is widely adopted to refer to the 'introduction of competition'. Deregulation often involves 'unbundling', which refers to disaggregating an electric utility service into its basic components and offering each component separately for sale with separate rates for each. As shown in Fig. 3.15, generation, transmission, and distribution could be unbundled and offered as discrete services.

The success of privatization of the airline and telecommunications industries has motivated the deregulation and restructuring of the electricity industry. In 1989, the United Kingdom became one of the pioneers in privatizing its vertically integrated electricity industry.

In many countries, a central independent body, usually called the independent system operator (ISO), is set up to cater to the demand, and the maintenance of system reliability

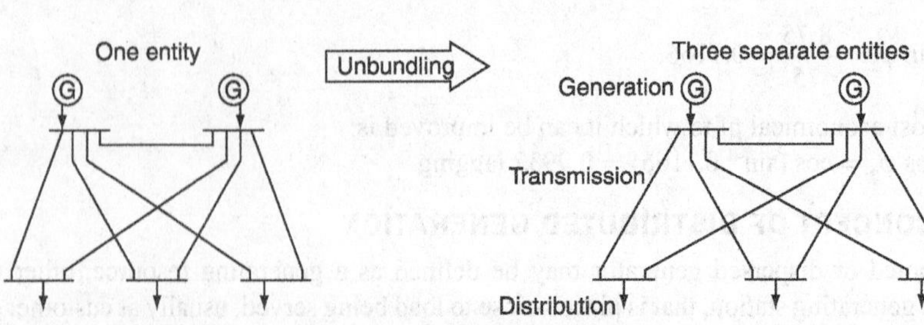

FIG. 3.15 *Unbundling of utilities*

and security. Sometimes, the system operator is also responsible for matching the bids of generators with the demand bids to facilitate exchange.

The restructuring of the utility into separate generation, transmission, and distribution companies has introduced competition in the generation and transmission of electricity. Several independent power producers and qualifying facilities produce electricity and the energy is traded on a real time basis to meet consumer requirements. The successful implementation of deregulation of electric utilities, in several developed countries, has motivated similar restructuring efforts in other developing countries. The effect of deregulation has had a great impact on the Indian power scenario, which has recently initiated deregulation and restructuring. There exist potential opportunities for the successful implementation of the principles of deregulation, as there is an abundance of dispersed sources of generation, such as renewable energy sources, for supplying energy at remote locations or load centers.

There are several motives for deregulation and restructuring of electric utility, important among them being the following factors, namely, (i) the break-up of entrenched bureaucracy, (ii) the reduction of public sector debt, (iii) the encouragement of private sector investment, (iv) the lower electricity prices, (v) the introduction of price competition, (vi) the improvement of efficiency, and (vii) the utilization of assets.

3.8 NEED FOR RESTRUCTURING

Restructuring promises an alternative to vertically integrated monopoly and works on the basic principle that transmission services should accommodate consumer choice and supply competition. In the past, the vertically integrated utilities have been monopolizing generation, transmission, and distribution services. A restructuring of this monopoly was required to provide reliable power at a lower cost.

3.8.1 Motivation for restructuring the power industry

A significant feature of restructuring the power industry is to allow for competition among generators and to create market conditions in the industry, which are seen as necessary for the reduction of the costs of energy production and distribution, the elimination of certain inefficiencies, the shedding of labor, and the increase of customer choice. Many factors such as technology advances, changes in political and ideological attitudes, regulatory failures, high tariffs, managerial inadequacy, global financial drives, and the rise of environmentalism contribute to the worldwide trend toward restructuring.

There are two potential benefits resulting from deregulation. Firstly, the advance of technology makes low-cost power plants owned by independent power producers very efficient. These independent power producers would not have emerged without the reform. Secondly, unbundling the services may result in fairer tariffs being assigned to individual services. Restructuring was done with the view that private organizations could do a better job of running the power industry, and that higher operating inefficiencies and reduction in labor could be reached by privatization. Private utilities also refuse to subsidize rates and have a greater interest in eliminating power thefts and managerial or workplace inefficiencies. A competitive power industry will provide rewards to risk takers and encourage the use of new technologies and business approaches. The regulated monopoly scheme was unable to provide incentives for innovation, since the utility had little motivation to use new ideas and technologies to lower costs under a regulated rate of return framework.

KEY NOTES

- A load curve is a plot of the load demand (on the *y*-axis) versus the time (on the *x*-axis) in the chronological order.
- The load duration curve is a plot of the load demands (in units of power) arranged in a descending order of magnitude (on *y*-axis) and the time in hours (on *x*-axis).
- In the operation of the hydro-electric plants, it is necessary to know the amount of energy between different load levels. This information can be obtained from the load duration curve.
- The integrated load duration curve is also the plot of the cumulative integration of area under the load curve starting at zero loads to the particular load.
- A base load station operates at a high-load factor while the peak-load plant operates at a low-load factor.
- Demand factor is the ratio of the maximum demand and the connected load.
- Load factor is the ratio of the average demand to the maximum demand. The higher the load factor of the power station, the lesser will be the cost per unit generated.
- Diversity factor is the ratio of the sum of the maximum demands of a group of consumers and the simultaneous maximum demand of the group of consumers.
- Base load is the unvarying load, which occurs almost the whole of the day on the station.
- Peak load is the various peak demands of load over and above the base load of the station.

SHORT QUESTIONS AND ANSWERS

(1) What is mean by connected load?

It is the sum of the ratings of the apparatus installed on the premises of a consumer.

(2) Define the maximum demand.

It is the maximum load used by a consumer at any time.

(3) Define the demand factor.

The ratio of the maximum demand and the connected load is called demand factor.

(4) Define the average load.

If the number of kWh supplied be a station in one day is divided by 24 hr, then the value so obtained is known as daily average load.

(5) Define the load factor.

It is the ratio of the average demand and maximum demand.

(6) Define the diversity factors.

It is the ratio of the sum of the maximum demands of a group of consumers and the simultaneous maximum demand of the group of consumers.

(7) Define the plant capacity.

It is the capacity or power for which a plant or station is designed.

(8) Define the utilization factor.

It is the ratio of kWh generated to the product of plant capacity and the number of hours for which the plant was in operation.

(9) What is mean by base load?

It is the unvarying load that occurs almost the whole of the day on the station.

(10) What is mean by peak load?

It is the various peak demands of load over and above the base load of the station.

(11) What is mean by load curve?

A load curve is a plot of the load demand versus the time in the chronological order.

(12) What is mean by load duration curve?

The load duration curve is a plot of the load demands arranged in a descending order of magnitude verses the time in hours.

MULTIPLE-CHOICE QUESTIONS

(1) In order to have a low cost of electrical generation:

(a) The load factor and diversity are high.

(b) The load factor should be low but the diversity factor should be high.

(c) The load factor should be high but the diversity factor should be low.

(d) The load factor and diversity factor should be low.

(2) Power plant having maximum demand more than installed capacity will have utilization factor:

(a) Less than 100%.

(b) Equal to 100%.

(c) More than 100%.

(d) None of these.

(3) The choice of number and size of units in a station is governed by best compromise between

(a) A plant load factor and a capacity factor.

(b) Plant capacity factor and plant use factor.

(c) Plant load factor and use factor.

(d) None of these.

(4) The load factor for domestic loads may be taken

(a) About 85%.

(b) 50–60%.

(c) 25–50%.

(d) 15–20%.

(5) If some reserve is available in a power plant,

(a) Its use factor is always greater than its capacity factor.

(b) Its use factor equals the capacity factor.

(c) Its use factor is always less than its capacity factor.

(d) None of these.

(6) A higher load factor means:

(a) Cost per unit is less.

(b) Less variation in load.

(c) The number of units generated is more.

(d) All of the above.

(7) The maximum demand of two power stations is same. The daily load factors of the stations are 10% and 20% the units generated by them are in the ratio:

(a) 2:1.

(b) 1:2.

(c) 3:3.

(d) 1:4.

(8) A plant had average load of 20 MW when the load factor is 50%. Its diversity factor is 20% sum of max. Demands of all loads amounts to:

(a) 12 MW.

(b) 8 MW.

(c) 6 MW.

(d) 4 MW.

(9) A peak load station

(a) Should have operating cost low.

(b) Should have low capital cost.

(c) Can have operating cost high.

(d) a and c.

(e) b and c.

(10) Two areas A and B have equal connected loads however load diversity in area A is more than in B then:

(a) Maximum demand of two areas is small.

(b) Maximum demand of A is greater than MD of B.

(c) MD of B is greater than MD of A.

(d) MD of A is lesser than that of B.

(11) Load curve helps in deciding:

(a) The total installed capacity of the plant.

(b) The size of the generating units.

(c) The operating schedule of the generating units.

(d) All of the above.

(12) The annual peak load on a 60-MW power station is 50 MW. The power station supplies loads having average demands of 9, 10, 17, and 20 MW. The annual load factor is 60% the average load on the plant is:

(a) 4,000 kW.

(b) 30,000 kW.

(c) 2,000 kW.

(d) 1,000 kW.

(13) A generating station has a connected load of 40 MW and a maximum demand of 20 MW. The demand factor is:

(a) 0.7.

(b) 0.6.

(c) 0.59.

(d) 0.4.

(14) A 100-MW power plant has a load factor of 0.5 and a utilization factor of 0.2; its average demand is:

(a) 10 MW.

(b) 5 MW.

(c) 7 MW.

(d) 6 MW.

(15) The value of demand factor is always:

(a) Less than one.

(b) Equal to one.

(c) Greater than one.

(d) None.

(16) If capacity factor = load factor then:

(a) Utilization factor is zero.

(b) Utilization capacity is zero

(c) Utilization factor is equal to one.

(d) None.

(17) If capacity factor = load factor then the plant's

(a) Reserve capacity is maximum.

(b) Reserve capacity is zero.

(c) Reserves capacity is less.

(d) None.

(18) Installed capacity of power plant is:

(a) More than MD.

(b) Less than MD.

(c) Equal to MD.

(d) a and c.

(19) In an interconnected system, diversity factor determining:

(a) Decreases.

(b) Increases.

(c) Zero.

(d) None.

(20) The knowledge of diversity factor helps in determining:

(a) Plant capacity.

(b) Reserve capacity.

(c) Maximum demand.

(d) Average demand.

(21) A power station has installed capacity 300 MW. Its capacity factor is 50% and its load factor is 75%. Its maximum demand is:

(a) 100 MW.

(b) 150 MW.

(c) 200 MW.

(d) 250 MW.

(22) The connected load of consumer is 2 kW and his or her MD is 1.5 kW. The load factor of the consumer is:

(a) 0.75.

(b) 0.375.

(c) 1.33.

(d) None.

(23) The maximum demand of a consumer is 2 kW and his or her daily energy consumption is 20 units. His or her load factor is:

(a) 10.15%.

(b) 41.6%.

(c) 50%.

(d) 52.6%.

(24) In a power plant, a reserve generating capacity that is not in service but its operation is known as:

(a) Hot reserve.

(b) Spinning reserve.

(c) Cold reserve.

(d) Firm power.

(25) The power intended to be always available is known as:

(a) Hot reserve.

(b) Spinning reserve.

(c) Cold reserve.

(d) Firm power.

(26) In a power plant, a reserve-generating capacity that is in service but it is not in operation is:

(a) Hot reserve.

(b) Spinning reserve.

(c) Cold reserve.

(d) Firm power.

(27) Power plant having MD more than the installed capacity will have utilization factor

(a) Less than 100%.

(b) Equal to 100%.

(c) More than 100%.

(d) None.

(28) The choice of the number and size of the units in a station is governed by best compromise between

(a) Plant load factor and capacity.

(b) Plant load factor and plant use factor.

(c) Plant capacity factor and plant use factor.

(d) None of the above.

(29) If some reserve capacity is available in power plant:

(a) Its use factor is always greater than its capacity factor.

(b) Its use factor is equal the capacity.

(c) Its use factor is always less than its capacity.

(d) None of the above.

(30) Which of the following is correct factor?

(a) Load factor = capacity × utilization factor.

(b) Utilization factor = capacity factor × load factor.

(c) Utilization factor = load factor/utilization factor.

(d) Capacity factor = load factor x utilization factor.

(31) If the rated plant capacity and the maximum load of generating station are equal, then:

(a) Load factor is 1.

(b) Capacity factor is 1.

(c) Load factor and capacity factor are equal.

(d) Utilization factor is poor.

(32) The capital cost of plant depends on:

(a) Total installed capacity only.

(b) Total number of units only.

(c) Both a and b.

(d) None.

(33) The reserve capacity in a system that is generally equal to:

(a) The capacity of the largest generating unit.

(b) The capacity of the two largest generating units.

(c) The total generating capacity.

(d) None of the above.

(34) The maximum demand of a consumer is 5 kW and his or her daily energy consumption is 24 units. His or her percentage load factor is:

(a) 5.

(b) 20.

(c) 24.

(d) 48.

(35) If load factor is poor,

(a) The electric energy produced is small.

(b) The charge per kWh is high.

(c) The fixed charges per kWh are high.

(d) All of the above.

(36) If a generating station had a maximum loads for a day 100 kW and a load factor of 0.2, its generation of that day was:

(a) 8.64 MWh.

(b) 21.6 units.

(c) 21.6 units.

(d) 2,160 kWh.

(37) The knowledge of maximum demand is important, as it helps in determining:

(a) The installed capacity of the plant.

(b) The connected load of the plant.

(c) The average demand of the plant.

(d) Either a or b.

(38) A power station is connected to 4.5 and 6 kW load. What is its maximum demand if its daily load factor was calculated as 0.2, where its generation of that day was 24 units.

(a) 2.6.

(b) 3.1.

(c) 5.

(d) 4.

(39) A 50-MW power station had produced 24 units in a day when its maximum demand was 50 Mw. Its plant load factor and capacity factor that day in percentage were

(a) 1 and 2.

(b) 2 and 3.

(c) 2 and 2.

(d) 4 and 3.

(40) The load curve of a power generation station is always:

(a) Negative.

(b) Zero slope.

(c) Positive.

(d) Any combination of (a), (b), and (c)

REVIEW QUESTIONS

(1) Explain the significance of daily load curve.

(2) Discuss the difference between load curve and load duration curve.

(3) Explain the differences in operations of peak load and base load stations.

(4) Explain the significance of load factor and diversity factor.

(5) Define the following:

(i) Load factor,
(ii) demand factor,
(iii) diversity factor,
(iv) plant capacity factor, and
(v) utilization factor.

Explain the load forecasting procedures.

EXERCISE PROBLEMS

(1) Calculate the diversity factor and the annual load factor of a generating station supplies loads to various consumers as follows:

Industrial consumer = 1,500 kW,
Establishment = 7,500 kW,

Domestic power = 100 kW, and
Domestic light = 400 kW.

And, assume the maximum demand on the station is 2,500 kW and the number of units produced per year is 40×10^5.

(2) A power station is to feed four regions of load whose peak loads are 10, 5, 14, and 6 MW. The diversity factor at the station is 1.3 and the average annual load factor is 60%, determine the (i) maximum demand on the station, (ii) annual energy supplied by the station, and (iii) suggest the installed capacity.

(3) The annual load duration curve of a certain power station is a straight line from 20 to 7 MW. To meet this load, three turbine-generator units, two rated, at 12 MW each, and one rated at 8 MW are installed. Calculate the following:

(i) Installed capacity,

(ii) Plant factor,

(iii) Units generated per annum, and

(iv) Utilization factor.

(4) A consumer is charged at the rate of Rs. 75/annum/kVA of maximum demand plus a flat rate per kWh. The phase-advancing plant can be purchased at a rate of Rs. 70/kVA. The rate of interest and depreciation on the capital is 12.5%. Find the most economical pf to which it can be improved.

ANSWERS

1. a	11. d	21. c	31. c
2. c	12. b	22. d	32. c
3. b	13. c	23. b	33. a
4. d	14. a	24. a	34. b
5. a	15. a	25. d	35. d
6. d	16. c	26. c	36. a
7. b	17. b	27. c	37. a
8. b	18. d	28. c	38. c
9. e	19. b	29. a	39. c
10. c	20. a	30. d	40. d

Electric Heating

OBJECTIVES

After reading this chapter, you should be able to:

- know the requirements of heating elements
- understand the causes of the failure of heating elements
- understand the different methods of electrical welding

4.1 INTRODUCTION

Heat plays a major role in everyday life. All heating requirements in domestic purposes such as cooking, room heater, immersion water heaters, and electric toasters and also in industrial purposes such as welding, melting of metals, tempering, hardening, and drying can be met easily by electric heating, over the other forms of conventional heating. Heat and electricity are interchangeable. Heat also can be produced by passing the current through material to be heated. This is called electric heating; there are various methods of heating a material but electric heating is considered far superior compared to the heat produced by coal, oil, and natural gas.

4.2 ADVANTAGES OF ELECTRIC HEATING

The various advantages of electric heating over other the types of heating are:

(i) *Economical*

Electric heating equipment is cheaper; they do not require much skilled persons; therefore, maintenance cost is less.

(ii) *Cleanliness*

Since dust and ash are completely eliminated in the electric heating, it keeps surroundings cleanly.

(iii) *Pollution free*

As there are no flue gases in the electric heating, atmosphere around is pollution free; no need of providing space for their exit.

(iv) *Ease of control*

In this heating, temperature can be controlled and regulated accurately either manually or automatically.

(v) *Uniform heating*

With electric heating, the substance can be heated uniformly, throughout whether it may be conducting or non-conducting material.

(vi) *High efficiency*

In non-electric heating, only 40–60% of heat is utilized but in electric heating 75–100% of heat can be successfully utilized. So, overall efficiency of electric heating is very high.

(vii) *Automatic protection*

Protection against over current and over heating can be provided by using fast control devices.

(viii) *Heating of non-conducting materials*

The heat developed in the non-conducting materials such as wood and porcelain is possible only through the electric heating.

(ix) *Better working conditions*

No irritating noise is produced with electric heating and also radiating losses are low.

(x) *Less floor area*

Due to the compactness of electric furnace, floor area required is less.

(xi) *High temperature*

High temperature can be obtained by the electric heating except the ability of the material to withstand the heat.

(xii) *Safety*

The electric heating is quite safe.

4.3 MODES OF TRANSFER OF HEAT

The transmission of the heat energy from one body to another because of the temperature gradient takes place by any of the following methods:

1. conduction,
2. convection, or
3. radiation.

4.3.1 Conduction

In this mode, the heat transfers from one part of substance to another part without the movement in the molecules of substance. The rate of the conduction of heat along the substance depends upon the temperature gradient.

The amount of heat passed through a cubic body with two parallel faces with thickness 't' meters, having the cross-sectional area of 'A' square meters and the temperature of its two faces T_1°C and T_2°C, during 'T' hours is given by:

$$Q = \frac{k\,A}{t}(T_1 - T_2)\ T \text{ MJ},$$

where k is the coefficient of the thermal conductivity for the material and it is measured in MJ/m^3/°C /hr.

Ex: Refractory heating, the heating of insulating materials, etc.

4.3.2 Convection

In this mode, the heat transfer takes place from one part to another part of substance or fluid due to the actual motion of the molecules. The rate of conduction of heat depends mainly on the difference in the fluid density at different temperatures.

Ex: Immersion water heater.

The mount of heat absorbed by the water from heater through convection depends mainly upon the temperature of heating element and also depends partly on the position of the heater.

Heat dissipation is given by the following expression.

$H = a\,(T_1 - T_2)^b$ W/m^2,

where 'a' and 'b' are the constants whose values are depend upon the heating surface and T_1 and T_2 are the temperatures of heating element and fluid in °C, respectively.

4.3.3 Radiation

In this mode, the heat transfers from source to the substance to be heated without heating the medium in between. It is dependent on surface.

Ex: Solar heaters.

The rate of heat dissipation through radiation is given by Stefan's Law.

$$\text{Heat dissipation, } H = 5.72 \times 10^4\, k\, e \left[\left(\frac{T_1}{1{,}000} \right)^4 - \left(\frac{T_2}{1{,}000} \right)^4 \right] \text{W/m}^2, \qquad \textbf{(4.1)}$$

where T_1 is the temperature of the source in kelvin, T_2 is the temperature of the substance to be heated in kelvin, and k is the radiant efficiency:

= 1, for single element
= 0.5–0.8, for several elements
e = emissivity = 1, for black body
= 0.9, for resistance heating element.

From Equation (4.1), the radiant heat is proportional to the difference of fourth power of the temperature, so it is very efficient heating at high temperature.

4.4 ESSENTIAL REQUIREMENTS OF GOOD HEATING ELEMENT

The materials used for heating element should have the following properties:

- **High-specific resistance**
  Material should have high-specific resistance so that small length of wire may be required to provide given amount of heat.
- **High-melting point**
  It should have high-melting point so that it can withstand for high temperature, a small increase in temperature will not destroy the element.
- **Low temperature coefficient of resistance**
  From Equation (4.1), the radiant heat is proportional to fourth powers of the temperatures, it is very efficient heating at high temperature.

  For accurate temperature control, the variation of resistance with the operating temperature should be very low. This can be obtained only if the material has low temperature coefficient of resistance.

- **Free from oxidation**
  The element material should not be oxidized when it is subjected to high temperatures; otherwise the formation of oxidized layers will shorten its life.
- **High-mechanical strength**
  The material should have high-mechanical strength and should withstand for mechanical vibrations.
- **Non-corrosive**
  The element should not corrode when exposed to atmosphere or any other chemical fumes.
- **Economical**
  The cost of material should not be so high.

4.5 MATERIAL FOR HEATING ELEMENTS

The selection of a material for heating element is depending upon the service conditions such as maximum operating temperature and the amount of charge to be heated, but no single element will not satisfy all the requirements of the heating elements. The materials normally used as heating elements are either alloys of nickel–chromium, nickel–chromium–iron, nickel–chromium–aluminum, or nickel–copper.

Nickel–chromium–iron alloy is cheaper when compared to simple nickel–chromium alloy. The use of iron in the alloy reduces the cost of final product but, reduces the life of the alloy, as it gets oxidized soon. We have different types of alloys for heating elements. Table 4.1 gives the relevant properties of some of the commercial heating elements.

The properties of some commercial heating element materials commonly employed for low and medium temperatures up to 1,200°C are Ni–Cr and an alloy of Ni–Cr–Fe composition

TABLE 4.1 *Properties of some heating elements*

S. No.	Type of alloy	Composition	Commercial name	Max. operating temperature	Resistivity at 20°C	Specific gravity
1	Nickel chromium (Ni–Cr)	80% Ni 20% Cr	Nichrome	1,150°C	1.03 μΩ-m	8.35
2	Nickel chromium iron (Ni–Cr–Fe)	60% Ni 16% Cr 24% Fe	—	950°C	1.06 μΩ-m	8.27
3	Nickel Copper (Ni–Cu)	45% Ni 55% Cu	Eureka or constantan	400°C	0.49 μΩ-m	8.88
4	Iron chromium aluminum (Fe–Cr–Al)	70% Fe 25% Cr 5% Al	Kanthal	1,200°C	1.4 μΩ-m	7.20

of these alloys are given in Table 4.1. For operating temperatures above 1,200°C, the heating elements are made up of silicon carbide, molebdenum, tungsten, and graphite. (Ni–Cu alloy is frequently used for heating elements operating at low temperatures. Its most important property is that it has virtually zero resistance and temperature coefficient.)

4.6 CAUSES OF FAILURE OF HEATING ELEMENTS

Heating element may fail due to any one of the following reasons.

1. Formation of hot spots.
2. Oxidation of the element and intermittency of operation.
3. Embrittlement caused by gain growth.
4. Contamination and corrosion.

4.6.1 Formation of hotspots

Hotspots are the points on the heating element generally at a higher temperature than the main body. The main reasons of the formation of hotspot in the heating element are the high rate of the local oxidation causing reduction in the area of cross-section of the element leading to the increase in the resistance at that spot. It gives rise to the damage of heating element due to the generation of more heat at spot. Another reason is the shielding of element by supports, etc., which reduces the local heat loss by radiation and hence the temperature of the shielded portion of the element will increase. So that the minimum number of supports should be used without producing the distortion of the element. The sagging and wrapping of the material arise due to the insufficient support for the element (or) selection of wrong fuse material may lead to the uneven spacing of sections thereby developing the hotspots on the element.

4.6.2 Oxidation and intermittency of operation

A continuous oxide layer is formed on the surface of the element at very high temperatures such layer is so strong that it prevents further oxidation of the inner metal of the element. If the element is used quite often, the oxide layer is subjected to thermal stresses; thus, the layer cracks and flakes off, thereby exposing fresh metal to oxidation. Thus, the local oxidation of the metal increases producing the hotspots.

4.6.3 Embrittlement causing grain growth

In general, most of the alloys containing iron tend to form large brittle grains at high temperatures. When cold, the elements are very brittle and liable to rupture easily on the slightest handling and jerks.

4.6.4 Contamination and corrosion

The heating elements may be subjected to dry corrosion produced by their contamination with the gases of the controlled atmosphere prevailing in annealing furnaces.

4.7 DESIGN OF HEATING ELEMENTS

By knowing the voltage and electrical energy input, the design of the heating element for an electric furnace is required to determine the size and length of the heating element. The wire employed may be circular or rectangular like a ribbon. The ribbon-type heating element permits the use of higher wattage per unit area compared to the circular-type element.

Circular-type heating element

Initially when the heating element is connected to the supply, the temperature goes on increasing and finally reaches high temperature.

Let V be the supply voltage of the system and R be the resistance of the element, then

electric power input, $P = \frac{V^2}{R}$ W.

If ρ is the resistivity of the element, l is the length, 'a' is the area, and d is the diameter of the element, then:

$$R = \rho \frac{l}{a} = \frac{\rho l}{\frac{\pi d^2}{4}}.$$

$$\text{Therefore, power input, } P = \frac{V^2 \pi d^2}{4 \rho l}. \tag{4.2}$$

By rearranging the above equation, we get:

$$\frac{l}{d^2} = \frac{\pi V^2}{4P\rho}, \tag{4.3}$$

where P is the electrical power input per phase (watt), V is the operating voltage per phase (volts), R is the resistance of the element (Ω), l is the length of the element (m), a is the area of cross-section (m^2), d is the diameter of the element (m), and ρ is the specific resistance (Ω-m).

According to Stefan's law, heat dissipated per unit area is:

$$H = 5.72 \times 10^4\ k\,e\left[\left(\frac{T_1}{1{,}000}\right)^4 - \left(\frac{T_2}{1{,}000}\right)^4\right] \text{W/m}^2, \tag{4.4}$$

where T_1 is the absolute temperature of the element (K), T_2 is the absolute temperature of the charge (K), e is the emissivity, and k is the radiant efficiency.

The surface area of the circular heating element:

$S = \pi dl.$

$\therefore$ Total heat dissipated = surface area $\times H$

$= H\pi dl.$

Under thermal equilibrium,

Power input = heat dissipated

$$P = H \times \pi dl.$$

Substituting P from Equation (4.2) in above equation:

$$\frac{V^2}{\rho l}\left(\frac{\pi d^2}{4}\right) = H \times \pi dl$$

$$\therefore \frac{d}{l^2} = \frac{4 \rho H}{V^2}. \tag{4.5}$$

By solving Equations (4.3) and (4.4), the length and diameter of the wire can be determined.

Ribbon-type element

Let 'w' be the width and 't' be the thickness of the ribbon-type heating element.

$$\text{Electrical power input } P = \frac{V^2}{R}. \quad (4.6)$$

We know that, $R = \frac{\rho l}{a} = \frac{\rho l}{w \times t}$ (for ribbon or rectangular element, $a = w \times t$)

$$\therefore P = \frac{V^2}{\left(\frac{\rho l}{w \times t}\right)}$$

$$\therefore \frac{l}{w} = \frac{V^2 t}{P \rho}. \quad (4.7)$$

The surface area of the rectangular element $(S) = 2\,l \times w$.

$\therefore$ Total heat dissipated $= H \times S$

$= H \times 2\,lw.$

$\therefore$ Under the thermal equilibrium,

Electrical power input = heat dissipated

$$P = H \times 2\,lw$$

$$lw = \frac{P}{2\,H}. \quad (4.8)$$

By solving Equations (4.7) and (4.8), the length and width of the heating element can be determined.

Example 4.1: A 4.5-kW, 200-V, and 1-ϕ resistance oven is to have nichrome wire heating elements. If the wire temperature is to be 1,000°C and that of the charge 500°C. Estimate the diameter and length of the wire. The resistivy of the nichrome alloy is 42.5 μΩ-m. Assume the radiating efficiency and the emissivity of the element as 1.0 and 0.9, respectively.

Solution:

Given data

Power input $(P) = 4.5$ kW

Supply voltage $(V) = 200$ V

Temperature of the source $(T_1) = 1{,}000 + 273$

$= 1{,}273$ K.

Temperature of the charge $T_2 = 500 + 273$

$= 773$ K.

According to the Stefan's law,

$$\text{The amount of heat dissipation } (H) = 5.72 \times 10^4 \times k\,e\left[\left(\frac{T_1}{1{,}000}\right)^4 - \left(\frac{T_2}{1{,}000}\right)^4\right] \text{W/m}^2$$

$$H = 5.72 \times 10^4 \times 0.1 \times 0.9\left[\left(\frac{1{,}273}{1{,}000}\right)^4 - \left(\frac{773}{1{,}000}\right)^4\right]$$

$$= 11.68 \times 10^3 \text{ W/m}^2.$$

$$\text{Power, } P = \frac{V^2}{R}$$

$$= \frac{V^2}{\frac{\rho l}{A}} \qquad \left(R = \frac{\rho l}{A}\right)$$

$$= \frac{V^2 A}{\rho l}$$

$$= \frac{V^2 \pi d^2}{4\rho l} \qquad \left[\because \text{The area of circular type element} = \frac{\pi}{4} d^2\right]$$

$$\frac{d^2}{l} = \frac{4 P\rho}{V^2 \pi}$$

$$= \frac{4 \times 42.5 \times 10^{-6} \times 4.5 \times 10^3}{(200)^2\, 3.14}$$

$$= 6.09 \times 10^{-9}. \qquad (1)$$

The heat dissipation is given by:

$$P = H \times S \qquad (S = \text{circular full-face area})$$

$$= H \times \pi dl$$

$$dl = \frac{P}{H\pi} = \frac{4.5 \times 10^3}{3.14 \times 11.68 \times 10^3}$$

$$l = 0.1226. \qquad (2)$$

By solving Equations (1) and (2):

$$d^3 = 0.7466$$

$$d = 0.907 \text{ mm.}$$

Substitute the value of 'd' in Equation (2):

$$l = 135.14 \text{ m.}$$

Example 4.2: A 20-kW, 230-V, and single-phase resistance oven employs nickel–chrome strip 25-mm thick is used, for its heating elements. If the wire temperature is not to exceed 1,200°C and the temperature of the charge is to be 700°C. Calculate the width and length of the wire. Assume the radiating efficiency as 0.6 and emissivity as 0.9. Determine also the temperature of the wire when the charge is cold.

Solution:

Power supplied, $P = 20 \times 10^3$ W.

Let 'w' be the width in meters, t be the thickness in meters, and 'l' be the length also in meters. Then:

$$P = \frac{V^2}{R}$$

$$= \frac{V^2}{\frac{\rho\, l}{A}}$$

$$= \frac{V^2 \times wt}{\rho l} \qquad \text{(since } A = w \times t\text{)}$$

$$\frac{w}{l} = \frac{P\rho}{V^2 t}$$

$$= \frac{20 \times 10^3 \times 1.016 \times 10^{-6}}{(230)^2 \times 0.25 \times 10^{-3}}$$

$$= 1.536 \times 10^{-3}. \qquad (1)$$

According to the Stefan's law of heat radiation:

$$H = 5.72 \times 10^4 \times ke\left[\left(\frac{T_1}{1{,}000}\right)^4 - \left(\frac{T_2}{1{,}000}\right)^4\right] \text{W/m}^2$$

$$H = 5.72 \times 10^4 \times 0.6 \times 0.9\left[\left(\frac{1{,}200+273}{1{,}000}\right)^4 - \left(\frac{700+273}{1{,}000}\right)^4\right]$$

$(\because T_1 = 1{,}200 + 273 = 1{,}473 \text{ K}, \quad T_2 = 700 + 273 = 973 \text{ K})$

$H = 117.714 \text{ kW/m}^2$.

The total amount of the heat dissipation × the surface area of strip = power supplied

$$P = H \times S$$

$$= H \times 2\, lw \qquad (S = \text{surface area of strip} = 2lw)$$

$$lw = \frac{P}{2H}$$

$$= \frac{20 \times 10^3}{2 \times 117.714 \times 10^3}$$

$$= 0.0849. \qquad (2)$$

From Equations (1) and (2):

$$\frac{w}{l} \times lw = 1.536 \times 10^{-3} \times 0.0849$$

$$w^2 = 1.304 \times 10^{-4}$$

$$w = 11.42 \text{ mm.}$$

Substitute the value of '*w*' in Equation (2) then:

$l = 7.435$ m.

When the charge is cold, it would be at normal temperature, say 25°C.

$$117.714 \times 10^3 = 5.72 \times 10^4 \times 0.6 \times 0.9\left[\left(\frac{T_1}{1{,}000}\right)^4 - \left(\frac{273+25}{1{,}000}\right)^4\right]$$

$$\left(\frac{T_1}{1{,}000}\right)^4 - 0.00788 = 3.8109$$

$$\left(\frac{T_1}{1{,}000}\right)^4 = 3.818$$

$$T_1 = 1{,}397.9169 \text{ K absolute}$$

Or, $$T_1 = 1{,}124.9°\text{C}.$$

Example 4.3 Determine the diameter and length of the wire, if a 17-kW, 220-V, and 1-ϕ resistance oven employs nickel–chrome wire for its heating elements. The temperature is not exceeding to 1,100°C and the temperature of the charge is to be 500°C. Assume the radiating efficiency as 0.5 and the emissivity as 0.9, respectively.

Solution:

For a circular element:

$$P = \frac{V^2}{R}$$

$$= \frac{V^2}{\frac{\rho l}{A}}$$

$$= \frac{V^2 A}{\rho l}$$

$$= \frac{V^2 \pi d^2}{\rho l 4} \quad \left[\because \text{The area of circular element } A = \frac{\pi}{4}d^2\right]$$

$$\frac{d^2}{l} = \frac{4P\rho}{V^2 \pi}$$

$$= \frac{4 \times 17 \times 10^3 \times 1.016 \times 10^{-6}}{(220)^2 \times 3.14}$$

$$= 4.545 \times 10^{-7}. \quad (1)$$

According to Stefan's law of heat dissipation:

$$H = 5.72 \times 10^4\, ke\left[\left(\frac{T_1}{1{,}000}\right)^4 - \left(\frac{T_2}{1{,}000}\right)^4\right] \text{W/m}^2$$

$$H = 5.72 \times 10^4 \times 0.5 \times 0.9 \left[\left(\frac{1{,}100+273}{1{,}000}\right)^4 - \left(\frac{500+273}{1{,}000}\right)^4\right]$$

$$= 82.28 \text{ kW/m}^2.$$

At steady temperature, crucial power input = heat output:

$$P = H \times \pi dl$$

$$dl = \frac{P}{H \times \pi}$$

$$= \frac{7 \times 10^3}{3.14 \times 62.28 \times 10^3}$$

$$= 0.0658.$$

Solving Equations (1) and (2), we get:

$$\frac{d^2}{l} \times dl = 4.545 \times 10^{-7} \times 0.0658$$

$$d^3 = 2.99 \times 10^{-8}$$

$$d = 3.1 \text{ mm}.$$

Substitute the value of 'd' in Equation (2) gives:

$$l = 21.198 \text{ m}.$$

Example 4.4: A 40-kW, 3-phase, and 400-V resistance oven is to employ Ni–Cr strip of 0.3 mm thickness. The heating elements are star commuted. If the temperature of the wire is to be 1,200°C and that of the charge is 700°C. Determine the length and width of the wire. Take the radiation efficiency 0.5 and the emissivity as 0.9. Take the specific resistance of Ni–Cr = 1.03×10^{-6} Ω-m.

Solution:

For the star connection, $V_L = \sqrt{3}\, V_{ph}$

$$V_{ph} = \frac{400}{\sqrt{3}} = 230.94 \text{ V}.$$

$$\text{Power, } P = \frac{V^2}{R}$$

$$= \frac{V^2 A}{\rho l}$$

$$= \frac{V^2 \times w \times t}{\rho l}$$

$$\frac{w}{l} = \frac{P\rho}{V^2 t}$$

$$= \frac{40 \times 10^3 \times 1.03 \times 10^{-6}}{(230.94)^2 \times 0.3 \times 10^{-6}}$$

$$= 2.575.$$

According to the Stefan's law of heat dissipation:

$$H = 5.72 \times 10^4 \ K_e \left[\left(\frac{T_1}{1{,}000}\right)^4 - \left(\frac{T_2}{1{,}000}\right)^4\right] \text{W/m}^2$$

$$= 5.72 \times 10^4 \times 0.5 \times 0.9\left[\left(\frac{1{,}200+273}{1{,}000}\right)^4 - \left(\frac{700+273}{1{,}000}\right)^4\right]$$

$$= 98.09 \text{ kW/m}^2.$$

At steady state:

Electrical power input = heat output

$$P = H \times S$$

$$= H \times 2\,wl \qquad (S = \text{the area of the strip} = 2\,wl)$$

$$wl = \frac{P}{2\,H}$$

$$= \frac{40 \times 10^3}{2 \times 98.09 \times 10^3}$$

$$= 0.2038. \tag{2}$$

Solving Equations (1) and (2):

$$= \frac{w}{l} \times wl = 2.575 \times 0.2038$$

$$w^2 = 0.5247$$

$$w = 0.724 \text{ m}.$$

Substitute the value of '*w*' in Equation (2):

$$l = 0.2812 \text{ m}.$$

4.8 METHODS OF ELECTRIC HEATING

Heat can be generated by passing the current through a resistance or induced currents. The initiation of an arc between two electrodes also develops heat. The bombardment by some heat energy particles such as α, γ, β, and x-rays or accelerating ion can produce heat on a surface.

Electric heating can be broadly classified as follows.

(i) *Direct resistance heating*

In this method, the electric current is made to pass through the charge (or) substance to be heated. This principle of heating is employed in electrode boiler.

(ii) *Indirect resistance heating*

In this method, the electric current is made to pass through a wire or high-resistance heating element, the heat so developed is transferred to charge from the heating element by convection or radiation. This method of heating is employed in immersion water heaters.

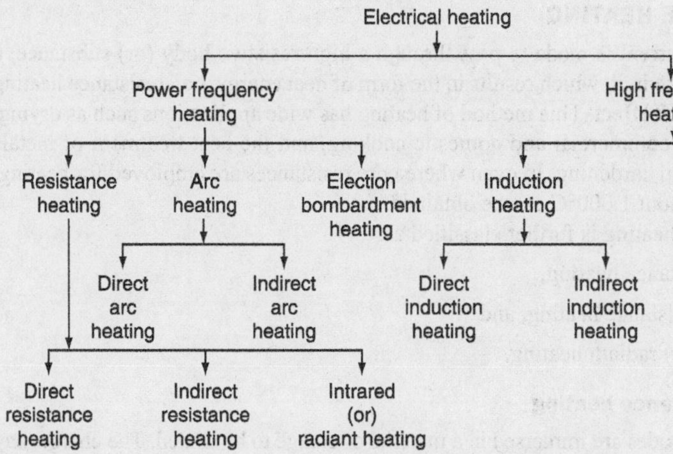

FIG. 4.1 *Classification of electrical heating*

Infrared (or) radiant heating

In this method of heating, the heat energy is transferred from source (incandescent lamp) and focused upon the body to be heated up in the form of electromagnetic radiations. Normally, this method is used for drying clothes in the textile industry and to dry the wet paints on an object.

Direct arc heating

In this method, by striking the arc between the charge and the electrode or electrodes, the heat so developed is directly conducted and taken by the charge. The furnace operating on this principle is known as direct arc furnaces. The main application of this type of heating is production of steel.

Indirect arc heating

In this method, arc is established between the two electrodes, the heat so developed is transferred to the charge (or) substance by radiation. The furnaces operating on this principle are known as indirect arc furnaces. This method is generally used in the melting of non-ferrous metals.

Direct induction heating

In this method of heating, the currents are induced by electromagnetic action in the charge to be heated. These induced currents are used to melt the charge in induction furnace.

Indirect induction heating

In this method, eddy currents are induced in the heating element by electromagnetic action. Thus, the developed heat in the heating element is transferred to the body (or) charge to be heated by radiation (or) convection. This principle of heating is employed in induction furnaces used for the heat treatment of metals.

Dielectric heating

In this method of electric heating, the heat developed in a non-metallic material due to inter-atomic friction, known as dielectric loss. This principle of heating usually employed for preheating of plastic performs, baking foundry cores, etc.

4.9 RESISTANCE HEATING

When the electric current is made to pass through a high-resistive body (or) substance, a power loss takes place in it, which results in the form of heat energy, i.e., resistance heating is passed upon the I^2R effect. This method of heating has wide applications such as drying, baking of potteries, commercial and domestic cooking, and the heat treatment of metals such as annealing and hardening. In oven where wire resistances are employed for heating, temperature up to about 1,000°C can be obtained.

The resistance heating is further classified as:

1. direct resistance heating,
2. indirect resistance heating, and
3. infrared (or) radiant heating.

4.9.1 Direct resistance heating

In this method, electrodes are immersed in a material or charge to be heated. The charge may be in the form of powder, pieces, or liquid. The electrodes are connected to AC or DC supply as shown in Fig. 4.1(a). In case of DC or 1-ϕ AC, two electrodes are immersed and three electrodes are immersed in the charge and connected to supply in case of availability of 3-ϕ supply. When metal pieces are to be heated, the powder of lightly resistive is sprinkled over the surface of the charge (or) pieces to avoid direct short circuit. The current flows through the charge and heat is produced in the charge itself. So, this method has high efficiency. As the current in this case is not variable, so that automatic temperature control is not possible. This method of heating is employed in salt bath furnace and electrode boiler for heating water.

(i) Salt bath furnace

This type of furnace consists of a bath and containing some salt such as molten sodium chloride and two electrodes immersed in it.

Such salt have a fusing point of about 1,000–1,500°C depending upon the type of salt used. When the current is passed between the electrodes immersed in the salt, heat is developed and the temperature of the salt bath may be increased. Such an arrangement is known as a salt bath furnace.

In this bath, the material or job to be heated is dipped. The electrodes should be carefully immersed in the bath in such a way that the current flows through the salt and not

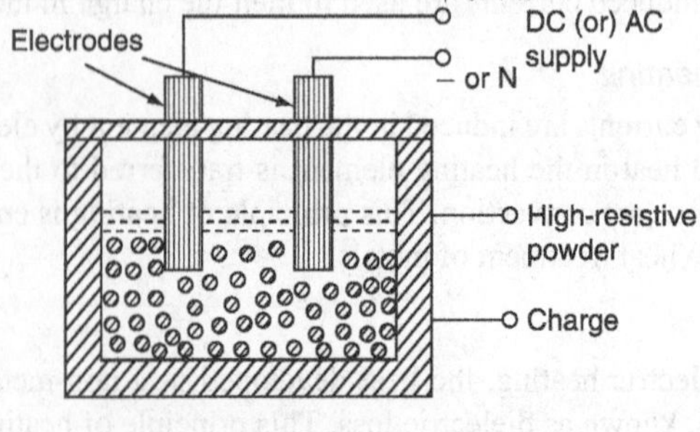

FIG. 4.1(a) *Direct resistance heating*

through the job being heated. As DC will cause electrolysis so, low-voltage AC up to 20 V and current up to 3,000 A is adopted depending upon the type of furnaces.

The resistance of the salt decreases with increase in the temperature of the salt, therefore, in order to maintain the constant power input, the voltage can be controlled by providing a tap changing transformer. The control of power input is also affected by varying the depth of immersion and the distance between the electrodes.

(II) *Electrode boiler*

It is used to heat the water by immersing three electrodes in a tank as shown in Fig. 4.2. This is based on the principle that when the electric current passed through the water produces heat due to the resistance offered by it. For DC supply, it results in a lot of evolution of H_2 at negative electrode and O_2 at positive electrode. Whereas AC supply hardly results in any evolution of gas, but heats the water. Electrode boiler tank is earthed solidly and connected to the ground. A circuit breaker is usually incorporated to make and break all poles simultaneously and an over current protective device is provided in each conductor feeding an electrode.

4.9.2 Indirect resistance heating

In the indirect resistance heating method, high current is passed through the heating element. In case of industrial heating, some times the heating element is placed in a cylinder which is surrounded by the charge placed in a jacket is known as heating chamber is shown in Fig. 4.3. The heat is proportional to power loss produced in the heating element is delivered to the charge by one or more of the modes of the transfer of heat viz. conduction, convection, and radiation. This arrangement provides uniform temperature and automatic temperature control. Generally, this method of heating is used in immersion water heaters, room heaters, and the resistance ovens used in domestic and commercial cooling and salt bath furnace.

Resistance ovens

According to the operating temperatures, the resistance furnaces may be classified into various types. Low-temperature heating chamber with the provision for ventilation is called as oven. For drying varnish coating, the hardening of synthetic materials, and commercial

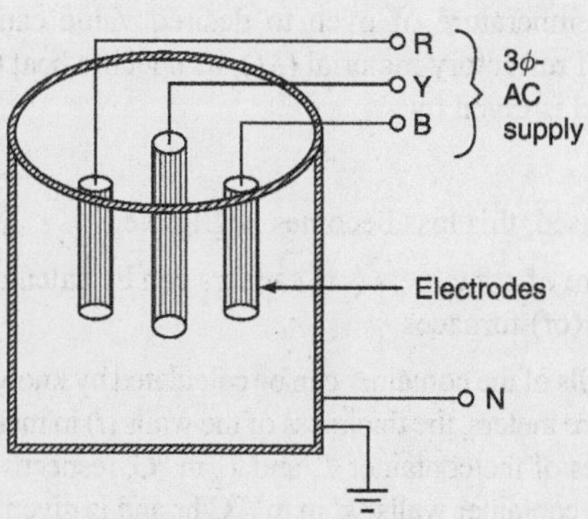

FIG. 4.2 *Electrode boiler*

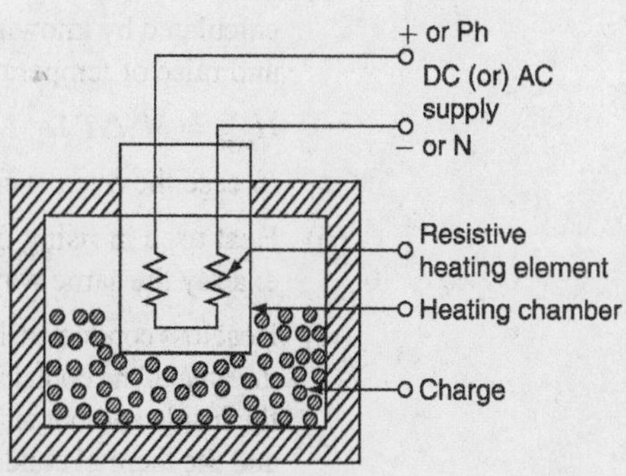

FIG. 4.3 *Indirect resistance heating*

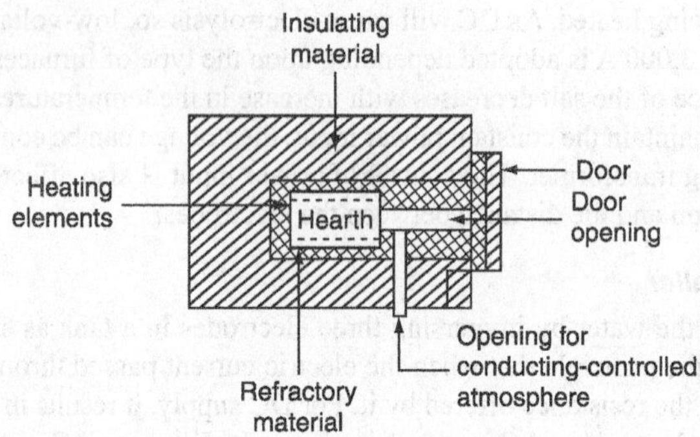

FIG. 4.4 *Resistance oven*

and domestic heating, etc., the resistance ovens are employed. The operating temperature of medium temperature furnaces is between 300°C and 1,050°C. These are employed for the melting of non-ferrous metals, stove (annealing), etc. Furnaces operating at temperature between 1,050°C and 1,350°C are known as high-temperature furnaces. These furnaces are employed for hardening applications. A simple resistance oven is shown in Fig. 4.4.

Resistance oven consists of a heating chamber in which heating elements are placed as shown in the Fig. 4.4. The inner surface of the heating chamber is made to suit the character of the charge and the type of furnace or oven. The type of insulation used for heating chamber is determined by the maximum temperature of the heating chamber.

Efficiency and losses of resistance ovens

The heat produced in the heating elements, not only raises the temperature of the charge to desired value, but also used to overcome the losses occurring due to:

(i) Heat used in raising the temperature of oven (or) furnace.

(ii) Heat used in raising the temperature of containers (or) carriers.

(iii) Heat conducted through the walls.

(iv) Heat loss due to the opening of oven door.

(i) The heat required to raise the temperature of oven to desired value can be calculated by knowing the mass of refractory material (M), its specific heat (S), and raise of temperature (ΔT) and is given by:

$$H_{oven} = MS\Delta T \text{ J}.$$

In case the oven is continuously used, this loss becomes negligible.

(ii) Heat used in rising the temperature of containers (or) carriers can be calculated exactly the same way as for oven (or) furnaces.

(iii) Heat loss conducted through the walls of the container can be calculated by knowing the area of the container (A) in square meters, the thickness of the walls (t) in meters, the inside and out side temperatures of the container T_1 and T_2 in °C, respectively, and the thermal conductivity of the container walls 'k' in m³/°C/hr and is given by:

$$\text{Heat loss by conduction} = \frac{k\,A(T_1 - T_2)}{t}\text{ W}.$$

Actually, there is no specific formula for the determination of loss occurring due to the opening of door for the periodic inspection of the charge so that this loss may be approximately taken as 0.58–1.15 MJ/m^2 of the door area, if the door is opened for a period of 20–30 sec.

The *efficiency of the oven* is defined as the ratio of the heat required to raise the temperature of the charge to the desired value to the heat required to raise the charge and losses.

The efficiency of the oven:

$$= \frac{\text{the heat required to raise the temperature of the charge}}{\text{the heat required to raise the temperature of the charge} + \text{total losses}}.$$

The efficiency of the resistance oven lies in between 60% and 80%.

4.9.3 Infrared or radiant heating

In this method of heating, the heat transfer takes place from the source to the body to be heated through radiation, for low and medium temperature applications. Whereas in resistance ovens, the heat transfers to the charge partly by convection and partly by radiation. In the radiant heating, the heating element consists of tungsten filament lamps together with reflector and to direct all the heat on the charge. Tungsten filament lamps are operating at 2,300°C instead of 3,000°C to give greater portion of infrared radiation and a longer life. The radiant heating is mainly used for drying enamel or painted surfaces. The high concentration of the radiant energy enables the heat to penetrate the coating of paint or enamel to a depth sufficient to dry it out without wasting energy in the body of the workpiece.

The main advantage of the radiant heating is that the heat absorption remains approximately constant whatever the charge temperature, whereas with the ordinary oven the heat absorption falls off very considerably as the temperature of the charge raises. The lamp ratings used are usually between 250 and 1,000 W and are operating at voltage of 115 V in order to ensure a robust filament.

4.10 TEMPERATURE CONTROL OF RESISTANCE HEATING

To control the temperature of a resistance heating at certain selected points in a furnace or oven, as per certain limits, such control may be required in order to hold the temperature constant or to vary it in accordance with a pre-determined cycle and it can be carried out by hand or automatically.

In resistance furnaces, the heat developed depends upon $I^2 R t$ (or) $\frac{V^2}{R} t$. Therefore, the temperature of the furnaces can be controlled either by:

(i) Changing the resistance of elements.

(ii) Changing the applied voltage to the elements (or) current passing through the elements.

(iii) Changing the ratio of the on-and-off times of the supply.

Voltage across the furnace can be controlled by changing the transformer tapings. Auto transformer or induction regulator can also be used for variable voltage supply. In addition to the above, voltage can be controlled by using a series resistance so that some voltage dropped across this series resistor. But this method is not economical as the power is continuously wasted in controlling the resistance. Hence, this method is limited to small furnaces. An on-off switch can be employed to control the temperature. The time for which the oven is connected to the supply and the time for which it is disconnected from supply will determine the temperature.

Temperature can be controlled by providing various combinations of groups of resistances used in the furnace and is given as follows:

(i) *Variable number of elements*

If 'R' be the resistance of one element and 'n' be the number of elements are connected in parallel, so that the equivalent resistance is R/n.

Heat developed in the furnace is:

$$H = \frac{V^2}{(R/n)} = \frac{V^2}{R} \times n$$

i.e., if the number of elements connected in parallel increases, the heat developed in the furnace also increased. This method does not provide uniform heating unless elements not in use are well distributed.

(ii) *Series parallel (or) star delta arrangement of elements*

If the available supply is single phase, the heating elements can be connected in series for the low temperatures and connected in parallel for the high temperature by means of a series–parallel switch.

In case, if the available supply is three phase, the heating elements can be connected in star for the low temperature and in delta for the high temperatures by using star–delta switch.

Example 4.5: Six resistances, each of 60 ohms, are used in a resistance; how much power is drawn for the following connections.

(a) Supply is 400 V, AC, and single phase and the connections are:
   (i) Three groups in parallel, each of two resistance units in series.
   (ii) Six groups are in parallel, each of one resistance unit.

(b) With the same three-phase supply, they are connected in delta fashion.
   (i) Two resistance units in parallel in each branch.
   (ii) Two resistance units in series in each branch.

(c) Supply is 400 V and three-phase while the connection is a star combination of:
   (i) Two resistance elements in series in each phase.
   (ii) Two resistance elements in parallel in each phase.

(d) If the supply is a 25% tapping with an auto transformer, calculate the output of the oven.

Solution:

(a) (i) The power consumption of the two resistances in series is:

$$P = \frac{V^2}{R} = \frac{(400)^2}{2 \times 60}$$

$$= 1{,}333.33 \text{ W}.$$

The power consumed by the three units in parallel is $P = 3 \times 1{,}333.33 = 4{,}000$ W.

(ii) The power consumed by each resistor is:

$$P = \frac{V^2}{R} = \frac{(400)^2}{60}$$

$$= 2{,}666.67 \text{ W}.$$

The power consumed by the six resistors in parallel is:

$$P = 6 \times 2{,}666.67$$
$$= 16{,}000 \text{ W.}$$

(b) Since in delta fashion, line voltage = phase voltage = 400 V:

(i) The power consumed by the each branch:

$$P = \frac{V^2}{R} = \frac{(400)^2}{30}$$
$$= 5{,}333.34 \text{ W.}$$

The power consumed by the three units is:

$$P = 3 \times 5{,}333.34$$
$$= 16{,}000 \text{ W.}$$

(ii) The power consumed by the each unit, when they are commuted in series is:

$$P = \frac{V^2}{R} = \frac{(400)^2}{60+60}$$
$$= 1{,}333.34 \text{ W.}$$

The power consumed by the three units is:

$$P = 4{,}000 \text{ W.}$$

(c) For the star connection, $V_L = \sqrt{3}\ V_{ph}$:

$$V_{ph} = \frac{400}{\sqrt{3}} = 230.94 \text{ V.}$$

(i) The power consumed by the two resistors in series is $P = \frac{V^2}{R} = \frac{(230.94)^2}{60+60}$:

$$P = 444.44 \text{ W.}$$

The power consumed by the three units is:

$$P = 1{,}333.33 \text{ W.}$$

(ii) The power consumed by the two resistors in parallel is:

$$P = \frac{(230.94)^2}{30}$$
$$P = 1{,}777.77 \text{ W.}$$

The power consumed by the three units in series is:

$$P = 3 \times 1{,}777.77$$
$$= 5{,}333.32 \text{ W.}$$

(d) The power is proportional to the square of the voltage. Hence, the voltage is 25%.

So that, the power loss will be $\frac{1}{16}$th of the values obtained as above.

4.11 ARC HEATING

If the high voltage is applied across an air gap, the air in the gap gets ionized under the influence of electrostatic forces and becomes conducting medium, current flows in the form of a continuous spark, known as *arc.* A very high voltage is required to establish an arc but very small voltage is sufficient to maintain it, across the air gap. The high voltage required for striking an arc can be obtained by using a step-up transformer fed from a variable AC supply.

Another method of striking the arc by using low voltage is by short circuiting the two electrodes momentarily and with drawing them back. Electrodes made up of carbon or graphite and are used in the arc furnaces when the temperature obtained is in the range of 3,000–3,500°C.

4.11.1 Electrodes used in the arc furnaces

Normally used electrodes in the arc furnaces are carbon electrodes, graphite electrodes, and self-baking electrodes. Usually the carbon and graphite electrodes are used and they can be selected based on their electrical conductivity insolubility, chemical inertness, mechanical strength, resistance to thermal shock, etc. The size of these electrodes may be 18–27 cm in diameter. The carbon electrodes are used with small furnaces for manufacturing of ferro-alloys, aluminum phosphorous, etc. The self-baking electrodes are employed in the electrochemical furnaces and in the electrolytic production of aluminum.

The salient features of carbon and graphite electrodes are:

(i) Resistivity: The graphite electrodes have low-specific resistance than the carbon electrodes, so the graphite required half in size for the same current resulting in easy replacement.

(ii) Oxidation: Graphite begins to oxides at 600°C where as carbon at 400°C.

(iii) Electrode consumption: For steel-melting furnaces, the consumption of the carbon electrodes is about 4.5 kg of electrodes per tonne of steel and 2.3–to 6.8 kg electrodes per tonne of steel for the graphite electrodes.

(iv) Cost: The graphite electrodes cost about twice as much per kg as the carbon electrodes. The choice of electrodes depends chiefly on the question of the total cost. In general, if the processes requiring large quantities of electrode, carbon is used but for other processes, the choice depends on local conditions.

4.11.2 Types of arc furnaces

There are two types of arc furnaces and they are:

(i) direct arc furnace and

(ii) indirect arc furnace.

(i) *Direct arc furnace*

When supply is given to the electrodes, two arcs are established and current passes through the charge, as shown in Fig. 4.5. As the arc is in direct contact with the charge and heat is also produced by current flowing through the charge itself, it is known as *direct arc furnace.*

If the available supply is DC or 1-ϕ, AC, two electrodes are sufficient, if the supply is 3-ϕ, AC, three electrodes are placed at three vertices of an equilateral triangle. The most important feature of the direct arc furnace is that the current flows through the charge, the stirring action is inherent due to the electromagnetic force setup by the current, such furnace is used for manufacturing alloy steel and gives purer product.

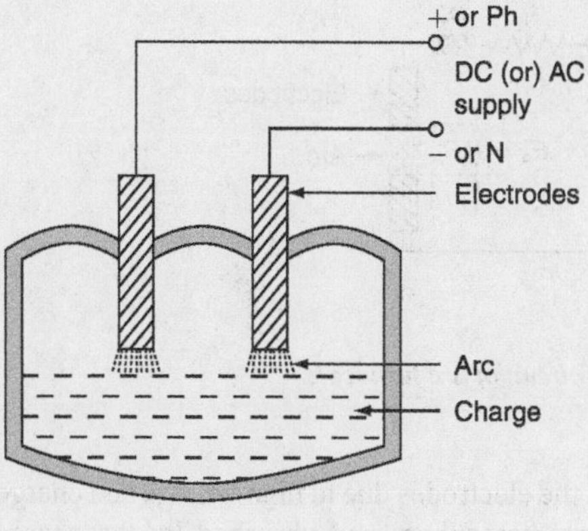

FIG. 4.5 *Direct arc furnace*

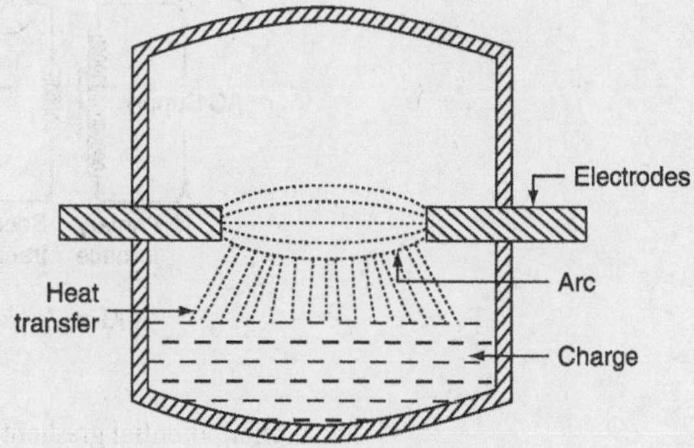

FIG. 4.6 *Indirect arc furnace*

It is very simple and easy to control the composition of the final product during refining process operating the power factor of arc furnace is 0.8 lagging. For 1-ton furnace, the power required is about 200 kW and the energy consumed is 1.0 MWh/ton.

(II) *Indirect arc furnace*

In indirect arc furnace, the arc strikes between two electrodes by bringing momentarily in contact and then with drawing them heat so developed, due to the striking of arc across air gap is transferred to charge is purely by radiation. A simple indirect arc furnace is shown in Fig. 4.6.

These furnaces are usually 1-ϕ and hence their size is limited by the amount of one-phase load which can be taken from one point. There is no inherent stirring action provided in this furnace, as current does not flow through the charge and the furnace must be rocked mechanically. The electrodes are projected through this chamber at each end along the horizontal axis. This furnace is also sometimes called as *rocking arc furnace*. The charge in this furnace is heated not only by radiation from the arc between electrode tips but also by conduction from the heated refractory during rocking action; so, the efficiency of such furnace is high. The arc is produced by bringing electrodes into solid contact and then withdrawing them; power input to the furnace is regulated by adjusting the arc length by moving the electrodes.

Even though it can be used in iron foundries where small quantities of iron are required frequently, the main application of this furnace is the melting of non-ferrous metals.

4.11.3 Power supply and control of arc furnace

As the arc voltage is of the order of 50–150 V in order to obtain large powers required for melting metals, the secondary current required will be of the order of several hundred amperes. Therefore, the power supply for the electric arc furnace is of low-voltage and high-current type. *This is due to the following reasons*:

- In order to obtain high temperatures, high currents are necessary since the heating effect is proportional to the square of the current.
- The maximum secondary voltage is limited to 275 V due to insulation and safety considerations.

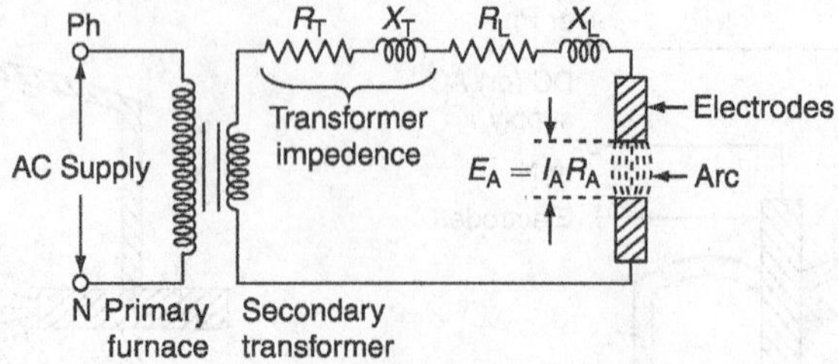

FIG. 4.7 *Equivalent circuit of arc furnace*

- The higher potential gradient between the electrodes due to high voltage and charge which ionizes the nitrogen of furnace atmosphere and absorbed by the charge produces embitterment.
- Generally, low-voltage and high-current electrodes are kept nearer to the charge thus arc remains away from the roof: therefore, the life of refractory is increased.

The equivalent circuit of an electric furnace is shown in Fig. 4.7. Here, the transformer used differs from the ordinary power transformer only in the provision of low-voltage and high-current secondary winding. The tappings for voltage regulation are provided on the primary side.

Equivalent circuit parameters are described as R_T is the equivalent resistance of transformer referred to secondary, X_T is the equivalent reactance of transformer referred to secondary, X_L is the reactance of load, R_L is the resistance of load, R_E is the resistance of electrodes, E_A is the arc voltage (voltage drop across arc), and R_A is the arc resistance.

Power input to the arc furnace can be controlled by varying the length of the arc by raising or lowering the electrodes thereby varying the resistance of the arc. Another way is by varying the applied voltage by the use of tappings provided on the primary side of furnace transformer.

In order to ensure the best operating conditions and to have the complete control of furnace temperature both voltage and electrode control must be employed.

Case-I: Condition for maximum power output

From Fig. 4.7:

The equivalent impedance of arc furnace (Z) $= \left[(R_T + R_L + R_A)^2 + (X_T + X_L)^2\right]^{\frac{1}{2}}$.

$$\therefore \text{Arc current, } I = \frac{V}{Z}$$

$$I = \frac{V}{\sqrt{(R_T + R_L + R_A)^2 + (X_T + X_L)^2}}. \quad \textbf{(4.9)}$$

Power loss in the arc $(P_L) = I^2 R_A$

$$\therefore P_L = \left(\frac{V}{\sqrt{(R_T + R_L + R_A)^2 + (X_T + X_L)^2}}\right)^2 R_A$$

$$= \frac{V^2 R_A}{(R_T + R_L)^2 + R_A{}^2 + 2(R_T + R_L)\, R_A + (X_T + X_L)^2}$$

$$= \frac{V^2}{R_A + 2\,(R_T + R_L) + \dfrac{(R_T + R_L)^2}{R_A} + \dfrac{(X_T + X_L)^2}{R_A}}.$$

Power loss 'P_L' is maximum when denominator is minimum.

i.e., $\dfrac{d\,(P_L)}{d\,R_A} = 0$

$$\therefore \frac{d}{dR_A}\left[R_A + 2\;(R_T + R_L) + \frac{(R_T + R_L)^2}{R_A} + \frac{(X_T + X_L)^2}{R_A}\right] = 0$$

(or) $\quad 1 + 0 - \dfrac{(R_T + R_L)^2}{R_A{}^2} - \dfrac{(X_T + X_L)^2}{R_A{}^2} = 0$

$$\therefore R_A{}^2 = (R_T + R_L)^2 + (X_T + X_L)^2$$

$$\therefore RA = \sqrt{(R_T + R_L)^2 \; + \; (X_T + X_L)^2}\,. \qquad \textbf{(4.10)}$$

From Equation (4.10), the power loss will be maximum when the arc resistance (R_A) will be equal to the impedance of whole electric circuit referred to the secondary of transformer excluding the arc resistance 'R_A'.

Case-II: Power factor at maximum power loss

From the Fig. 4.7, the power factor:

$$\cos\phi = \frac{R_A + R_T + R_L}{Z}$$

$$= \frac{R_A + R_T + R_L}{\sqrt{(R_A + R_T + R_L)^2 + (X_T + X_L)^2}}$$

$$= \frac{R_A + R_T + R_L}{\sqrt{R_A{}^2 + (R_T + R_L)^2 + 2\,R_A(R_T + R_L) + (X_T + X_L)^2}}$$

$$= \frac{R_A + R_T + R_L}{\sqrt{R_A{}^2 + 2\,R_A(R_T + R_L) + (R_T + R_L)^2 + (X_T + X_L)^2}}.$$

But, from the maximum power loss condition, replace $(R_T + R_L)^2 + (X_T + X_L)^2 = R_A{}^2$ in the above equation:

$$\therefore \cos\phi = \frac{R_A + R_T + R_L}{\sqrt{R_A{}^2 \; + 2\,R_A(R_T + R_L) + R_A{}^2}}$$

$$= \frac{R_A + R_L + R_L}{\sqrt{2\,R_A{}^2 + 2\,R_A(R_T + R_L)}}$$

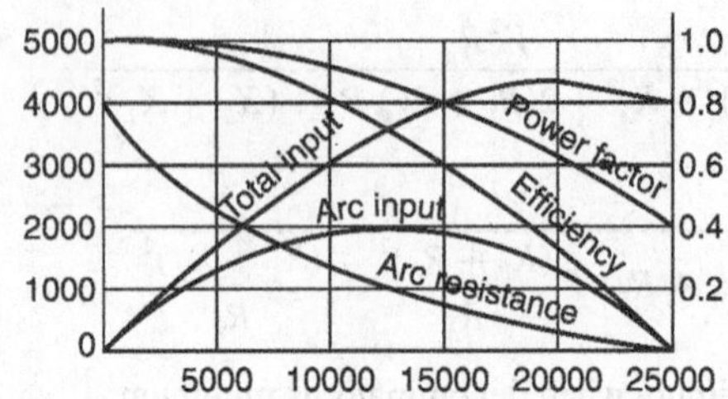

FIG. 4.8 *Performance characteristics of typical arc furnace*

$$= \frac{R_A + R_T + R_L}{\sqrt{2\,R_A\,(R_A + R_T + R_L)}}$$

$$= \sqrt{\frac{R_A + R_T + R_L}{2\,R_A}}$$

$$= \frac{1}{\sqrt{2}}\sqrt{1 + \frac{R_T + R_L}{R_A}}.$$

Assuming the value of $\frac{R_T + R_L}{R_A}$ is small and hence neglected.

$$\therefore \cos\phi = \frac{1}{\sqrt{2}} = 0.707.$$

That is, maximum power loss takes place when the power factor at secondary side is 0.707. And, it is not economical to operate an arc furnace with primary side power factor below 0.8. Figure 4.8 illustrates the performance characteristics of typical arc furnace.

When electrodes are short circuited, the total input to the furnace is almost zero. And, if electrodes are separated by some distance apart, arc is extinguished and then power drawn from the supply is also zero. In between these two limits, the power input to the furnace will be maximum at any particular loading.

Example 4.6: Calculate the time taken to melt 5 ton of steel in three-phase arc furnace having the following data.

Current = 8,000 A	Resistance = 0.003 Ω
Arc voltage = 50 V	Reactance = 0.005 Ω
Latent heat = 8.89 kcal/kg	Specific heat = 0.12
Initial temperature = 18°C	Melting point = 1,370°C

The overall efficiency is 50%. Find also the power factor and the electrical efficiency of the furnace.

Solution:

The equivalent circuit of the furnace is shown in Fig. P.4.1.

$$\text{Arc resistance per phase} = \frac{50}{8{,}000}$$

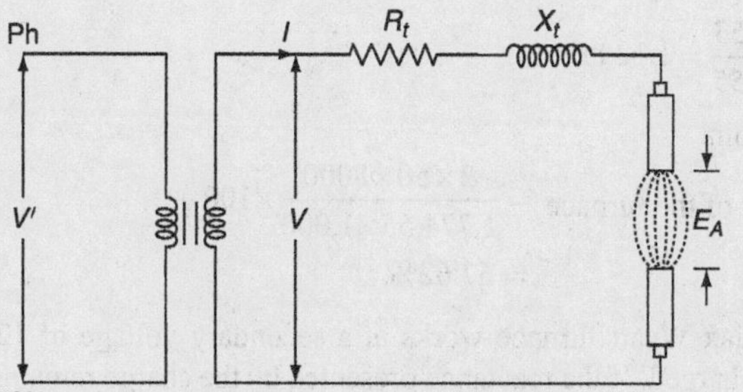

FIG. P.4.1 *Equivalent circuit of arc furnace*

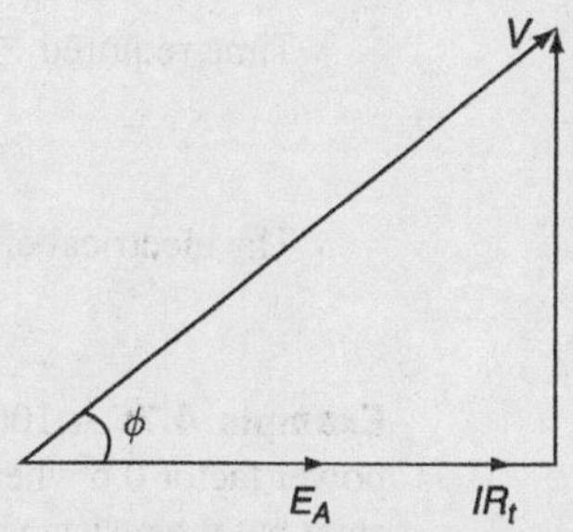

FIG. P.4.2 *Phasor diagram*

$R_A = 0.00625\ \Omega$.

Drop due to the resistance of transformer, $I R_t = 8{,}000 \times 0.003 = 24$ V and drop due to the reactance, $I X_t = 8{,}000 \times 0.005 = 40$ V.

From the phasor diagram (Fig. P.4.2):

$$V = \sqrt{(E_A + I R_t)^2 + (I X_t)^2}$$

$$= \sqrt{(50+24)^2 + (40)^2}$$

$$= 84.118 \text{ V.}$$

From the phasor diagram:

$$\cos\phi = \frac{E_A + IR_t}{V}$$

$$= \frac{50+24}{84.118}$$

$$= 0.879 \text{ lag.}$$

The amount of heat required per kg of steel:
= Specific heat × $(t_2 - t_1)$ + latent heat
= 0.12 × (1,370 – 18) + 8.89
= 171.13 kcal.

The heat required for 5 ton = 5,000 × 171.13
= 855,650 kcal.

$$\text{The actual heat required} = \frac{855{,}650 \times 1.162 \times 10^{-3}}{0.5}$$

= 1,988.53 kWh [∵ 1 kcal = 1.162×10^{-3} kWh].

Power input = 3 V $I \cos\phi \times 10^{-3}$ kW
= $3 \times 84.118 \times 8{,}000 \times 0.879 \times 10^{-3}$ kW
= 1,774.55 kW.

$$\text{Time required} = \frac{1,988.53}{1,774.55} = 1.12 \text{ hr}$$

$$= 67.2 \text{ min.}$$

$$\text{The electrical efficiency of the furnace} = \frac{3\times 50\times 8000}{1,774.55\times 1,000}\times 100$$

$$= 67.62\%.$$

Example 4.7: A 100-kW Ajax Wyatt furnace works at a secondary voltage of 12 V at power factor 0.6 when fully charged. If the reactance presented by the charge remains constant but the resistance varies invert as the charge depth in the furnace; calculate the charge depth that produces maximum heating effect when the furnace is fully charged.

Solution:

Secondary power, $P = V_2 I_2 \cos\phi$

$$I_2 = \frac{P}{V_2 \times \cos\phi}$$

$$= \frac{100\times 10^3}{12\times 0.6}$$

$$= 13.88 \text{ k A.}$$

When the crucible is fully charged, then the secondary impedance is:

$$Z_2 = \frac{V_2}{I_2}$$

$$= \frac{12}{13.88\times 10^3}$$

$$= 0.864 \text{ m}\Omega.$$

From the impedance triangle:

$$\cos\phi = \frac{R_2}{Z_2}$$

$$= Z_2 \cos\phi.$$

$$= 0.864 \times 10^{-3} \times 0.6$$

$$= 0.5184 \text{ m}\Omega.$$

$$\text{The secondary reactance } X_2 = \sqrt{(Z_2)^2 - (R_2)^2}$$

$$X_2 = \sqrt{(0.864\times 10^{-3})^2 - (0.5184\times 10^{-3})^2}$$

$$X_2 = 0.69 \text{ mm.}$$

Let 'H' be the height of the crucible when the crucible is full of charge and 'H_m' be the height of the charge at which maximum heating effect is possible.

$$\frac{H_m}{H} = h.$$

Given that the height of the charge is inversely proportional to the resistance. Let 'R_m' be the maximum resistance at which maximum heating effect will be possible.

At $R_m = X_2$, the heat produced will be maximum.

$$\frac{H_m}{H} = \frac{R_2}{R_m} = h \qquad \left[\because H_m \propto \frac{1}{R_m} H \propto \frac{1}{R_2}\right]$$

$$\frac{H_m}{H} = \frac{R_2}{X_2} = h$$

$$h = \frac{0.5184\times10^{-3}}{0.69\times10^{-3}}$$

$$= 0.75$$

$$\frac{H_m}{H} = 0.75$$

$$H_m = 0.75H.$$

4.12 HIGH-FREQUENCY HEATING

The main difference between the power-frequency and the high-frequency heating is that in the conventional methods, the heat is transferred either by conduction convection or by radiation, but in the high-frequency heating methods, the electromagnetic energy converted into the heat energy in side the material.

The high-frequency heating can be applied to two types of materials. The heating of the conducting materials, such as ferro-magnetic and non-ferro-magnetic, is known as *induction heating*. The process of heating of the insulating materials is known as *dielectric heating*. The heat transfer by the conventional method is very low of the order of 0.5–20 W/sq. cm. And, the heat transfer rate by the high-frequency heating either by induction or by dielectric heating is as much as 10,000 W/sq. cm. Thus, the high-frequency heating is most importance for tremendous speed of production.

4.13 INDUCTION HEATING

The induction heating process makes use of the currents induced by the electromagnetic action in the material to be heated. To develop sufficient amount of heat, the resistance of the material must be low $\left(\because \text{power drawn} = \frac{V^2}{R}\right)$, which is possible only with the metals, and the voltage must be higher, which can be obtained by employing higher flux and higher frequency. Therefore, the magnetic materials can be heated than non-magnetic materials due to their high permeability.

In order to analyze the factors affecting induction heating, let us consider a circular disc to be heated carrying a current of '*I*' amps at a frequency '*f*' Hz. As shown in Fig. 4.9.

Heat developed in the disc is depending upon the following factors.

- Primary coil current.
- The number of the turns of the coil.
- Supply frequency.

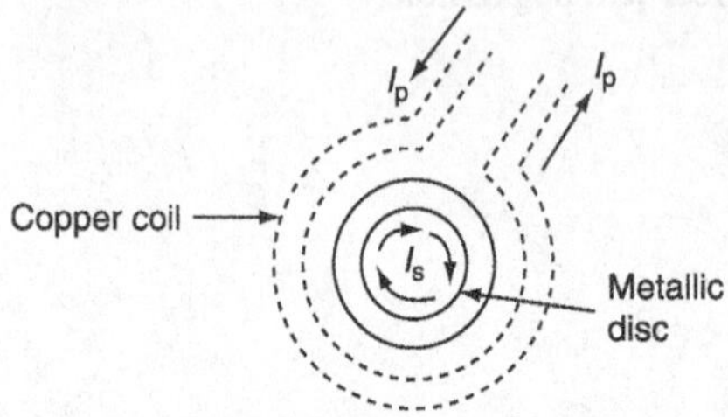

FIG. 4.9 *Induction heating*

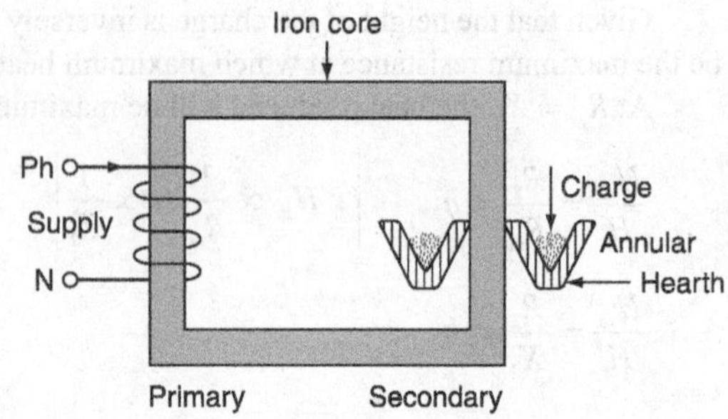

FIG. 4.10 *Direct core type furnace*

- The magnetic coupling between the coil and the disc.
- The high electrical resistivity of the disc.

If the charge to be heated is non-magnetic, then the heat developed is due to eddy current loss, whereas if it is magnetic material, there will be hysteresis loss in addition to eddy current loss. Both hysteresis and eddy current loss are depended upon frequency, but at high-frequency hysteresis, loss is very small as compared to eddy currents.

The depth of penetration of induced currents into the disc is given by:

$$d = \frac{1}{2\pi}\sqrt{\frac{\rho \times 10^9}{\mu f}} \text{ cm}$$

$$\text{i.e., } d \, \mu \frac{1}{\sqrt{f}},$$

where ρ is the specific resistance in Ω-cm, f is the frequency in Hz, and μ is the permeability of the charge.

There are basically two types of induction furnaces and they are:

1. Core type or low-frequency induction furnace.
2. Coreless type or high-frequency induction furnace.

4.13.1 Core type furnace

The operating principle of the core type furnace is the electromagnetic induction. This furnace is operating just like a transformer. It is further classified as:

(i) Direct core type.
(ii) Vertical core type.
(iii) Indirect core type.

(i) *Direct core type induction furnace*

The core type furnace is essentially a transformer in which the charge to be heated forms single-turn secondary circuit and is magnetically coupled to the primary by an iron core as shown in Fig. 4.10.

The furnace consists of a circular hearth in the form of a trough, which contains the charge to be melted in the form of an annular ring. This type of furnace has the following characteristics:

- This metal ring is quite large in diameter and is magnetically interlinked with primary winding, which is energized from an AC source. The magnetic coupling between primary and secondary is very weak; it results in high leakage reactance and low pf. To overcome the increase in leakage reactance, the furnace should be operated at low frequency of the order of 10 Hz.
- When there is no molten metal in the hearth, the secondary becomes open circuited thereby cutting of secondary current. Hence, to start the furnace, the molten metal has to be taken in the hearth to keep the secondary as short circuit.
- Furnace is operating at normal frequency, which causes turbulence and severe stirring action in the molten metal to avoid this difficulty, it is also necessary to operate the furnace at low frequency.
- In order to obtain low-frequency supply, separate motor-generator set (or) frequency changer is to be provided, which involves the extra cost.
- The crucible used for the charge is of odd shape and inconvenient from the metallurgical viewpoint.
- If current density exceeds about 500 A/cm^2, it will produce high-electromagnetic forces in the molten metal and hence adjacent molecules repel each other, as they are in the same direction. The repulsion may cause the interruption of secondary circuit (formation of bubbles and voids); this effect is known as *pinch effect.*

The pinch effect is also dependent on frequency; at low frequency, this effect is negligible, and so it is necessary to operate the furnace at low frequency.

(ii) *Vertical core type induction furnace*

It is an improvement over the direct core type furnace, to overcome some of the disadvantages mentioned above. This type of furnace consists of a vertical core instead of horizontal core as shown in Fig. 4.11. It is also known as *Ajax–Wyatt induction furnace.*

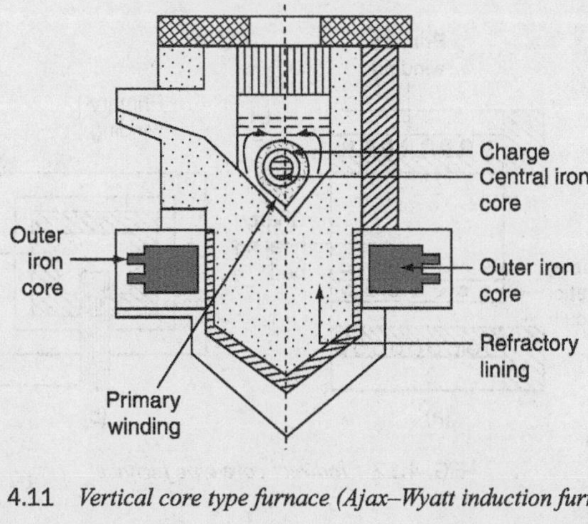

FIG. 4.11 *Vertical core type furnace (Ajax–Wyatt induction furnace)*

Vertical core avoids the pinch effect due to the weight of the charge in the main body of the crucible. The leakage reactance is comparatively low and the power factor is high as the magnetic coupling is high compared to direct core type.

There is a tendency of molten metal to accumulate at the bottom that keeps the secondary completed for a vertical core type furnace as it consists of narrow V-shaped channel.

The inside layer of furnace is lined depending upon the type charge used. Clay lining is used for yellow brass and an alloy of magnesia and alumina is used for red brass.

The top surface of the furnace is covered with insulating material, which can be removed for admitting the charge. Necessary hydraulic arrangements are usually made for tilting the furnace to take out the molten metal. Even though it is having complicated construction, it is operating at power factor of the order of 0.8–0.83. This furnace is normally used for the melting and refining of brass and non-ferrous metals.

Advantages

- Accurate temperature control and reduced metal losses.
- Absence of crucibles.
- Consistent performance and simple control.
- It is operating at high power factor.
- Pinch effect can be avoided.

(III) *Indirect core type furnace*

This type of furnace is used for providing heat treatment to metal. A simple induction furnace with the absence of core is shown in Fig. 4.12.

The secondary winding itself forms the walls of the container or furnace and an iron core links both primary and secondary windings.

The heat produced in the secondary winding is transmitted to the charge by radiation. An oven of this type is in direct competition with ordinary resistance oven.

It consists of a magnetic circuit AB is made up of a special alloy and is kept inside the chamber of the furnace. This magnetic circuit loses its magnetic properties at certain temperature and regains them again when it is cooled to the same temperature.

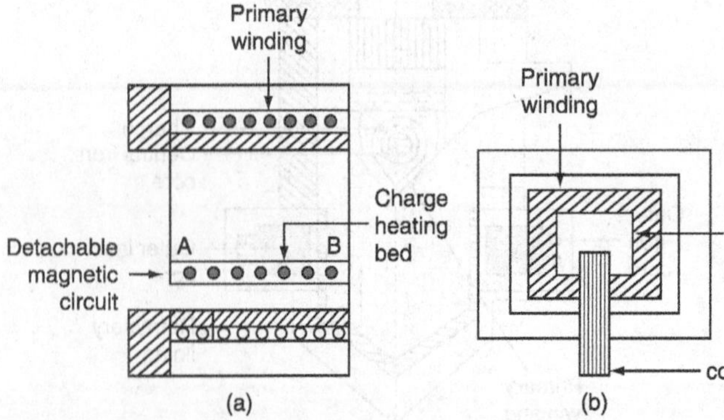

FIG. 4.12 *Indirect core type furnace*

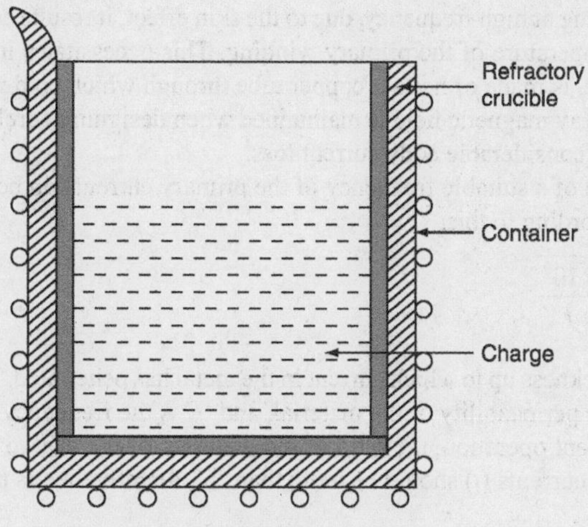

FIG. 4.13 *Coreless induction furnace*

When the oven reaches to critical temperature, the reluctance of the magnetic circuit increases many times and the inductive effect decreases thereby cutting off the supply heat. Thus, the temperature of the furnace can be effectively controlled. The magnetic circuit 'AB' is detachable type that can be replaced by the other magnetic circuits having critical temperatures ranging between 400°C and 1,000°C. The furnace operates at a pf of around 0.8.

The main advantage of such furnace is wide variation of temperature control is possible.

4.13.2 Coreless type induction furnace

It is a simple furnace with the absence core is shown in Fig. 4.13. In this furnace, heat developed in the charge due to eddy currents flowing through it.

The furnace consists of a refractory or ceramic crucible cylindrical in shape enclosed within a coil that forms primary of the transformer. The furnace also contains a conducting or non-conducting container that acts as secondary.

If the container is made up of conducting material, charge can be conducting or non-conducting; whereas, if the container is made up of non-conducting material, charge taken should have conducting properties.

When primary coils are excited by an alternating source, the flux set up by these coils induce the eddy currents in the charge. The direction of the resultant eddy current is in a direction opposite to the current in the primary coil. These currents heat the charge to melting point and they also set up electromagnetic forces that produce a stirring action to the charge.

∴ The eddy currents developed in any magnetic circuit are given as:

$$W_e \propto B_m^2 f^2,$$

where B_m is the maximum flux density (tesla), f is the frequency in (Hz), and W_e is the eddy current loss (watts).

In coreless furnace, the flux density will be low as there is no core. Hence, the primary supply should have high frequency for compensating the low flux density.

If it is operating at high frequency, due to the skin effect, it results copper loss, thereby increasing the temperature of the primary winding. This necessitates in artificial cooling. The coil, therefore, is made of hollow copper tube through which cold water is circulated.

Minimum stray magnetic field is maintained when designing coreless furnace, otherwise there will be considerable eddy current loss.

The selection of a suitable frequency of the primary current can be given by penetration formula. According to this:

$$t = \frac{1}{2\pi}\sqrt{\frac{\rho \times 10^9}{\mu f}}, \tag{4.11}$$

where 't' is the thickness up to which current in the metal has penetrated, 'ρ' is the resistivity in Ω-cm, 'μ' is the permeability of the material, and 'f' is the frequency in Hz.

For the efficient operation, the ratio of the diameter of the charge (d) to the depth of the penetration of currents (t) should be more than '6', therefore let us take:

$$\frac{d}{t} = 8.$$

Substitute above in Equation (4.11).

$$f = \frac{16 \times \rho \times 10^9}{\pi^2 \mu d^2}. \tag{4.12}$$

Following are the advantages of coreless furnace over the other furnaces:

- Ease of control.
- Oxidation is reduced, as the time taken to reach the melting temperature is less.
- The eddy currents in the charge itself results in automatic stirring.
- The cost is less for the erection and operation.
- It can be used for heating and melting.
- Any shape of crucible can be used.
- It is suitable for intermittent operation.

Example 4.8: Determine the amount of energy required to melt 2 ton of zinc in 1 hr, if it operates at an efficiency of 70% specific heat of zinc is equals to 0.1. The latent heat of zinc = 26.67 kcal/kg, the melting point is 480°C, and the initial temperature is 25°C.

Solution:

Weight of zinc = 2 × 1,000 = 2,000 kg.

The heat required raising the temperature from 25°C to 480°C:

$$H = w \times S \times (t_2 - t_1)$$
$$= 2{,}000 \times 0.1 \times (480 - 25)$$
$$= 91{,}000 \text{ kcal}.$$

The heat required for melting:

$$= w \times l$$
$$= 2{,}000 \times 26.67$$
$$= 53{,}340 \text{ kcal}.$$

$\therefore$ Total heat required = 91,000 + 53,340
= 144,340 kcal.

Since 4.18 J = 1 cal and 1 J/sec = 1 W.
So, 1 cal = 4.18 W-sec.

$$\text{Energy input} = \frac{144,340 \times 10^3 \times 4.18}{10^3 \times 3,600 \times 0.70}$$

= 239.42 kWh.

Energy = $I^2 R t$.

$$\text{Power} = \frac{\text{energy}}{\text{time}} = \frac{239.42 \text{ kW}}{1}$$

= 239.42 kW.

Example 4.9: A high-frequency induction furnace that takes 20 min to melt 1.9 kg of aluminum, the input to the furnace being 3 kW, and the initial temperature is 25°C. Then, determine the efficiency of the furnace.

The specific heat of aluminum = 0.212.
Melting point = 660°C.
The latent heat of the fusion of aluminum = 76.8 kcal/kg.

Solution:

Total heat required = 1.90 × 0.212 × (60 – 25)+1.9 × 76.8
= 401.698 kcal.

$$\text{Heat required per hour} = 401.698 \times \frac{60}{20}$$

= 1,205.094 kcal.

$$\text{The power delivered to the charge} = \frac{1,205.094}{860}$$

= 1.401 kW.

$$\text{The efficiency of the furnace } \% \, \eta = \frac{1.401}{3} \times 100.$$

Example 4.10: Determine the equivalent resistance of the charge and the current in the primary winding that is required to counter balance the mmf due to secondary current. If the power input to the charge is a 0.5-ton and 960-Hz induction, furnace have 20 turns on the primary winding is 340 kW. The cylindrical crucible has an internal diameter of 47 cm and the depth of the charge in it is 50 cm. Take resistivity of charge = 200 μΩ-cm.

Solution:

The depth of penetration of eddy currents through the charge is given by:

$$t = \frac{1}{2\pi}\sqrt{\frac{\rho \times 10^9}{\mu f}}$$

$$= \frac{1}{2\pi}\sqrt{\frac{200\times10^{-6}\times10^{9}}{1\times960}} \qquad (\mu = 1 \text{ for molten steel})$$

$t = 2.29$ cm.

The area of the cylinder in which current flows is $= 50 \times 2.29$

$= 114.5$ cm^2.

The mean diameter of the cylinder $= 47 + 2.29$

$= 49.29$ cm.

The mean length of the current flows $= \pi \times 49.29$

$= 154.77.$

The resistance of the cylinder $= \dfrac{\rho\, l}{A}$

$$= \frac{200\times10^{-6}\times154.77}{114.5}$$

$$R = 270.34\times10^{-6}\Omega.$$

The power loss, $P_L = I^2 R$

$$I = \sqrt{\frac{P}{R}} = \sqrt{\frac{340\times10^{3}}{270.34\times10^{-6}}}$$

$= 35.46$ kA.

Now, in case of transformer:

$N_1 I_1 = N_2 I_2.$

Assume secondary is a single-turn winding, the current in the primary:

$$I_1 = \frac{35.46\times10^{3}}{20}$$

$= 1{,}773$ A.

Example 4.11: A low-frequency induction furnace has a secondary voltage of 20 V and takes 600 kW at 0.5 pf when the hearth is full. If the secondary voltage is maintained at 20 V, determine the power absorbed and the power factor when the hearth is half-full. Assume the resistance of the secondary circuit to be doubled and the reactance to remain the same.

Solution:

Case-I: when the hearth is full:

The power input:

$P = VI\cos\phi$

$$I = \frac{600\times10^{3}}{20\times0.5}$$

$= 60 \times 10^3$ A.

The impedance, $Z = \dfrac{V}{I}$

$$= \frac{20}{60 \times 10^3}$$

$$= 0.33 \times 10^{-3} \Omega$$

$$\cos\phi = \frac{R}{Z}$$

$$R = Z \times \cos\phi$$

$$= 0.33 \times 10^{-3} \times 0.5$$

$$= 0.167 \times 10^{-3}\ \Omega.$$

Reactance, $X = \sqrt{(Z)^2 - (R)^2}$

$$= \sqrt{(0.33 \times 10^{-3})^2 - (0.167 \times 10^{-3})^2}$$

$$= 0.2846 \times 10^{-3} \Omega.$$

Case-II: when the hearth is half-full:

Now, $R = 2R$ and $X = X$

$$R = 2 \times 0.167 \times 10^{-3}$$

$$= 0.334 \times 10^{-3}\ \Omega.$$

Impedance, $Z = \sqrt{R^2 + X^2}$

$$= \sqrt{(0.334 \times 10^{-3})^2 + (0.2846 \times 10^{-3})^2}$$

$$= 0.438 \times 10^{-3}\ \Omega.$$

From the impedance triangle:

$$\cos\phi = \frac{R}{Z}$$

$$= \frac{0.334 \times 10^{-3}}{0.438 \times 10^{-3}}$$

$$= 0.762.$$

current, $I = \frac{V}{Z}$

$$= \frac{20}{(0.438 \times 10^{-3})}$$

$$= 45.66 \text{ kA}.$$

The power absorbed, $P = VI\cos\phi$

$$= 20 \times 45.66 \times 10^3 \times 0.762$$

$$= 695.85 \text{ kW}.$$

4.14 DIELECTRIC HEATING

When non-metallic materials i.e., insulators such as wood, plastics, and china glass are subjected to high-voltage alternating electric field, the atoms get stresses, and due to inter-atomic friction caused by the repeated deformation and the rotation of atomic structure (polarization), heat is produced. This is known as dielectric loss. This dielectric loss in insulators corresponds to hysteresis loss in ferro-magnetic materials. This loss is due to the reversal of magnetism or magneto molecular friction. These losses developed in a material that has to be heated.

An atom of any material is neutral, since the central positive charge is equals to the negative charge. So that, the centers of positive and negative charges coincide as long as there is no external field is applied, as shown in Fig. 4.14(a). When this atom is subjected to the influence of the electric field, the positive charge of the nucleus is acted upon by some force in the direction of negative charges in the opposite direction. Therefore, the effective centers of both positive and negative charges no longer coincident as shown in Fig. 4.14(b). The electric charge of an atom equivalent to Fig. 4.14(b) is shown in Fig. 4.14(c).

This gives raise to an electric dipole moment equal to $P = q\,d$, where d is the distance between the two centers and q is the charge on the nucleus.

Now, the atom is said to be polarized atom. If we apply alternating voltage across the capacitor plate, we will get alternating electric field.

Electric dipoles will also try to change their orientation according to the direction of the impressed electric field. In doing so, some energy will be wasted as inter-atomic friction, which is called dielectric loss.

As there is no perfect conductor, so there is no perfect insulator. All the dielectric materials can be represented by a parallel combination of a leakage resistor 'R' and a capacitor 'C' as shown in Fig. 4.15 (a) and (b).

If an AC voltage is applied across a piece of insulator, an electric current flows; total current 'I' supposed to be made up of two components I_C and I_R, where I_C is the capacitive current leading the applied voltage by 90° and I_R is in phase with applied voltage as shown in Fig. 4.15(c).

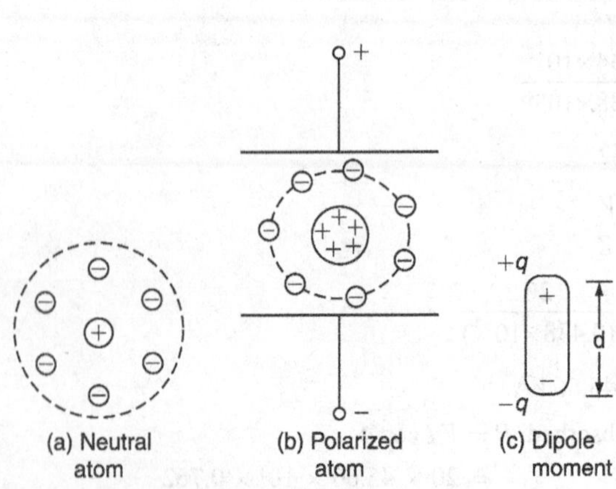

FIG. 4.14 *Polarization*

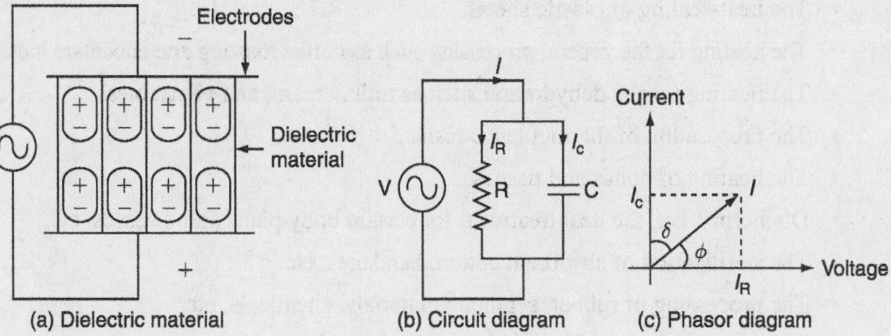

FIG. 4.15 *Dielectric heating*

$$\text{Dielectric loss, } P_L = VI\cos\phi$$

$$= VI_R \qquad [\because I_R = I\cos\phi]$$

$$= VI_C \tan\delta \qquad \left[\because \tan\delta = \frac{I_R}{I_C}\right].$$

$$V\cdot\left(\frac{V}{X_C}\right)\tan\delta \qquad \left[QI_C = \frac{V}{X_C}\right]$$

$$= V^2\,\omega\,C\tan\delta \tag{4.13}$$

$$= V^2 \times 2\pi f \times \frac{\varepsilon_O\,\varepsilon_r\,A}{d} \times \delta \text{ W} \tag{4.14}$$

where 'V' is the applied voltage in volts, 'f' is the supply frequency in Hz, ε_0 is the absolute permittivity of the medium $= 8.854 \times 10^{-12}$ F/m, ε_r is the relative permittivity of the medium = 1 for free space, A is the area of the plate or electrode (m²), d is the thickness of the dielectric medium, and δ is the loss angle in radian.

From Equation (4.14):

$$P_L \propto V^2 \quad \text{and} \quad P_L \propto f. \tag{4.15}$$

Normally frequency used for dielectric heating is in the range of 1–40 MHz. The use of high voltage is also limited due to the breakdown voltage of thin dielectric that is to be heated, under normal conditions; the voltage gradient used is limited to 18 kV/cm.

The advantages of the dielectric heating

- The heating of the non-conducting materials is very rapid.
- The uniform heating of material is possible.
- Heat is produced in the whole mass of the material.

The applications of the dielectric heating

- The drying of paper, wood, etc.
- The gluing of wood.

- The heat-sealing of plastic sheets.
- The heating for the general processing such as coffee roasting and chocolate industry.
- The heating for the dehydration such as milk, cream, and vegetables.
- The preparation of thermoplastic resins.
- The heating of bones and tissues.
- Diathermy, i.e., the heat treatment for certain body pains and diseases, etc.
- The sterilization of absorbent cotton, bandages, etc.
- The processing of rubber, synthetic materials, chemicals, etc.

Example 4.12: A piece of insulating material is to be heated by dielectric heating. The size of the piece is 10 × 10 × 3 cm³. A frequency of 30 mega cycles is used and the power absorbed is 400 W. Determine the voltage necessary for heating and the current that flows in the material. The material has a permittivity of 5 and a power factor of 0.05.

Solution:

The capacitance offered by the material is given by:

$$C = \frac{\varepsilon_o\ \varepsilon_r\ A}{d},$$

where ε_o is 8.854×10^{-12}, ε_r is 5, and A is area in $m^2 = 10 \times 10 \times 10^{-4} = 0.01\ m^2$.

$$\therefore C = \frac{8.854 \times 10^{-12} \times 5 \times 0.01}{3 \times 10^{-2}}$$

$$= 14.75\ \text{pF}.$$

In the phasor diagram, δ is called the dielectric loss angle and ϕ is called the power factor angle.

From the phasor diagram (Fig. P.4.3):

$$\tan \delta = \frac{I_R}{I_C}$$

$$= \frac{V/R}{V\ \omega\ C}$$

$$\frac{V}{R} = V\omega C \tan \delta.$$

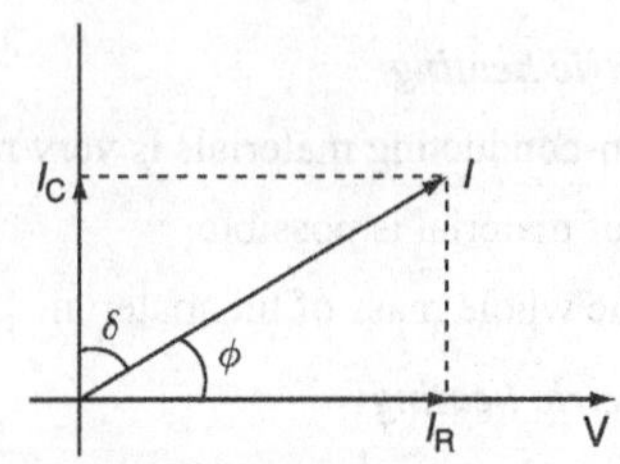

FIG. P.4.3 *Phasor diagram*

The power loss, $P_L = \frac{V^2}{R}$

$= V^2\, \omega\, C \tan \delta$ (or)

$= V^2\, \omega\, C \cos \phi.$

$\therefore 400 = V^2 \times 2 \times 3.14 \times 30 \times 10^6 \times 14.75 \times 10^{-12} \times 0.05$

$V = 1{,}696.71$ V.

The total current:

$I = I_C + I_R$

$= I_C \qquad (I_R \cong 0)$

$= V\omega C$

$= 4.71$ A.

Example 4.13: A piece of an insulating material 2-cm thick and 120 cm² in area is to be heated by the dielectric heating. The material has a permittivity of 5 and a power factor of 0.05. The power at 800 V is 300 W. Determine the cycles per second.

Solution:

The capacitance offered by the dielectric material is:

$$C = \frac{\varepsilon_o\, \varepsilon_r\, A}{d}$$

$$= \frac{8.854 \times 10^{-12} \times 5 \times 120 \times 10^{-4}}{2 \times 10^{-12}}$$

$$= 26.56 \text{ pF.}$$

From the phasor diagram shown in Fig. P.4.4:

$$\tan \delta = \frac{I_R}{I_C}$$

$$= \frac{V/R}{V\, \omega\, C}$$

$$\frac{V}{R} = V\omega C \tan\delta.$$

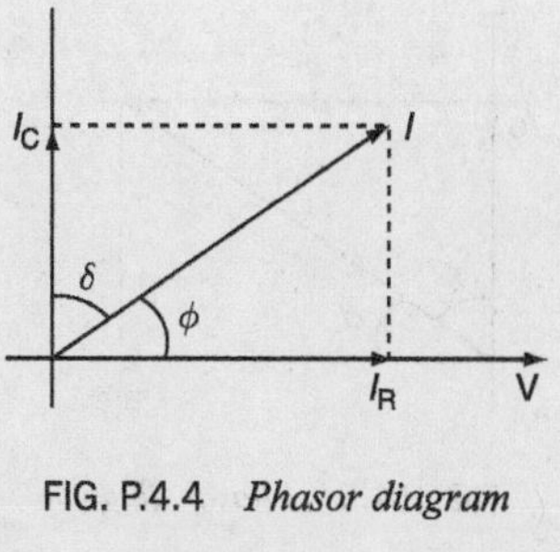

FIG. P.4.4 *Phasor diagram*

The power loss, $P_L = \frac{V^2}{R}$

$$= V^2 \omega C \tan \delta \quad \text{(or)}$$
$$= V^2 \omega C \cos \phi$$
$$= V^2 \pi f C \cos \phi.$$

$$f = \frac{P_L}{V^2 2\pi C \cos\phi}$$

$$= \frac{300}{(800)^2 \times 2 \times 3.14 \times 26.56 \times 10^{-12} \times 0.05}$$

$$= 56.2 \text{ MHz.}$$

Example 4.14: The power required for dielectric heating of a slab of resin 150 cm² in area and 2-cm thick is 200 W, at a frequency of 30 MHz. The material has a relative permittivity of 5 and power factor 0.05. Find the voltage necessary and the current flowing through the material. If the voltage is limited to 700 V, what will be the frequency to obtain the same heating?

Solution:

The capacitance offered by the dielectric slab is:

$$C = \frac{\varepsilon_0 \varepsilon_r A}{d}$$

$$= \frac{8.854 \times 10^{-12} \times 5 \times 150 \times 10^{-4}}{2 \times 10^{-2}}$$

$$= 33.2 \text{ pF.}$$

The equivalent circuit and the phasor diagram of the dielectric heating is shown in Fig. P.4.5.

From the phasor diagram:

$$\tan \delta = \frac{I_R}{I_C}$$

$$= \frac{V/R}{VWC}$$

$$\frac{V}{R} = VWC \tan \delta.$$

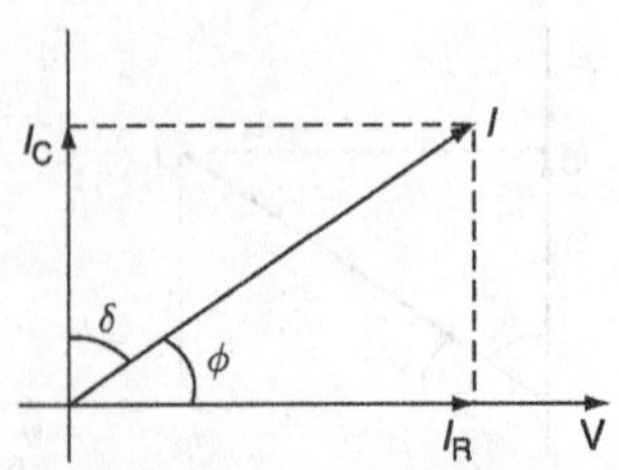

FIG. P.4.5 *Phasor diagram*

The power loss, $P_L = \dfrac{V^2}{R}$

$= V^2\omega C \tan\delta$ (or)

$= V^2\omega C \cos\phi \qquad [\because I = I_R + I_C; \quad I = I_C]$

$200 = V^2 \times 2 \times 3.14 \times 30 \times 10^6 \times 33.2 \times 10^{-12} \times 0.05 \qquad [\omega = 2\pi f]$

$V = 799.68$ V.

From the equivalent circuit:

$$I = I_R + I_C$$
$$= I_C \qquad [I_R \cong \text{small}]$$
$$= VWC$$
$$= 799.68 \times 2 \times 3.14 \times 30 \times 10^6 \times 33.2 \times 10^{-12}$$
$$= 5\text{ A}.$$

The voltage, $V \alpha \dfrac{1}{\sqrt{f}}$.

Let V_1 be the voltage at the frequency 30 MHz.
Let V_2 be the voltage at the frequency f_2.
So that:

$$\frac{V_1}{V_2} = \sqrt{\frac{f_2}{f_1}}$$

$$f_2 = \left(\frac{V_1}{V_2}\right)^2 \times f_1$$

$$f_2 = \left(\frac{799.68}{700}\right)^2 \times 30 \times 10^6$$

$$= 39.15\text{ mHz}.$$

Example 4.15 A piece of plastic material of length 5 cm, width 2 cm, and thickness 1 cm is placed in between two electrodes having dimensions: length 25 cm, width 2 cm, and with 2-cm distance between them. The frequency of voltage impressed across the electrodes is 20 mHz. If the power consumed is 80 W, find the voltage applied across the electrodes and the current through the material. Assume relative permittivity as 5 and power factor 0.05.

Solution:

The arrangement of the heating material is shown in the Fig. P.4.6.
The capacitance offered by the parallel plate capacitor is:

$$C = \frac{\varepsilon_0\, \varepsilon_r\, A_1}{t} + \frac{A_2}{\dfrac{t_1}{\varepsilon_{r1}} + \dfrac{t_2}{\varepsilon_{r2}}},$$

where, $A_1 = (25 - 2) \times 2 = 0.004\ \text{m}^2$

$A_2 = 5 \times 2 = 0.001\ \text{m}^2$.

The separation of the distance between two plates = 2 cm = 0.02 m.

Similarly:

$t_1 = 0.01$ m

$t_2 = 2 - 1 = 1$ cm $= 0.01$ m

$\varepsilon_{r1} = 5$ for wood

$\varepsilon_{r2} = 1$ for air

$$\therefore C = 8.854 \times 10^{-12} \left[\frac{0.004 \times 1}{0.02} + \frac{0.001}{\frac{0.01}{5} + \frac{0.01}{1}} \right]$$

$$= 2.509\ \text{pF}.$$

The power loss, $P_L = V^2 \omega C \cos\phi$.

$$\therefore V = \sqrt{\frac{P}{2\pi f C \cos\theta}}$$

$$= \sqrt{\frac{80}{2 \times 3.14 \times 20 \times 10^6 \times 2.509 \times 10^{-12} \times 0.05}}$$

$$= 2.25\ \text{kV}.$$

Power, $P = VI\cos\phi$

$$I = \frac{P}{V\cos\phi}$$

$$= \frac{80}{2.25 \times 10^3 \times 0.05}$$

$$= 0.71\ \text{A}.$$

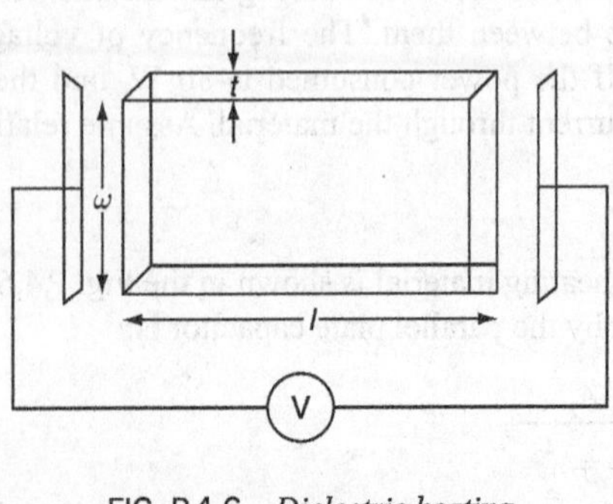

FIG. P.4.6 *Dielectric heating*

KEY NOTES

- The modes of the transfer of heat are:
  (i) Conduction.
  (ii) Convection.
  (iii) Radiation.
- **Conduction** is the process of the transfer of heat from one part of a substance to another part without movement in the molecules of substance.
- **Convection** is the process of the transfer of heat takes place from one part to another part of a substance or a fluid due to the actual motion of the molecules.
- **Radiation** is the process of the transfer of heat from the source to the substance to be heated without heating the medium in between the source and the substance.
- **Stefan's Law** for heat dissipation is:

$$H = 5.72\times10^4\ ke\left[\left(\frac{T_1}{1{,}000}\right)^4 - \left(\frac{T_2}{1{,}000}\right)^4\right]\text{W/m}^2$$

- **Resistance heating** is the process of heating the charge or substance by the heat produced due to the resistance offered by the charge or heating element.
- **Induction heating** is the process of heating the material due to the heat developed by the currents induced in the material by electromagnetic induction process.
- **Dielectric heating is** the process of heating non-metallic materials due to the heat developed by the process of polarization.
- **Oven** means that a low-temperature heating chamber with provision for ventilation.
- **Pinch effect** is the formation of bubbles and voids in the charge to be heated by the electromagnetic induction due to the high electromagnetic forces, which causes the interruption of the secondary circuit.
- **High frequency eddy current heating** is the process of heating any material by the heat developed due to the conversion of electromagnetic energy into heat energy.

SHORT QUESTIONS AND ANSWERS

(1) Give any two advantages of electric heating.

(i) Electric heating equipment is cheaper; it does not require much skilled persons so maintenance cost is less.

(ii) In this heating, the temperature can be controlled and regulated accurately either manually or automatically.

(2) What are the modes of the transfer of heat?

The modes of the transfer of heat are:

(i) Conduction.

(ii) Convection.

(iii) Radiation.

(3) What is an oven?

Oven is mean that a low-temperature heating chamber with provision for ventilation.

(4) Define conduction.

The process of heat transfers from one part of a substance to another part without movement in the molecules of substance. The rate of conduction of heat along the substance depends upon temperature gradient.

(5) Define convection.

The process of heat transfer takes place from one part to another part of a substance or a fluid due to the actual motion of the molecules. The rate of conduction of the heat depends mainly on the difference in the fluid density at different temperatures.

(6) Define radiation.

The process of heat transfers from the source to the substance to be heated without heating the medium in between the source and the substance.

(7) What are the essentials requirements of heating elements?

The materials used for heating element should have:

(i) High-specific resistance.

(ii) High-melting point.

(iii) High-mechanical strength.

(iv) Free from oxidation.

(8) What is the Stefan's formula for heat dissipation?

Stefan's law for heat dissipation is:

$$H = 5.72\times10^4\ ke\left[\left(\frac{T_1}{1{,}000}\right)^4 - \left(\frac{T_2}{1{,}000}\right)^4\right] \text{W/m}^2.$$

(9) What are the causes of the failure of the heating elements?

The failure of the heating element may cause due to:

- The formation of hotspots.
- The oxidation of the element and the intermittency of operation.
- The embitterment caused by gain growth
- Contamination and corrosion.

(10) What is meant by resistance heating?

The process of heating the charge or substance by the heat produced due to the resistance offered by the charge or heating element.

(11) What is meant by induction heating?

The process of heating the material due to the heat developed by the currents induced in the material by electromagnetic induction process.

(12) What is meant by dielectric heating?

The process of heating non-metallic materials, i.e., the insulators such as wood, plastics, and china clay due to the heat developed in the material when they are subjected to high voltage alternating electric field, the atoms get stresses and due to inter-atomic friction caused by the repeated deformation and rotation of atomic structure.

(13) What are the various losses occurring in resistance oven?

The heat produced in the heating elements, not only raises the temperature of charge to desired value, but also used to overcome the losses occurring due to:

(i) The heat used in raising the temperature of oven (or) furnace.

(ii) The heat used in raising the temperature of containers (or) carriers.

(iii) The heat conducted through the walls.

(iv) The heat loss due to the opening of oven door.

(14) List out various methods of controlling the temperature of resistance heating.

The temperature of the furnaces can be controlled either by:

(i) Varying the resistance of elements.

(ii) Varying the applied voltage to the elements or the current flowing through the elements

(iii) Varying the ratio of the on-and-off times of supply.

(15) What are the types of arc furnaces?

There are two types of arc furnaces and they are:

(i) Direct arc furnace.

(ii) Indirect arc furnace.

(16) What is the condition for the maximum power output of electric arc furnace?

The condition for the maximum power output of electric arc furnace is:

$$R_A = \sqrt{(R_T + R_L)^2 + (X_T + X_L)^2}.$$

(17) What is pinch effect?

The formation of bubbles and voids in the charge to be heated by the electromagnetic induction due to high-electromagnetic forces, which causes the interruption of secondary circuit. This effect is known as pinch effect.

(18) Write any two advantages of dielectric heating?

The advantages of dielectric heating are:

(i) The heating of non-conducting materials is very rapid.

(ii) The uniform heating of material is possible.

(iii) Heat is produced in the whole mass of the material.

(19) List out some of the applications of dielectric heating?

(i) The drying of paper, wood, etc.

(ii) The gluing of wood.

(iii) The heat-sealing of plastic sheets.

(iv) The heating for general processing such as coffee roasting and chocolate industry.

(v) The heating for dehydration such as milk, cream, and vegetables.

(20) What is high-frequency eddy current heating?

The process of heating any material by the heat developed due to the conversion of electromagnetic energy into heat energy.

(21) How amount of heat is controlled in high-frequency eddy current heating?

The amount of heat is controlled by controlling the supply frequency and the flux density in high-frequency eddy current heating.

(22) How can the rate of dielectric heating be varied?

The rate of dielectric heating can be varied by varying either supply voltage or supply frequency.

MULTIPLE-CHOICE QUESTIONS

(1) Electric heating is considered advantageous over the other systems of heating (coal, oil, or gas heating) as:
- (a) It is economical, clean, hygienic, efficient, and safe in operation.
- (b) It provides better working conditions, automatic protection against overheating, and simple, accurate, and reliable temperature control.
- (c) There is no upper limit to the temperature obtainable except the ability of the material to withstand the heat.
- (d) All of the above.

(2) A perfect black body is one which:
- (a) Absorbs all incident radiations.
- (b) Reflects all incident radiations.
- (c) Transmits all in incident radiations.
- (d) All of the above.

(3) For the transmission of heat from one body to another it is essential that:
- (a) Both bodies are solids.
- (b) The two bodies are at different temperatures.
- (c) Both bodies are in contact.
- (d) At least one of the bodies has some source of heating.

(4) Heat is transferred simultaneously by conduction, convection, and radiation:
- (a) During the melting of ice.
- (b) From refrigerator coils to refrigerator freezer.
- (c) Inside boiler furnaces.
- (d) Through the surface of the insulated pipe carrying steam.

(5) Heat transfer by conduction will not take place when the:
- (a) Two bodies are at the same temperatures.
- (b) Bodies are kept in vacuum.
- (c) Bodies are immersed in water.
- (d) Bodies are exposed to thermal radiations.

(6) Thermal conductivity is measured in:
- (a) MJ/m^2/m/°C/hr.
- (b) MJ/m/°C/hr.
- (c) MJ/m^2/°C/hr.
- (d) MJ/m^2/°C.

(7) The highest value of thermal conductivity is for:
- (a) Aluminum.
- (b) Brass.
- (c) Copper.
- (d) Iron.

(8) Radiations from a black body are proportional to:
- (a) T^2.
- (b) T^3.
- (c) T^4.
- (d) $1/T^4$.

(9) A body reflecting entire radiations incidenting on it is called the:
- (a) White body.
- (b) Gray body.
- (c) Black body.
- (d) Transparent body.

(10) The insulating material suitable for the low temperature applications is:
- (a) Cork.
- (b) Diatomaceous earth.
- (c) Asbestos paper.
- (d) 75% magnesia.

(11) The quantity of heat absorbed from the heater by convection depends upon:
- (a) The temperature of heating element above the surroundings.
- (b) The surface area of the heater.

(c) The position of the heater.

(d) All of the above.

(12) In case of immersion type water heater, the heat is transferred by:

(a) Radiation.

(b) Conduction.

(c) Convection

(d) All of the above.

(13) The material used as the heating element for a furnace should have:

(a) High resistivity.

(b) High-melting point.

(c) Low temperature coefficient.

(d) All of the above.

(14) The material of the heating element should be:

(a) Such that it may withstand the required temperature without getting oxidized.

(b) Of low resistivity.

(c) Of low melting point.

(d) Of high temperature coefficient.

(15) The material to be used for the heating element should be of high resistivity so as to:

(a) Increase the life of the heating element.

(b) Reduce the length of the heating element.

(c) Reduce the effect of oxidation.

(d) Produce large amount of heat.

(16) The material to be used for the heating element should be of low temperature coefficient so as to:

(a) Avoid initial rush of current.

(b) Avoid change in kW rating with temperature.

(c) Reduce the effect of oxidation

(d) Both (a) and (b) above.

(17) Which of the following heating element will have the least temperature range?

(a) Eureka.

(b) Silicon carbon.

(c) Nichrome.

(d) Kanthal.

(18) Which of the following heating element can give the highest temperature in resistance heating?

(a) Nichrome.

(b) Silicon carbide.

(c) Copper.

(d) Nickel–Cr–Fe alloy.

(19) The heat element to be used in a furnace employed for heating around 1,600°C should be of the material:

(a) Nichrome.

(b) Eureka.

(c) Molybdenum.

(d) Silicon-carbide.

(20) In an electric press, mica is used:

(a) For induction heating.

(b) For dielectric heating.

(c) As an insulator.

(d) For the improvement of power factor.

(21) In a resistance furnace, the temperature is controlled by:

(a) The variation of operating voltage.

(b) The variation of the resistance of heating circuit.

(c) Switching on and off the supply periodically.

(d) All of the above.

(22) The simplest and the most commonly used method for temperature control is:

(a) The external series resistance in the heating circuit.

(b) The change of connections of heating circuit.

(c) The use of variable number of heating elements.

(d) Transformer tappings.

(23) In a domestic baking oven, the temperature is controlled by:

(a) Voltage.

(b) Series–parallel operation.

(c) Thermostat.

(d) Star-delta connections.

(24) The device necessarily used for automatic temperature control in a furnace is:

(a) Thermostat.

(b) Auto-transformer.

(c) Thermo-couple.

(d) Any of the above.

(25) The control of power input to salt-bath furnace is affected by:

(a) Varying the depth of immersion of electrodes.

(b) Varying the distance between the electrodes.

(c) Both (a) and (b).

(d) None of (a) and (b).

(26) The temperature inside a furnace is usually measured by:

(a) Mercury thermometer.

(b) Optical pyrometer.

(c) Alcohol thermometer.

(d) Any of the above.

(27) In a resistance furnace, the atmosphere is:

(a) Oxidizing.

(b) Deoxidizing.

(c) Reducing.

(d) Neutral.

(28) In the direct resistance heating method, the maximum heat transfer takes place by:

(a) Convection.

(b) Radiation.

(c) Conduction.

(d) Any of the above.

(29) Radiant heating is used for:

(a) The malting of ferrous metals.

(b) The annealing of metals.

(c) The drying of paints and varnishes.

(d) Any of the above.

(30) Direct resistance heating is used in:

(a) Electrode boiler.

(b) Salt-bath furnace.

(c) Resistance welding.

(d) All of the above.

(31) Resistance ovens are used for:

(a) Domestic and commercial heating.

(b) The vulcanizing and hardening of synthetic materials.

(c) The drying of varnish coatings, drying, and baking of potteries.

(d) All of the above.

(32) In direct arc furnace, which of the following is of high value?

(a) Current.

(b) Voltage.

(c) Power factor.

(d) All of the above.

(33) The power factor at which the direct arc furnace operates is:

(a) Low lagging.

(b) Low leading.

(c) Unity.

(d) High leading.

(34) For arc heating, the electrodes used are made of:

(a) Copper.

(b) Graphite.

(c) Tungsten.

(d) Aluminum.

(35) In an arc furnace, the choke is provided to:

(a) Reduce the surge severity.

(b) Stabilize the arc.

(c) Improve the power factor.

(d) All of the above.

(36) It is desirable to operate the arc furnaces at a power factor of:

(a) Zero.

(b) Unity.

(c) 0.707 lagging.

(d) 0.707 leading.

(37) It is desirable to keep the arc length short in order to:

(a) Have better heating.

(b) Have better stirring action and reduce oxidation problem.

(c) Increase the life of roof refractory.

(d) All of the above.

(38) Usually arc furnaces are of:

(a) Cylindrical or conical shapes.

(b) Rectangular shape.

(c) Spherical shape.

(d) V-Shape.

(39) In submerged arc furnaces, the power is controlled by:

(a) Varying the spacing between the electrodes.

(b) Varying the voltage applied to the electrodes.

(c) Either (a) or (b).

(d) Varying the arc length.

(40) In induction heating:

(a) Heat is produced due to the currents induced in the charge by the electromagnetic action.

(b) The resistance of the charge must be low and the voltage applied must be high in order to produce sufficient heat.

(c) Magnetic materials can be easily treated in comparison to non-magnetic materials.

(d) All of the above.

(41) Induction heating takes place in:

(a) Insulating materials.

(b) Conducting and magnetic materials.

(c) Conducting but non-magnetic materials.

(d) Conducting materials may be magnetic or non-magnetic.

(42) Low-frequency supply is necessary for direct core type induction furnaces because:

(a) The magnetic coupling between the primary and secondary circuit is poor.

(b) With the normal frequency supply, the electromagnetic forces cause severe stirring action in the molten metal.

(c) Both (a) and (b).

(d) None of (a) and (b).

(43) In induction heating, which of the following is of high value?

(a) Frequency.

(b) Current.

(c) Voltage.

(d) Power factor.

(44) Induction furnaces are used for:

(a) The heat treatment of castings.

(b) The heating of insulators.

(c) The melting of aluminum.

(d) All of the above.

(45) In induction heating, the depth up to which the current will penetrate is proportional to:

(a) $1/(\text{Frequency})^{1/2}$.

(b) 1/Frequency.

(c) Frequency.

(d) $(\text{Frequency})^2$.

(46) The supply frequency usually employed for high-frequency eddy current heating is:

(a) 10 MHz.

(b) 10–400 KHz.

(c) 5 KHz.

(d) 1 KHz.

(47) In dielectric heating, the current flows through:

(a) Air.

(b) Dielectric.

(c) Metallic conductor.

(d) The ionic discharge between dielectric medium and metallic conductor.

(48) Dielectric loss is proportional to:

(a) Frequency.

(b) $(\text{Frequency})^2$.

(c) $(\text{Frequency})^3$.

(d) $(\text{Frequency})^{1/2}$.

(49) The dielectric loss in a dielectric is proportional to:

(a) The voltage impressed on the dielectric.

(b) The square of the voltage impressed on the dielectric.

(c) The square root of the voltage impressed on the dielectric.

(d) None of the above.

(50) For heating of plywood, the frequency should be:

(a) 1–2 MHz.

(b) 10–25 khz.

(c) 1 khz.

(d) 100 Hz.

(51) The power factor will be maximum in case of:

(a) Electric arc heating.

(b) Resistance heating.

(c) Induction heating.

(d) Dielectric heating.

(52) Which of the following methods of heating is independent of supply frequency?

(a) Electric heating.

(b) Induction heating.

(c) Electric resistance heating.

(d) Dielectric heating.

(53) The furnaces used for cremation are:

(a) Electric resistance heating.

(b) Electric arc heating.

(c) Dielectric heating.

(d) High-frequency eddy current heating.

(54) In an electric room heat convector, the method of heating used is:

(a) Arc heating.

(b) Resistance heating.

(c) Induction heating.

(d) Infrared heating.

(55) Hysteresis loss and eddy current loss are used in:

(a) Resistance heating.

(b) Dielectric heating.

(c) Induction heating of steel.

(d) Induction heating of brass.

(56) The power factor will be leading in case of:

(a) Dielectric heating.

(b) Induction heating.

(c) Electric arc heating.

(d) Resistance heating.

(57) The method of heating used for the non-conducting material is:

(a) Induction heating.

(b) Dielectric heating.

(c) Electric resistance heating.

(d) Electric arc heating.

(58) The method appropriate for the heating of non-ferrous metals is:

(a) Indirect resistance heating.

(b) Radiant heating.

(c) Indirect arc heating

(d) Dielectric heating.

(59) The method suitable for the heating of conducting medium is:

(a) Induction heating.

(b) Indirect arc heating.

(c) Eddy current heating.

(d) Resistance heating.

(60) The most modern method for the food processing is:

(a) Induction heating.

(b) Resistance heating.

(c) Dielectric heating.

(d) Eddy current heating.

REVIEW QUESTIONS

(1) Discuss the various modes of heat dissipation.

(2) What are the advantages of electric heating?

(3) What are the causes of the failure of heating element?

(4) What are the advantages and the disadvantages of direct and indirect arc furnaces?

(5) Explain the principle of high-frequency eddy current heating.

(6) Explain with a neat sketch the principle of Ajax–Wyatt induction furnace.

(7) Explain with a neat sketch the principle of core type induction furnaces.

(8) Explain the principle of arc heating.

(9) What are the applications of induction heating?

(10) Explain the principle of dielectric heating.

(11) What are the applications of dielectric heating?

(12) Compare high-frequency and power frequency coreless furnaces.

EXERCISE PROBLEMS

(1) A 30-kW, 220-V, and single-phase resistance oven employs nickel–chrome strip of 35-mm thick is used, for its heating elements. If the wire temperature is not exceed 1,300°C and the temperature of the charge is to be 600°C. Calculate its width and length of the wire. Assume the radiating efficiency as 0.75 and emissivity as 0.9.

(2) A 150-kW Ajax–Wyatt furnace works at a secondary voltage of 15 V at power factor 0.7 when fully charged. If the reactance presented by the charge remains constant but the resistance varies invert as the charge depth in the furnace, calculate the charge depth that produces the maximum heating effect when the furnace is fully charged.

(3) Determine also the temperature of the wire when the charge is cold. Calculate the time taken to melt 5 ton of steel in three-phase arc furnace having the following data.

Current = 7,000 A	Resistance = 0.002 Ω
Arc voltage = 45 V	Reactance = 0.004 Ω
Latent heat = 9 kcal/kg	Specific heat = 0.12
Initial temperature = 17°C	Melting point = 1,400°C

The overall efficiency is 60%. Find also the power factor and the electrical efficiency of the furnace.

(4) Determine the amount of energy required to melt 2 tons of zinc in 1 hr, if it operates at an efficiency of 60%, the specific heat of zinc is equals to 0.2. The latent heat of zinc = 25.67 kcal/kg, the melting point is 500° C, and the initial temperature is 35° C.

(5) A piece of insulating material is to be heated by dielectric heating. The size of the piece is 12 × 12 × 5 cm³. A frequency of 30 mega cycles is used and the power absorbed is 500 W. Determine the voltage necessary for heating and the current that flows in the material. The material has a permittivity of 6 and a power factor of 0.04.

(6) A piece of an insulating material 3-cm thick and 110 cm² in area is to be heated by dielectric heating. The material has a permittivity of 5 and power factor of 0.05. The power at 700 V is 400 W. Determine the cycles per second.

ANSWERS

1. d	12. c	23. c	34. b
2. a	13. d	24. a	35. b
3. b	14. a	25. c	36. c
4. c	15. b	26. d	37. d
5. a	16. d	27. a	38. a
6. b	17. a	28. b	39. c
7. c	18. b	29. c	40. d
8. c	19. c	30. d	41. d
9. a	20. c	31. d	42. c
10. b	21. d	32. a	43. a
11. d	22. b	33. a	44. a

45. a	49. b	53. a	57. b
46. b	50. a	54. b	58. c
47. b	51. b	55. c	59. a
48. a	52. c	56. a	60. c

Electric Welding

OBJECTIVES

After reading this chapter, you should be able to:

- understand the various methods of electrical welding
- know the importance of the choice of welding time
- discuss the different welding equipments
- study the salient features of the electrical welding equipments

5.1 INTRODUCTION

Welding is the process of joining two pieces of metal or non-metal together by heating them to their melting point. Filler metal may or may not be used to join two pieces. The physical and mechanical properties of a material to be welded such as melting temperature, density, thermal conductivity, and tensile strength take an important role in welding. Depending upon how the heat applied is created; we get different types of welding such as thermal welding, gas welding, and electric welding. Here in this chapter, we will discuss only about the electric welding and some introduction to other modern welding techniques. Welding is nowadays extensively used in automobile industry, pipe-line fabrication in thermal power plants, machine repair work, machine frames, etc.

5.2 ADVANTAGES AND DISADVANTAGES OF WELDING

Some of the advantages of welding are:

- Welding is the most economical method to permanently join two metal parts.
- It provides design flexibility.
- Welding equipment is not so costly.
- It joins all the commercial metals.
- Both similar and dissimilar metals can be joined by welding.
- Portable welding equipment are available.

Some of the disadvantages of welding are:

- Welding gives out harmful radiations and fumes.
- Welding needs internal inspection.
- If welding is not done carefully, it may result in the distortion of workpiece.
- Skilled welding is necessary to produce good welding.

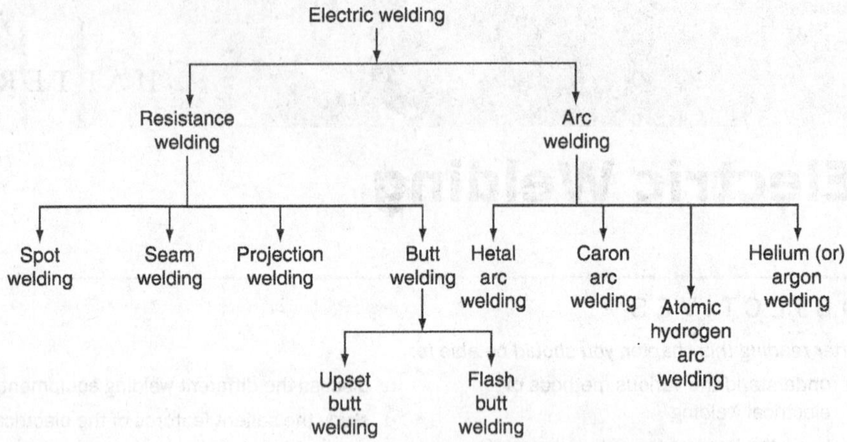

FIG. 5.1 *Classification of electric welding*

5.3 ELECTRIC WELDING

It is defined as the process of joining two metal pieces, in which the electrical energy is used to generate heat at the point of welding in order to melt the joint.

The classification of electric welding process is shown in Fig. 5.1.

The selection of proper welding process depends on the following factors.

- The type of metal to be joined.
- The techniques of welding adopted.
- The cost of equipment used.
- The nature of products to be fabricated.

5.4 RESISTANCE WELDING

Resistance welding is the process of joining two metals together by the heat produced due to the resistance offered to the flow of electric current at the junctions of two metals. The heat produced by the resistance to the flow of current is given by:

$$H = I^2Rt,$$

where I is the current through the electrodes, R is the contact resistance of the interface, and t is the time for which current flows.

Here, the total resistance offered to the flow of current is made up of:

(i) The resistance of current path in the work.

(ii) The resistance between the contact surfaces of the parts being welded.

(iii) The resistance between electrodes and the surface of parts being welded.

In this process of welding, the heat developed at the contact area between the pieces to be welded reduces the metal to plastic state or liquid state, then the pieces are pressed under high mechanical pressure to complete the weld. The electrical voltage input to the welding varies in between 4 and 12 V depending upon area, thickness, composition, etc. and usually power ranges from about 60 to 180 W for each sq. mm of area.

Any desired combination of voltage and current can be obtained by means of a suitable transformer in AC; hence, AC is found to be most suitable for the resistance welding. The magnitude of current is controlled by changing the primary voltage of the welding transformer, which can be done by using an auto-transformer or a tap-changing transformer. Automatic arrangements are provided to switch off the supply after a pre-determined time from applying the pressure, why because the duration of the current flow through the work is very important in the resistance welding.

The electrical circuit diagram for the resistance welding is shown in Fig. 5.2. This method of welding consists of a tap-changing transformer, a clamping device for holding the metal pieces, and some sort of mechanical arrangement for forcing the pieces to form a complete weld.

Advantages

- Welding process is rapid and simple.
- Localized heating is possible, if required.
- No need of using filler metal.
- Both similar and dissimilar metals can be welded.
- Comparatively lesser skill is required.
- Maintenance cost is less.
- It can be employed for mass production.

However, the resistance welding has got some drawbacks and they are:

- Initial cost is very high.
- High maintenance cost.
- The workpiece with heavier thickness cannot be welded, since it requires high input current.

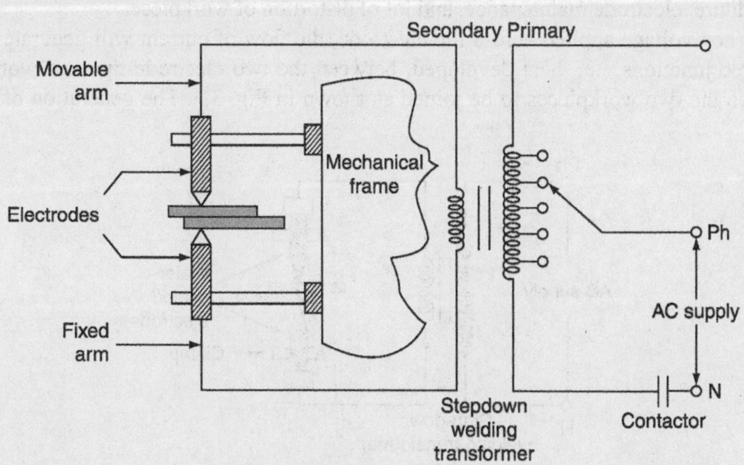

FIG. 5.2 *Electric circuit for resistance welding*

Applications

- It is used by many industries manufacturing products made up of thinner gauge metals.
- It is used for the manufacturing of tubes and smaller structural sections.

5.4.1 Types of resistance welding

Depending upon the method of weld obtained and the type of electrodes used, the resistance welding is classified as:

1. Spot welding.
2. Seam welding.
3. Projection welding.
4. Butt welding.

(i) *Spot welding*

Spot welding means the joining of two metal sheets and fusing them together between copper electrode tips at suitably spaced intervals by means of heavy electric current passed through the electrodes as shown in Fig. 5.3.

This type of joint formed by the spot welding provides mechanical strength and not air or water tight, for such welding it is necessary to localize the welding current and to apply sufficient pressure on the sheet to be welded. The electrodes are made up of copper or copper alloy and are water cooled. The welding current varies widely depending upon the thickness and composition of the plates. It varies from 1,000 to 10,000 A, and voltage between the electrodes is usually less than 2 V. The period of the flow of current varies widely depending upon the thickness of sheets to be joined. A step-down transformer is used to reduce a high-voltage and low-current supply to low-voltage and high-current supply required. Since the heat developed being proportional to the product of welding time and square of the current. Good weld can be obtained by low currents for longer duration and high currents for shorter duration; longer welding time usually produces stronger weld but it involves high energy expenditure, electrode maintenance, and lot of distortion of workpiece.

When voltage applied across the electrode, the flow of current will generate heat at the three junctions, i.e., heat developed, between the two electrode tips and workpiece, between the two workpieces to be joined as shown in Fig. 3.3. The generation of heat at

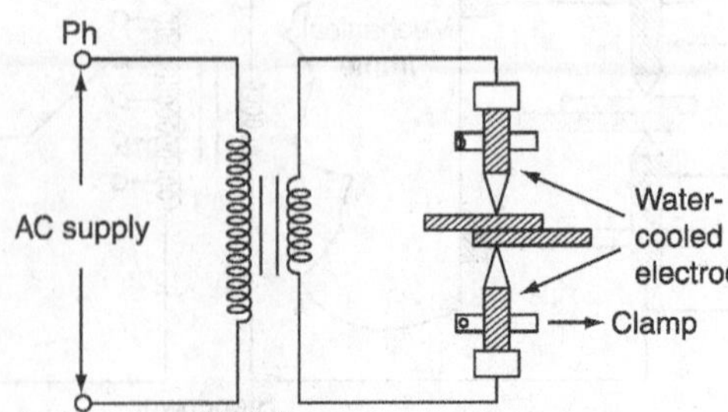

FIG. 5.3 *Spot welding*

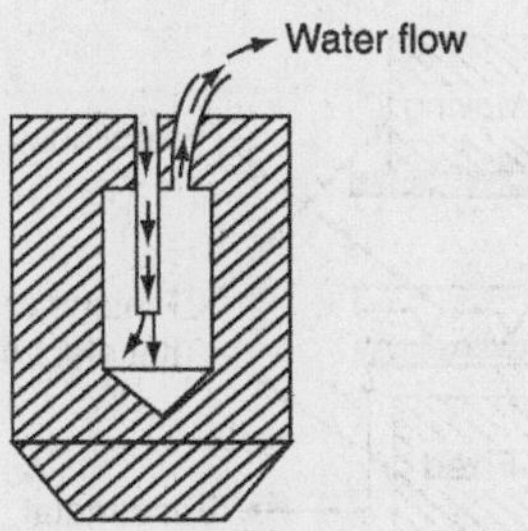

FIG. 5.4 *Water cooled electrode*

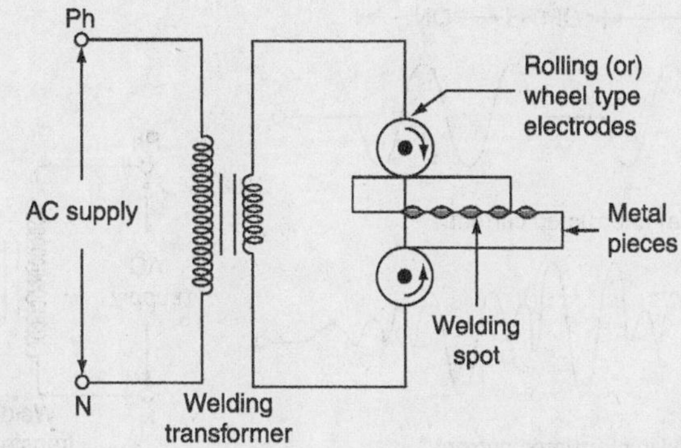

FIG. 5.5 *Seam welding*

junctions 1 and 3 will effect electrode sticking and melt through holes, the prevention of electrode striking is achieved by:

(i) Using water-cooled electrodes shown in Fig. 5.4. By avoiding the heating of junctions 1 and 3 electrodes in which cold water circulated continuously as shown in Fig. 5.3.

(ii) The material used for electrode should have high electrical and thermal conductivity. Spot welding is widely used for automatic welding process, for joining automobile parts, joining and fabricating sheet metal structure, etc.

(ii) *Seam welding*

Seam welding is nothing but the series of continuous spot welding. If number spots obtained by spot welding are placed very closely that they can overlap, it gives rise to seam welding.

In this welding, continuous spot welds can be formed by using wheel type or roller electrodes instead of tipped electrodes as shown in Fig. 5.5.

Seam welding is obtained by keeping the job under electrodes. When these wheel type electrodes travel over the metal pieces which are under pressure, the current passing between them heats the two metal pieces to the plastic state and results into continuous spot welds.

In this welding, the contact area of electrodes should be small, which will localize the current pressure to the welding point. After forming weld at one point, the weld so obtained can be cooled by splashing water over the job by using cooling jets.

In general, it is not satisfactory to make a continuous weld, for which the flow of continuous current build up high heat that causes burning and wrapping of the metal piece. To avoid this difficulty, an interrupter is provided on the circuit which turns on supply for a period sufficient to heat the welding point. The series of weld spots depends upon the number of welding current pulses.

The two forms of welding currents are shown in Fig. 5.6(a) and (b).

Welding cannot be made satisfactorily by using uninterrupted or un-modulated current, which builds up high heat as the welding progress; this will over heat the workpiece and cause distortion.

Seam welding is very important, as it provides leak proof joints. It is usually employed in welding of pressure tanks, transformers, condensers, evaporators, air craft tanks, refrigerators, varnish containers, etc.

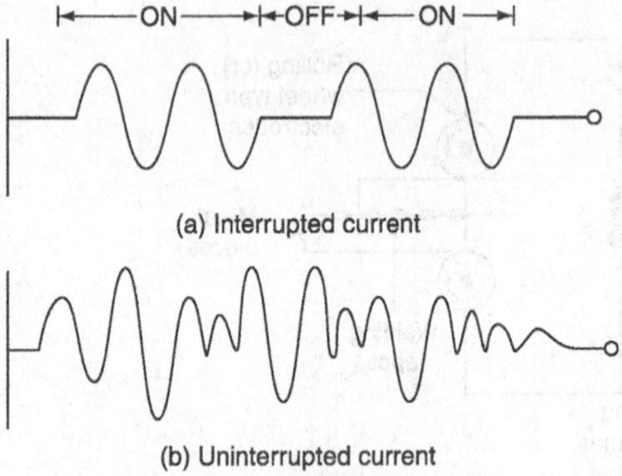

FIG. 5.6 *Welding current*

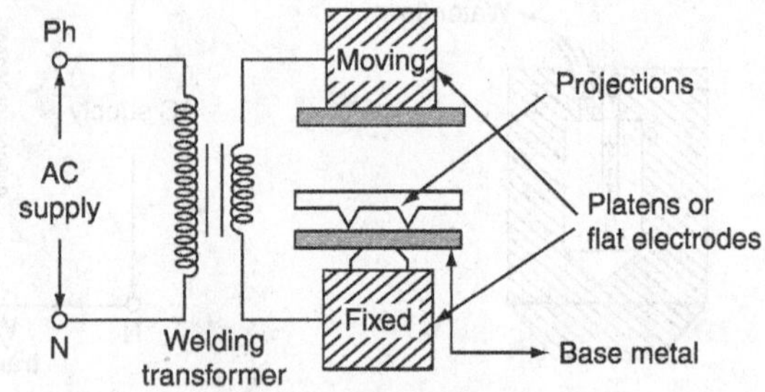

FIG. 5.7 *Projection welding*

(III) *Projection welding*

It is a modified form of the spot welding. In the projection welding, both current and pressure are localized to the welding points as in the spot welding. But the only difference in the projection welding is the high mechanical pressure applied on the metal pieces to be welded, after the formation of weld. The electrodes used for such welding are flat metal plates known as *platens*. The two pieces of base metal to be weld are held together in between the two platens, one is movable and the other is fixed, as shown in Fig. 5.7.

One of the two pieces of metal is run through a machine that makes the bumps or projections of required shape and size in the metal. As current flows through the two metal parts to be welded, which heat up and melt. These weld points soon reach the plastic state, and the projection touches the metal then force applied by the two flat electrodes forms the complete weld.

The projection welding needs no protective atmosphere as in the spot welding to produce successful results. This welding process reduces the amount of current and pressure in order to join two metal surfaces, so that there is less chance of distortion of the surrounding areas of the weld zone. Due to this reason, it has been incorporated into many manufacturing process.

The projection welding has the following advantages over the spot welding.

- Simplicity in welding process.
- It is easy to weld some of the parts where the spot welding is not possible.
- It is possible to join several welding points.
- Welds are located automatically by the position of projection.
- As the electrodes used in the projection welding are flat type, the contact area over the projection is sufficient.

This type of welding is usually employed on punched, formed, or stamped parts where the projection automatically exists. The projection welding is particularly employed for mass production work, i.e., welding of refrigerators, condensers, crossed wire welding, refrigerator racks, grills, etc.

(iv) *Butt welding*

Butt welding is similar to the spot welding; however, the only difference is, in butt welding, instead of electrodes the metal parts that are to be joined or butted together are connected to the supply.

The three basic types of the butt welding process are:

(a) Upset butt welding.

(b) Flash butt welding.

(c) Percussion butt welding.

(a) Upset butt welding

In upset welding, the two metal parts to be welded are joined end to end and are connected across the secondary of a welding transformer as shown in Fig. 5.8.

Due to the contact resistance of the metals to be welded, heating effect is generated in this welding. When current is made to flow through the two electrodes, heat will develop due to the contact resistance of the two pieces and then melts. By applying high mechanical pressure either manually or by toggle mechanism, the two metal pieces are pressed. When jaw-type electrodes are used that introduce the high currents without treating any hot spot on the job.

This type of welding is usually employed for welding of rods, pipes, and wires and for joining metal parts end to end.

(b) Flash butt welding

Flash butt welding is a combination of resistance, arc, and pressure welding. This method of welding is mainly used in the production welding. A simple flash butt welding arrangement is shown in Fig. 5.9.

In this method of welding, the two pieces to be welded are brought very nearer to each other under light mechanical pressure. These two pieces are placed in a conducting movable clamps. When high current is passed through the two metal pieces and they are separated by some distance, then arc established between them. This arc or flashing

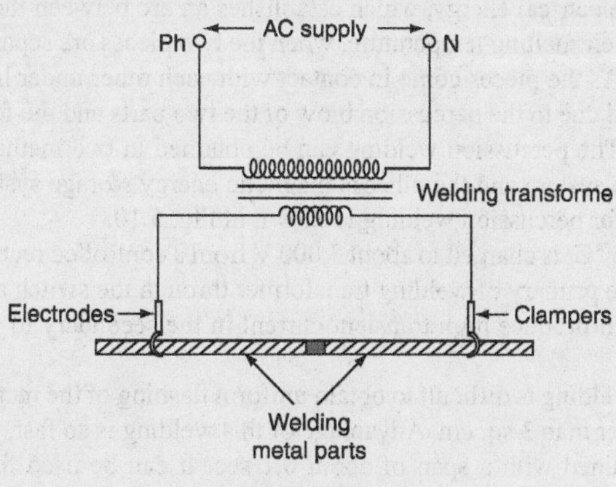

FIG. 5.8 *Upset butt welding*

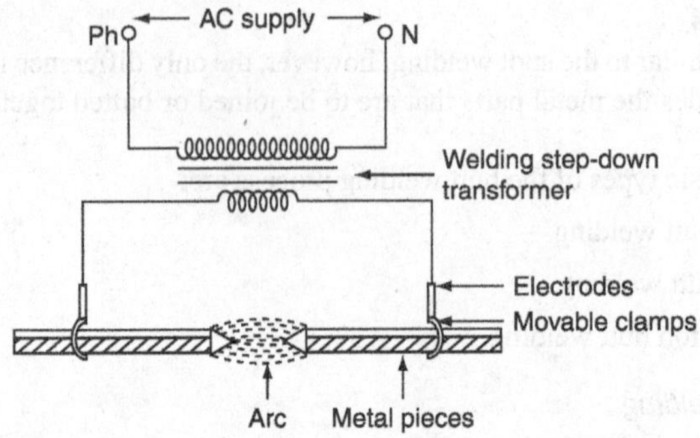

FIG. 5.9 *Flash butt welding*

is allowed till the ends of the workpieces reach melting temperature, the supply will be switched off and the pieces are rapidly brought together under light pressure. As the pieces are moved together, the fused metal and slag come out of the joint making a good solid joint.

Following are the advantages of the flash butt welding over the upset welding.

- Less requirement of power.
- When the surfaces being joined, it requires only less attention.
- Weld obtained is so clean and pure; due to the foreign metals appearing on the surfaces will burn due to flash or arc.

(c) Percussion welding

It is a form of the flash butt welding, where high current of short duration is employed using stored energy principle. This is a self-timing spot welding method.

Percussion welding arrangement consists of one fixed holder and the other one is movable. The pieces to be welded are held apart, with the help of two holders, when the movable clamp is released, it moves rapidly carrying the piece to be welded. There is a sudden discharge of electrical energy, which establishes an arc between the two surfaces and heating them to their melting temperature, when the two pieces are separated by a distance of 1.5 mm apart. As the pieces come in contact with each other under heavy pressure, the arc is extinguished due to the percussion blow of the two parts and the force between them affects the weld. The percussion welding can be obtained in two methods; one is capacitor energy storage system and the other is magnetic energy storage system. The capacitor discharge circuit for percussion welding is shown in Fig. 5.10.

The capacitor 'C' is charged to about 3,000 V from a controlled rectifier. The capacitor is connected to the primary of welding transformer through the switch and will discharge. This discharge will produce high transient current in the secondary to join the two metal pieces.

Percussion welding is difficult to obtain uniform flashing of the metal part areas of the cross-section grater than 3 sq. cm. Advantage of this welding is so fast, extremely shallow of heating is obtained with a span of about 0.1 sec. It can be used for welding a large number of dissimilar metals.

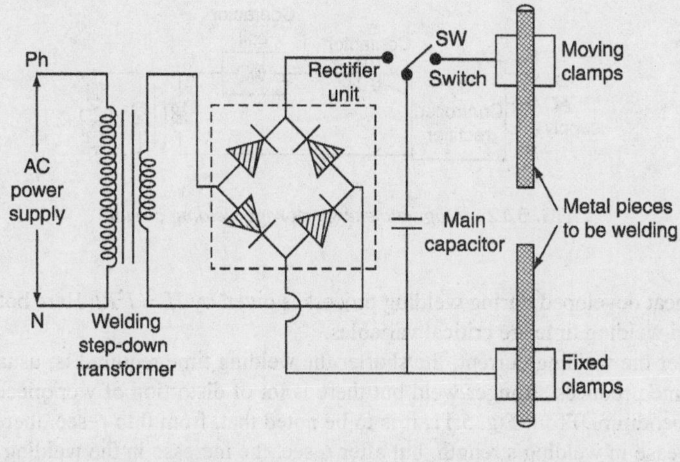

FIG. 5.10 *Capacitor discharge circuit for percussion welding*

Applications

- It is useful for welding satellite tips to tools, sliver contact tips to copper, cast iron to steel, etc.
- Commonly used for electrical contacts.
- The metals such as copper alloys, aluminum alloys, and nickel alloys are percussion welded.

5.5 CHOICE OF WELDING TIME

The successful welding operation mainly depends upon three factors and they are:

1. Welding time.
2. Welding current.
3. Welding pressure.

Figure 5.11 shows how the energy input to the welding process, welding strength, and welding current vary with welding time.

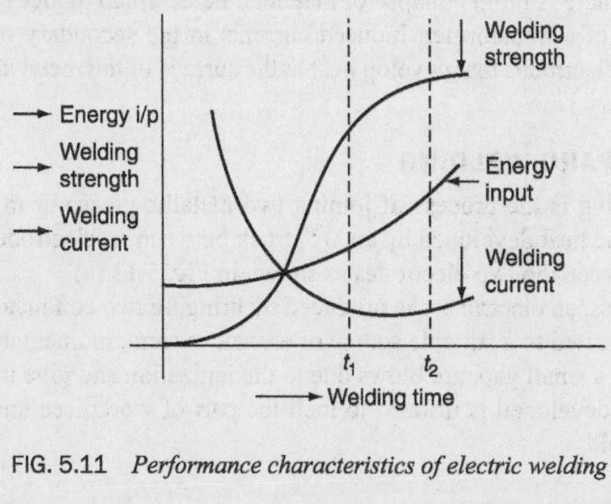

FIG. 5.11 *Performance characteristics of electric welding*

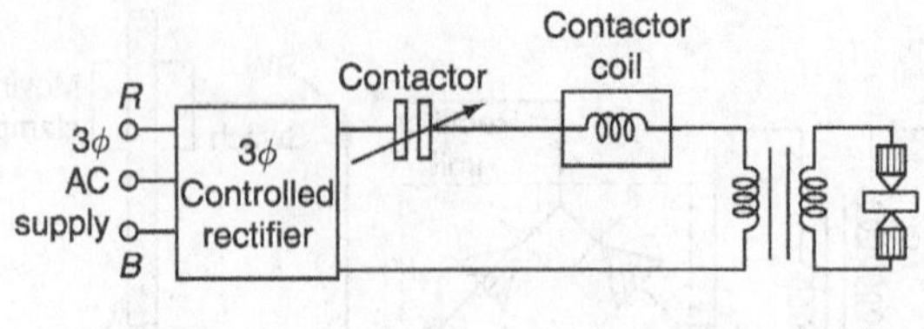

FIG. 5.12 *Magnetic energy storage welding circuit*

The heat developed during welding process is given by $H = I^2Rt$. Here both welding current and welding time are critical variables.

Greater the welding current, the shorter the welding time required is; usually longer welding time produces stronger weld but there is lot of distortion of workpiece and high energy expenditure. From Fig. 5.11, it is to be noted that, from 0 to t_1 sec, there is appreciable increase in welding strength, but after t_2 sec, the increase in the welding time does not appreciably result in the increase in strength; therefore, 't_2' is the optimum welding time. This optimum time varies with the thickness of the material. The optimum times of material (sheet steel) with different thickness are given as:

Dimensions of material	Optimum time
2 × 24 SWG	8 cycles
2 × 14 SWG	20 cycles
2¼"	2 sec

Therefore, from the above discussion, it is observed that shorter welding times with strength and economy are always preferable.

Electromagnetic storage welding circuit is shown in Fig. 5.12. In this type of welding, the energy stored in the magnetic circuit is used in the welding operation.

In this system, rectifier is fed from AC supply, which is converted to DC, the DC voltage of rectifier is controlled in such a way that, voltage induced in the primary without causing large current in the secondary of transformer on opening the contactor switch, DC on longer flows, there is rapid collapse of magnetic field, which induces very high current in the secondary of a transformer. Induced currents in the secondary of the transformer flow through the electrodes that develop heat at the surface of the metal and so forming the complete weld.

5.6 ELECTRIC ARC WELDING

Electric arc welding is the process of joining two metallic pieces or melting of metal is obtained due to the heat developed by an arc struck between an electrode and the metal to be welded or between the two electrodes as shown in Fig. 5.13 (a).

In this process, an electric arc is produced by bringing two conductors (electrode and metal piece) connected to a suitable source of electric current, momentarily in contact and then separated by a small gap, arc blows due to the ionization and give intense heat.

The heat so developed is utilized to melt the part of workpiece and filler metal and thus forms the weld.

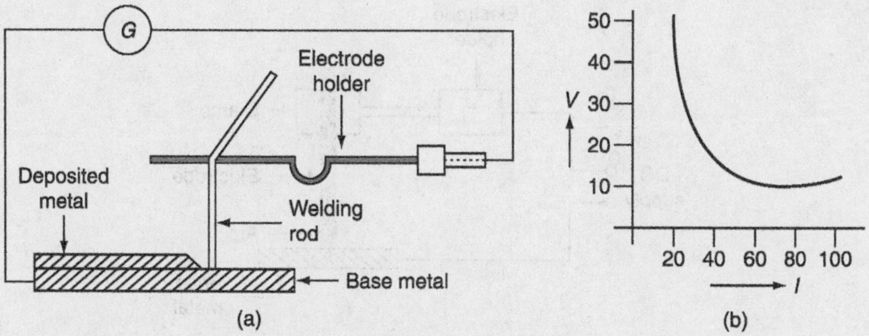

FIG. 5.13 *Arrangement of electric welding equipment*

In this method of welding, no mechanical pressure is employed; therefore, this type of welding is also known as '*non-pressure welding*'.

The length of the arc required for welding depends upon the following factors:

- The surface coating and the type of electrodes used.
- The position of welding.
- The amount of current used.

When the supply is given across the conductors separated by some distance apart, the air gap present between the two conductors gets ionized, as the arc welding is in progress, the ionization of the arc path and its surrounding area increases. This increase in ionization decreases the resistance of the path. Thus, current increases with the decrease in voltage of arc. This *V–I* characteristic of an arc is shown in Fig. 5.13(b), it also known as *negative resistance characteristics of an arc.* Thus, it will be seen that this decrease in resistance with increase in current does not remain the arc steadily. This difficulty cab be avoided, with the supply, it should fall rapidly with the increase in the current so that any further increase in the current is restricted.

For the arc welding, the temperature of the arc should be 3,500°C. At this temperature, mechanical pressure for melting is not required. Both AC and DC can be used in the arc welding. Usually 70–100 V on AC supply and 50–60 V on DC supply system is sufficient to struck the arc in the air gap between the electrodes. Once the arc is struck, 20–30 V is only required to maintain it.

However, in certain cases, there is any danger of electric shock to the operator, low voltage should be used for the welding purpose. Thus, DC arc welding of low voltage is generally preferred.

Electric arc welding is extensively used for the joining of metal parts, the repair of fractured casting, and the fillings by the deposition of new metal on base metal, etc.

Various types of electric arc welding are:

1. Carbon arc welding.
2. Metal arc welding.
3. Atomic hydrogen arc welding.
4. Inert gas metal arc welding.
5. Submerged arc welding.

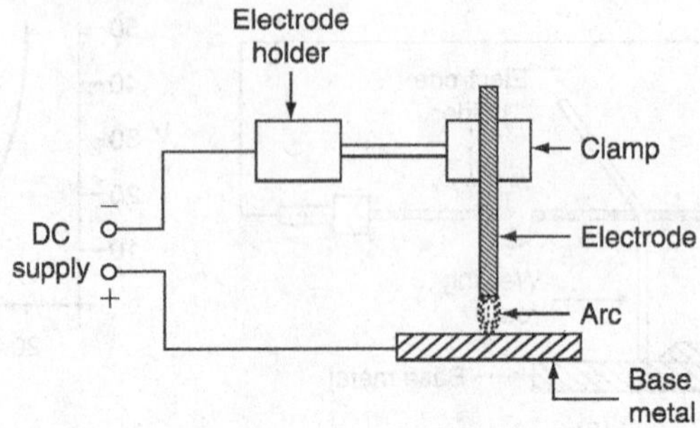

FIG. 5.14 *Carbon arc welding*

5.6.1 Carbon arc welding

It is one of the processes of arc welding in which arc is struck between two carbon electrodes or the carbon electrode and the base metal. The simple arrangement of the carbon arc welding is shown in Fig. 5.14.

In this process of welding, the electrodes are placed in an electrode holder used as negative electrode and the base metal being welded as positive. Unless, the electrode is negative relative to the work, due to high temperature, there is a tendency of the particles of carbon will fuse and mix up with the base metal, which causes brittleness; DC is preferred for carbon arc welding since there is no fixed polarity maintained in case of AC.

In the carbon arc welding, carbon or graphite rods are used as electrode. Due to longer life and low resistance, graphite electrodes are used, and thus capable of conducting more current. The arc produced between electrode and base metal; heat the metal to the melting temperature, on the negative electrode is 3,200°C and on the positive electrode is 3,900°C.

This process of welding is normally employed where addition of filler metal is not required. The carbon arc is easy to maintain, and also the length of the arc can be easily varied. One major problem with carbon arc is its instability which can be overcome by using an inductor in the electrode of 2.5-cm diameter and with the current of about of 500–800 A employed to deposit large amount of filler metal on the base metal.

Filler metal and flux may not be used depending upon the type of joint and material to be welded.

Advantages

- The heat developed during the welding can be easily controlled by adjusting the length of the arc.
- It is quite clean, simple, and less expensive when compared to other welding process.
- Easily adoptable for automation.
- Both the ferrous and the non-ferrous metals can be welded.

Disadvantages

- Input current required in this welding, for the workpiece to rise its temperature to melting/welding temperature, is approximately double the metal arc welding.
- In case of the ferrous metal, there is a chance of disintegrating the carbon at high temperature and transfer to the weld, which causes harder weld deposit and brittlement.
- A separate filler rod has to be used if any filler metal is required.

Applications

- It can be employed for the welding of stainless steel with thinner gauges.
- Useful for the welding of thin high-grade nickel alloys and for galvanized sheets using copper silicon manganese alloy filler metal.

5.6.2 Metal arc welding

In metal arc welding, the electrodes used must be of the same metal as that of the workpiece to be welded. The electrode itself forms the filler metal. An electric arc is stuck by bringing the electrode connected to a suitable source of electric current, momentarily in contract with the workpieces to be welded and withdrawn apart. The circuit diagram for the metal arc welding is shown in Fig. 5.15.

The arc produced between the workpiece and the electrode results high temperature of the order of about 2,400°C at negative metal electrode and 2,600°C at positive base metal or workpiece.

This high temperature of the arc melts the metal as well as the tip of the electrode, then the electrode melts and deposited over the surface of the workpiece, forms complete weld.

Both AC and DC can be used for the metal arc welding. The voltage required for the DC metal arc welding is about 50–60 V and for the AC metal arc welding is about 80–90 V.

In order to maintain the voltage drop across the arc less than 13 V, the arc length should be kept as small as possible, otherwise the weld will be brittle. The current required for the welding varies from 10 to 500 A depending upon the type of work to be welded.

The main disadvantage in the DC metal arc welding is the presence of arc blow, i.e., distortion of arc stream from the intended path due to the magnetic forces of the non-uniform magnetic field with AC arc blow is considerably reduced. For obtaining good weld, the flux-coated electrodes must be used, so the metal which is melted is covered with slag produces a non-oxidizing gas or a molten slag to cover the weld, and also stabilizes the arc.

5.6.3 Atomic hydrogen arc welding

In atomic hydrogen arc welding, shown in Fig. 5.16, the heat for the welding process is produced from an electric arc struck between two tungsten electrodes in an atmosphere of hydrogen. Here, hydrogen serves mainly two functions; one acts as a protective screen for

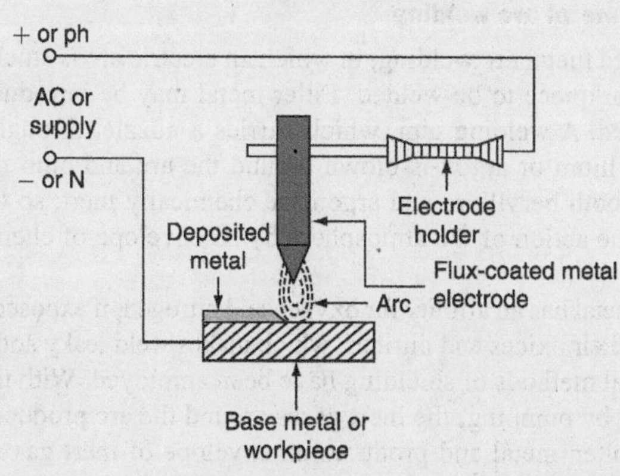

FIG. 5.15 *Metal arc welding*

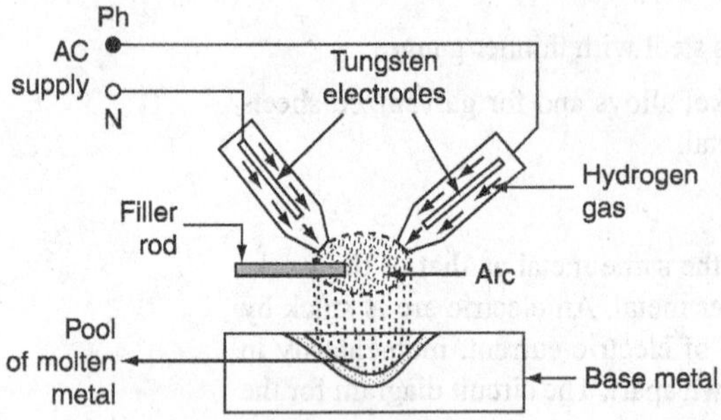

FIG. 5.16 *Atomic hydrogen arc welding*

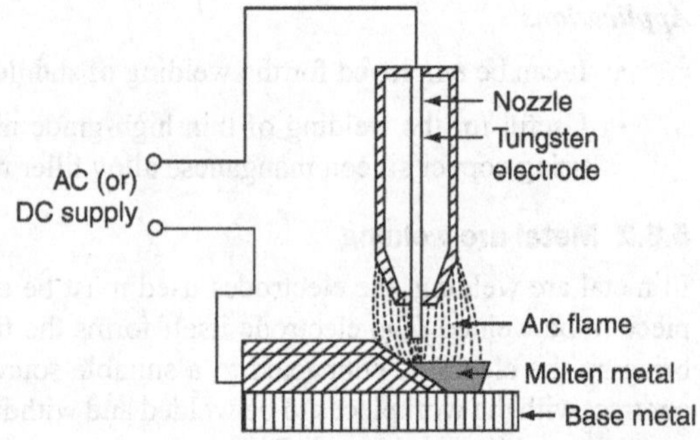

FIG. 5.17 *Inert gas metal are welding*

the arc and the other acts as a cooling agent for the glowing tungsten electrode tips. As the hydrogen gas passes through the arc, the hydrogen molecules are broken up into atoms, absorbs heat from the glowing tungsten electrodes so that these are cooled.

But, when the atoms of hydrogen recombine into molecules outside the arc, a large amount of heat is liberated. This extraheat is added to the intense heat of arc, which produces a temperature of about 4,000°C that is sufficient to melt the surfaces to be welded, together with the filler rod if used. Moreover hydrogen includes oxygen and some other gases that might combine with the molten metal and forms oxides and other impurities. Hydrogen also removes oxides from the surface of workpiece. Thus, this process is capable of producing strong, uniform, smooth, and ductile welds.

In the atomic hydrogen arc welding, the arc is maintained between the two non-consumable tungsten electrodes under a pressure of about 0.5 kg/cm^2. In order to obtain equal consumption of electrodes, AC supply is used. Arc currents up to 150 A can be used. High voltage about 300 V is applied for this welding through a transformer. For striking the arc between the electrodes the open circuit voltage required varies from 80 to 100 V.

As the atomic hydrogen welding is too expensive, it is usually employed for welding alloy steel, carbon steel, stainless steel, aluminum, etc.

5.6.4 Inert gas metal arc welding

It is a gas-shielded metal arc welding, in which an electric arc is stuck between tungsten electrode and workpiece to be welded. Filler metal may be introduced separately into the arc if required. A welding gun, which carries a nozzle, through this nozzle, inert gas such as beryllium or argon is blown around the arc and onto the weld, as shown in Fig. 5.17. As both beryllium and argon are chemically inert, so the molten metal is protected from the action of the atmosphere by an envelope of chemically reducing or inert gas.

As molten metal has an affinity for oxygen and nitrogen, if exposed to the atmosphere, thereby forming their oxides and nitrides, which makes weld leaky and brittle.

Thus, several methods of shielding have been employed. With the use of flux coating electrodes or by pumping, the inert gases around the arc produces a slag that floats on the top of molten metal and produces an envelope of inert gas around the arc and the weld.

Advantages

- Flux is not required since inert gas envelope protects the molten metal without forming oxides and nitrates so the weld is smooth, uniform, and ductile.
- Distortion of the work is minimum because the concentration of heat is possible.

Applications

- The welding is employed for light alloys, stainless steel, etc.
- The welding of non-ferrous metal such as copper, aluminum, etc.

5.7 SUBMERGED ARC WELDING

It is an arc welding process, in which the arc column is established between above metal electrode and the workpiece. Electric arc and molten pool are shielded by blanket of granular flux on the workpiece. Initially to start an arc, short circuit path is provided by introducing steel wool between the welding electrode and the workpiece. This is due to the coated flux material, when cold it is non-conductor of the electricity but in molten state, it is highly conductive. Welding zone is shielded by a blanket of flux, so that the arc is not visible. Hence, it is known as '*submerged arc welding*'. The arc so produced, melts the electrode, parent the metal and the coated flux, which forms a protective envelope around both the arc and the molten metal.

As the arc in progress, the melted electrode metal forms globules and mix up with the molten base metal, so that the weld is completed. In this welding, the electrode is completely covered by flux. The flux may be made of silica, metal oxides, and other compounds fused together and then crushed to proper size. Therefore, the welding takes place without spark, smoke, ash, etc. Thus, there is no need of providing protective shields, smoke collectors, and ventilating systems. Figure 5.18 shows the filling of parent metal by the submerged arc welding.

Voltage required for the submerged arc welding varies from 25 to 40 V. Current employed for welding depends upon the dimensions of the workpiece. Normally, if DC supply is used employing current ranging from 600 to 1,000 A, the current for AC is usually 2,000 A.

Advantages

- Deep penetration with high-quality weld is possible.
- Job with heavy thickness can be welded.
- The weld so obtained has good ductility, impact strength, high corrosion resistance, etc.
- The submerged arc welding can be done manually or automatically.

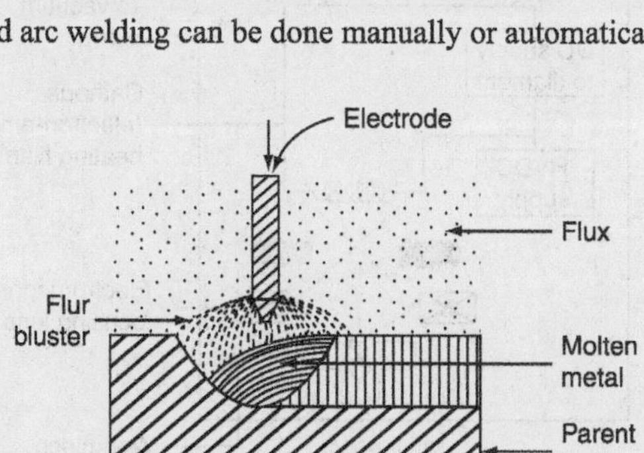

FIG. 5.18 *Submerged arc welding*

Applications

- The submerged arc welding is widely used in the heavy steel plant fabrication work.
- It can be employed for welding high strength steel, corrosion resistance steel, and low carbon steel.
- It is also used in the ship-building industry for splicing and fabricating subassemblies, manufacture of vessels, tanks, etc.

5.8 ELECTRON BEAM WELDING

It is one of the processes of the electric welding, in which the heat required for carrying out the welding operation is obtained by the electron bombardment heating.

In the electron bombardment heating, continuous stream of electron is produced between the electron emitting material cathode and the material to be heated. The electrons released from cathode possess KE traveling with high velocity in vacuum of 10^{-3}–10^{-5} mmHg. When the fast moving electrons hit, the material or workpiece releases their KE as heat in the material to be heated. This heat is utilized to melt the metal.

If this process is carried out in high vacuum, without providing any electrodes, gasses, or filler metal, pure weld can be obtained. Moreover, high vacuum is maintained around the (filament) cathode. So that, it will not burn up and also produces continuous stable beam. If a vacuum was not used, the electron would strike the small partials in the atmosphere, reducing their velocity and also the heating ability. Thus, the operation should be performed in vacuum to present the reduction of the velocity of electron. That's why this is also called as '*vacuum electron beam welding*'. The power released by the electron beam is given by:

$$P = nqv \text{ watts,}$$

where n is the number of charged particles, q is the charge in coulombs per meter, and v is the voltage required to accelerate the electrum from rest.

The electron beam welding (Fig. 5.19) process requires electron-emitting heating filament as cathode, focusing lens, etc.

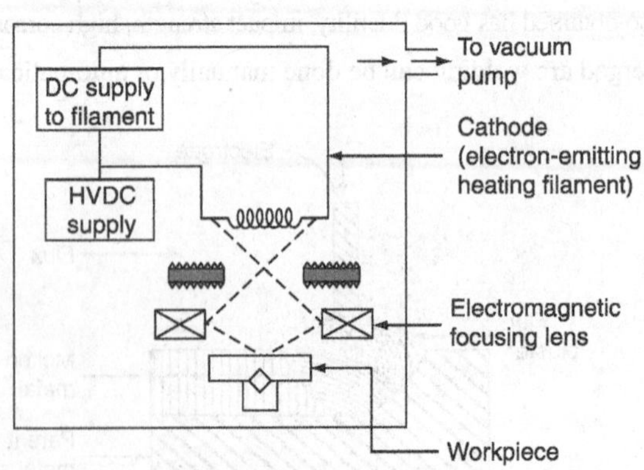

FIG. 5.19 *Electron beam welding*

Advantages

- Heat input to the electron beam welding can be easily controlled by varying beam current, voltage, the position of filament, etc.
- The electron beam welding can be used to join high temperature metals such as columbium.
- It can be employed for the welding of thick sections, due to high penetration to width ratio.
- It eliminates contamination of both weld zone and weld metal.
- Narrow electron beam reduces the distortion of workpiece.

Disadvantages

- The pressure build up in the vacuum chamber due to the vapor of parent metal causes electrical break down.
- Most of the super alloys, refractory metals, and combinations of dissimilar metals can also be welded.

5.9 LASER BEAM WELDING

The word laser means '*light amplification stimulated emission of radiation*'. It is the process of joining the metal pieces by focusing a monochromatic light into the extremely concentrated beams, onto the weld zone.

This process is used without shielding gas and without the application of pressure. The laser beam is very intense and unidirectional but can be focused and refracted in the same way as an ordinary light beam. The focus of the laser beam can be controlled by controlling the lenses, mirrors, and the distance to the workpiece. A block diagram of the laser beam welding system is shown in Fig. 5.20.

In laser beam welding system, flash tube is designed to give thousands of flashes per second. When capacitor bank is triggered, the electrical energy is injected into the flash tube through trigger wire. Flash tube consists of thick xenon material, which produces

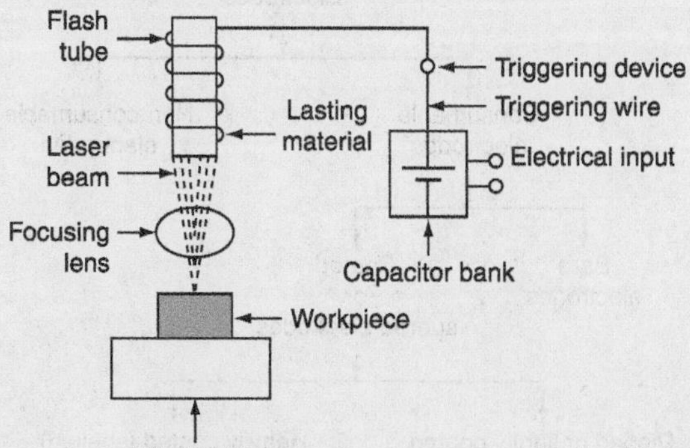

FIG. 5.20 *Laser beam welding*

high power levels for very short period. If the bulb is operated in this manner, it becomes an efficient device, which converts electrical energy to light energy. The laser is then activated.

The laser beam emitting from the flash tube, passing through the focusing lens, where it is pinpointed on the workpiece. The heat so developed by the laser beam melts the work-piece and the weld is completed. The welding characteristics of the laser are similar to the electron beam.

The laser beam has been used to weld carbon steel, low-alloy steel, aluminum, etc. The metals with relatively high-electrical resistance and the parts of different sizes and mass can be welded.

5.10 TYPES OF WELDING ELECTRODES

An electrode is a piece of metal in the form of wire or rod that is either bare or coated uniformly with flux. Electrode carries current for the welding operation. One contact end of the electrode must be clean and is inserted into the electrode holder, an arc is set up at the other end.

The electrodes used for the arc welding are classified as follows (Fig. 5.21).

5.10.1 Non-consumable electrodes

Electrodes, which do not consume or fuse during the welding process, are called non-consumable electrodes.
Ex: Electrodes made up of carbon, graphite, or tungsten do not consume during welding.

5.10.2 Consumable electrodes

Electrodes, which are consumed during the welding operation, are consumable electrodes. These are made up of various materials depending upon their purpose and the chemical composition of metal to be welded.

The consumable electrodes are made in the form of rod having diameter of about 2–8 mm and length of about 200–500 mm. They act as filler rod and are consumed during welding operation.

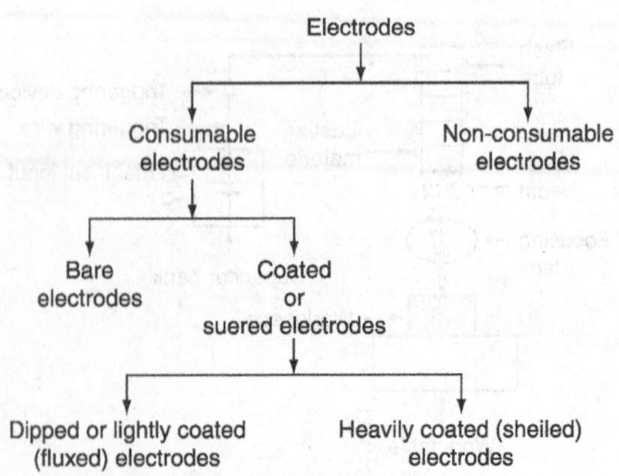

FIG. 5.21 *Classification of electrods*

5.10.2.1 Bare electrodes

These are the consumable electrodes, which are not coated with any fluxing material. Bare electrodes are in the form of wire. During welding operation, an arc is struck between the workpiece and the electrode wire, then the electrode is melted down into the weld.

When the molten metal electrode and the workpiece are exposed to the atmosphere of oxygen and nitrogen, they form their oxides and nitrides and cause the formation of some non-metallic constituent, which reduces the strength and ductility of the deposited weld. The bare electrodes are usually employed in automatic and semiautomatic welding. With bare electrode, the welding can be done satisfactorily with DC supply only if the electrode should be connected to the negative terminal of the supply.

5.10.2.2 Coated electrodes

Depending upon the thickness of flux coating, the coated electrode may classified into:

(i) lightly coated electrodes and

(ii) heavily coated electrodes.

For obtaining good weld, the coated electrodes are always preferred.

(i) Lightly coated electrodes

These electrodes are coated with thin layer of coating material up to less than 1 mm. This coating is usually consists of lime mixed with soluble glass which serves as a binder. These electrodes are considered as improvement over bare electrodes.

The main purpose of using the light coating layer on the electrode is to increase the arc stability, so they are also called as stabilizing electrodes. The mechanical strength of the weld increased because slag layer will not formed on the molten weld. For this reason, lightly coated electrodes may only be used for welding non-essential workpieces.

(ii) Heavily coated electrodes

These electrodes have coating layer with heavy thickness. The heavily coated electrodes sometimes referred to as the shielded arc electrodes. The materials commonly used for coating the electrodes are titanium oxide, ferromanganese, silica, flour, asbestos clay, calcium carbonate, etc. This electrode coating helps in improving the quality of weld, as if the coating layer of the electrodes burns in the heat of the arc provides gaseous shield around the arc, which prevents the formation oxides and nitrites.

Advantages

- Arc is stabilized due to the flux compounds of sodium and potassium.
- The weld metal can be protected from the oxidizing action of oxygen and the nitrifying action of nitrogen due to the gas shielded envelope.
- The impurities present on the surface being welded are fluxed away.
- The electrode coating increases deposition efficiency and weld metal deposition rate through iron powder and ferro alloy addition.
- In case of AC supply arc cools at zero current and there is a tendency of deionizing the arc path. Covering gases keep the arc space ionized.
- The welding operation becomes faster due to the increased melting rate.
- The coated electrodes help to deoxidize and refine the weld metal.

The type of electrode used for the welding process depends upon the following factors.

- The nature of the electric supply, either AC or DC.
- The type of the metal to be welded.
- The welding position.
- The polarity of the welding machine.

5.11 COMPARISON BETWEEN RESISTANCE AND ARC WELDINGS

	Resistance welding	*Arc welding*
1	The source of supply is AC only.	The source of supply is either AC (1-ϕ or 3-ϕ) or DC.
2	The head developed is mainly due to the flow of contact resistance.	The heat developed is mainly due to the striking of arc between electrodes or an electrode and the workpiece.
3	The temperature attained by the workpiece is not so high.	The temperature of the arc is so high, so proper care should be taken during the welding.
4	External pressure is required.	No external pressure is required hence the welding equipment is more simple and easy to control.
5	Filler metal is not required to join two metal pieces.	Suitable filler electrodes are necessary to get proper welding strength.
6	It cannot be used for repair work; it is suitable for mass production.	It is not suitable for mass production. It is most suitable for repair works and where more metal is to be deposited.
7	The power consumption is low.	The power consumption is high.
8	The operating power factor is low.	The operating power factor is high.
9	Bar, roller, or flat type electrodes are used (not consumable).	Bare or coated electrodes are used (consumable or non-consumable).

5.12 ELECTRIC WELDING EQUIPMENT

Electric welding accessories required to carry out proper welding operation are:

i. Electric welding power sets.
ii. Electrode holder to hold the electrodes.
iii. Welding cable for connecting electrode and workpiece to the supply.
iv. Face screen with colored glass.
v. Chipping hammers to remove slag from molten weld.
vi. Wire brush to clean the weld.
vii. Earth clamp and protective clothing.

5.12.1 Electric welding power sets

Welding power sets may be of different types and they can be selected depending upon the nature of available power supply (either DC or 1-ϕ AC). Sometimes diesel driven engine may be used under the absence of power supply, initial and running costs, the location of operation, required output and the type of work, and based on the available floor space.

Based on the nature of available supply, commonly used welding sets are:

(i) DC welding sets and

(ii) AC welding sets.

(i) *DC welding sets*

Commonly used DC welding sets are:

(a) welding generator and

(b) rectifier.

(a) Welding generator

A DC generator is driven by a primer mover (electric motor or diesel engine) which produces DC current in either or reversed polarity. The current supplied by DC generator is alternating that can be converted to direct quantity by the use of a commutator. The differential compound DC generator is used as a welding generator as shown in Fig. 5.22, since it has drooping volt–amp characteristics. As the load current increases, the net flux due to the series and the shunt fields in opposition decrease and hence the generated EMF also decreases.

This drooping characteristics is important in view of arc stability and this steep characteristics of the differential compound generator is shown in Fig. 5.23.

(b) Rectifier set

Rectifier set is a thyristor control electronic circuit or device, converts AC to DC supply. The power supply to the rectifier set is fed through a transformer. The rectifier consists of silicon diodes or metal plates coated with selenium compound. This allows unidirectional current. It is a fast controlled device and is reliable, but the cost of the set is too high.

The rectifier type welders are said to combine some of the desirable arcing characteristics of the DC welding. Such as easy arc starting, with those of welding transformers such as reduced no losses. In this case, the DC voltage can be controlled by regulating the transformer output.

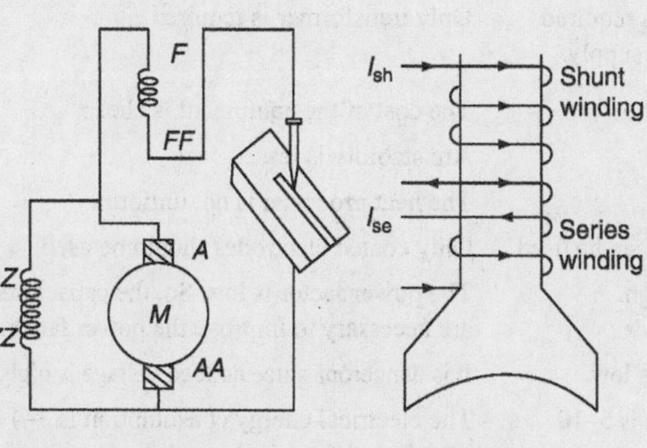

FIG. 5.22 *Welding generator*

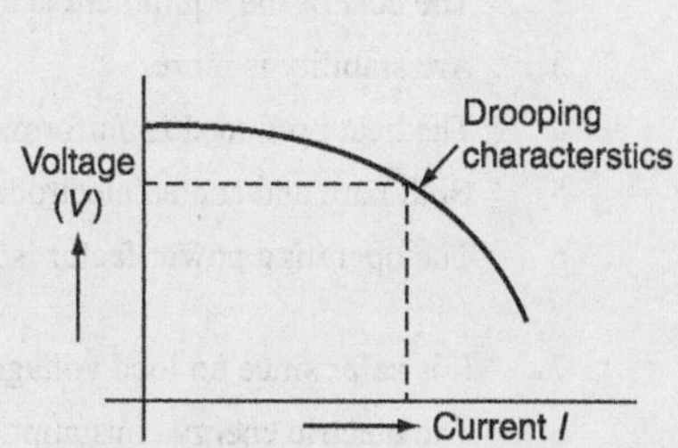

FIG. 5.23 *Drooping characteristics*

(II) *AC welding set*

Commonly used AC welding sets are single-phase or three-phase step-down transformers under the availability of AC supply. These transformers provide low voltage of the order of 80–100 V on open circuit for the welding operation. The welding transformers may be of air- or oil-cooled types, since AC supply passes through zero twice for every cycle, at which the welding arc would extinguish twice, which can be prevented by using coated electrodes, produce more complete ionization in the arc stream even though current passes through zero.

In the AC welding, i.e., the transformer welding set, the current control is achieved by using (a) magnetic shunt and (b) a choke coil or reactor connected in series with primary and secondary winding. The transformers are mostly used for the flux-shielded metal arc welding, the production welding on heavy gauge steel, some industrial welding operations, etc.

5.12.2 Electrode holder

It is a device used to insert or hold the electrodes for carrying out the welding operation. Electrode jaws used to hold the electrode in holder must be completely insulated against thermal and electric shocks. Electric holders must be mechanically strong.

5.12.3 Welding cables

Welding cables are conductors to carry current throughout the welding operation. Two electrode cables are necessary to connect the electrode and workpiece to the welding power source. These cables must be insulated with rubber, and needs periodic inspection for proper welding operation.

5.12.4 Chipping hammer and wire brush

A chipping hammer is a chisel-shaped device that is used to remove the slag formed over the molten weld.

A wire brush is made up of stiff steel wire, surrounded by wood layer that removes the remaining slag articles and clean the weld after chipping hammer has done its job.

Protective clothing (apron, gloves, etc.) are necessary to protect welder from the hot spattering particles, against the thermal shocks, etc.

5.13 COMPARISON BETWEEN AC AND DC WELDINGS

	AC welding	*DC welding*
1	Motor generator set or rectifier is required in case of the availability of AC supply.	Only transformer is required.
2	The cost of the equipment is high.	The cost of the equipment is cheap.
3	Arc stability is more.	Arc stability is less.
4	The heat produced is uniform.	The heat produced is not uniform.
5	Both bare and coated electrodes can be used.	Only coated electrodes should be used.
6	The operating power factor is high.	The power factor is low. So, the capacitors are necessary to improve the power factor.
7	It is safer since no load voltage is low.	It is dangerous since no load voltage is high.
8	The electric energy consumption is 5–10 kWh/kg of deposited metal.	The electrical energy consumption is 3–4 kWh/kg of deposited metal

9	Arc blow occurs due to the presence of non-uniform magnetic field.	Arc blow will not occur due to the uniform magnetic field.
10	The efficiency is low due to the rotating parts.	The efficiency is high due to the absence of rotating parts.

KEY NOTES

- *Electric welding* is the process of joining two metal pieces, in which electrical energy is used to generate heat at the point of welding in order to melt the joint.
- *Resistance welding* is the process of joining two metals together by the heat produced due to the resistance offered to the flow of electric current at the junctions of two metals.
- Resistance welding are:
  - (i) Spot welding.
  - (ii) Seam welding.
  - (iii) Projection welding.
  - (iv) Butt welding.
- *Spot welding* is the process of joining of two metal sheets and fusing them together between copper electrode tips at suitably spaced intervals by means of heavy electric current passed through electrodes.
- *Seam welding* is nothing but series of continuous spot welding. If number spots obtained by spot welding are placed very closely, they can overlap.
- *Projection welding* is the modified form of spot welding, in which two metallic parts are joined together with high mechanical pressure, which is applied on the metal pieces to be welded.
- *Butt welding* is similar to the spot welding; the only difference is, in butt welding instead of electrodes, the metal parts, which are to be joined or butted together, are connected to the supply.
- A successful welding operation mainly depends upon:
  - (i) welding time,
  - (ii) welding current, and
  - (iii) welding pressure.
- *Electric arc welding* is the process of joining two metallic pieces. It is obtained by the heat developed by an arc struck between two electrodes. It is also known as '*non-pressure welding*'.
- *Various types of electric arc welding are:*
  - (i) carbon arc welding,
  - (ii) metal arc welding,
  - (iii) atomic hydrogen arc welding,
  - (iv) inert gas metal arc welding, and
  - (v) submerged arc welding.
- *Electron beam welding* is the process of electric welding, in which heat required for carrying out the welding operation is obtained by electron bombardment heating. It is also known as '*vacuum electron beam welding*'.
- The word laser means '*light amplification stimulated emission of radiation*'. It is the process of joining the metal pieces by focusing a monochromatic light into extremely concentrated beams, onto the weld zone.

SHORT QUESTIONS AND ANSWERS

(1) What is meant by electrical welding?

It is defined as the process of joining two metal pieces, in which the electrical energy is used to generate the heat at the point of welding in order to melt the joint.

(2) What are the advantages of electrical welding?

The advantages of electric welding are:

(i) Welding is the most economical method to permanently join two metal parts.

(ii) It provides design flexibility.

(iii) The welding equipment is not so costly.

(3) List out the disadvantages of electric welding.

The disadvantages of electric welding are:

(i) The welding gives out harmful radiations and fumes.

(ii) The welding needs internal inspection.

(iii) If the welding is not done carefully, it may result in the distortion of workpiece.

(iv) Skilled welding is necessary to produce good welding.

(4) List out the applications of the electrical welding.

The welding is nowadays extensively used in the automobile industry, pipe-line fabrication in thermal power plants, machine repair work, machine frames, etc.

(5) List out the factors based on which the electric welding can be carried out properly.

The selection of proper welding process depends on the following factors.

(i) The type of metal to be joined.

(ii) The techniques of welding adopted.

(iii) The cost of equipment used.

(iv) The nature of products to be fabricated.

(6) What is resistance welding?

Resistance welding is the process of joining two metals together by the heat produced due to the resistance offered to the flow of electric current at the junctions of two metals.

(7) What are the types of resistance welding?

Depending upon method of weld obtained and type of electrodes used, the resistance welding classified as:

(i) spot welding,

(ii) projection welding,

(iii) seam welding, and

(iv) butt welding.

(8) What is meant by spot welding?

The process of joining of two metal sheets and fusing them together between copper electrode tips at suitably spaced intervals by means of the heavy electric current passed through the electrodes is known as the spot welding.

(9) What is meant by seam welding?

The seam welding is nothing but series of continuous spot welding. If number spots obtained by spot welding are placed very closely that they can overlap, it gives rise to seam welding.

(10) What are the types of butt welding?

The basic types of butt welding process are:

(i) upset butt welding,

(ii) flash butt welding, and

(iii) percussion butt welding.

(11) List out the factors which effect the successful welding operation.

A successful welding operation mainly depends upon three factors and they are:

(1) welding time,

(2) welding current, and

(3) welding pressure.

(12) What are the various types of electric arc welding?

The various types of electric arc welding are:

(i) carbon arc welding,

(ii) metal arc welding,

(iii) atomic hydrogen arc welding,

(iv) inert gas metal arc welding, and

(v) submerged arc welding.

(13) What is meant by electron beam welding?

It is one of the processes of the electric welding, in which the heat required for carrying out welding operation is obtained by the electron bombardment heating.

(14) What is meant by laser beam welding?

The word laser means '*light amplification stimulated emission of radiation*'. The laser beam welding is the process of joining the metal pieces by focusing a monochromatic light into extremely concentrated beams onto the weld zone.

(15) What is the fundamental difference between the electric arc welding and the resistance welding?

The resistance welding processes differ from arc welding by the pressure is applied but not by the filler metal or fluxes.

(16) Why AC is more suitable for the resistance welding?

AC is more suitable for the resistance welding, as it can provide any desired combination of current and voltage by means of transformer

MULTIPLE-CHOICE QUESTIONS

(1) Which of the followings falls under the category of the plastic or non-fusion welding?

(a) Resistance welding.

(b) Electron beam welding.

(c) Electro-slag welding.

(d) Arc welding.

(2) Which of the following falls under the category of the fusion or non-pressure welding?

(a) Resistance welding.

(b) Metal arc welding.

(c) Ultrasonic welding.

(d) Explosive welding.

(3) The proper selection of welding depends upon, in addition to cost involved:

(a) The kinds of metals to be joined.

(b) The nature of products to be fabricated.

(c) The production technique used.

(d) All of the above.

(4) During the resistance welding, the heat produced at the joint is proportional to:

(a) Current.

(b) Voltage.

(c) I^2R.

(d) Volt–amperes.

(5) The metal surfaces for the electrical resistance welding must be:

(a) Cleaned.

(b) Lubricated.

(c) Moistened.

(d) Rough.

(6) The resistance welding cannot be used for:

(a) Ferrous materials.

(b) Non-ferrous materials.

(c) Dielectrics .

(d) Any of the above.

(7) In the electrical resistance welding, the greatest resistance is offered by:

(a) Metal surface.

(b) The contact layer of metals to be welded.

(c) The contact point of electrode with metal top.

(d) The contact point of electrode with metal bottom.

(8) In the electric resistance welding:

(a) The current required exceeds 100 A.

(b) The voltage ranges from 4 to 12 V.

(c) The amount of power supplied to the weld usually ranges from 60 to 80 W for each square mm of area.

(d) All of the above.

(9) The resistance to the flow of current is made of:

(a) The resistance of current path in the work.

(b) The resistance between the contact surfaces of the parts being welded.

(c) The resistance between the electrodes and the surface of the parts being welded.

(d) All of the above.

(10) The electric resistance welding has the advantages of:

(a) Reduced distortion, higher production rates, suitability for large quantity production, and comparatively lesser skill need.

(b) Heat is localized where required.

(c) No filler material is required.

(d) All of the above.

(11) The main drawbacks of the resistance welding are:

(a) High initial as well as maintenance cost.

(b) Difficult shapes and sections cannot welded.

(c) Only similar metals can be welded.

(d) Parent metal is affected.

(12) Plain and butt welds may be used on materials up to thickness of about:

(a) 5 mm.

(b) 10 mm.

(c) 25 mm.

(d) 40 mm.

(13) In the upset butt welding:

(a) The faces of the metal pieces to be joined are prepared for even contact.

(b) Heating is obtained by the contact resistance of the metal pieces to be welded.

(c) The voltage required is 2–8 V and the current required ranges from 50 A to several hundred amperes depending upon material and the area to be welded at a time.

(d) All of the above.

(14) In the flash butt welding:

(a) No special preparation of the faces to be welded is necessary.

(b) Clean and pure weld is obtained.

(c) Power requirement is less.

(d) All of the above.

(15) The spot welding process basically depends on:

(a) The generation of heat.

(b) The application of forging pressure.

(c) Both (a) and (b).

(d) Ohmic resistance.

(16) In the spot welding, the composition and thickness of the base metal determines:

(a) The holding time.

(b) The amount of weld current.

(c) The amount of squeeze pressure.

(d) All of the above.

(17) The tips of the electrodes, for the spot welding are made of:

(a) Carbon.

(b) Copper alloy or pure copper.

(c) Mica.

(d) Porcelain.

(18) The power factor of a spot welding machine is expected to be about:

(a) 0.3–0.5 lagging.

(b) 0.8–0.85 lagging.

(c) 0.75–0.85 lagging.

(d) Unity.

(19) During the spot welding, the current flows for:

(a) Fraction of a minute.

(b) Fraction of a second to several seconds.

(c) Few milliseconds.

(d) Few microseconds.

(20) The spot welding is employed for:

(a) Thin metal sheets (thickness being usually limited to 10–12 mm).

(b) Castings only.

(c) Thick sections.

(d) Rough and irregular surfaces.

(21) The spot welding:

(a) Makes the weld air tight.

(b) Makes the weld water tight.

(c) Provides mechanical strength.

(d) All of the above.

(22) In the spot welding:

(a) It is desirable to clean the sheets thoroughly before welding.

(b) The workpieces being welded are pressed together by mechanical pressure exerted through electrodes.

(c) The current required is above 5,000 A and the voltage between the electrodes is usually less than 2 V (open-circuit voltage less than 12 V).

(d) All of the above.

(23) The projection welding can be considered as a mass production form of:

(a) Seam welding.

(b) Spot welding.

(c) Upset welding.

(d) Flash welding.

(24) The basic electrical requirement in the arc welding is that there should be:

(a) High open circuit voltage.

(b) No arc blow.

(c) DC power supply.

(d) Coated electrodes.

(25) In the arc welding, the temperature of the arc produced is of the order of:

(a) 1,000°C.

(b) 3,500–4,000°C.

(c) 5,000–7,500°C.

(d) 7,500–10,000°C.

(26) The electric arc has:

(a) Linear resistance characteristic.

(b) Positive resistance characteristic.

(c) Negative resistance characteristic.

(d) Highly inductive characteristic.

(27) In an electric arc welding, the voltage required to strike DC arc is about:

(a) 50–60 V.

(b) 80–90 V.

(c) 100–120 V.

(d) 220 V.

(28) In an electric arc welding, the voltage required to strike AC arc is about:

(a) 50–60 V.

(b) 80–90 V.

(c) 100–120 V.

(d) 230 V.

(29) In an electric arc welding, the voltage required to maintain the arc will be:

(a) 250–500 V.

(b) 150–250 V.

(c) 20–30 V.

(d) Below 10 V.

(30) For an electric arc welding, the current range is usually:

(a) 50–1,000 A.

(b) 30–50 A.

(c) 20–30 A.

(d) Below 20 A.

(31) An arc blow is a welding defect that is countered in:

(a) The arc welding using DC supply.

(b) The arc welding using AC supply.

(c) The thermit welding.

(d) The resistance welding.

(32) For the electric arc welding DC supply is obtained from:

(a) Motor-generator set.

(b) AC rectified welding unit.

(c) Either from motor-generator set or from AC rectified unit.

(d) None of the above.

(33) In an electric welding, the major personal hazards are:

(a) Weld spatter.

(b) Flying sparks.

(c) Harmful infra-red and ultraviolet rays from the arc.

(d) All of the above.

(34) During the electric arc welding, as the thickness of the metal to be welded increases:

(a) The voltage is increased keeping current the same.

(b) The current is increased keeping voltage unchanged.

(c) Both the current and the voltage are increased.

(d) Both the current and the voltage are reduced.

(35) The length of arc required depends on:

(a) The kind of electrode used, its coating and its diameter.

(b) The magnitude of current used.

(c) The position of welding.

(d) All of the above.

(36) The overhead welding position is thought to be the most:

(a) Hazardous.

(b) Economical.

(c) Useful.

(d) Difficult.

(37) The electrode is coated in order to:

(a) Improve the bead quality.

(b) Cleanse the base metal.

(c) Provide the shielding to weld pool.

(d) Prevent the atmospheric contamination.

(38) A 10-swg electrode has approximate diameter of:
(a) 0.8 mm.
(b) 1.0 mm.
(c) 3.3 mm.
(d) 10 mm.

(39) The purpose of coating on the arc welding electrodes is to:
(a) Provide a protective covering.
(b) Provide slag for the protection of the molten metal.
(c) Stabilize the arc.
(d) All of the above.

(40) In the carbon arc welding:
(a) Electrode is +ve wrt work in case of DC supply.
(b) Electrode is −ve wrt work in case of DC supply.
(c) Electrode is connected to neutral in case of AC supply.
(d) None of the above.

(41) The carbon arc welding has the advantages of:
(a) Easy control of molten pool temperature simply by varying the arc length.
(b) Easily adaptable to automation.
(c) Excellent heat source for brazing, braze welding, soldering, etc.
(d) All of the above.

(42) For the metal arc welding:
(a) Both DC and AC can be used but AC is preferred.
(b) Bare electrodes are no longer used except for the automatic welding having arrangement to protect the weld area from the atmosphere.
(c) Correcting welding current, voltage, and speed are very important.
(d) All of the above.

(43) In the argon arc welding, the electrode is made of:
(a) Carbon.
(b) Graphite.
(c) Tungsten.
(d) Steel.

(44) Argon is:
(a) An inert gas.
(b) A rare gas.
(c) An inactive gas.
(d) An oxidizing agent.

(45) The MiG welding is becoming more and more popular as it:
(a) Is easy in operation.
(b) Has high metal deposit rate.
(c) Both (a) and (b)
(d) It can be used for both ferrous and non-ferrous metals.

(46) In the electro-slag welding, theoretically there is no limit to:
(a) The rate of metal deposit.
(b) The thickness of weld bead.
(c) The temperature of salt bath.
(d) The rate of slag consumption.

(47) In the ultrasonic welding, the frequency range is usually:
(a) 20–60 kHz.
(b) 50–100 kHz.
(c) 100–200 kHz.
(d) Above 250 kHz.

(48) Welding is not done directly from the supply mains as:
(a) Its voltage is too high.
(b) It is impracticable to draw heavy currents directly from the supply mains.
(c) Its voltage remains fluctuating.
(d) None of the above.

(49) The AC welding machine cannot be used for:
(a) The resistance welding.
(b) The submerged arc welding.
(c) The MIG welding.
(d) The atomic hydrogen welding.

(50) In the electric welding, the arc blow can be avoided by:
(a) Using AC machines.
(b) Increasing arc length.
(c) Using bare electrodes.
(d) Welding away from ground connections.

(51) Welding leads have:
(a) High current carrying capacity.
(b) High flexibility.

(c) Both (a) and (b).

(d) None of the above.

(52) The transformer used in a welding set is:

(a) Step-up transformer.

(b) Step-down transformer.

(c) Constant current transformer.

(d) Booster transformer.

(53) A rectifier used for welding has voltage current characteristic as:

(a) Drooping.

(b) Rising.

(c) Straight line.

(d) None of the above.

(54) The load power factor using the welding transformer depends on:

(a) Arc length.

(b) Material to be welded.

(c) Type of electrode to be used.

(d) All of the above.

(55) For power factor correction in a welding circuit, a capacitor is usually connected:

(a) Across the mains.

(b) Across the secondary side of the welding transformer.

(c) Across the primary side of the welding transformer.

(d) Across the arcing electrodes.

(56) In a welded joint, poor fusion is on account of:

(a) Improper current.

(b) High welding speed.

(c) Uncleaned metal surface.

(d) Lack of flux.

(57) The method recommended for the welding of aluminum alloy is:

(a) DC arc welding.

(b) AC arc welding.

(c) Acetylene–oxygen gas welding.

(d) Tungsten arc welding.

(58) Steel pipes are manufactured by:

(a) The arc welding.

(b) The argon arc welding.

(c) The resistance welding.

(d) The thermit welding.

(59) Steel rails are welded by:

(a) The thermit welding.

(b) The argon arc welding.

(c) The gas welding.

(d) The resistance welding.

(60) Electronic components are joined by:

(a) Spot welding.

(b) Soldering.

(c) Brazing.

(d) None of the above.

(61) Air craft body is:

(a) Riveted.

(b) Seam welded.

(c) Gas welded.

(d) Spot welded.

REVIEW QUESTIONS

(1) Explain the resistance welding and its application.

(2) Explain the principle of spot and seam weldings.

(3) What are the differences between resistance welding and arc welding?

(4) Discuss the difference between carbon and metallic arc weldings. Give their relative merits and demerits.

(5) Explain varies types of arc welding processes.

(6) Compare AC and DC weldings.

(7) Compare flash and upset butt weldings.

(8) What are the types of electrodes used for welding operation? Give the advantages of coated electrodes.

(9) List out the equipment used for the welding operations.

(10) Compare resistance and arc weldings.

ANSWERS

1. a
2. b
3. d
4. c
5. a
6. c
7. b
8. d
9. d
10. d
11. a
12. c
13. d
14. d
15. c
16. d
17. b
18. a
19. b
20. a
21. c
22. d
23. b
24. a
25. b
26. c
27. a
28. b
29. c
30. a
31. a
32. c
33. d
34. b
35. d
36. a
37. a
38. c
39. d
40. b
41. d
42. d
43. c
44. a
45. c
46. b
47. a
48. b
49. c
50. a
51. c
52. b
53. a
54. b
55. c
56. a
57. b
58. c
59. a
60. b
61. a

Fundamentals of Illumination

OBJECTIVES

After reading this chapter, you should be able to:

- understand the terms used in illumination
- discuss the various laws of illumination
- analyze of photometry for the measurement of candle power of a source

6.1 INTRODUCTION

Study of illumination engineering is necessary not only to understand the principles of light control as applied to interior lighting design such as domestic and factory lighting but also to understand outdoor applications such as highway lighting and flood lighting. Nowaday, the electrically produced light is preferred to the other source of illumination because of an account of its cleanliness, ease of control, steady light output, low cost, and reliability. The best illumination is that it produces no strain on the eyes. Apart from its esthetic and decorative aspects, good lighting has a strictly utilitarian value in reducing the fatigue of the workers, protecting their health, increasing production, etc. The science of illumination engineering is therefore becoming of major importance.

6.1.1 Nature of light

Light is a form of electromagnetic energy radiated from a body and human eye is capable of receiving it. Light is a prime factor in the human life as all activities of human being ultimately depend upon the light.

Various forms of incandescent bodies are the sources of light and the light emitted by such bodies depends upon their temperature. A hot body about 500–800°C becomes a red hot and about 2,500–3,000°C the body becomes white hot. While the body is red-hot, the wavelength of the radiated energy will be sufficiently large and the energy available in the form of heat. Further, the temperature increases, the body changes from red-hot to white-hot state, the wavelength of the radiated energy becomes smaller and enters into the range of the wavelength of light. The wavelength of the light waves varying from 0.0004 to 0.00075 mm, i.e. 4,000–7,500 Å (1 Angstrom unit = 10^{-10} mm).

The eye discriminates between different wavelengths in this range by the sensation of color. The whole of the energy radiated out is not useful for illumination purpose. Radiations of very short wavelength varying from 0.0000156×10^{-6} m to 0.001×10^{-6} m are not in the visible range are called as rontgen or x-rays, which are having the property of penetrating through opaque bodies.

6.2 TERMS USED IN ILLUMINATION

The following terms are generally used in illumination.

Color: The energy radiation of the heated body is monochromatic, i.e. the radiation of only one wavelength emits specific color. The wavelength of visible light lies between 4,000 and 7,500 Å. The color of the radiation corresponding to the wavelength is shown in Fig. 6.1.

Relative sensitivity: The reacting power of the human eye to the light waves of different wavelengths varies from person to person, and also varies with age. The average relative sensitivity is shown in Fig. 6.2.

The eye is most sensitive for a wavelength of 5,500 Å. So that, the relative sensitivity according to this wavelength is taken as unity.

Referred from Fig. 6.1, blue and violet corresponding to the short wavelengths and red to the long wavelengths, orange, yellow, and green being in the middle of the visible region of wavelength. The color corresponding to 5,500 Å is not suitable for most of the applications since yellowish green. The relative sensitivity at any particular wavelength (λ) is known as relative luminous factor (K_λ).

Light: It is defined as the radiant energy from a hot body that produces the visual sensation upon the human eye. It is expressed in lumen-hours and it analogous to watt-hours, which denoted by the symbol 'Q'.

Luminous flux: It is defined as the energy in the form of light waves radiated per second from a luminous body. It is represented by the symbol 'ϕ' and measured in lumens.

***Ex*:** Suppose the luminous body is an incandescent lamp.

The total electrical power input to the lamp is not converted to luminous flux, some of the power lost through conduction, convection, and radiation, etc. A fraction of the remaining radiant flux is in the form of light waves lies in between the visual range of wavelength, i.e. between 4,000 and 7,000 Å, as shown in Fig. 6.3.

6.2.1 Radiant efficiency

When an electric current is passed through a conductor, some heat is produced to I^2R loss, which increases its temperature of the conductor. At low temperature, conductor radiates energy in the form of heat waves, but at very high temperatures, radiated energy will be in the form of light as well as heat waves.

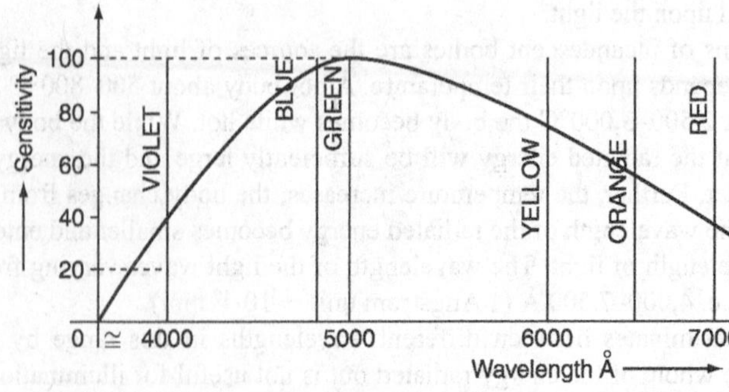

FIG. 6.1 *Wavelength*

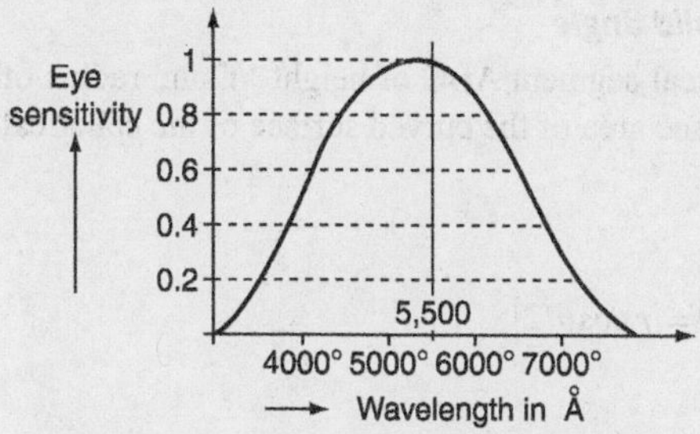

FIG. 6.2 *The average relative sensitivity*

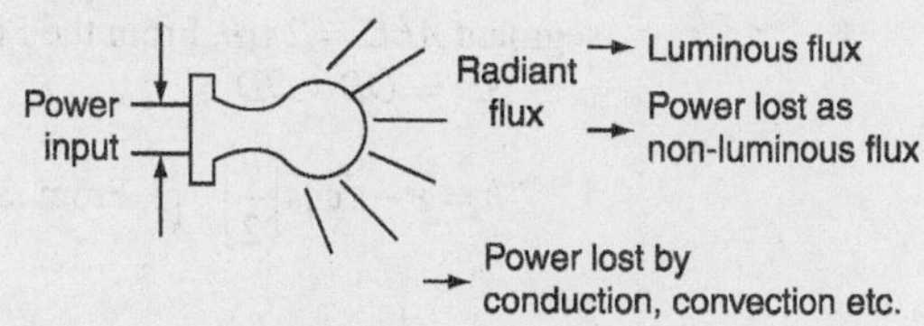

FIG. 6.3 *Flux diagram*

'Radiant efficiency is defined as the ratio of energy radiated in the form of light, produces sensation of vision to the total energy radiated out by the luminous body'.

$$\text{Radiant efficiency} = \frac{\text{energy radiated in the form of light}}{\text{total energy radiated by the body}}.$$

6.2.2 Plane angle

A plane angle is the angle subtended at a point in a plane by two converging lines (Fig. 6.4). It is denoted by the Greek letter 'θ' (theta) and is usually measured in degrees or radians.

$$\therefore \text{Plane angle } (\theta) = \frac{\text{arc}}{\text{radius}}. \qquad \textbf{(6.1)}$$

One radian is defined as the angle subtended by an arc of a circle whose length by an arc of a circle whose length is equals to the radius of the circle.

6.2.3 Solid angle

Solid angle is the angle subtended at a point in space by an area, i.e., the angle enclosed in the volume formed by numerous lines lying on the surface and meeting at the point (Fig. 6.5). It is usually denoted by symbol 'ω' and is measured in steradian.

$$\therefore \text{Solid angle } (\omega) = \frac{\text{area}}{\text{radius}^2}. \qquad \textbf{(6.2)}$$

The largest solid angle subtended at the center of a sphere:

$$\omega = \frac{\text{area of sphere}}{\text{radius}^2} = \frac{4\pi r^2}{R^2} = 4\pi \text{ steradians.}$$

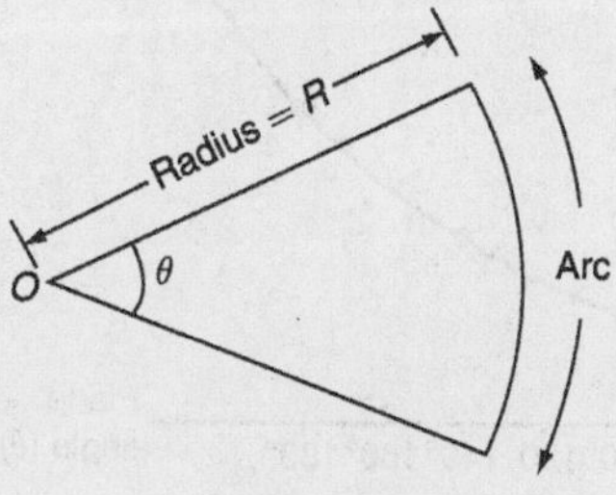

FIG. 6.4 *Plane angle*

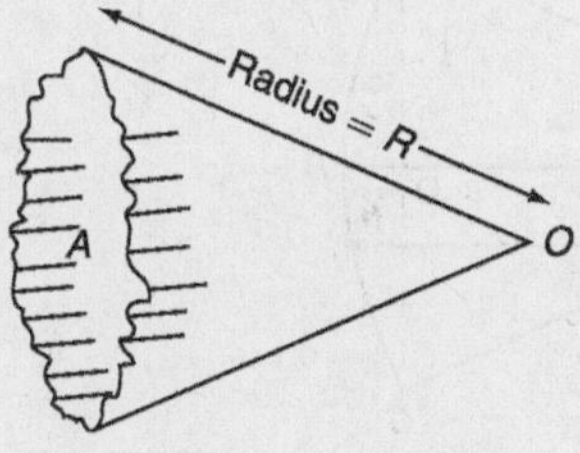

FIG. 6.5 *Solid angle*

Relationship between plane angle and solid angle

Let us consider a curved surface of a spherical segment ABC of height '*h*' and radius of the sphere '*r*' as shown in Fig. 6.6. The surface area of the curved surface of the spherical segment $ABC = 2\pi rh$. From the Fig. 6.6:

$$BD = OB - OD$$

$$h = r - r\cos\left(\frac{\theta}{2}\right) \quad [\therefore \text{From } \Delta ODA, OD = r\cos\theta/2]$$

$$= r\left(1 - \cos\frac{\theta}{2}\right).$$

$\therefore$ The surface area of the segment $= 2\pi rh$

$$= 2\pi r^2\left[r - \cos\frac{\theta}{2}\right].$$

$$\text{We know solid angle } (\omega) = \frac{\text{area}}{(\text{radius})^2}$$

$$= \frac{2\pi r^2\left(1 - \cos\frac{\theta}{2}\right)}{r^2}$$

$$= 2\pi\left(1 - \cos\frac{\theta}{2}\right). \quad (6.3)$$

From the Equation (6.3), the curve shows the variation of solid angle with plane angle is shown in Fig. 6.7.

Luminous intensity

Luminous intensity in a given direction is defined as the luminous flux emitted by the source per unit solid angle (Fig. 6.8).

It is denoted by the symbol '*I*' and is usually measured in 'candela'.

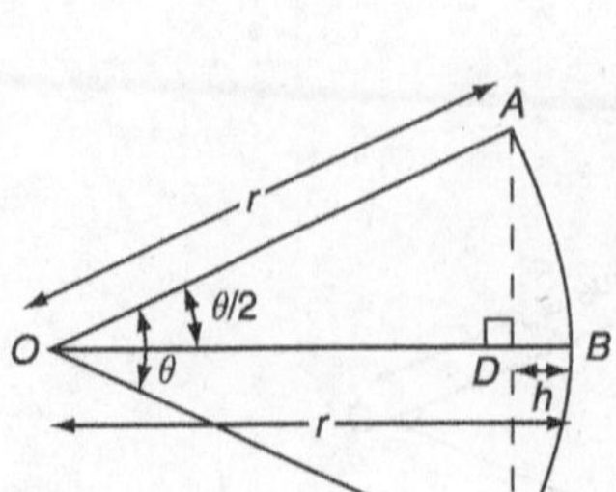

FIG. 6.6 *Sectional view for solid angle*

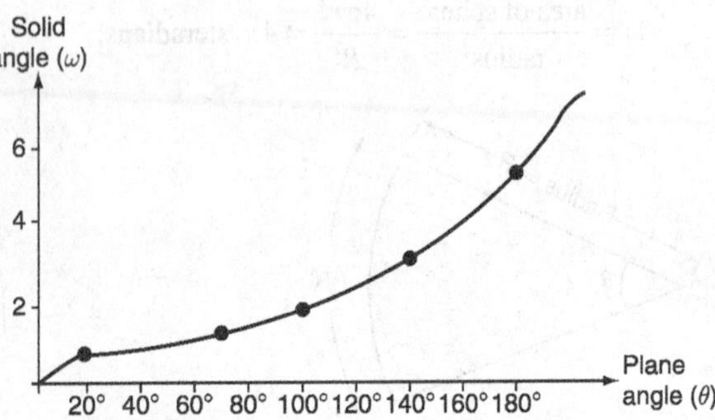

FIG. 6.7 *Relation between solid angle and plane angle*

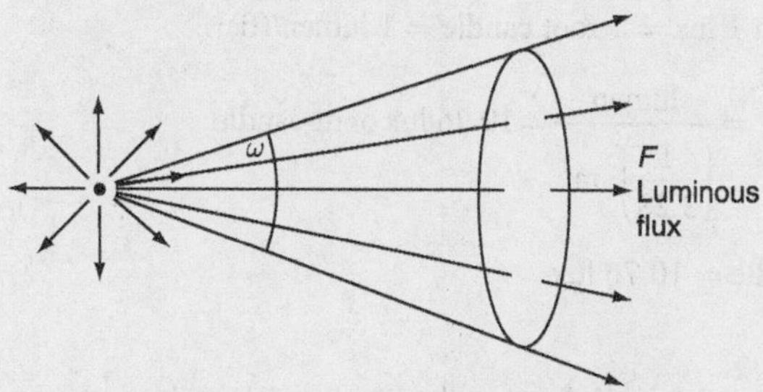

FIG. 6.8 *Luminous flux emitting from the source*

Let '*F*' be the luminous flux crossing a spherical segment of solid angle 'ω'. Then luminous intensity $(I) = \frac{\phi}{\omega}$ lumen/steradian or candela.

Lumen: It is the unit of luminous flux.

It is defined as the luminous flux emitted by a source of one candle power per unit solid angle in all directions.

Lumen = candle power of source × solid angle.

Lumen = CP × ω

Total flux emitted by a source of one candle power is 4π lumens.

Candle power (CP)

The CP of a source is defined as the total luminous flux lines emitted by that source in a unit solid angle.

$$CP = \frac{\text{lumen}}{\omega} \text{ lumen/steradian or candela.}$$

Illumination

Illumination is defined as the luminous flux received by the surface per unit area.

It is usually denoted by the symbol '*E*' and is measured in lux or lumen/m^2 or meter candle or foot candle.

$$\text{Illumination, } E = \frac{\text{luminous flux}}{\text{area}}$$

$$= \frac{\phi}{A} = \frac{CP \times \omega}{A} \text{ lux.}$$

Lux or meter candle

It is defined as the illumination of the inside of a sphere of radius 1 m and a source of 1 CP is fitted at the center of sphere.

Foot candle

It is the unit of illumination and is defined as the illumination of the inside of a sphere of radius 1 foot, and a source of 1 CP is fitted at the center of it.

We know that 1 lux = 1 foot candle = 1 lumen/(ft)^2

$$1 \text{ foot candle} = \frac{\text{lumen}}{\left(\frac{1}{3.28}\right)^2 \text{m}^2} = 10.76 \text{ lux or m-candle}$$

$\therefore$ 1 foot candle = 10.76 lux.

Brightness

Brightness of any surface is defined as the luminous intensity pen unit surface area of the projected surface in the given direction. It is usually denoted by symbol 'L'.

If the luminous intensity of source be 'I' candela on an area A, then the projected area is $A \cos \theta$.

$$\therefore \text{Brightness}, L = \frac{I}{A\cos\theta}$$

The unit of brightness is candela/m^2 or candela/cm^2 or candela/(ft)^2.

Relation between I, E, and L

Let us consider a uniform diffuse sphere with radius r meters, at the center a source of 1 CP, and luminous intensity I candela.

$$\therefore \text{Brightness } (L) = \frac{I}{\pi r^2}$$

$$\text{and Illumination } (E) = \frac{\phi}{A} = \frac{CP \times \omega}{A}$$

$$= \frac{I}{4\pi r^2} \times 4\pi = \frac{I}{r^2}$$

$$\therefore E = \frac{I}{r^2} = \frac{I}{\pi r^2} \times \pi = \pi L$$

$$\therefore E = \pi L = \frac{I}{r^2}. \tag{6.4}$$

Mean horizontal candle power (MHCP)

MHCP is defined as the mean of the candle power of source in all directions in horizontal plane.

Mean spherical candle power (MSCP)

MSCP is defined as the mean of the candle power of source in all directions in all planes.

Mean hemispherical candle power (MHSCP)

MHSCP is defined as the mean of the candle power of source in all directions above or below the horizontal plane.

Reduction factor

Reduction factor of the source of light is defined as the ratio of its mean spherical candle power to its mean horizontal candle power.

$$\text{i.e., reduction factor} = \frac{MSCP}{MHCP}.$$

Lamp efficiency

It is defined as the ratio of the total luminous flux emitting from the source to its electrical power input in watts.

$$\therefore \text{Lamp efficiency} = \frac{\text{luminous flux}}{\text{power input}}.$$

It is expressed in lumen/W.

Specific consumption

It is defined as the ratio of electric power input to its average candle power.

Space to height ratio

It is defined as ratio of horizontal distance between adjacent lamps to the height of their mountings.

$$\text{Space to height ratio} = \frac{\text{horizontal distance between two adjacent lamps}}{\text{mounting height of lamps above the working plane}}.$$

Coefficient of utilization or utilization factor

It is defined as the ratio of total number of lumens reaching the working plane to the total number of lumens emitting from source.

$$\text{Utilization factor} = \frac{\text{total lumens reaching the working plane}}{\text{total lumens emitting from source}}.$$

Maintenance factor

It is defined as the ratio of illumination under normal working conditions to the illumination when everything is clean.

$$\text{Maintanance factor} = \frac{\text{illumination under normal working condition}}{\text{illumination under every thing is clean}}.$$

Its value is always less than 1, and it will be around 0.8. This is due to the accumulation of dust, dirt, and smoke on the lamps that emit less light than that they emit when they are so clean. Frequent cleaning of lamp will improve the maintenance factor.

Depreciation factor

It is defined as the ratio of initial illumination to the ultimate maintained illumination on the working plane.

$$\therefore \text{Depreciation factor} = \frac{1}{\text{maintenance factor}}.$$

Its values is always more than 1.

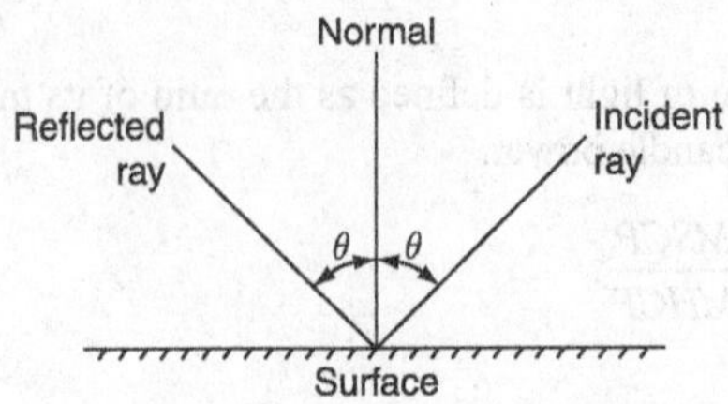

FIG. 6.9 *Reflected ray*

Waste light factor

When a surface is illuminated by several numbers of the sources of light, there is certain amount of wastage due to overlapping of light waves; the wastage of light is taken into account depending upon the type of area to be illuminated. Its value for rectangular area is 1.2 and for irregular area is 1.5 and objects such as statues, monuments, etc.

Absorption factor

Normally, when the atmosphere is full of smoke and fumes, there is a possibility of absorption of light. Hence, the total lumens available after absorption to the total lumens emitted by the lamp are known as absorption factor.

$$\text{Absorption factor} = \frac{\text{the total lumens available after absorption}}{\text{the total lumens given out by the lamp}}.$$

Reflection factor or coefficient of reflection

When light rays impinge on a surface, it is reflected from the surface at an angle of incidence shown in Fig. 6.9. A portion of incident light is absorbed by the surface.

The ratio of luminous flux leaving the surface to the luminous flux incident on it is known as reflection factor.

$$\text{Reflection factor} = \frac{\text{reflected light}}{\text{incident light}}.$$

Its value will be always less than 1.

Beam factor

It is defined as the ratio of 'lumens in the beam of a projector to the lumens given out by lamps'. Its value is usually varies from 0.3 to 0.6. This factor is taken into account for the absorption of light by reflector and front glass of the projector lamp.

Example 6.1: A 200-V lamp takes a current of 1.2 A, it produces a total flux of 2,860 lumens. Calculate:

(i) the MSCP of the lamp and

(ii) the efficiency of the lamp.

Solution:

Given V = 200 V

$I = 1.2$ A, flux = 2,860 lumens.

(i) MSCP $= \dfrac{\text{total flux}}{4\pi} = \dfrac{2860}{4\pi} = 227.59.$

(ii) Lamp efficiency $(\eta) = \dfrac{\text{total flux output}}{\text{electrical input}} = \dfrac{\text{lumen}}{VI\cos\phi}$

$$= \frac{21,860}{200 \times 1.2 \times 1}[\cos\phi = 1]$$

$$= 11.9 \text{ lumens/W.}$$

Example 6.2: A room with an area of 6 × 9 m is illustrated by ten 80-W lamps. The luminous efficiency of the lamp is 80 lumens/W and the coefficient of utilization is 0.65. Find the average illumination.

Solution:

Room area = 6 × 9 = 54 m^2.
Total wattage = 80 × 10 = 800 W.
Total flux emitted by ten lamps = 80 × 800 = 64,000 lumens.
Flux reaching the working plane = 64,000 × 0.65 = 41,600 lumens.

$\therefore$ Illumination, $E = \dfrac{\phi}{A} = \dfrac{41,600}{54} = 770.37$ lux.

Example 6.3: The luminous intensity of a lamp is 600 CP. Find the flux given out. Also find the flux in the hemisphere containing the source of light and zero above the horizontal.

Solution:

Flux emitted by source (lumen)
= Intensity (I) × solid angle (ω)
= 600 × 2π = 3,769.911 lumens
$\therefore$ Flux emitted in the lower hemisphere = 3,769.911 lumens.

Example 6.4: The flux emitted by 100-W lamp is 1,400 lumens placed in a frosted globe of 40 cm diameter and gives uniform brightness of 250 milli-lumens/m^2 in all directions. Calculate the candel power of the globe and the percentage of light absorbed by the globe.

Solution:

Flux emitted by the globe
= brightness × globe area

$$= \left[\frac{250}{1,000}\right] \times \left[4\pi\left(\frac{40}{2}\right)^2\right]$$

= 1,256.63 lumens
Flux absorbed by the globe
= flux emitted by source − flux emitted by globe
= 1,400 − 1,256.63
= 143.36 lumens.

$\therefore$ The percentage of light absorbed by the globe $= \dfrac{143.36}{1,400} \times 100 = 10.24\%.$

Example 6.5: A surface inclined at an angle 40° to the rays is kept 6 m away from 150 candle power lamp. Find the average intensity of illumination on the surface.

Solution:

From the Fig. P.6.1:

$\theta = (90° - 40°) = 50°.$

$\therefore$ Average illumination:

$$E = \frac{I}{d^2} \times \cos\theta$$

$$= \frac{150}{(4)^2} \times \cos 50°$$

$$= 6.026 \text{ lux.}$$

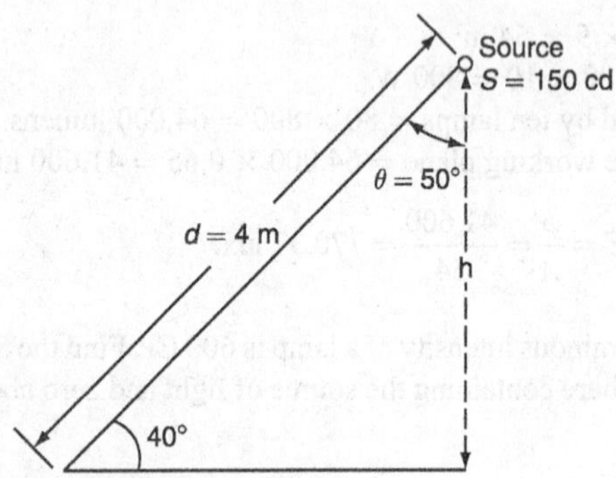

FIG. P.6.1

6.3 LAWS OF ILLUMINATION

Mainly there are two laws of illumination.

1. Inverse square law.
2. Lambert's cosine law.

6.3.1 Inverse square law

This law states that 'the illumination of a surface is inversely proportional to the square of distance between the surface and a point source'.

Proof:

Let, '*S*' be a point source of luminous intensity '*I*' candela, the luminous flux emitting from source crossing the three parallel plates having areas A_1, A_2, and A_3 square meters, which are separated by a distances of *d*, 2*d*, and 3*d* from the point source respectively as shown in Fig. 6.10.

For area A_1, solid angle $\omega = \frac{A_1}{d^2}$.

Luminous flux reaching the area A_1 = luminous intensity × solid angle

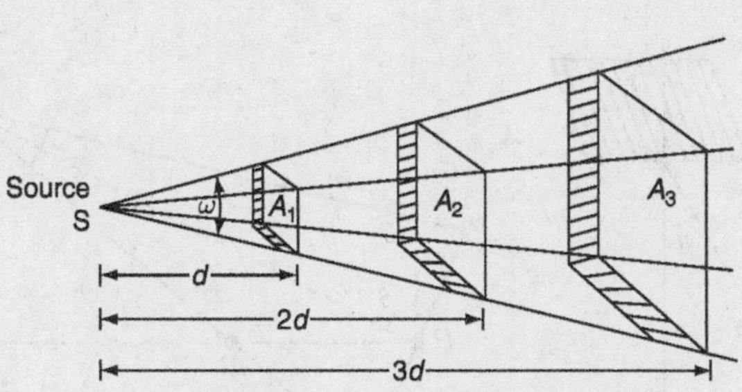

FIG. 6.10 *Inverse square law*

$$= I \times \omega = I \times \frac{A_1}{d^2}.$$

∴ Illumination 'E_1' on the surface area 'A_1' is:

$$E_1 = \frac{\text{flux}}{\text{area}} = \frac{IA_1}{d^2} \times \frac{1}{A_1}$$

$$\therefore E_1 = \frac{I}{d^2} \text{ lux.} \tag{6.5}$$

Similarly, illumination 'E_2' on the surface area A_2 is:

$$E_2 = \frac{I}{(2d)^2} \text{ lux} \tag{6.6}$$

and illumination 'E_3' on the surface area A_3 is:

$$E_3 = \frac{\text{I}}{(3d)^2} \text{ lux.} \tag{6.7}$$

From Equations (6.5), (6.6), and (6.7)

$$E_1 : E_2 : E_3 = \frac{1}{d^2} : \frac{1}{(2d)^2} : \frac{1}{(3d)^2}. \tag{6.8}$$

Hence, from Equation (6.8), illumination on any surface is inversely proportional to the square of distance between the surface and the source.

6.3.2 Lambert's cosine law

This law states that 'illumination, E at any point on a surface is directly proportional to the cosine of the angle between the normal at that point and the line of flux'.

Proof:

While discussing, the Lambert's cosine law, let us assume that the surface is inclined at an angle 'θ' to the lines of flux as shown in Fig. 6.11.

Let PQ = The surface area normal to the source and inclined at 'θ' to the vertical axis.

RS = The surface area normal to the vertical axis and inclined at an angle θ to the source 'O'.

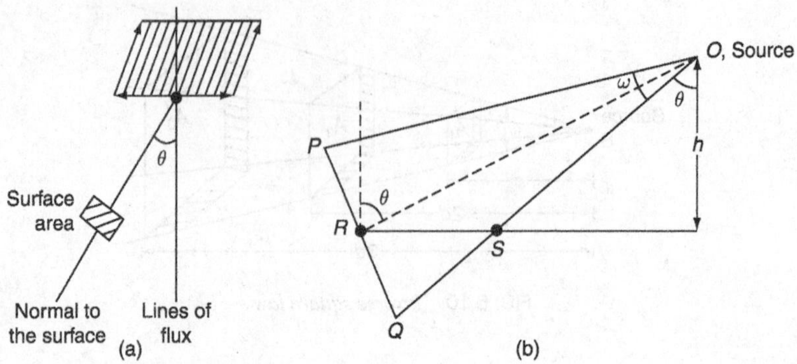

FIG. 6.11 *Lambert's cosine law*

Therefore, from Fig. 6.11:

$$PQ = RS\cos\theta.$$

$$\therefore \text{The illumination of the surface } PQ, E_{PQ} = \frac{\text{flux}}{\text{area of } PQ}$$

$$= \frac{I \times \omega}{\text{area of } PQ} = \frac{I}{\text{area of } PQ} \times \frac{\text{area of } PQ}{d^2} \quad \left[\because \omega = \text{area/(radius)}^2\right]$$

$$= \frac{I}{d^2}. \tag{6.9}$$

$$\therefore \text{The illumination of the surface } RS, E_{RS} = \frac{\text{flux}}{\text{area of } RS} = \frac{\text{flux}}{\text{area of } PQ/\cos\theta}$$

$$[\because PQ = RS\cos\theta]$$

$$= \frac{I}{d^2}\cos\theta. \tag{6.10}$$

From Fig. 6.11(b):

$$\cos\theta = \frac{h}{d}$$

$$\text{or } d = \frac{h}{\cos\theta}.$$

Substituting '*d*' from the above equation in Equation (6.10):

$$\therefore E_{RS} = \frac{I}{(h/\cos\theta)^2} \times \cos\theta = \frac{I}{h^2}\cos^3\theta \tag{6.11}$$

$$\therefore E_{RS} = \frac{I}{d^2}\cos\theta = \frac{I}{h^2}\cos^3\theta \tag{6.12}$$

where d is the distance between the source and the surface in m, h is the height of source from the surface in m, and I is the luminous intensity in candela.

Hence, Equation (6.11) is also known as 'cosine cube' law. This law states that the 'illumination at any point on a surface is dependent on the cube of cosine of the angle between line of flux and normal at that point'.

Note:

*From the above laws of illumination, it is to be noted that inverse square law is only applicable for the surfaces if the surface is normal to the line of flux. And Lambert's cosine law is applicable for the surfaces if the surface is inclined an angle 'θ' to the line of flux.

Example 6.6: The illumination at a point on a working plane directly below the lamp is to be 60 lumens/m^2. The lamp gives 130 CP uniformly below the horizontal plane. Determine:

(i) The height at which lamp is suspended.

(ii) The illumination at a point on the working plane 2.8 m away from the vertical axis of the lamp.

Solution:

Given data:

Candle power of the lamp = 130 CP.

The illumination just below the lamp, E = 60 lumen/m^2.

(i) From the Fig. P.6.2, the illumination just below the lamp, i.e., at point A:

$$E_A = \frac{I}{h^2}$$

$$\therefore h = \sqrt{\frac{I}{EA}} = \sqrt{\frac{130}{60}} = 1.471\,\text{m}.$$

(ii) The illumination at point 'B':

$$E_B = \frac{I}{h^2}\cos^3\theta$$

$$= \frac{130}{(2.8)^2}\left\{\frac{2.8}{\sqrt{2.8^2+1.471^2}}\right\}^3 = 11.504\,\text{lux}.$$

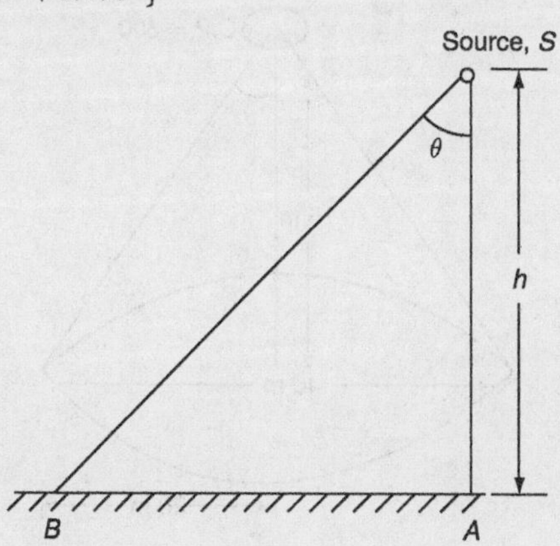

FIG. P.6.2

Example 6.7: A lamp having a candle power of 300 in all directions is provided with a reflector that directs 70% of total light uniformly on a circular area 40-m diameter. The lamp is hung at 15 m above the area.

(i) Calculate the illumination.

(ii) Also calculate the illumination at the center.

(iii) The illumination at the edge of the surface without reflector.

Solution:

Given data:

Candle power of the lamp = 300 CP.
Circular area diameter (D) = 40 m.
Height of mounting = 15 m.

(i) The illumination on the circular area (Fig. P.6.3):

$$E = \frac{\text{flux}}{\text{area}} = \frac{CP \times \omega}{A}.$$

Here, $A = \frac{\pi}{4} D^2 = \frac{\pi}{4} \times 40^2 = 400\ \pi\text{m}^2.$

Solid angle 'ω' $= 2\pi\,(1 - \cos\theta)$

$$= 2\pi \left(1 - \frac{15}{\sqrt{15^2 + 20^2}}\right)$$

$$= 0.8\ \pi \text{ steradians.}$$

$$\therefore \text{Illumination } E = \frac{\text{flux}}{\text{area}} = \frac{CP \times \omega}{A}$$

$$= \frac{300 \times 0.8\pi}{400\pi}$$

$$= 0.6 \text{ lux.}$$

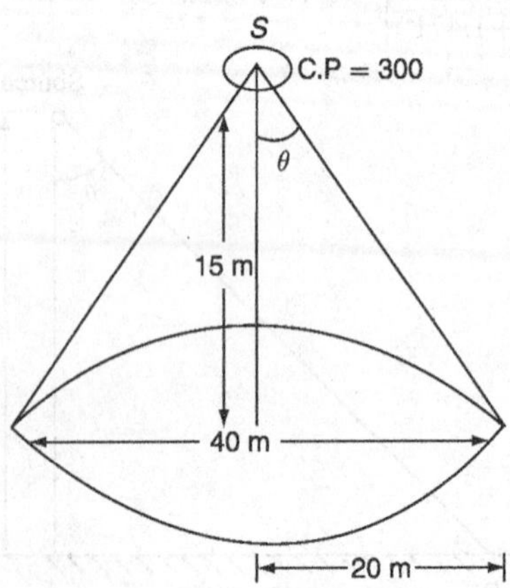

FIG. P.6.3

(ii) The illumination at the center with reflector 70%:

$$= \frac{\phi}{A} \times 0.7 = \frac{CP \times \omega}{A} \times 0.7$$

$$= \frac{300 \times 4\pi}{400\pi} \times 0.7$$

$$= 2.1 \text{ lux.}$$

(iii) The illumination at the edge without reflector:

$$= \frac{CP}{d^2} \times \cos\theta$$

$$= \frac{300}{(\sqrt{15^2 + 10^2})^2} \times \frac{15}{\sqrt{15^2 + 10^2}}$$

$$= 0.768 \text{ lux.}$$

Example 6.8: The luminous intensity of a source is 600 candela is placed in the middle of a 10 × 6 × 2 m room. Calculate the illumination:

(i) At each corner of the room.

(ii) At the middle of the 6-m wall.

Solution:

Given data:

Luminous intensity, $(I) = 600$ cd.

Room area = 10 × 6 × 2 m.

(i) From the Fig. P.6.4:

$$OB = BD = \frac{\sqrt{10^2 + 6^2}}{2} = 5.83 \text{ m}$$

$$\text{BS} = \text{d} = \sqrt{2^2 + (5.38)^2} = 6.163 \text{ m.}$$

∴ The illumination at the corner '*B*':

$$E_B = E_A = E_C = E_D$$

$$\frac{I}{d^2}\cos\theta = \frac{600}{(6.163)^2} \times \frac{2}{(6.163)}$$

$$= 5.126 \text{ lux.}$$

(ii) From Fig. P.6.5:

$$PS = \sqrt{2^2 + 5^2}$$

$$= 5.385 \text{ m.}$$

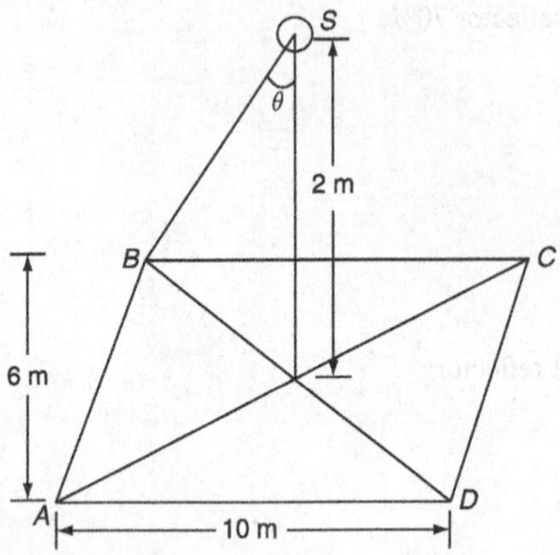

FIG. P.6.4

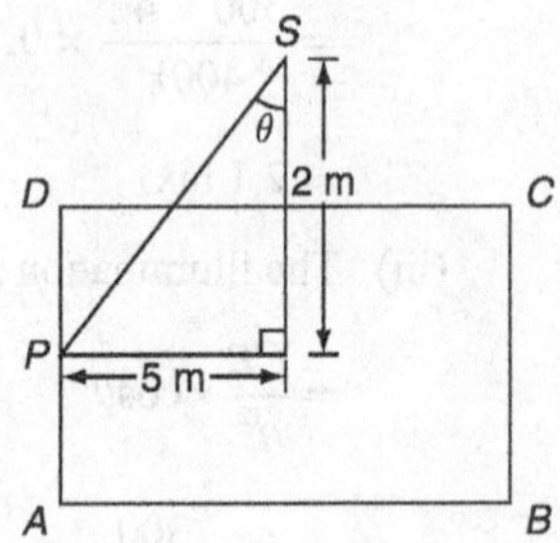

FIG. P.6.5

The illumination at the point '*P*', $E_p = \frac{I}{d^2}\cos\theta$

$$= \frac{600}{(5.385)^2} \times \frac{2}{(5.385)}$$

$$= 7.684 \text{ lux.}$$

Example 6.9: The candle power of a source is 200 candela in all directions below the lamp. The mounting height of the lamp is 6 m. Find the illumination:

(i) Just below the lamp.

(ii) 3 m horizontally away from the lamp on the ground.

(iii) The total luminous flux in an area of 1.5-m diameter around the lamp on the ground.

Solution:

The candle power of the source, $I = 200$ candela.
Mounting height $(h) = 6$ m.

(i) The illumination just below the lamp, i.e., at point '*A*':

$$E_A = \frac{I}{h^2}$$

$$= \frac{200}{6^2} = 5.55 \text{ lux.}$$

(ii) From Fig. P.6.6:

$$d = \sqrt{3^2 + 6^2} = 6.708.$$

The illumination 3 m away from the lamp on the ground, i.e., at point '*B*' (Fig. P.6.7):

$$E_B = \frac{I}{d^2}\cos\theta$$

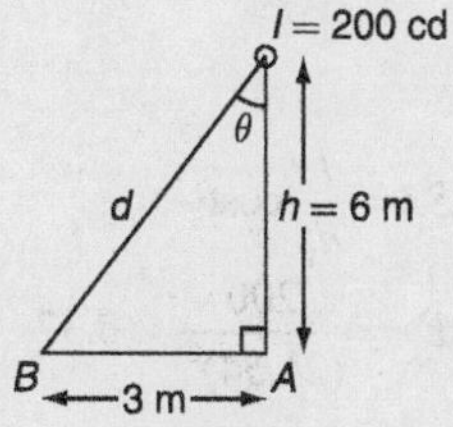

FIG. P.6.6

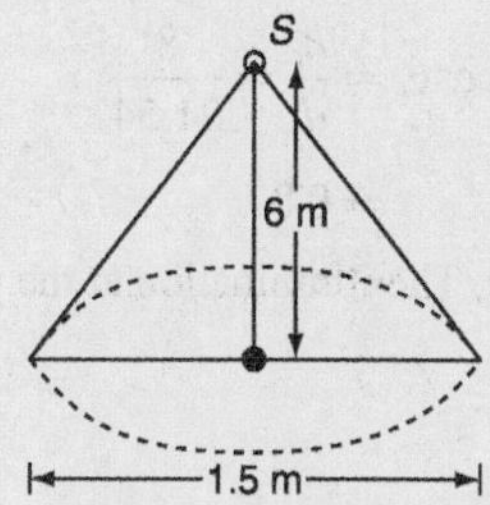

FIG. P.6.7

$$= \frac{200}{(6.708)^2} \times \frac{6}{(6.708)}$$

$$= 3.975 \text{ lux.}$$

(iii) Surface area $= \frac{\pi}{4} d^2$

$$= \frac{\pi}{4} \times (1.5)^2$$

$$= 1.767 \text{ m}^2.$$

The total flux reaching the area around the lamp:

$= E_A \times$ surface area

$= 5.55 \times 1.767$

$= 9.80$ lumens.

Example 6.10: Two sources of candle power or luminous intensity 200 candela and 250 candela are mounted at 8 and 10 m, respectively. The horizontal distance between the lamp posts is 40 m, calculate the illumination in the middle of the posts.

Solution:

From Fig. P.6.8:

$$d_1 = \sqrt{8^2 + 20^2}$$

$$= 21.54.$$

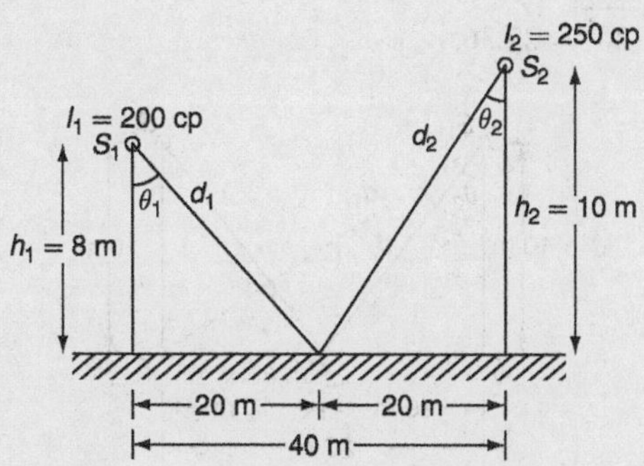

FIG. P.6.8

$$\cos\theta_1 = \frac{h_1}{d_1} = \frac{8}{21.54}$$

$$= 0.37.$$

$\therefore$ The illumination at the point 'P' due to the source 'S_1'$= \frac{I_1}{d_1^2}\cos\theta_1$

$$E_1 = \frac{200}{(21.54)^2} \times 0.37$$

$$= 0.159 \text{ lux}.$$

and $d_2 = \sqrt{10^2 + 20^2} = 22.36$

$$\cos\theta_2 = \frac{h_2}{d_2} = \frac{10}{22.36} = 0.447.$$

The illumination at the point 'P' due to the source 'S_2':

$$E_2 = \frac{I_2}{d_2^2} \times \cos\theta_2$$

$$= \frac{250}{(22.36)^2} \times 0.447 = 0.2235 \text{ lux}.$$

$\therefore$ The total illumination at 'P' due to both the sources S_1 and $S_2 = E_1 + E_2$

$$= 0.159 + 0.2235$$

$$= 0.3825 \text{ lux}.$$

Example 6.11: Two sources of having luminous intensity 400 candela are hung at a height of 10 m. The distance between the two lamp posts is 20 m. Find the illumination (i) beneath the lamp and (ii) in the middle of the posts.

Solution:

Given data:

Luminous intensity = 400 CP.

Mounting height = 10 m.

Distance between the lamp posts = 20 m.

(i) From Fig. P.6.9:

$$d_1 = \sqrt{10^2 + 20^2} = 22.36.$$

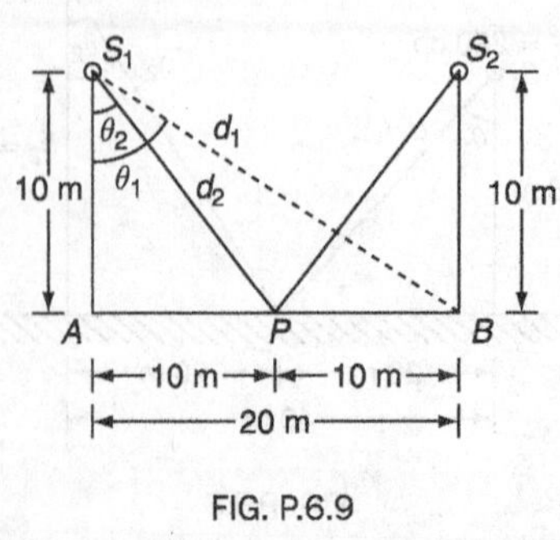

FIG. P.6.9

$$\cos\theta_1 = \frac{h}{d_1} = \frac{10}{22.36} = 0.4472.$$

The illumination at 'B' due to 'S_1':

$$E_1 = \frac{I}{d_1^2}\cos\theta_1$$

$$= \frac{400}{(22.36)^2} \times 0.4472$$

$$= 0.35778 \text{ lux.}$$

The illumination at 'B' due to 'S_2':

$$E_2 = \frac{400}{10^2} = 4 \text{ lux.}$$

$\therefore$ The total illumination at 'B' $= E_1 + E_2$

$= 0.3577 + 4$

$= 4.3577$ lux.

$$d_2 = \sqrt{10^2 + 10^2} = 14.14.$$

$$\cos\theta_2 = \frac{10}{14.14} = 0.707.$$

The illumination at 'P' due to S_1 is:

$$E_1 = \frac{I}{d_2^2} \times \cos\theta_2$$

$$= \frac{400}{(14.14)^2} \times 0.707 = 1.414 \text{ lux.}$$

The illumination at 'P' due to S_2, 'E_2' will be same as E_1.

$\therefore$ The illumination at 'P' due to both S_1 and S_2:

$= E_1 + E_2 = E_1 + E_1$

$= 2E_1 = 2 \times 1.414$

$= 2.828$ lux.

Example 6.12: In a street lighting, two lamps are having luminous intensity of 300 candela, which are mounted at a height of 6 and 10 m. The distance between lamp posts is 12 m. Find the illumination, just below the two lamps.

Solution:

(i) The illumination at 'B' = the illumination due to L_1 + the illumination due to L_2. Form Fig. P.6.10:

$$d_1 = \sqrt{6^2 + 12^2} = 13.416 \text{ m.}$$

$$\cos\theta_1 = \frac{h_1}{d_1} = \frac{6}{13.416} = 0.447.$$

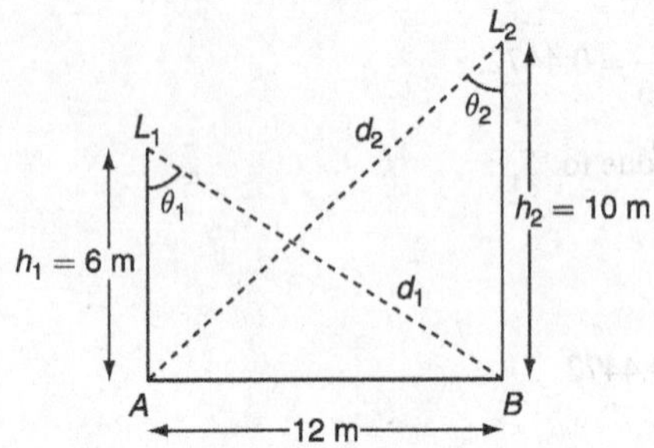

FIG. P.6.10

$\therefore$ The illumination at 'B' due to $L_1 = \frac{I}{d_1^2}\cos\theta_1$

$$= \frac{300}{(13.416)^2} \times 0.447$$

$$= 0.745 \text{ lux.}$$

Illumination at 'B' due to $L_2 = \frac{I}{h_2^2}$

$$= \frac{300}{10^2}$$

$$= 3 \text{ lux.}$$

$\therefore$ The total illumination at 'B' due to the two lamps = 0.745 + 3 = 3.745 lux.

(ii) The illumination at 'A' = the illumination due to L_1+ the illumination due to L_2.

$$d_2 = \sqrt{10^2 + 12^2} = 15.62 \text{ m.}$$

$$\cos\theta_2 = \frac{h_2}{d_2} = \frac{10}{15.62} = 0.64.$$

$\therefore$ The illumination at 'A' due to lamp $L_1 = \frac{I}{d_2^2}\cos\theta_2$

$$= \frac{300}{(15.62)^2} \times 0.64$$

$$= 0.786 \text{ lux.}$$

Illumination at A due to lamp 'L_2' $= \frac{I}{h_1^2}$

$$= \frac{300}{6^2}$$

$$= 8.33 \text{ lux.}$$

$\therefore$ The total illumination at 'A' due to both lamps = 0.786 + 8.33 = 9.116 lux.

Example 6.13: Four lamps 15 m apart are arranged to illuminate a corridor. Each lamp is suspended at a height of 8 m above the floor level. Each lamp gives 450 CP in all directions below the horizontal; find the illumination at the second and the third lamp.

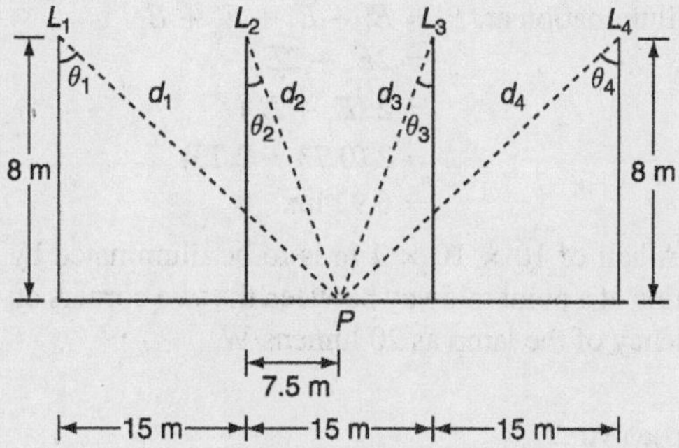

FIG. P.6.11

Solution:

Given data:

Luminous intensity = 450 CP.

Mounting height = 8 m.

Distance between the adjacent lamps = 15 m (Fig. P.6.11).

The illumination at 'P' = the illumination due to L_1 + the illumination due to L_2 + the illumination due to L_3 + the illumination due to L_4.

The illumination at 'P' due to L_1, $E_1 = \dfrac{I}{d_1^2}\cos\theta_1$.

But, $d_1 = \sqrt{8^2+15^2} = 17.$

$$\cos\theta_1 = \frac{h}{d_1} = \frac{8}{17} = 0.470.$$

$$\therefore \mathrm{E}_1 = \frac{I}{d_1^2}\cos\theta_1$$

$$= \frac{450}{(17)^2}\times 0.47$$

$$= 0.73 \text{ lux.}$$

The illumination at 'P' due to lamp 'L_2' is:

$$\mathrm{E}_2 = \frac{I}{d_2^2}\cos\theta_2$$

$$= \frac{450}{\left[\sqrt{8^2+(7.5)^2}\right]^2}\times\frac{8}{\sqrt{8^2+7.5^2}}$$

$$= 2.73 \text{ lux.}$$

Similarly, the illumination at 'P' due to the lamp L_3, 'E_3' = the illumination at 'P' due to the lamp 'L_2', 'E_2',

and the illumination at 'P' due to the lamp L_4, 'E_4'= illumination at 'P' due to the lamp 'L_1', 'E_1.'

$$\therefore \text{The total illumination at 'P'} = E_1 + E_2 + E_3 + E_4$$
$$= 2E_1 + 2E_2$$
$$= 2\,(E_1 + E_2)$$
$$= 2\,(0.73 + 2.73)$$
$$= 6.92 \text{ lux.}$$

Example 6.14: A hall of 10 × 10 × 4 m is to be illuminated by four lamps each 60 W. Find the illumination at a point midway between the two corners on the floor along the side. Assume the efficiency of the lamp as 20 lumens/W.

Solution:

Given data (Fig. P.6.12):

Hall dimensions = 10 × 10 × 4 m.

The number of lamps = 4.

The wattage of each lamp = 60 W.

The efficiency of the lamp, $\eta = 20$.

The luminous flux emitted by each lamp is:

$\phi = \text{wattage} \times \eta$

$= 60 \times 20 = 1{,}200$ lumens.

$$\text{Luminous intensity, } I = \frac{\text{flux}}{4\pi}$$
$$= \frac{1200}{4\pi} = 95.49 \text{ CP.}$$

From Fig. P.6.13:

$$d_1 = \sqrt{10^2 + 4^2} = 10.77 \text{ m.}$$
$$\cos\theta_1 = \frac{h}{d_1 10.77} = 0.3714.$$

The illumination at 'P' due to L_1 is:

$$E_1 = \frac{I}{d_2^2}\cos\theta_1$$
$$= \frac{95.49}{(10.77)^2} \times 0.371$$
$$= 0.3054 \text{ lux.}$$

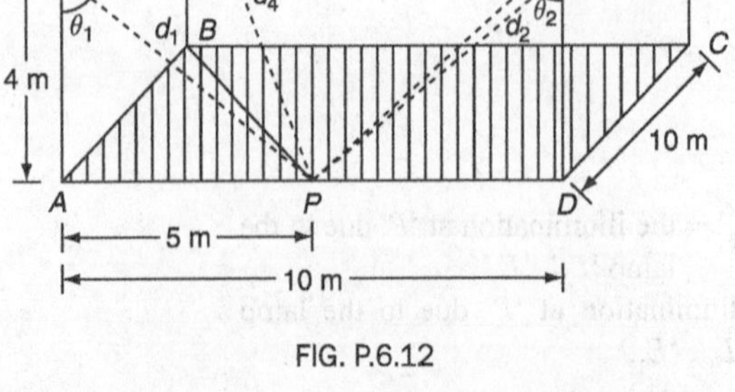

FIG. P.6.12

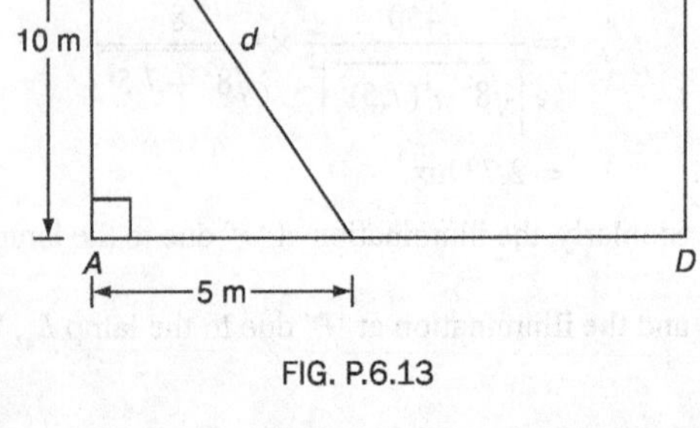

FIG. P.6.13

The illumination E_2, at 'P' due to 'L_2' is will be same as $E_1 = E_2 = 0.3054$ lux.

$$d = \sqrt{10^2 + 5^2}$$
$$= 11.180 \text{ m.}$$
$$\therefore d_4 = \sqrt{4^2 + (11.18)^2}$$
$$= 11.874 \text{ m.}$$
$$\cos\theta_4 = \frac{4}{11.874} = 0.3368.$$

$\therefore$ The illumination at the point 'P' due to L_4 is:

$$E_4 = \frac{I}{d_4^2}\cos\theta_4$$
$$= \frac{95.49}{(11.874)^2} \times 0.3368$$
$$= 0.228 \text{ lux.}$$

The illumination 'E_3' at 'P' due to 'L_3' will be same as to 'E_4'.

$$\therefore E_3 = E_4 == 0.228 \text{ lux.}$$

$\therefore$ The total illumination at the point 'P' $= E_1 + E_2 + E_3 + E_4$
$= 2E_1 + 2E_3$
$= 2\,(E_1 + E_3)$
$= 2(0.3054 + 0.228)$
$= 1.0668$ lux.

Example 6.15: Two lamps of each 500 CP are suspended 10 m from the ground and are separated by a distance of 20 m apart. Find the intensity of illumination at a point on the ground in line with the lamps and 12 m from the base on both sides of the lamps.

Solution:

Given data:

Luminous intensity, $I = 500$ CP.

Mounting height, $h = 10$ m.

Case (i) :

From Fig. P.6.14:

$$d_1 = \sqrt{10^2 + 12^2}$$
$$= 15.62 \text{ m.}$$
$$\cos\theta_1 = \frac{h}{d_1} = \frac{10}{15.62} = 0.64.$$

The illumination at 'P' due to lamp L_1 is:

$$E_1 = \frac{I}{d_1^2}\cos\theta_1$$
$$= \frac{500}{(5.62)^2} \times 0.64$$
$$= 1.3115 \text{ lux.}$$

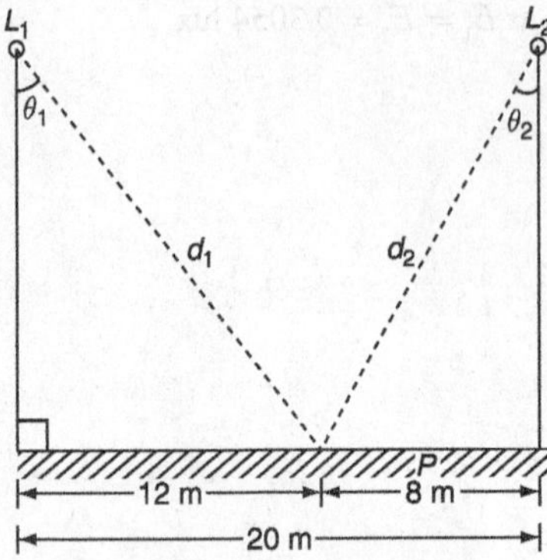

FIG. P.6.14

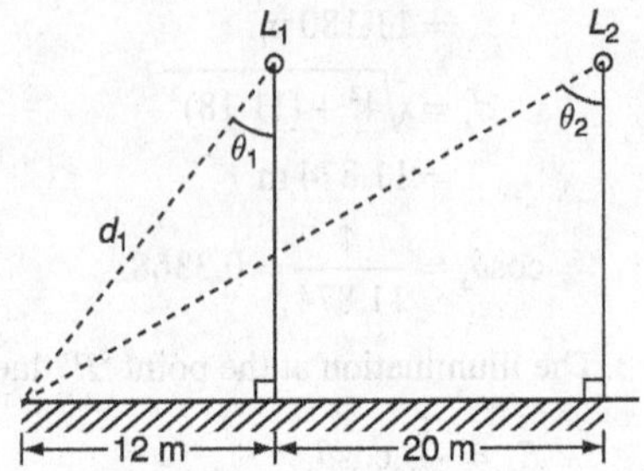

FIG. P.6.15

$$d_2 = \sqrt{8^2 + 10^2} = 12.806 \text{ m.}$$

$$\cos\theta_2 = \frac{h}{d_2} = \frac{10}{12.806} = 0.780.$$

∴ The illumination at 'P' due to lamp 'L_2' is:

$$E_2 = \frac{I}{d_2^2}\cos\theta_2$$

$$= \frac{500}{(12.806)^2} \times 0.78$$

$$= 2.378 \text{ lux.}$$

∴ The total illumination at the point 'P' $= E_1 + E_2$

$$= 1.3115 + 2.378$$

$$= 3.689 \text{ lux.}$$

Case (ii):
From Fig. P.6.15:

$$d_1 = \sqrt{10^2 + 12^2} = 15.62 \text{ m.}$$

$$\cos\theta_1 = \frac{h}{d_1} = \frac{10}{15.62} = 0.64.$$

The illumination at 'P' due to lamp L_1 is:

$$E_1 = \frac{I}{d_1^2} \times \cos\theta_1$$

$$= \frac{500}{(15.62)^2} \times 0.64$$

$$= 1.3115 \text{ lux.}$$

$$d_2 = \sqrt{10^2 + 32^2} = 33.52 \text{ m.}$$

$$\cos\theta_2 = \frac{I}{d_2} = \frac{10}{33.52} = 0.298.$$

The illumination at 'P' due to the lamp 'L_2' is:

$$E_2 = \frac{I}{d_2^2}\cos\theta_2$$

$$= \frac{500}{(33.52)^2} \times 0.298$$

$$= 0.1326 \text{ lux.}$$

$\therefore$ The total illumination at 'P' due to both lamps $= E_1 + E_2$

$$= 1.3115 + 0.1326$$

$$= 1.44 \text{ lux.}$$

Example 6.16: Two similar lamps having luminous intensity 500 CP in all directions below horizontal are mounted at a height of 8 m. What must be the spacing between the lamps so that the illumination on the ground midway between the lamps shall be at least one-half of the illumination directly below the lamp.

Solution:

Given data:

The candle power of lamp, $I = 600$ CP.

The mounting height of lamps form the ground, $H = 8$ m.

Let, the maximum spacing between the lamps $= x$ *m*.

From Fig. P.6.16:

The illumination at 'C' due to the lamp 'L_1' is:

$$E_1 = \frac{I}{h^2}\cos^3\theta_1$$

$$= \frac{600}{8^2} \times \frac{(8)^3}{\left[8^2 + (x/2)^2\right]^{3/2}}.$$

The illumination 'E_2' at 'C' due to the lamp 'L_2' is same as to 'E_1'.

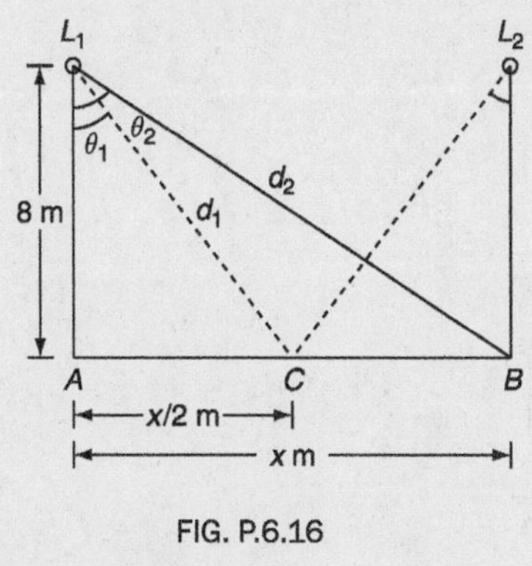

FIG. P.6.16

∴ The total illumination at 'C' due to the lamps, L_1 and L_2 is:

$$E_C = 2\,E_1$$

$$= 2 \times \left[\frac{600}{8^2} \times \frac{8^3}{\left[8^2 + (x/2)^2\right]^{3/2}} \right]$$

$$= \frac{9{,}600}{\left[8^2 + (x/2)^2\right]^{3/2}}.$$

The illumination just below the lamp, L_2 is:

E_B = the illumination due to lamp L_1 + the illumination due to lamp L_2:

$$= \frac{600}{8^2} \times \frac{8^3}{\left[8^2 + x^2\right]^{3/2}} + \frac{600}{8^2}.$$

But, given $E_C = \frac{1}{2} E_B$.

$$\therefore \frac{9{,}600}{\left[8^2 + \left(\frac{x}{2}\right)^2\right]^{3/2}} = \frac{1}{2}\left[\frac{4{,}800}{\left[8^2 + x^2\right]^{3/2}} + 9.375 \right]$$

$$\frac{9{,}600}{\left[8^2 + \left(\frac{x}{2}\right)^2\right]^{3/2}} = \frac{2400}{\left[8^2 + x^2\right]^{3/2}} + 4.6875.$$

Example 6.17: Find the height at which a light source having uniform spherical distribution should be placed over a floor in order that the intensity of horizontal illumination at a given distance from its vertical line may be greatest.

Solution:

Let the luminous intensity of the lamp = 'I' CP.

The illumination at the point 'A' due to source is:

$$E_A = \frac{I}{\sqrt{h^2 + x^2}}.\cos\theta$$

$$= \frac{I}{h^2}\cos^3\theta.$$

But, from Fig. P.6.17:

$$\cos\theta = \frac{h}{\sqrt{h^2 + x^2}}.$$

$$\therefore E_A = \frac{I}{h^2} \times \left[\frac{h}{h^2 + x^2} \right]^3$$

$$= I \times \frac{h}{(h^2 + x^2)^{3/2}}.$$

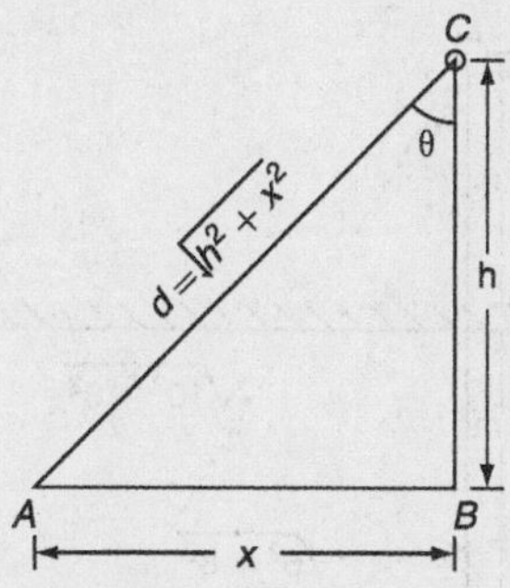

FIG. P.6.17

Given that, the illumination at a point away from the base of lamp may be the greatest:

$$\therefore \frac{dE_A}{dh}=0$$

$$=I\left[\frac{h}{dh}\left[\frac{h}{(h^2+x^2)^{3/2}}\right]\right]=0$$

$$=\frac{(h^2+x^2)^{3/2}\cdot 1-h\cdot\frac{3}{2}(h^2+x^2)^{1/2}\cdot 2h}{\left\{(h^2+x^2)^{3/2}\right\}^2}=0$$

$$=(h^2+x^2)^{1/2}\cdot\left[(h^2+x^2)-3h^2\right]=0$$

$$=x^2-2h^2=0$$

$$\Rightarrow x^2=2h^2$$

$$\Rightarrow h=\frac{x}{\sqrt{2}}=0.707x$$

$$\therefore h=0.707x.$$

Example 6.18: A lamp of 250 candela is placed 2 m below a plane mirror that reflects 60% of light falling on it. The lamp is hung at 6 m above ground. Find the illumination at a point on the ground 8 m away from the point vertically below the lamp.

Solution:

Figure P.6.18 shows the lamp and the mirror arrangements. Here, the lamp '*L*' produces an image '*L*', then the height of the image from the ground = 8 + 2 = 10 m.

And L^1 acts as the secondary sources of light whose candle power is equals to 0.85 × CP of the lamp '*L*'.

i.e., 0.85 × 250 = 212.5 CP.

∴ The illumination at the point '*B*', '8' m away from the lamp = illumination at '*B*' due to *L* + the illumination at '*B*' due to L^1:

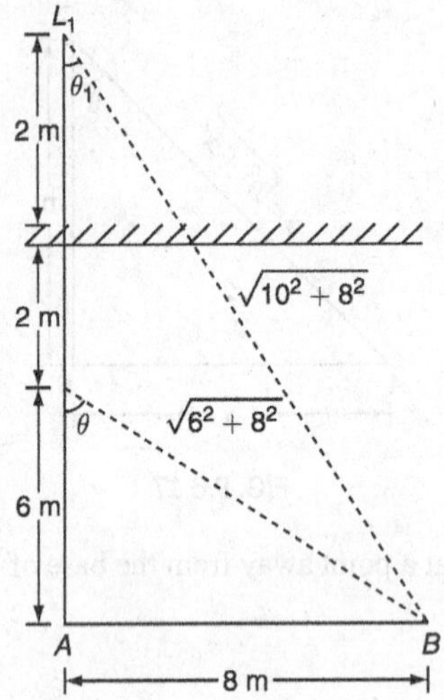

FIG. P.6.18

$$= \frac{250}{\left(\sqrt{6^2+8^2}\right)^2} \times \frac{6}{\sqrt{6^2+8^2}} + \frac{212.5}{\left(\sqrt{10^2+8^2}\right)^2} \times \frac{10}{\sqrt{10^2+8^2}}$$

$$= \frac{1500}{(6^2+8^2)^{3/2}} + \frac{2125}{(10^2+8^2)^{3/2}}$$

$$= 1.5 + 1.0117$$

$$= 2.5117 \text{ lux.}$$

Example 6.19: A light source with an intensity uniform in all direction is mounted at a height of 20 ms above a horizontal surface. Two points '*A*' and '*B*' both lie on the surface with point *A* directly beneath the source. How far is *B* from *A* if the illumination at '*B*' is only 1/15th as great as *A*?

Solution:
Let the luminous intensity of the lamp '*L*' be '*I*' candela and the distance of the point of illumination from the base of the lamp is '*x*' m (Fig. P.6.19).
The illumination at the point '*A*', due to the lamp '*L*' is:

$$E_A = \frac{I}{h^2} = \frac{I}{20^2} = \frac{I}{400}.$$

The illumination at the point '*B*', due to the lamp '*L*' is:

$$E_B = \frac{I}{h^2}\cos^3\theta$$

$$E_B = \frac{I}{(20)^2}\left[\frac{20}{\sqrt{(20^2+x^2)}}\right]^3.$$

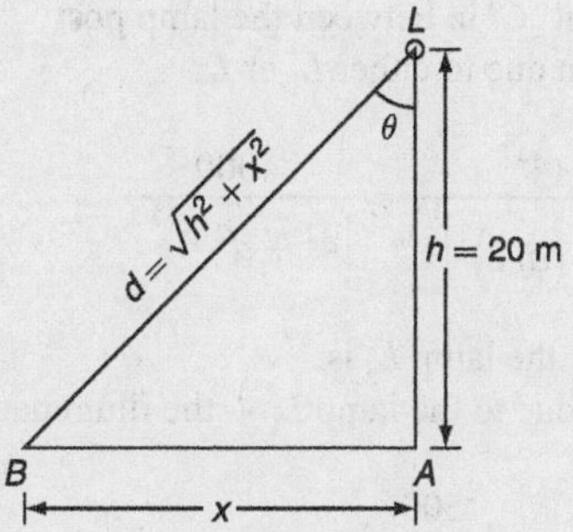

FIG. P.6.19

Given, $E_B = \frac{1}{15} E_A$

$$\frac{20I}{(20^2 + x^2)^{3/2}} = \frac{1}{15} \times \frac{I}{400}$$

$$20 \times 15 \times 400 = (20^2 + x^2)^{3/2}$$

$$2143.98 = 20^2 + x^2.$$

$$x^2 = 1743.98$$

$$x = 41.76 \text{ m}.$$

Example 6.20: Two similar lamps having uniform intensity 500 CP in all directions below the horizontal are mounted at a height of 4 m. What must be the maximum spacing between the lamps so that the illumination on the ground midway between the lamps shall be at least one-half the illuminations directly under the lamps?

Solution:

The candle power of the lamp = 500 CP (Fig. P.6.20).
The height of the lamps from the ground, $h = 4$ m.
Let the maximum spacing between the lamps be of 'd' meters.

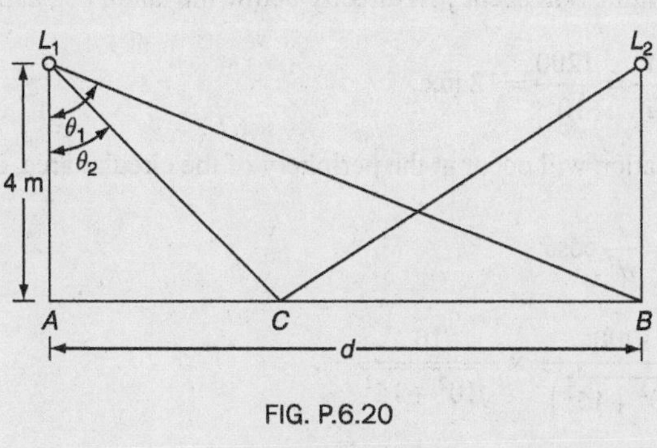

FIG. P.6.20

The illumination at the point 'C' in between the lamp post

$= 2 \times$ Illumination due to either L_1 or L_2

$$E_C = 2 \times \frac{500}{4^2} \times \frac{4^3}{\left[4^2 + (d/2)^2\right]^{3/2}} = \frac{4000}{\left[4^2 + d^2/4\right]^{3/2}}.$$

The illumination just below the lamp L_2 is:

$E_B =$ the illumination due to the lamp L_1 + the illumination due to the lamp L_2:

$$= \frac{500}{4^2} \times \frac{4}{\left[4^2 + d^2/2\right]^{3/2}} + \frac{500}{4^2}$$

$$= \frac{2{,}000}{(4^2 + d^2)^{3/2}} + 31.25.$$

Given:

$$E_C = \frac{1}{2} E_B$$

$$\frac{4000}{\left[4^2 + d^2/4\right]^{3/2}} = \frac{1}{2}\left[31.25 + \frac{200}{(4^2 + d^2)^{3/2}}\right]$$

$$\frac{4{,}000}{(4^2 + d^2/4)^{3/2}} = 15.625 + \frac{1{,}000}{(4^2 + d^2)^{3/2}}$$

$$\therefore d = 9.56 \text{ m}.$$

Example 6.21: A lamp with a reflector is mounted 10 m above the center of a circular area of 30-m diameter. If the combination of lamp and reflector gives a uniform CP of 1,200 over circular area, determine the maximum and minimum illumination produced.

Solution:

The mounting height of the lamp $h = 10$ m (Fig. P.6.21, P.6.22).
The diameter of the circular area = 30 m.
The candle power of the lamp $I = 1{,}200$ CP.

The maximum illumination occur just directly below the lamp, i.e., at point 'C' is:

$$E_C = \frac{I}{d^2} = \frac{I}{h^2} = \frac{1200}{10^2} = 12 \text{ lux}.$$

Minimum Illumination will occur at the periphery of the circular area, i.e., at A (or) B.

$$\therefore E_A = E_B = \frac{I}{d^2}\cos\theta$$

$$= \frac{1200}{\left(\sqrt{10^2 + 15^2}\right)^2} \times \frac{10}{\sqrt{10^2 + 15^2}}$$

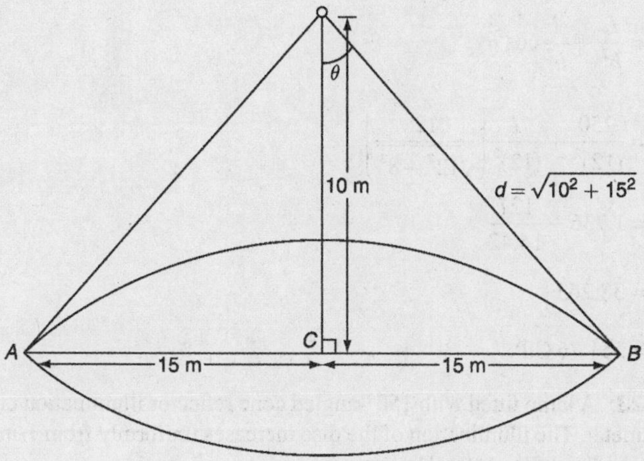

FIG. P.6.21

$$= \frac{12,000}{(10^2 + 15^2)^{3/2}}$$

$= 2.048$ lux.

Example 6.22: Two lamps hung at a height of 12 m from the floor level. The distance between the lamps is 8 m. Lamp one is of 250 CP. If the illumination on the floor vertically below this lamp is 40 lux, find the CP of the second lamp.

Solution:

Given data:

The candle power of the lamp, $I = 250$ CP.

The intensity of L_1 illumination just below the lamp $L_1 = 40$ lux.

Let CP of $L_2 = I$ CP.

$\therefore$ The illumination at the point A = the illumination due to the lamp L_1 + the illumination due to the lamp L_2:

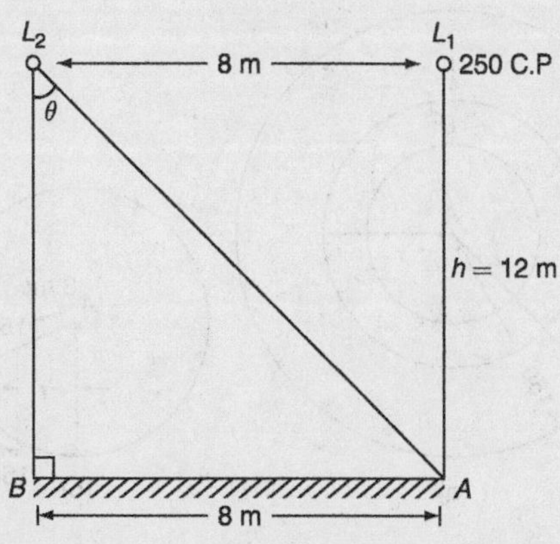

FIG. P.6.22

$$40 = \frac{I_1}{h^2} + \frac{I}{h^2}\cos^3\theta$$

$$= \frac{250}{(12)^2} + \frac{I}{(12)^2}\left(\frac{12}{\sqrt{12^2+8^2}}\right)^3$$

$$= 1.736 + \frac{12I}{14.42}$$

$$\frac{12I}{14.42} = 38.263$$

$$I = 551.76 \text{ C.P.}$$

Example 6.23: A lamp fitted with 150°-angled cone reflector illumination circular area of 300 m in diameter. The illumination of the disc increases uniformly from 1-m candle at the edge to 3-m candle at the center. Determine:

(i) The total light received.

(ii) The average illumination of the disc.

(iii) The average CP of the source.

Solution:

Area illuminated from the source and the variation of meter candle from edge to the center.

(i) Let us consider a circular strip of width '*dr*' and radius '*r*' from the center as shown in Fig. P.6.23(a):

∴ The area of the strop = $2\pi r \times dr$.

The intensity of illumination at radius '*r*'

$$= 1 + \frac{150 - r}{150} \times 2.$$

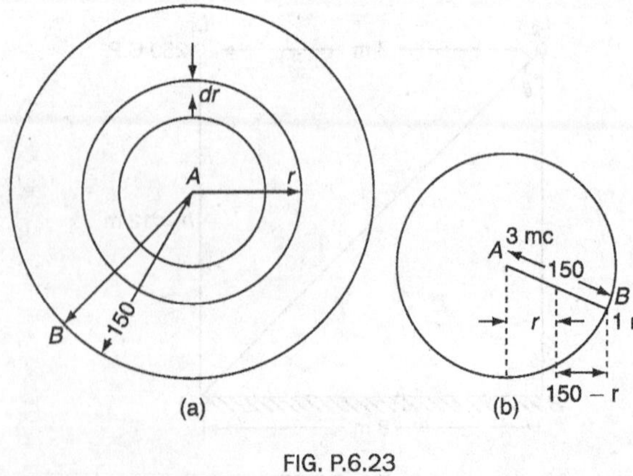

FIG. P.6.23

The total flux falling on the strop is:

= area × intensity of illumination

$$= 2\pi r \times dr \times 1 + \frac{150-r}{150} \times 2 \text{ lumens.}$$

∴ The total light falling on the circular area:

$$= \int_0^{150} 2\pi r\left(\frac{450-2r}{100}\right)dr$$

$$= \int_0^{100} (9\pi r - 0.04\pi r^2)\,dr$$

$$= (9\pi)\frac{r^2}{2}\int_0^{100} - (0.04\pi)\frac{r^3}{3}\int_0^{100}$$

$$= 4.5\pi(100^2) - 0.04\pi(100)^3$$

=15,707.96 lumens.

(ii) The average illumination:

$$= \frac{\text{total illumination}}{\text{area}} = \frac{15707.96}{\pi \times 150^2} = 0.2222.$$

(iii) The average CP of the source:

$$= \frac{\text{total illumination}}{\text{solid angle}}.$$

Consider a small circular strip with an angle $d\theta$.
From Fig. P.6.23(b):

The radius of the strip = $r \sin\theta$.
The circumference of the circular strip = $2\pi r \sin\theta$.
The width of the strip = $(2\pi r \sin\theta)\, r\, d\theta$.
The area of the strip = $(2\, \pi r \sin\theta)\, r\, d\theta$
= $2\pi r^2 \sin\theta\, d\theta$.

The total area with a cone angle of 150° or for θ varying from 0 to 30° (or) from 0 to $\frac{\pi}{6}$

$$= \int_0^{\pi/6} 2\pi r^2 \sin\theta \cdot d\theta.$$

$$\therefore \text{ The solid angle } = \frac{\text{area}}{r^2} = \int_0^{\pi/6} \frac{2\pi r^2 \sin\theta}{r^2}\,d\theta$$

$$= \int_0^{\pi/6} 2\pi \sin\theta . d\theta = -2\pi(\cos\theta)\int_0^{\pi/6}$$

$$= -2\pi\left(\sqrt{3}/2 - 1\right) = 0.8417 \text{ steradians.}$$

$$\text{The average CP of the source} = \frac{15{,}707.96}{0.8417} = 18{,}662.18 \text{ candela.}$$

6.4 POLAR CURVES

The luminous flux emitted by a source can be determined using the intensity distribution curve. Till now we assumed that the luminous intensity or the candle power from a source is distributed uniformly over the surrounding surface. But due to its unsymmetrical shape, it is not uniform in all directions. The luminous intensity or the distribution of the light can be represented with the help of the polar curves.

The polar curves are drawn by taking luminous intensities in various directions at an equal angular displacement in the sphere. A radial ordinate pointing in any particular direction on a polar curve represents the luminous intensity of the source when it is viewed from that direction. Accordingly, there are two different types of polar curves and they are:

(i) A curve is plotted between the candle power and the angular position, if the luminous intensity, i.e., candle power is measured in the horizontal plane about the vertical axis, called '*horizontal polar curve*'.

(ii) A curve is plotted between the candle power, if it is measured in the vertical plane and the angular position is known as '*vertical polar curve*'.

Figure 6.12 shows the typical polar curves for an ordinary lamp.

Depression at 180° in the vertical polar curve is due to the lamp holder. Slight depression at 0° in horizontal polar curve is because of coiled coil filament.

Polar curves are used to determine the actual illumination of a surface by employing the candle power in that particular direction as read from the vertical polar curve. These are also used to determine mean horizontal candle power (MHCP) and mean spherical candle power (MSCP).

The mean horizontal candle power of a lamp can be determined from the horizontal polar curve by considering the mean value of all the candle powers in a horizontal direction.

The mean spherical candle power of a symmetrical source of a light can be found out from the polar curve by means of a Rousseau's construction.

6.4.1 Rousseau's construction

Let us consider a vertical polar curve is in the form of two lobes symmetrical about XOX^1 axis. A simple Rousseau's curve is shown in Fig. 6.13.

Rules for constructing the Rousseau's curve are as follows:

(i) Draw a circle with any convenient radius and with '*O*' as center.

(ii) Draw a line '*AF*' parallel to the axis XOX^1 and is equal to the diameter of the circle.

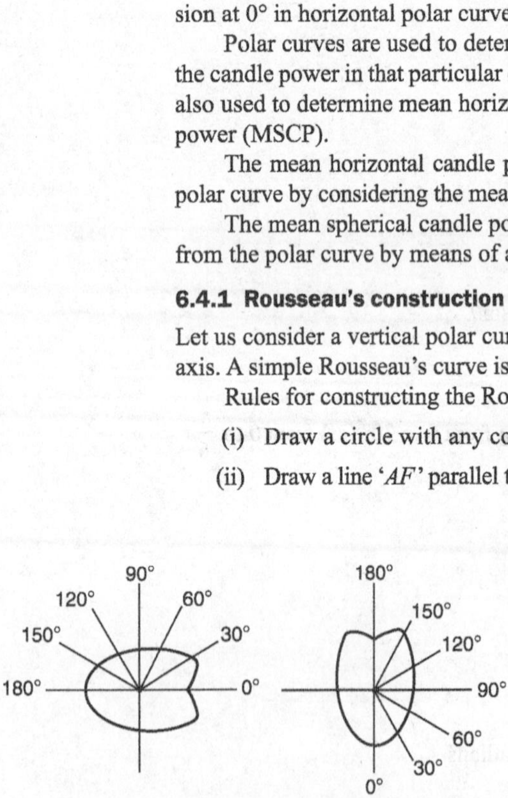

(a) Horizontal polar curves (b) Vertical polar curves

FIG. 6.12 *Polar curves*

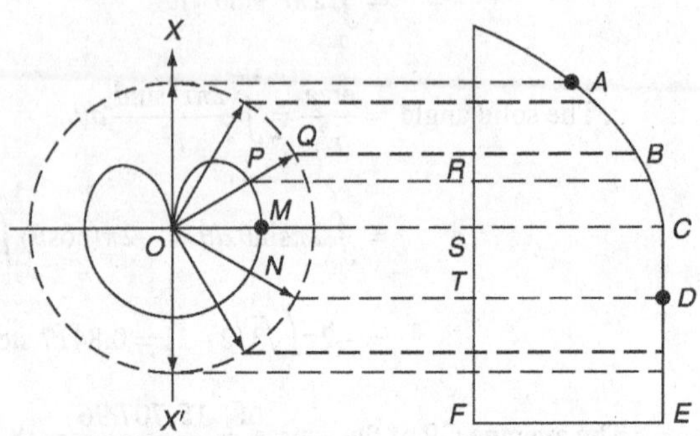

FIG. 6.13 *Rousseau's curve*

(iii) Draw any line '*OPQ*' in such a way that the line meeting the circle at point '*Q*'. Now let the projection be '*R*' onto the parallel line '*AF*'.

(iv) Erect an ordinate at '*R*' as, $RB = OP$.

(v) Now from this line '*AF*' ordinate equals to the corresponding radius on the polar curve are setup such as $SC = OM$, $TD = ON$, and so on.

(vi) The curve *ABC DEFA* so obtained by joining these ordinates is known as Rousseau's curve.

The mean ordinate of this curve gives the mean spherical candle power (MSCP) of the lamp having polar curve given in Fig. 6.13.

The mean ordinate of the curve:

$$= \frac{\text{area of } ABCDEFA}{\text{length of } AF}.$$

The area under the Rousseau's curve can be determined by Simpson's rule.

6.5 PHOTOMETRY

Photometry involves the measurement of candle power or luminous intensity of a given source. Now, we shall discuss the comparison and measurement of the candle powers.

The candle power of a given source in a particular direction can be measured by the comparison with a standard or substandard source. In order to eliminate the errors due to the reflected light, the experiment is conducted in a dark room with dead black walls and ceiling. The comparison of the test lamp with the standard lamp can be done by employing a photometer bench and some form of photometer.

6.5.1 Principle of simple photometer

The photometer bench essentially consists of two steel rods with 2- to 3-m long. This bench carries stands or saddles for holding two sources (test and standard lamps), the carriage for the photometer head and any other apparatus employed in making measurements. Graduated scale in centimeters or millimeters in one of the bar strips. The circular table is provided with a large graduated scale in degrees round its edge so that the angle of the rotation of lamp from the axis of bench can be measured.

The photometer bench should be rigid so that the source being compared may be free from vibration. The photometer head should be capable of moving smoothly and the photometer head acts as screen for the comparison of the illumination of the standard lamp and the test lamp.

The principle methods of measurement are based upon the inverse square law.

The photometer bench consists of two sources, the standard source '*S*' whose candle power is known and the other source '*T*' whose candle power is to be determined. The photometer head acts as screen is moved in between the two fixed sources until the illumination on both the sides of screen is same. A simple arrangement for the measurement of the candle power of the test source is shown in Fig. 6.14.

If the distances of the standard source '*S*' and the test source '*T*' from the photometer head are L_1 and L_2, respectively, then, according to the inverse square law, if the illumination on both the sides of screen are equal then the candle power of the source is proportional to the square of the distance between the source and the photometer head.

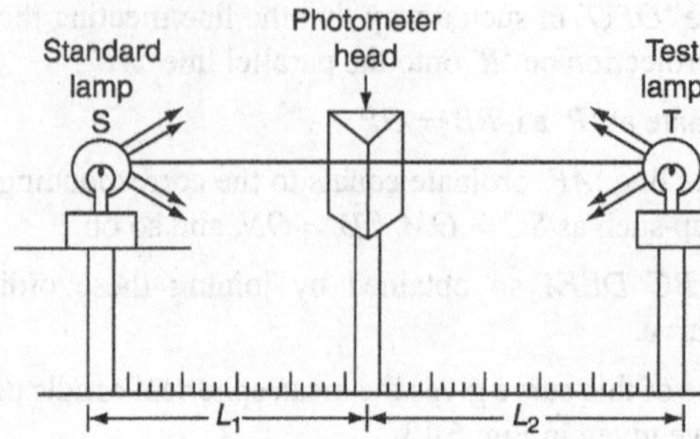

FIG. 6.14 *Measurement of candle power*

The CP of standard source $\propto L_1^2$.
The CP of test source $\propto L_2^2$.

$$\therefore \frac{\text{CP of test source}}{\text{CP of standard source}} = \frac{L_2^2}{L_1^2}$$

$$\therefore \text{CP of test source} = S \times \frac{L_2^2}{L_1^2}.$$

In order to obtain the accurate candle power of test source, the distance of the sources from the photometer head should be measured accurately.

6.5.2 Photometer heads

The photometer heads that are most common in use are:

(i) Bunsen grease spot photometer.
(ii) Lumer–Brodhun photometer.
(iii) Flicker photometer.

The first two are best suited for use, if the two sources to be compared give the light of same or approximately similar colors. Increase the light from the two sources to be compared differ in color, a flicker photometer is best suited.

(I) Bunsen grease spot photometer

Bunsen photometer consists of a tissue paper, with a spot of grease or wax at its center. It held vertically in a carrier between the two light sources to be compared. The central spot will appear dark on the side, having illumination in excess when seen from the other side. Then, the observer will adjust the position of photometer head in such a way that until the semitransparent spot and the opaque parts of the paper are equally bright then the grease spot is invisible, i.e., same contrast in brightness is got between the spot and the disc when seen from each sides as shown in Fig. 6.15. The distance of the photometer from the two sources is measured. Hence, the candle power of test source is then determined by using relation:

$$\text{The CP of the test lamp} = \text{the CP of the standard lamp} \times \left(\frac{L_2}{L_1}\right)^2.$$

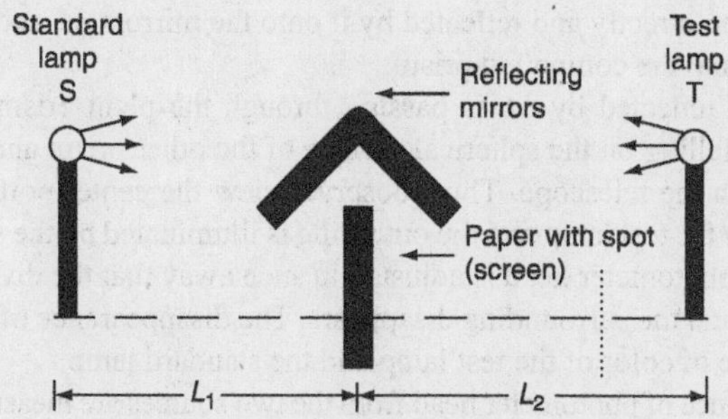

FIG. 6.15 *Bunsen grease spot photometer*

The use of two reflecting mirrors above the photometer head makes it perhaps the accurate method, since the two sides of spot and position of the head can be viewed simultaneously.

(ii) Lumer–Brodhun photometer

There are two types of Lumen–Brodhun photometer heads.

(a) Equality of brightness type.

(b) Contrast type.

The Contrast type is more accurate and therefore, extensively used in the photometric measurements.

(a) Equality of brightness type photometer head

The photometer head essentially consists of screen made of plaster of Paris, two mirrors M_1 and M_2, glass cube or compound prism, and a telescope.

The compound prism made up of two right-angled glass prisms held together, one of which has sand blasted pattern on its face, i.e., principal surface as spherical with small flat portion at the center and the other is perfectly plain. A typical Lumer–Brodhun photometer head is shown in Fig. 6.16.

The two sides of the screen are illuminated by two sources such as the standard and test lamps as shown in Fig. 6.16. The luminous flux lines emitting from the two sources are

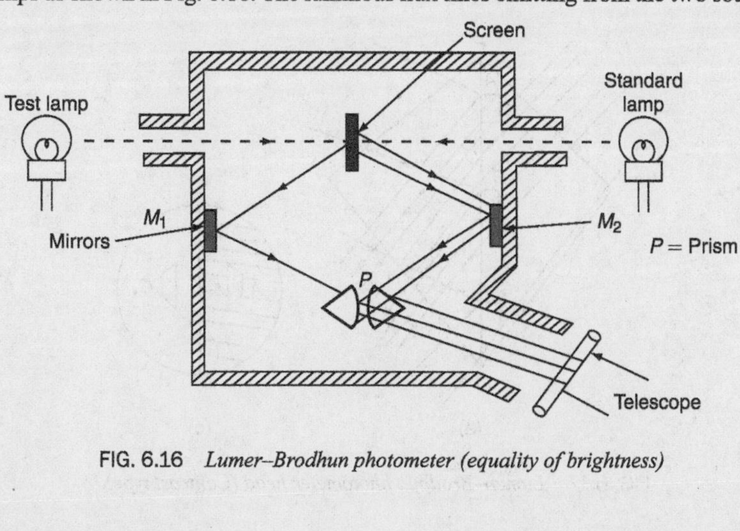

FIG. 6.16 *Lumer–Brodhun photometer (equality of brightness)*

falling on the screen directly and reflected by it onto the mirrors M_1 and M_2, which in turn reflects the same onto the compound prism.

The light ray reflected by M_1 is passing through the plain prism and the light ray reflected by M_2 is falling on the spherical surface of the other prism and is reflected again which pass through the telescope. Thus, observer view the center portion of the circular area illuminated by the test lamp and the outer ring is illuminated by the standard lamp. The positioning of the photometer head is adjusted in such away that the dividing line between the center portion and the surrounding disappears. The disappearance of dividing line indicates the same type of color of the test lamp and the standard lamp.

Now, the distance of photometer head from the two sources are measured and the candle power or luminous intensity of test lamp can be calculated by using inverse square law.

(b) Contrast type photometer head

Similar to the equal brightness type photometer, it consists of a compound prism, which is made up of two right-angled glass prism. The joining surfaces of the two right-angled glass prisms are flat, but one of the prism has its hypotenuses surface etched away at *A*, *B*, and *C* to get pattern of the type shown in Fig. 6.17.

As in case of equal brightness type, the light falling on the both sides of the screen passes through the unetched portion of the joining surface and gets reflected at the etched surfaces (*A*, *B*, and *C*). *P* and *Q* are the sheets of glass that give little reflected light to maintain the difference between the illuminations of both the etched and the unetched portions. If the illumination of the surfaces of the prism is different, then the etched portion will have difference in illumination as compared to unetched portion.

If the balance is got, the difference in illuminations of both etched and unetched portions are same and equal to half of the circular area; then, the photometer head is said to be in a balance position. When the balance position is altered, the difference or the contrast in the illumination of area '*C*' and its surrounding area decreases. In addition, the contrast illumination area AB and the inner trapezium will increase. Generally, in balanced position, the contrast is about 8%. The position of photometer head is adjusted in such a way that the equal contrast is obtained between the etched and the unetched portions. This contrast type of the head gives accuracy within 1%.

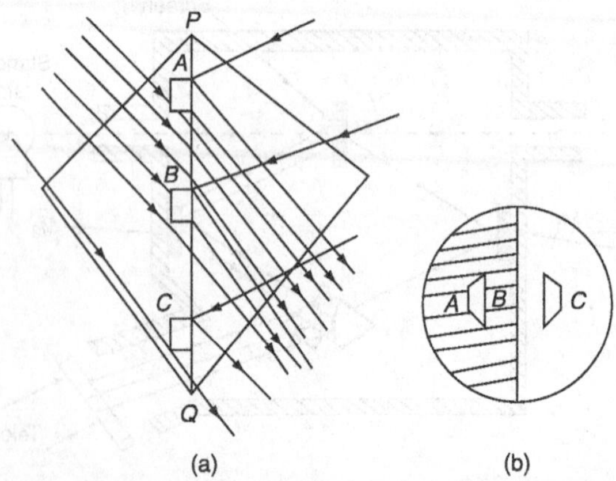

FIG. 6.17 *Lumen–Brodhun photometer head (Contrast type)*

(iii) Flicker photometer

The flicker photometers are employed when two sources giving light of different colors to be compared. The color contrast between two lights do not affect their working is the unit feature of the flicker photometer. This is because the color contrast between the two alternating fields of the light disappears at a lower speed of alternation than does a contrast of brightness.

A typically used flicker photometer is a Simmance–Abady flicker photometer, where used rotating disc made up of plaster of Paris. The dick is in the form of a double-truncated cone as shown in Fig. 6.18. The truncated portions of cone are fitted together to form the disc. The disc is continuously rotated at the required minimum speed by small motor in between the two sources to be compared. Each half of the disc is illuminated from one source and the eye is presented with the two fields of the light to be compared alternately. When the two halves are having unequal illuminations a flicker appears. Now, the disc is rotated to that position where the flicker disappears. When the two halves of the disc are illuminated equally and then the candle power of the test source can be calculated by measuring the distances of the disc from the two sources in the usual manner.

6.6 PHOTO CELLS (FOR PHOTOMETRIC MEASUREMENTS)

Photo cells are employed for photometric measurements. Except for the measurement of the luminance of source, the measurements by visual comparison have been replaced by these photocells.

Photo cells have various advantages over the conventional methods of photometric measurements are:

- These cells give more accurate and faster measurements.
- The more complex procedures and apparatus involved in the visual measurements are thus avoided.
- The measurement by photocells is consistent.

Generally used photocells for the photometric measurements are:

(i) photo voltaic cell and

(ii) photo emissive cell.

The photo voltaic cell is most widely used one because of its simplicity and associated circuits.

6.6.1 Photo voltaic cell

Photo voltaic cell is also known as *barrier layer or rectified cell*. The construction of a photo voltaic cell is shown in Fig. 6.19. It consists of a base metal plate and it is made up of either steel or aluminum over which a metallic selenium layer is situated which is light sensitive.

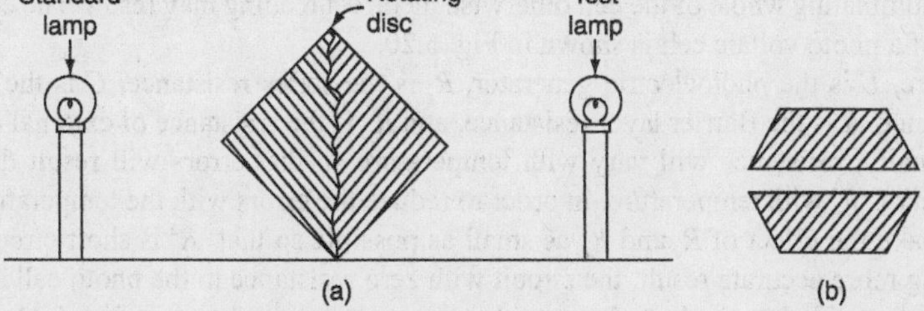

FIG. 6.18 *Flicker photometer*

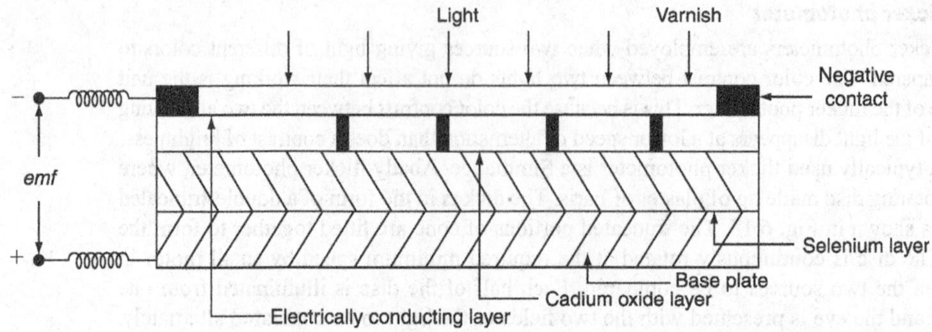

FIG. 6.19 *Photo voltaic cell*

An electrically conducting layer of cadmium oxide is applied by sputtering over the selenium layer. The layer is sufficiently thin in order to allow light to reach the selenium and is electrically continuous which acts as the negative pole. A strip of wood's metal sprayed onto the edge of the top surface forms the negative contact, and the base plate forms the positive contact. The transparent varnish is used to protect the front surface of the cell. When light falls on the surface of selenium layer through cadmium oxide layer, selenium compound releases the electrons that are sufficient to maintain the flow of current through the external circuit connected between the positive and the negative contacts. Some important points should be remember while taking the photometric measurements by the photocell are:

(i) In photometry, the current output of a photo cell should be proportional to the illumination that can be achieved by keeping the external resistance to a low value.

(ii) Illumination should not be allowed to exceed 25 lumens/sq. ft.

(iii) The better linear relation is obtained between current and illumination, with small size of the cell of course compatible, by measuring the sufficient current properly. This is because, for such type of cell, the resistance of the electrically conducting filament is minimum. And also, since the current is small, the voltage drop due to the circuit resistance is minimum.

Main drawback of the photo voltaic cell is if light incident on the surface of cell at angle 60°, the laquer or varnish tends to reflect a significant amount of light, so that the reflected light does not reach the selenium layer. Thus, reading is less what it should be according to the cosine law of illumination. This can be prevented by using a matt laquer or the cell is covered with a hemispherical dome of transparent plastic. More care should be taken while illuminating whole of the cell otherwise incorrect reading may result. The equivalent circuit of a photo voltaic cell is shown in Fig. 6.20.

Here, E is the photoelectric generator, R_s is the series resistance, C is the effective capacitance, R is the Barrier layer resistance, and R_L is the resistance of external circuit.

Usually, resistance will vary with temperature so that errors will result due to the variation of 'R' with temperature. In order to reduce the errors with the temperature variations, make the effect of R_s and R_L as small as possible so that 'R' is short circuited. For obtaining more accurate result, the circuit with zero resistance to the photo cell should be used. The modified equivalent circuit with zero resistance is shown in Fig. 6.21.

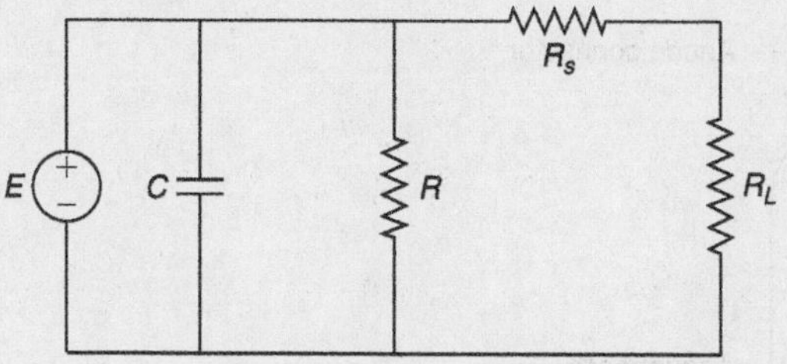

FIG. 6.20 *Equivalent circuit of photo voltaic cell*

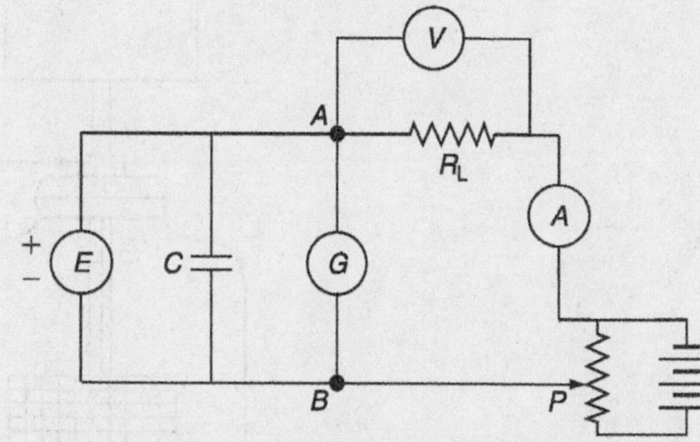

FIG. 6.21 *Modified equivalent circuit*

Zero resistance to the photo voltaic cell is obtained by adjusting the potentiometer in such a way that the potentials at the points *A* and *B* should be zero, across which the galvanometer is balanced, i.e., *A* and *B* are short circuited and hence '*R*' is short circuited. It is necessary for obtaining the linear relation between illumination and the current since no current is diverted through '*R*'.

6.6.2 Photo emissive cell

In order to get the greater precision in terms of linearity and stability, the photo emissive cell is used rather than photo voltaic cell. But the circuit involved in photo emissive cell is more complex and requires some sort of valve amplifiers.

Simple construction of a photo emissive cell is shown in Fig. 6.22. It consists of anode made up of a cylindrical wire mesh with cathode placed in its axis. The cathode is made up Bi-O-Ag-Cs compound, when light falls on it, which tend to emit electrons. The anode is normally 30–50 V at a higher potential as compared to the cathode. In order to match the spectral response of the cell with the human eye, a suitable solution of potassium dichromate and cupric chloride are mixed to get the blue and red ends of the response.

Here, the circuitry involved in photo emissive cell is more complex and requires some sort of valve amplifier. A typical DC amplifier to be used along with the photo cell is shown in Fig. 6.23.

Normally, the amplifier used in a photo cell is DC amplifier and is basically a Wheatstone bridge circuit. V_1 and V_2 are the two valve amplifiers with a very high input resistance to prevent the diverting current from the grid leak resistors R_1 and R_2, galvanometer '*G*' indicates the balance position of bridge circuit. The dark current or the out of balance current will disturb the grid bias of V_1, thereby changing its effective resistance is indicated by the galvanometer '*G*', this can be prevented by adjusting the variable resistance 'R_3' with '*P*' set to zero. The bridge balance is restored by adjusting *P*, when light falls on the photo cell.

6.7 INTEGRATING SPHERE

The integrating sphere consists of a hollow sphere whose diameter is one or more than one meter. The inner surface of the sphere is to be coated with white paint of reflection factor 'ρ' has uniform diffusing properties so that illumination is produced over the whole surface.

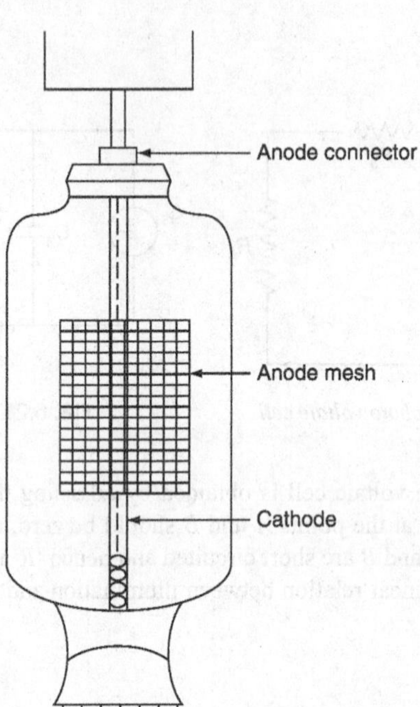

FIG. 6.22 *Photo emissive cell (vacuum type)*

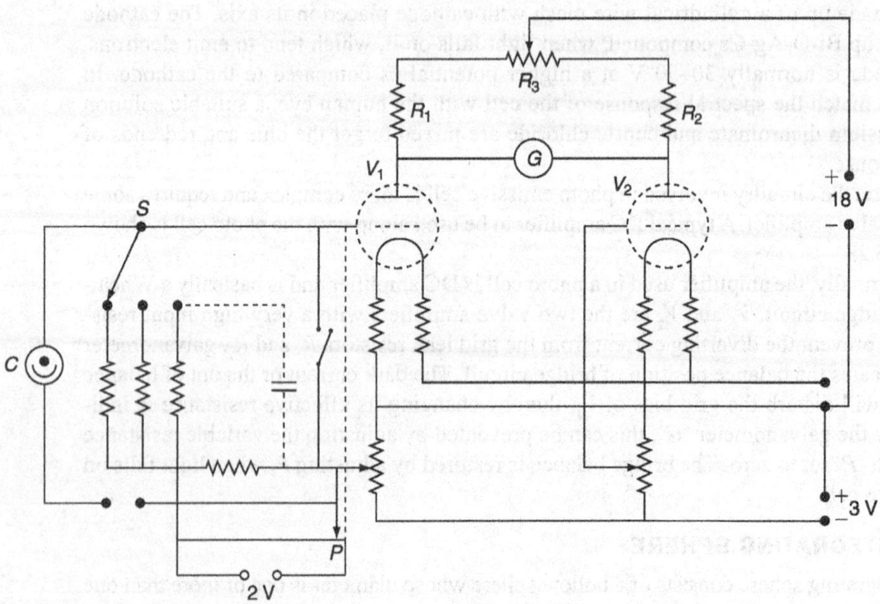

FIG. 6.23 *DC amplifier used in photo cell*

A source '*S*' is long inside the sphere, with a small window of translucent glass provided at one side of the sphere is illuminated by reflected light from the inner surface of the sphere. A small screen known as baffle is inserted between the lamp and window so that the light does not reach the window directly. A typical arrangement is shown in Fig. 6.24.

'The operation of integrating sphere used for photometry is based on the principle that the illumination received on one area of a sphere from another part is independent of the relative positions of two parts'. For an integrating sphere, the illumination at any point on the spherical surface is proportional to the flux emitted and the MSCP of the lamp but is independent of the position of source or fitting. Hence, the luminous intensity of the source may not be the same in all directions and no need of mounting the source only at the center of the sphere.

Let us assume the initial brightness of a differential element dA of the sphere wall be 'L', 'ρ' be the reflection factor of the wall, then the total flux reflected by the sphere wall is given as:

$$\phi = \frac{1}{\rho}\int LdA. \tag{6.13}$$

The luminous flux reflected by the wall after first reflection will illuminate all parts of the sphere.

So, the illumination of all parts of the sphere due to the reflected light from dA is:

$$\frac{1}{A}\cdot LdA.$$

In addition, the illumination of all parts of the sphere due to the reflected light from all parts of the sphere will be:

$$E_{11} = \frac{1}{A}\int L\cdot dA$$

$$E_{11} = \frac{\phi\times\rho}{A}. \tag{6.14}$$

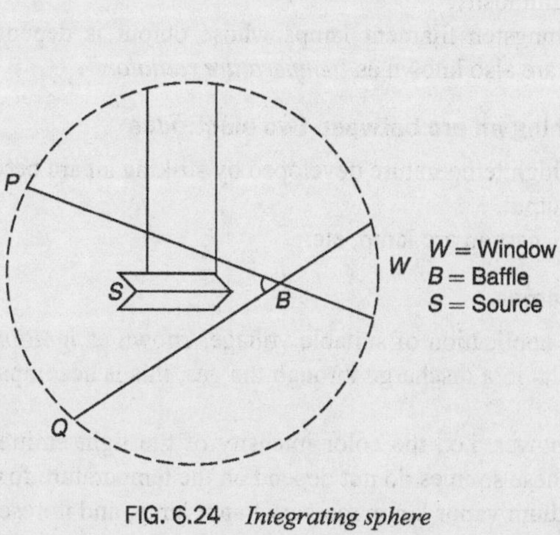

FIG. 6.24 *Integrating sphere*

Similarly, the illumination due to second reflection is:

$$E_{12} = \left(\frac{\rho^2}{A}\right) \times \phi. \tag{6.15}$$

∴ The total illumination of the sphere due to infinity reflections is given by:

$$E = E_{11} + E_{12} + \cdots + E_{1\infty}$$

$$= \frac{\rho \times \phi}{A} + \frac{\rho^2 \phi}{A} + \cdots + \frac{\rho^K \phi}{A} + \cdots$$

$$= \frac{\rho\phi}{A}\left[1 + \rho + \rho^2 + \cdots \infty\right]$$

$$= \frac{\rho\phi}{A}\left[\frac{1}{1-\rho}\right]. \tag{6.16}$$

Hence from Equation (6.16), it is proved that the illumination of the sphere is independent of the distribution from the source or fitting.

6.8 SOURCES OF LIGHT

Light is the prime factor in the human life, as all the activities of human being ultimately depend upon the light. The natural source of light is the 'Sun', which emits both heat and light energy at a very tremendous rate. Where there is no natural light, artificial light is made. The artificial lighting produced by electricity is playing an important role due to its cleanliness, ease of control, reliability, etc. In a boarder sense, the different methods of producing light by electricity may be divided into the following three groups.

6.8.1 By temperature effect

Here, in this method, an electric current is made to pass through a fine metallic filament of thin wire placed in vacuum or inert gas. The current develops enough heat to raise the temperature of filament that emits light waves, which falls in the visible region of wavelength, i.e., which causes luminosity.
Ex: Incandescent tungsten filament lamps whose output is depending on its filament temperature. These are also known as '*temperature radiators*'.

6.8.2 By establishing an arc between two electrodes

In this method, the high temperature developed by striking an arc between two electrodes, which cause light output.
Ex: Flame arc lamp, carbon arc lamp, etc.

6.8.3 Discharge lamps

In this method, the application of suitable voltage, known as *ignition voltage*, across the two electrodes results in a discharge through the gas, this is accompanied by electromagnetic radiation.

Here, candle power, i.e., the color intensity of the light emitted depends upon the nature of the gas. These sources do not depend on the temperature for higher efficiencies.
Ex: Neon lamp, sodium vapor lamp, mercury vapor lamp, and florescent lamp.

KEY NOTES

- **Light** is a form of electromagnetic energy radiated from a body, which is capable of being received by the human eye.
- **Luminous flux** is defined as the energy in the form of light waves radiated per second from a luminous body.
- **Radiant efficiency** is defined as the ratio of energy radiated in the form of light, produces the sensation of vision to the total energy radiated out by the luminous body.
- **Plane angle** is the angle subtended at a point in a plane by two converging lines.
- **Solid angle** is the angle subtended at a point in space by an area, i.e., the angle enclosed in the volume formed by numerous lines lying on the surface and meeting at the point.
- **Lamp efficiency** is defined as the ratio of the total luminous flux emitting from the source to its electrical power input in watts.
- **MHCP** is defined as the mean of candle power of the source in all directions in the horizontal plane.
- **MSCP** is defined as the mean of candle power of the source in all directions in all planes.
- **MHSCP** is defined as the mean of candle power of source in all directions above or below the horizontal plane.
- **Inverse square law** states that 'the illumination of a surface is inversely proportional to the square of distance between the surface and a point source'.
- **Lambert's cosine law** states that 'illumination, E at any pint on a surface is directly proportional to the cosine of the angle between the normal at that point and the line'.
- **Photometry** involves the measurement of the candle power or luminous intensity of a given source. The photometer heads used for photometry are:
  - (i) Bunsen grease spot photometer.
  - (ii) Lumer–Brodhun photometer.
  - (iii) Flicker photometer.
- **Photo cells** are employed for photometric measurements are:
  - (i) Photo voltaic cell.
  - (ii) Photo emissive cell.
- **Integrating sphere** is used for the photometry and the illumination of the sphere is independent of the distribution from the source or fitting.

SHORT QUESTIONS AND ANSWERS

(1) What is light?

It is defined as the radiant energy from a hot body that produces the visual sensation upon the human eye. It is expressed in lumen-hours and it analogous to watt-hours, which denoted by the symbol '*Q*'.

(2) Write the expression that shows the relation between solid angle and plane angle.

$$\omega = 2\pi\left(1-\cos\frac{\theta}{2}\right).$$

(3) States the inverse square law of illumination.

This law states that 'the illumination of a surface is inversely proportional to the square of distance between the surface and a point source'.

(4) States the Lambert's cosine law of illumination.

This law states that 'illumination, *E* at any pint on a surface is directly proportional to the cosine of the angle between the normal at that point and the line of flux'.

(5) Define the MSCP.

It is defined as the mean of the candle power of the source in all directions in horizontal plane.

(6) Define the MHCP.

It is defined as the mean of the candle power of the source in all directions in all planes.

(7) Define the MHSCP.

It is defined as the mean of the candle power of the source in all directions above or below the horizontal plane.

(8) What is the need of polar curves?

The luminous flux emitted by a source can be determined from the intensity distribution curve. But the luminous intensity or the candle power of any practical lamp is not uniform in all directions due to its unsymmetrical shape. The luminous intensity or the distribution of such sources can be determined by polar curves.

(9) List out the types of photometers used for the photometric measurements.

The photometer heads that are most common in use are:

(i) Bunsen grease spot photometer.

(ii) Lumer–Brodhun photometer.

(iii) Flicker photometer.

(10) What is photometry?

Photometry means the measurement of the candle power or the luminous intensity of a given source. The candle power of any test source is measured with the comparison of a standard source.

(11) List out the various photocells used for photometric measurements.

Generally used photocells for photometric measurements are:

- photo voltaic cell and
- photo emissive cell.

The photo voltaic cell is most widely used one because of its simplicity and associated circuits.

(13) Define plane angle.

A plane angle is the angle subtended at a point in a plane by two converging lines. It is denoted by the Greek letter 'θ' (theta) and is usually measured in degrees or radians.

$$\therefore \text{Plane angle}(\theta) = \frac{\text{arc}}{\text{radius}}.$$

(14) Define solid angle.

Solid angle is the angle subtended at a point in space by an area, i.e., the angle enclosed in the volume formed by numerous lines lying on the surface and meeting at the point. It is usually denoted by symbol 'ω', and is measured in steradian.

$$\therefore \text{Solid angle}(\omega) = \frac{\text{area}}{(\text{radius})^2}.$$

(15) Define luminous flux.

It is defined as the energy in the form of light waves radiated per second from a luminous body. It is represented by the symbol 'ϕ' and measured in lumens.

(16) Define luminous intensity.

Luminous intensity in a given dissection is defined as the luminous flux emitted by the source per unit solid angle.

$$\text{Luminous intensity } (I) = \frac{\phi}{\omega} \text{ lumen/steradian or candela.}$$

(17) Define illumination.

Illumination is defined as the luminous flux received by the surface per unit area.

$$\text{Illumination, } E = \frac{\text{luminous flux}}{\text{area}}$$

$$= \frac{\phi}{A} = \frac{CP \times \omega}{A} \text{ lux.}$$

(18) Define lamp efficiency.

It is defined as the ratio of total luminous flux emitting from the source to its electrical power input in watts.

$$\therefore \text{Lamp efficiency} = \frac{\text{luminous flux}}{\text{power input}}.$$

(19) What for an integrating sphere is used?

An integrated sphere is a piece of apparatus that is commonly used for the measurement of MSCP.

MULTIPLE-CHOICE QUESTIONS

(1) Light:
 (a) Is a form of heat energy.
 (b) Is a form of electrical energy.
 (c) Consists of electromagnetic waves.
 (d) Consists of shooting particles.

(2) Radiant efficiency of the luminous source depends on:
 (a) The temperature of the source.
 (b) The wavelength of light rays.
 (c) The shape of the source.
 (d) All of the above.

(3) One Angstrom is equal to:
 (a) 10^{-8} m.
 (b) 10^{-10} m.
 (c) 10^{-8} cm.
 (d) 10^{-8} mm.

(4) A substance that changes its electrical resistance when illuminated by light is called:
 (a) Photoconductive.
 (b) Photovoltaic.
 (c) Photoelectric.
 (d) None of the above.

(5) Visible spectrum of light has a wavelength in the range of:
 (a) 1,000–4,000 Å.
 (b) 4,000–7,000 Å.
 (c) 7,500–4,000 Å.
 (d) 12,500–30,000 Å.

(6) Materials that reflect all wavelength in the spectrum of light appear to be:
 (a) Opaque.
 (b) Transparent.
 (c) Black to white.
 (d) Green to red.

(7) An object that appears red to the eyes absorbs:
 (a) Blue radiations.
 (b) Violet radiations.
 (c) Green radiations.
 (d) All of the above.

(8) The amount of light entering the eye is controlled by:
 (a) Pupil.
 (b) Retina.
 (c) Lens.
 (d) Pupil and lens.

(9) The eye lens focuses the image on:
 (a) Membrane.
 (b) Retina.
 (c) Corona.
 (d) Pupil.

(10) The color temperature of day-light is about:
 (a) 100 K.
 (b) 200 K.
 (c) 500 K.
 (d) 600 K.

(11) Light waves travel with a velocity of:
 (a) 3×10^6 m/s.
 (b) 3×10^8 m/s.
 (c) 3×10^{10} m/s.
 (d) 3×10^{12} m/s.

(12) The color of light depends upon:
 (a) The velocity of light.
 (b) Frequency.
 (c) Wavelength.
 (d) Both (b) and (c).

(13) The color having the shortest wavelength is:
 (a) Yellow.
 (b) Blue.
 (c) Orange.
 (d) Green.

(14) The color having the longest wavelength is:
 (a) Blue.
 (b) Green.

(c) Red.

(d) Violet.

(15) Wavelength for red color is:

(a) 4,000 Å.

(b) 5,000 Å.

(c) 6,000 Å.

(d) 7,000 Å.

(16) The wavelength of 5,500 Å will give light of:

(a) Green color.

(b) Red color.

(c) Orange color.

(d) Yellow–green color.

(17) Candle power is:

(a) The luminous flux emitted by the source per unit solid angle.

(b) The light radiating capacity of a source in a given direction.

(c) The unit of illumination.

(d) None of the above.

(18) The unit of illumination is:

(a) Lux.

(b) Lumen.

(c) Cd/m^2.

(d) Candela.

(19) Candela is the unit of:

(a) Illumination.

(b) Luminous intensity.

(c) Luminance.

(d) Light.

(20) The luminous flux is measured in:

(a) Candela.

(b) Lux.

(c) Stilb.

(d) Lumens.

(21) The unit of solid angle is:

(a) Radian.

(b) Steradian.

(c) Degree.

(d) Radian per meter.

(22) The solid angle subtended at the center of a hemisphere of radius r will be:

(a) $2\pi\gamma$.

(b) 2π.

(c) 4π.

(d) $2\pi/\gamma$.

(23) The illumination at a surface due to a source of light placed at a distance 'd' from the surface varies as:

(a) $1/d^2$.

(b) $1/d$.

(c) D.

(d) d^2.

(24) The illumination of various points on a horizontal surface illuminated by the same source varies as:

(a) $\cos\theta$.

(b) $\cos^2\theta$.

(c) $\cos^3\theta$.

(d) $1/\cos\theta$.

REVIEW QUESTIONS

(1) Discuss the inverse square law and the cosine cube law of illumination.

(2) Define the following terms:

(i) Mean horizontal candle power.

(ii) Mean spherical candle power.

(iii) Mean hemispherical candle power.

(iv) Luminous flux.

(3) Explain the measurement techniques for luminous intensity.

(4) Write short note the following:

(i) Bunsen photometer head.

(ii) Lumen–Brodhun photometer head.

(iii) Flicker photometer head.

(5) Write short note on polar curves.

(6) What is meant by photometry?

(7) Explain in detail about the photo voltaic and the photo conductive cells.

(8) Define the following terms:

(i) Luminous flux.

(ii) Luminous intensity.

(iii) Illumination.

(9) What do you understand by polar curves? Explain Rousseau's construction for calculating MSCP of a lamp.

EXERCISE PROBLEMS

(1) A room of 15 × 20 m is illustrated by ten 150-W lamps. The luminous efficiency of the lamp is 100 lumens/W and the coefficient of utilization is 0.7. Find the average illumination.

(2) The flux emitted by 150-W lamp is 1,600 lumens placed in a frosted globe of 50-cm diameter and gives uniform brightness of 350 milli-lumen/m^2 in all directions. Calculate the candel power of the globe and the percentage of light absorbed by the globe.

(3) A lamp having a candle power of 500 in all directions is provided with a reflector that directs 80% of total light uniformly on a circular area 50-m diameter. The lamp is hung at 25 m above the area.

(i) Calculate the illumination.

(ii) Calculate the illumination at the center.

(iii) Calculate the illumination at the edge of the surface without reflector.

(4) The luminous intensity of a source is 900 candela is placed in the middle of a 12 × 8 × 4 m room. Calculate the illumination,

(i) At each corner of the room.

(ii) At the middle of the 10-m wall.

(5) Two sources of having luminous intensity of 600 candela are hung at a height of 12 m.
The distance between two lamp posts is 25 m. Find the illumination (i) beneath the lamp and (ii) in the middle of the posts.

(6) Four lamps 25 m apart are arranged to illuminate a corridor. Each lamp is suspended at a height of 10 m above the floor level. If each lamp gives 500 CP in all directions below the horizontal, find the illumination at second and third lamps.

(7) A hall of 12 × 12 × 6 m is to be illuminated by four lamps each 100 W. Find the illumination at a point midway between the two corners on the floor along the side. Assume the efficiency of the lamp as 25 lumens/W.

(8) Two similar lamps having luminous intensity of 600 CP in all directions below horizontal are mounted at a height of 10 m. What must be the spacing between the lamps so that the illumination on the ground midway between the lamps shall be at least one-half of the illumination directly below the lamp?

ANSWERS

1. c	7. d	13. b	19. b
2. a	8. a	14. c	20. d
3. b	9. b	15. d	21. b
4. a	10. d	16. b	22. b
5. b	11. b	17. b	23. a
6. c	12. d	18. a	24. c

CHAPTER 7

Various Illumination Methods

OBJECTIVES

After reading this chapter, you should be able to:

- study the various sources of illumination
- understand the stroboscopic effect of fluorescent lamp
- understand the various principles of light control
- design different lighting schemes
- analyze the various schemes of lighting calculations.

7.1 INTRODUCTION

Light plays major role in human life. Natural light restricted for some duration in a day, it is very difficult to do any work by human being without light. So, it is necessary to have substitute for natural light. Light from incandescent bodies produced electrically, which playing important role in everyday life due to its controlled output, reliability, and clean-liness nowadays; various sources are producing artificial light. Each source has its own characteristics and specific importance.

7.2 TYPES OF SOURCES OF ILLUMINATION

Usually in a broad sense, based upon the way of producing the light by electricity, the sources of light are classified into following four types.

7.2.1 Electric arc lamps

The ionization of air present between the two electrodes produces an arc and provides intense light.

7.2.2 Incandescent lamps

When the filaments of these lamps are heated to high temperature, they emit light that falls in the visible region of wavelength. Tungsten-filament lamps are operating on this principle.

7.2.3 Gaseous discharge lamps

When an electric current is made to pass through a gas or metal vapor, it produces visible radiation by discharge takes place in the gas vapor. Sodium and mercury vapor lamps operate on this principle.

7.2.4 Fluorescent lamps

Certain materials like phosphor powders exposed to ultraviolet rays emits the absorbed energy into visible radiations fall in the visible range of wavelength. This principle is employed in fluorescent lamps.

7.3 Arc LAMPS

In arc lamps, the electrodes are in contact with each other and are separated by some distance apart; the electric current is made to flow through these two electrodes. The discharge is allowed to take place in the atmosphere where there are the production of a very intense light and a considerable amount of UV radiation, when an arc is struck between two electrodes.

The arcs maintain current and is very efficient source of light. They are used in search lights, projection lamps, and other special purpose lamps such as those in flash cameras.

Generally, used arc lamps are:

1. carbon arc lamp,
2. flame arc lamp, and
3. magnetic arc lamp.

7.3.1 Carbon arc lamp

Carbon arc lamp consists of two hard rod-type electrodes made up of carbon. Two electrodes are placed end to end and are connected to the DC supply. The positive electrode is of a large size than that of the negative electrode. The carbon electrodes used with AC supply are of the same size as that of the DC supply. The DC supply across the two electrodes must not be less than 45 V. When electric current passes through the electrodes are in contact and then withdrawn apart about 2–3 mm an arc is established between the two rods.

The two edges of the rods becomes incandescence due to the high resistance offered by rods as shown in Fig. 7.1 by transfer of carbon particles from one rod to the other. It is observed that carbon particles transfer from the positive rod to the negative one. So that the positive electrode gets consumed earlier than the negative electrode. Hence, the positive electrode is of twice the diameter than that of the negative electrode.

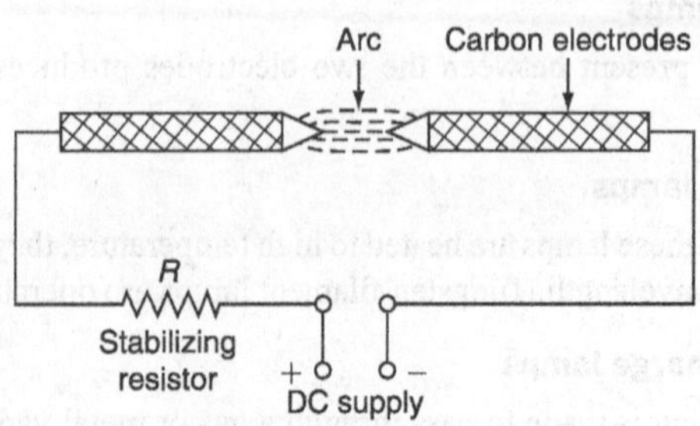

FIG. 7.1 *Carbon arc lamp*

In case of AC supply, the rate of consumption of the two electrodes is same; therefore, the cross-section of the two electrodes is same. *A* resistance '*R*' is connected in series with the electrode for stabilizing the arc. As current increases, the vaporizing rate of carbon increases, which decreases the resistance so much, then voltage drop across the arc decreases. So, to maintain the arc between the two electrodes, series resistance should be necessarily connected.

For maintaining the arc, the necessary voltage required is:

$$V = (39 + 2.8\, l)\ \text{V},$$

where l is the length of the arc. The voltage drop across the arc is 60 V, the temperature of the positive electrode is 3,500–4,200°C, and the temperature of the negative electrode is 2,500°C. The luminous efficiency of such lamps is 9–12 lumens/W. This low luminous efficiency is due to the service resistance provided in DC supply while in case of AC supply, an inductor is used in place of a resistor. In carbon arc lamps, 85% of the light is given out by the positive electrode, 10% of the light is given out by the negative electrodes, and 5% of the light is given out by the air.

7.3.2 Flame arc lamp

The electrodes used in flame arc lamp are made up of 85% of carbon and 15% of fluoride. This fluoride is also known as flame material; it has the efficient property that radiates light energy from high heated arc stream. Generally, the core type electrodes are used and the cavities are filled with fluoride. The principle of operation of the flame arc lamp is similar to the carbon arc lamp. When the arc is established between the electrodes, both fluoride and carbon get vaporized and give out very high luminous intensities. The color output of the flame arc lamps depends upon the flame materials. The luminous efficiency of such lamp is 8 lumens/W. A simple flame arc lamp is shown in Fig. 7.2. Resistance is connected in service with the electrodes to stabilize the arc.

7.3.3 Magnetic arc lamp

The principle of the operation of the magnetic arc lamp is similar to the carbon arc lamp. This lamp consists of positive electrode that is made up of copper and negative electrode that is made up of magnetic oxide of iron. Light energy radiated out when the arc is struck between the two electrodes. These are rarely used lamps.

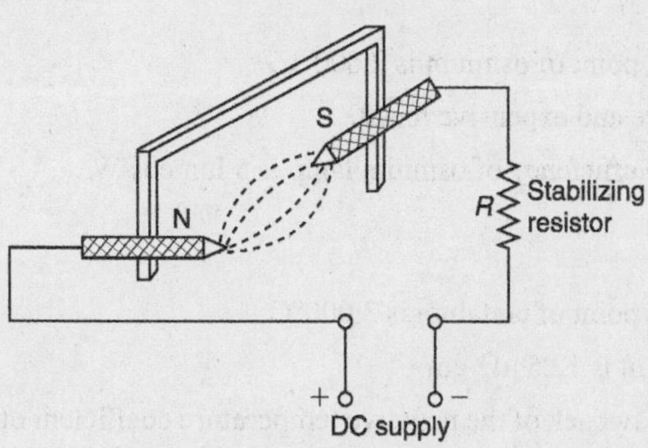

FIG. 7.2 *Flame arc lamp*

7.4 INCANDESCENT LAMP

These lamps are temperature-dependent sources. When electric current is made to flow through a fine metallic wire, which is known as *filament,* its temperature increases. At low temperatures, it emits only heat energy, but at very high temperature, the metallic wire emits both heat and light energy. These incandescent lamps are also known as *temperature radiators.*

7.4.1 Choice of material for filament

The materials commonly used as filament for incandescent lamps are carbon, tantalum, tungsten, and osmium.

The materials used for the filament of the incandescent lamp have the following properties.

- The melting point of the filament material should be high.
- The temperature coefficient of the material should be low.
- It should be high resistive material.
- The material should possess good mechanical strength to withstand vibrations.
- The material should be ductile.

7.4.2 Comparisons of carbon, osmium, tantalum, and tungsten used for making the filament

Carbon

- Carbon has high melting point of 3,500°C; even though, its melting point is high, carbon starts disintegration at very fast rate beyond its working temperature of 1,800°C.
- Its resistance decreases with increase in temperature, i.e., its temperature coefficient of resistivity is negative, so that it draws more current from the supply. The temperature coefficient (α) is –0.0002 to –0.0008.
- The efficiency of carbon filament lamp is low; because of its low operating temperature, large electrical input is required. The commercial efficiency of carbon lamp is 3–4.5 lumens/W approximately.
- Carbon has high resistivity (ρ), which is about 1,000–7,000 μΩ-cm and its density is 1.7–3.5.

Osmium

- The melting point of osmium is 2,600°C.
- It is very rare and expensive metal.
- The average efficiency of osmium lamp is 5 lumens/W.

Tantalum

- The melting point of tantalum is 3,000°C.
- Resistivity (ρ) is 12.5 μΩ-cm.
- The main drawback of the negative temperature coefficient of carbon is overcome in tantalum. It has positive temperature coefficient (α) and its value is 0.0036.

- The density of tantalum is 16.6.
- The efficiency of tantalum lamp is 2 lumens/W.

Tungsten

- The working temperature of tungsten is 2,500–3,000°C.
- Its resistance at working temperature is about 12–15 times the cold resistance.
- It has positive temperature coefficient of resistance of 0.0045.
- Its resistivity is 5.6 12.5 μΩ-cm.
- The density of tungsten is 19.3.
- The efficiency of tantalum when working at 2,000°C is 18 lumens/W.
- Its vapor pressure is low when compared to carbon.

In fact, the carbon lamp is the first lamp introduced by Thomas Alva Edison in 1879, owing to two drawbacks, tungsten radiates more energy in visible spectrum and somewhat less in infrared spectrum so that there was a switch over in infrared spectrum so that there was a switch over from carbon filament to tungsten filament. Nowadays, tungsten filament lamps are widely used incandescent lamps.

The chemically pure tungsten is very strong and fragile. In order to make it into ductile, tungsten oxide is first reduced in the form of gray power in the atmosphere of hydrogen and this powder is pressed in steel mold for small bars; the mechanical strength of these bars can be improved by heating them to their melting point and then hammered at red-hot position and re-rolled into wires.

Construction

Figure 7.3 shows the construction of the pure tungsten filament incandescent lamp. It consists of an evacuated glass bulb and an aluminum or brass cap is provided with two pins to insert the bulb into the socket. The inner side of the bulb consists of a tungsten filament and the support wires are made of molybdenum to hold the filament in proper position. A glass button is provided in which the support wires are inserted. A stem tube forms an air-tight seal around the filament whenever the glass is melted.

Operation

When electric current is made to flow through the fine metallic tungsten filament, its temperature increases. At very high temperature, the filament emits both heat and light radiations, which fall in the visible region. The maximum temperature at which the filament can be worked without oxidization is 2,000°C, i.e., beyond this temperature, the tungsten filament blackens the inside of the bulb. The tungsten filament lamps can be operated efficiently beyond 2,000°C, it can be attained by inserting a small quantity of inert gas nitrogen with small quantity of organ. But if gas is inserted instead of vacuum in the inner side of the bulb, the heat of the lamp is conducted away and it reduces the efficiency of the lamp. To reduce this loss of heat by conduction and convection, as far as possible, the filament should be so wound that it takes very little space. This is achieved by using a single-coil filament instead of a straight wire filament as shown in Fig. 7.4(a). This single-coil filament is used in vacuum bulbs up to 25 W and gas filled bulbs from 300 to 1,000 W.

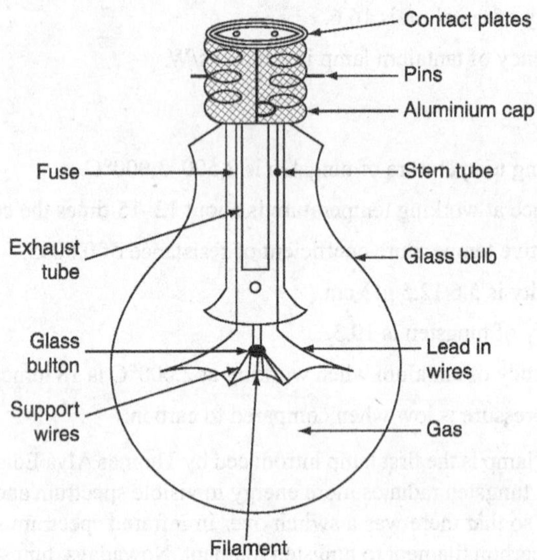

FIG. 7.3 *Incandescent lamp*

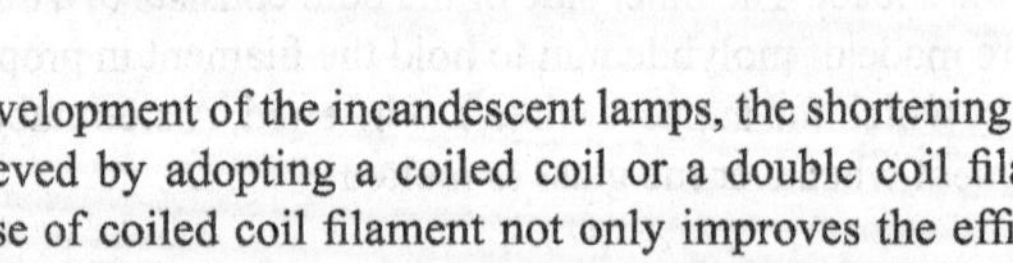

FIG. 7.4 *Various filaments used in incandescent lamps*

On further development of the incandescent lamps, the shortening of the length of the filament was achieved by adopting a coiled coil or a double coil filament as shown in Fig. 7.4(b). The use of coiled coil filament not only improves the efficiency of the lamp but also reduces the number of filament supports and thus simplified interior construction because the double coil reduces the filament mounting length in the ratio of 1:25 as compared to the straight wire filaments.

Usually, the tungsten filament lamp suffers from 'aging effect', the output of the light an incandescent lamp decreases as the lamp ages. The output of the light of the lamp decreases due to two reasons.

- At very high temperature, the vaporization of filament decreases the coil diameter so that resistance of the filament increases and hence its draws less current from the supply, so the temperature of the filament and the light output of the bulb decrease.
- The current drawn from the mains and the power consumed by the filament decrease, which decrease the efficiency of the lamp with the passage of time. In addition, the evaporation of the filament at high temperature blackens the inside of the bulb.

The effects of voltage variations

The variations in normal supply voltages will affect the operating characteristics of incandescent lamps. The performance characteristic of an incandescent lamp, when it is subjected to voltage other than normal voltage, is shown in Fig. 7.5.

With an increase in the voltage owing to the increase in the temperature, the luminous output of the incandescent lamps, and the efficiency and power consumption, but its life span decreases.

The depreciation in the light output is around 15% over the useful life of the lamp. The above-stated factors are related to the variations of voltage are given as:

- Lumens output $\propto$ (voltage)$^{3.55}$.
- Power consumption $\propto$ (voltage)$^{1.55}$.
- Luminous efficiency $\propto$ (voltage)^{2}.
- Life $\propto$ (voltage)$^{-13}$ (for vacuum lamps).
- Life $\propto$ (voltage)$^{-14}$ (for gas filled lamps).

The advantages of the incandescent lamps

- These lamps are available in various shapes and sizes.
- These are operating at unity power factor.
- These lamps are not affected by surrounding air temperature.
- Different colored light output can be obtained by using different colored glasses.

Filament dimensions

Let us consider a lamp, which is connected to the mains, is given the steady light output, i.e., whatever the heat produced, it is dissipated and the filament temperature is not going to

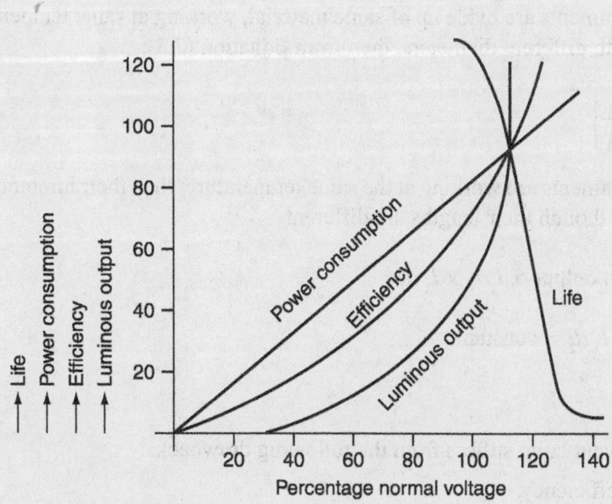

FIG. 7.5 *Performance characteristics of incandescent lamp*

be increase further. It is found to be the existence of a definite relation between the diameter of a given filament and the current through it.

The input wattage to the lamp is expressed as:

$$I^2R = I^2\frac{\rho l}{a} \qquad \left(\because\ R = \rho\frac{l}{a}\right)$$

$$= \frac{I^2 \times \rho l}{(\pi d^2/4)}$$

$$= I^2 \times \frac{4\rho l}{\pi d^2}, \tag{7.1}$$

where I is the current taken by the lamp A, a is the filament cross-section, sq. m, ρ is the resistivity of the filament at working temperature Ω-m, l is the length of the filament m, and d is the diameter of the filament.

Let the emissivity of the material be 'e'. Total heat dissipated will depend upon the surface area and the emissivity of the material

∴ Heat dissipated ∝ surface area × emissivity:

$$\propto \pi dl \times e. \tag{7.2}$$

At the steady state condition, the power input should be equal to the heat dissipated. From Equations (7.1) and (7.2), we can write that:

$$I^2\frac{4\rho l}{\pi d^2} \propto \pi dl \times e$$

$$I^2 \propto d^3 \quad \text{or} \quad I \propto d^{3/2}. \tag{7.3}$$

If two filaments are made up of same material, working at same temperature and efficiency but with different diameters, then from Equation (7.3):

$$\frac{I_1}{I_2} = \left(\frac{d_1}{d_2}\right)^{3/2} \tag{7.4}$$

If two filaments are working at the same temperature, then their luminous output must be same even though their lengths are different.

$$\therefore \text{Lumen output} \propto l_1 d_1 \propto l_2\, d_2$$

$$\therefore l_1 d_1 \propto l_2\, d_2 = \text{constant}. \tag{7.5}$$

Limitations

The incandescent lamp suffers from the following drawbacks:

- Low efficiency.
- Colored light can be obtained by using different colored glass enclosures only.

7.5 DISCHARGE LAMPS

Discharge lamps have been developed to overcome the drawbacks of the incandescent lamp. The main principle of the operation of light in a gaseous discharge lamp is illustrated as below.

In all discharge lamps, an electric current is made to pass through a gas or vapor, which produces its illuminance. Normally, at high pressures and atmospheric conditions, all the gases are poor conductors of electricity. But on application of sufficient voltage across the two electrodes, these ionized gases produce electromagnetic radiation. In the process of producing light by gaseous conduction, the most commonly used elements are neon, sodium, and mercury. The wavelength of the electromagnetic radiation depends upon the nature of gas and the gaseous pressure used inside the lamp. A simple discharge lamp is shown in Fig. 7.6.

The production of light in the gaseous discharge lamps is based on the phenomenon of excitation and ionization of gas or metal vapor present between the two electrodes of a discharge tube.

When the potential between the two electrodes is equals to ionizing potential, gas or metal vapor starts ionizing and an arc is established between the two electrodes. Volt–ampere characteristics of the arc is negative, i.e., gaseous discharge lamp possess a negative resistance characteristics. A choke or ballast is provided to limit high currents to a safe value. Here, the choke serves two functions.

- It provides ignition voltage initially.
- Limits high currents.

The use of choke will reduce the power factor (0.3–0.4) of all the gaseous lamps so that all the discharge lamps should be provided with a condenser to improve the power factor. The nature of the gas and vapor used in the lamp will affect the color affected of light.

7.5.1 Types of discharge lamps

Generally used discharge lamps are of two types. They are:

1. The lamps that emit light of the color produced by discharge takes place through the gas or vapor present in the discharge tube such as neon gas, sodium vapor, mercury vapor, etc.

   Ex: Neon gas, sodium vapor lamp, and mercury vapor lamp.

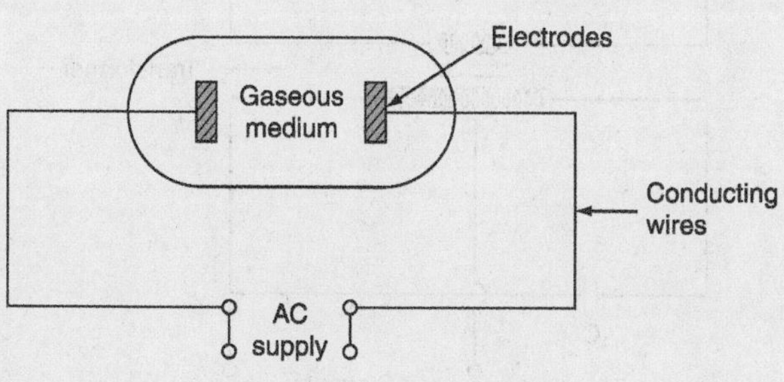

FIG. 7.6 *Discharge lamps*

2. The lamp that emits light of color depends upon the type of phosphor material coated inside the walls of the discharge tube. Initially, the discharge takes place through the vapor produces UV radiation, then the invisible UV rays absorbed by the phosphors and radiates light energy falls in the visible region. This UV light causes fluorescence in certain phosphor materials, such lamps are known as fluorescent lamps.

   Ex: Fluorescent mercury vapor tube.

   In general, the gaseous discharge lamps are superior to the tungsten filament lamps.

7.5.2 Drawbacks

The discharge lamps suffer from the following drawbacks.

1. The starting of the discharge lamps requires starters and transformers; therefore, the lamp circuitry is complex.
2. High initial cost.
3. Poor power factor; therefore, the lamps make use of the capacitor.
4. Time required to give its full output brilliancy is more.
5. These lamps must be placed in particular position.
6. These lamps require stabilizing choke to limit current since the lamps have negative resistance characteristics.

7.6 NEON DISCHARGE LAMP

This is a cold cathode lamp, in which no filament is used to heat the electrode for starting.

Neon lamp consists of two electrodes placed at the two ends of a long discharge tube is shown in Fig. 7.7.

The discharge tube is filled with neon gas. A low voltage of 150 V on DC or 110 V on AC is impressed across the two electrodes; the discharge takes place through the neon

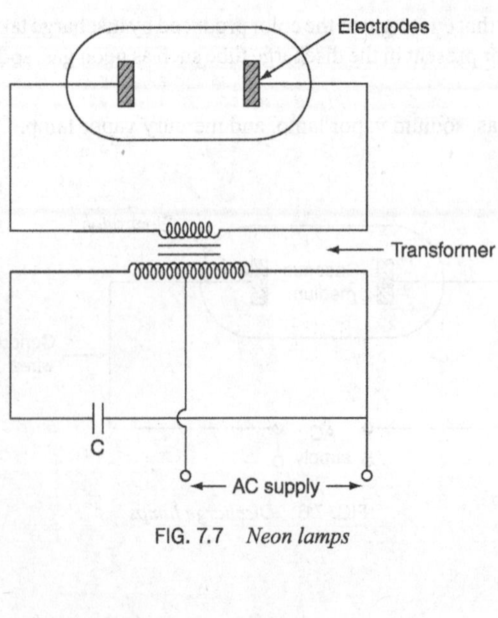

FIG. 7.7 *Neon lamps*

gas that emits light or electro magnetic radiation reddish in color. The sizes of electrodes used are equal for both AC and DC supplies. On DC, neon glow appear nearer to the negative electrode; therefore, the negative electrode is made larger in size. Neon lamp electric circuit consists of a transformer with high leakage reactance in order to stabilize the arc. Capacitor is used to improve the power factor. Neon lamp efficiency is approximately 15–40 lumens/W. The power consumption of the neon lamp is 5 W.

If the helium gas is used instead of neon, pinkish white light is obtained. These lamps are used as night lamps and as indicator lamps and used for the determination of the polarity of DC mains and for advertising purpose.

7.7 SODIUM VAPOR LAMP

A sodium vapor lamp is a cold cathode and low-pressure lamp. A sodium vapor discharge lamp consists of a *U*-shaped tube enclosed in a double-walled vacuum flask, to keep the temperature of the tube within the working region. The inner *U*-tube consists of two oxide-coated electrodes, which are sealed with the ends. These electrodes are connected to a pin type base construction of sodium vapor lamp is shown in Fig. 7.8.

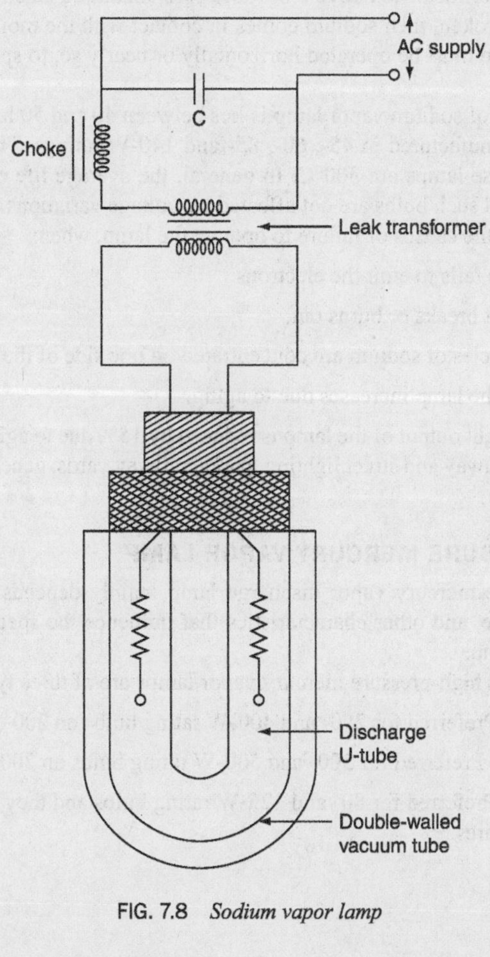

FIG. 7.8 *Sodium vapor lamp*

This sodium vapor lamp is low luminosity lamp, so that the length of the lamp should be more. In order to get the desired length, it is made in the form of a *U*-shaped tube. This long *U*-tube consists of a small amount of neon gas and metallic sodium. At the time of start, the neon gas vaporizes and develops sufficient heat to vaporize metallic sodium in the *U*-shaped tube.

7.7.1 Working

Initially, the sodium is in the form of a solid, deposited on the walls of inner tube. When sufficient voltage is impressed across the electrodes, the discharge starts in the inert gas, i.e., neon; it operates as a low-pressure neon lamp with pink color. The temperature of the lamp increases gradually and the metallic sodium vaporizes and then ionizes thereby producing the monochromatic yellow light. This lamp takes 10–15 min to give its full light output. The yellowish output of the lamp makes the object appears gray.

In order to start the lamp, 380–450 V of striking voltage required for 40- and 100-W lamps. These voltages can be obtained from a high reactance transformer or an auto transformer. The operating power factor of the lamp is very poor, so that a capacitor is placed to improve the power factor to above 0.8. More care should be taken while replacing the inner tube, if it is broken, then sodium comes in contact with the moisture; therefore, fire will result. The lamp must be operated horizontally or nearly so, to spread out the sodium well along the tube.

The efficiency of sodium vapor lamp is lies between 40 and 50 lumens/W. Normally, these lamps are manufactured in 45-, 60-, 85- and 140-W ratings. The normal operating temperatures of these lamps are 300°C. In general, the average life of the sodium vapor lamp is 3,000 hr and such bulbs are not affected by voltage variations.

Following are the causes of failure to operate the lamp, when:

- The cathode fails to emit the electrons.
- The filament breaks or burns out.
- All the particles of sodium are concentrated on one side of the inner tube.
- The life of the lamp increases due to aging.

The average light output of the lamp is reduced by 15% due to aging. These lamps are mainly used for highway and street lighting, parks, railway yards, general outdoor lighting, etc.

7.8 HIGH-PRESSURE MERCURY VAPOR LAMP

The working of the mercury vapor discharge lamp mainly depends upon the pressure, voltage, temperature, and other characteristics that influence the spectral quality and the efficiency of the lamp.

Generally used high-pressure mercury vapor lamps are of three types. They are:

1. MA type: Preferred for 250- and 400-W rating bulbs on 200–250-V AC supply.
2. MAT type: Preferred for 300- and 500-W rating bulbs on 200–250-V AC supply.
3. MB type: Preferred for 80- and 125-W rating bulbs and they are working at very high pressures.

7.8.1 MA type lamp

It is a high-pressure mercury vapor discharge lamp that is similar to the construction of sodium vapor lamp. The construction of MA type lamp is shown in Fig. 7.9

MA type lamp consists of a long discharge tube in '*U*' shape and is made up of hard glass or quartz. This discharge tube is enclosed in an outer tube of ordinary glass. To prevent the heat loss from the inner bulb, by convection, the gap between the two tubes is completely evacuated. The inner tube contains two main electrodes and an auxiliary starting electrode, which is connected through a high resistance of about 50 kΩ. It also contains a small quantity of argon gas and mercury. The two main electrodes are tungsten coils coated with electron emitting material (such as thorium metal).

Working

Initially, the tube is cold and hence the mercury is in condensed form. Initially, when supply is given to the lamp, argon gas present between the main and the auxiliary electrodes gets ionized,

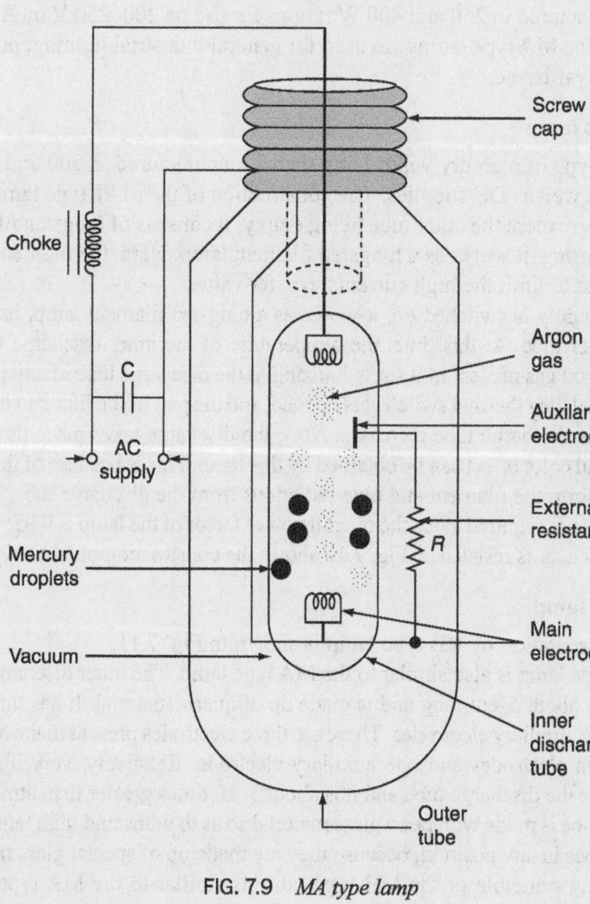

FIG. 7.9 *MA type lamp*

and an arc is established, and then discharge takes place through argon for few minutes between the main and the auxiliary electrodes. As a result, discharge takes place through argon for few minutes in between the main and the auxiliary electrodes. The discharge can be controlled by using high resistance that is inserted in-series with the auxiliary electrode. After few minutes, the argon gas, as a whole, gets ionized between the two main electrodes. Hence, the discharge shifts from the auxiliary electrode to the two main electrodes. During the discharge process, heat is produced and this heat is sufficient to vaporize the mercury. As a result, the pressure inside the discharge tube becomes high and the voltage drop across the two main electrodes will increases from 20 to 150 V. After 5–7 min, the lamp starts and gives its full output.

Initially, the discharge through the argon is pale blue glow and the discharge through the mercury vapors is greenish blue light; here, choke is provided to limit high currents and capacitor is to improve the power factor of the lamp.

If the supply is interrupted, the lamp must cool down and the vapor pressure be reduced before it will start. It takes approximately 3–4 min. The operating temperature of the inner discharge tube is about 600°C. The efficiency of this type of lamp is 30–40 lumens/W. These lamps are manufactured in 250 and 400 W ratings for use on 200–250 V on AC supply.

Generally, the MA type lamps are used for general industrial lighting, ports, shopping centers, railway yards, etc.

7.8.2 MAT type lamp

This is another type of mercury vapor lamp that is manufactured in 300 and 500 W rating for use on AC as well as DC supplies. The construction of the MAT type lamp is similar to the MA type lamp except the outer tube being empty; it consists of tungsten filament so that at the time of starting, it works as a tungsten filament lamp. Here, the filament itself acts as a choke or ballast to limit the high currents to safer value.

When the supply is switched on, it works as a tungsten filament lamp, its full output is given by the outer tube. At this time, the temperature of the inner discharge tube increases gradually, the argon gas present in it starts ionizing in the discharge tube at any particular temperature is attained then thermal switch gets opened, and the part of the filament is detached and voltage across the discharge tube increases. Now, the discharge takes place through the mercury vapor. Useful color effect can be obtained by this lamp. This is because of the combination of light emitted form the filament and blue radiations from the discharge tube. In this type of lamp, capacitor is not required since the overall power factor of the lamp is 0.95; this is because the filament itself acts as resistance. Fig. 7.10 shows the construction of MAT type lamp.

7.8.3 MB type lamp

Schematic representation of MB type lamp is shown in Fig. 7.11.

The MB type lamp is also similar to the MA type lamp. The inner discharge tube for the MB type lamp is about 5-cm long and is made up of quartz material. It has three electrodes; two main and one auxiliary electrodes. There are three electrodes present in the MB type lamp, namely two main electrodes and one auxiliary electrode. Relatively, very high pressure is maintained inside the discharge tube and it is about 5–10 times greater than atmospheric pressure. The outer tube is made with pearl glass material so as to withstand high temperatures. We can use these tubes in any position, because they are made up of special glass material.

The working principle of the MB type lamp is similar to the MA type lamp. These lamps are manufactured in 300 and 500 W rating for use in AC as well as DC supplies. An MB type lamp consists a bayonet cap with three pins, so it may not be used in an ordinary sense. A choke coil and a capacitor are necessary for working with these types of lamps.

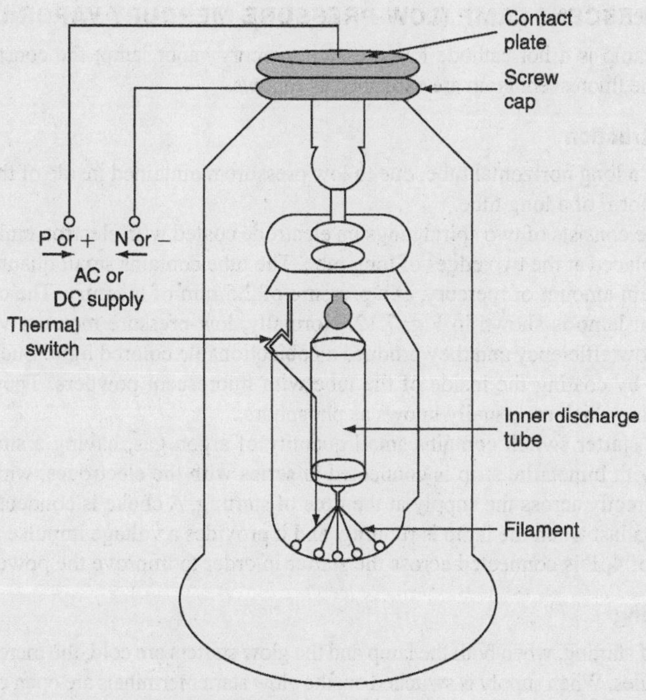

FIG. 7.10 *MAT type lamp*

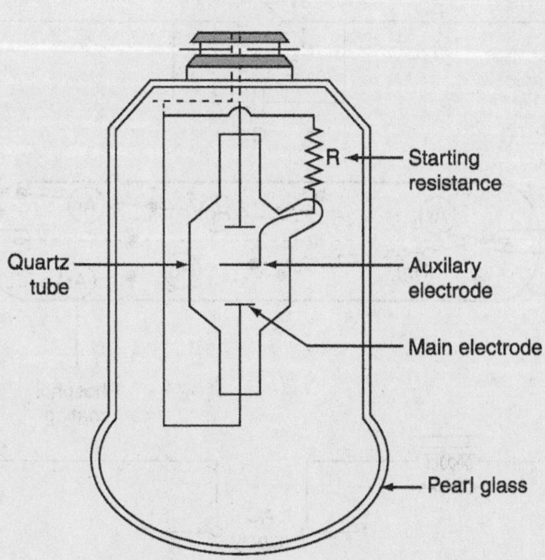

FIG. 7.11 *MB type lamp*

7.9 FLUORESCENT LAMP (LOW-PRESSURE MERCURY VAPOR LAMP)

Fluorescent lamp is a hot cathode low-pressure mercury vapor lamp; the construction and working of the fluorescent lamp are explained as follows.

7.9.1 Construction

It consists of a long horizontal tube, due to low pressure maintained inside of the bulb; it is made in the form of a long tube.

The tube consists of two spiral tungsten electrode coated with electron emissive material and are placed at the two edges of long tube. The tube contains small quantity of argon gas and certain amount of mercury, at a pressure of 2.5 mm of mercury. The construction of fluorescent lamp is shown in Fig. 7.12. Normally, low-pressure mercury vapor lamps suffer from low efficiency and they produce an objectionable colored light. Such drawback is overcome by coating the inside of the tube with fluorescent powders. They are in the form of solids, which are usually knows as phosphors.

A glow starter switch contains small quantity of argon gas, having a small cathode glow lamp with bimetallic strip is connected in series with the electrodes, which puts the electrodes directly across the supply at the time of starting. A choke is connected in series that acts as ballast when the lamp is running, and it provides a voltage impulse for starting. A capacitor of 4μF is connected across the starter in order to improve the power factor.

7.9.2 Working

At the time of starting, when both the lamp and the glow starters are cold, the mercury is in the form of globules. When supply is switched on, the glow starter terminals are open circuited and full supply voltage appeared across these terminals, due to low resistance of electrodes and choke coil. The small quantity of argon gas gets ionized, which establishes an arc with a starting glow.

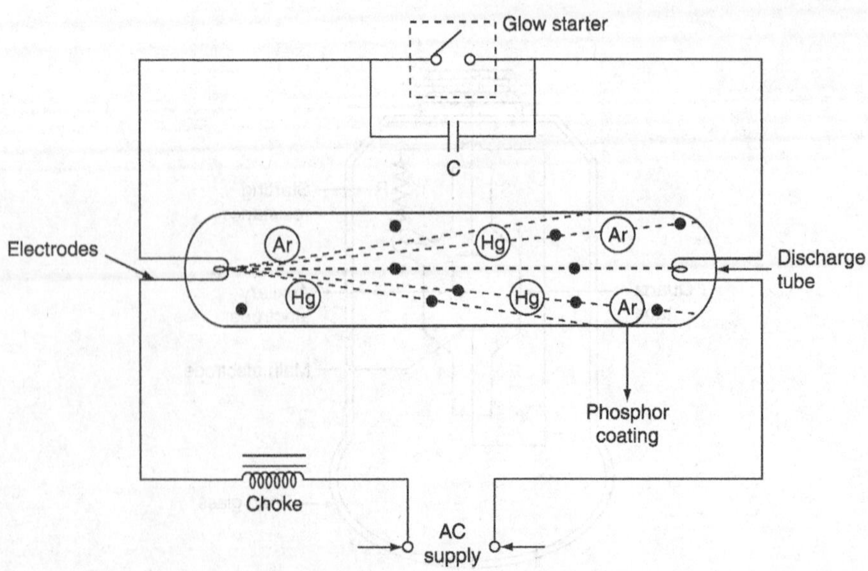

FIG. 7.12 *Fluorescent lamp*

This glow warms up the bimetallic strip thus glow starts gets short circuited. Hence, the two electrodes come in series and are connected across the supply voltage. Now, the two electrodes get heated and start emitting electrons due to the flow of current through them. These electrons collide with the argon atoms present in the long tube discharge that takes place through the argon gas. So, in the beginning, the lamp starts conduction with argon gas as the temperature increases, the mercury changes into vapor form and takes over the conduction of current.

In the mean time, the starter potential reaches to zero and the bimetallic strip gets cooling down. As a result, the starter terminals will open. This results breaking of the series circuit. A very high voltage around 1,000 V is induced, because of the sudden opening of starter terminals in the series circuit. But in the long tube, electrons are already present; this induced voltage is quite sufficient to break down the long gap. Thus, more number of electrons collide with argon and mercury vapor atoms. The excited atom of mercury gives UV radiation, which will not fall in the visible region.

Meanwhile, these UV rays are made to strike phosphor material; it causes the re-emission of light of different wavelengths producing illumination. The phenomenon of the re-emission is called as *luminescence.*

This luminescence is classified into two ways. They are:

(i) **Fluorescence:** In this case, the excitation presents for the excited periods only.

(ii) **Phosphorescence:** In this case, even after the exciting source is removed, the excitation will present.

In a lamp, the re-emission of light causes fluorescence, then such lamp is known as *fluorescent lamp.*

Depending upon the type of phosphor material used, we get light of different colors as given in Table. 7.1.

Advantages of fluorescent lamp

The fluorescent lamp has the following advantages:

- High efficiency.
- The life of the lamp is three times of the ordinary filament lamp.
- The quality of the light obtained is much superior.
- Less chances of glare.
- These lamps can be mounted on low ceiling, where other light sources would be unsatisfactory.

TABLE 7.1 *Colors of light*

	Phosphor material	*Color effect*
1.	Zinc silicate	Green
2.	Calcium tungstate	Green
3.	Magnesium tungstate	Bluish while
4.	Cadmium silicate	Yellowish pink
5.	Zinc beryllium silicate	Yellowish while
6.	Cadmium borate	Pink

Although the fluorescent lamp has the above advantages, it sufferers form the following disadvantages:

- The initial cost is high because of choke and starter.
- The starting time as well as the light output of the lamp will increases because of low ambient temperature.
- Because of the presence of choke, these lamps suffer from magnetic humming and may cause disturbance.
- The stroboscopic effect of this lamp is objectionable.

Stroboscopic effect

We all know that because of 'the alternating nature of supply, it crosses zero two times in a cycle'. For 50-Hz frequency supply of the alternating current, a discharge lamp will be extinguished twice in a cycle and 100 times per second (for 50-Hz supply). A human eye cannot identify this extinguish phenomenon, because of the persistence of vision. If this light falls upon a moving object, the object appearing like slow moving or fast moving or moving in reverse direction, sometimes stationary. This effect is due to the extinguishing nature of the light of the lamp. This effect is called as '*stroboscopic effect*'.

This effect can be avoided by employing any of the two techniques listed below.

(i) If we have three-phase supply, then the fluorescent lamps that are adjacent should be fed from different phases. Then, no two lamps will not be in same phase at zero instant of AC supply, so light is present at any instant.

(ii) If the available supply is single phase, then twin tube circuitry as shown in Fig. 7.13, we can eliminate stroboscopic effect.

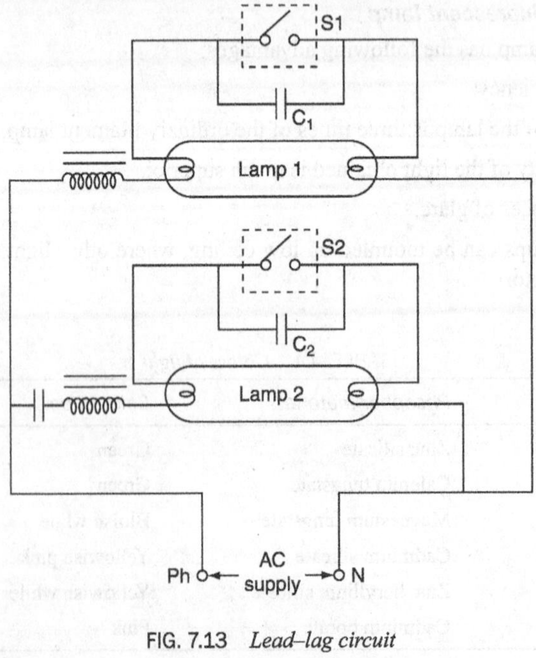

FIG. 7.13 *Lead–lag circuit*

Twin tube circuit is also known as *lead–lag circuit.* Here two tubes are connected in parallel. One of the two tubes provided with a capacitor in series with the choke coil. The current through the lamps is almost 90° out of phase and under these conditions, the light output of one of the lamps is at maximum. Moreover, the overall power factor of lamps is unity. In this lead–lag arrangement, one of the lamps is operating at 0.5 lagging, the other, provided with capacitor, is operating at 0.5 leading.

In general, the life of a fluorescent lamp is about 7,500 hr. Based on the operating conditions, the lamp's actual life can be varied from 5,000 to 10,000 hr. It is recommended to replace a lamp after 4,000–5,000 of its working hours.

7.9.3 Startless fluorescent lamp

A fluorescent lamp without a starter is commercially called as 'quick start' or 'instant start' fluorescent lamp. Figure 7.14 shows the circuit of a startless fluorescent lamp. This lamp does not need a starter.

In the present days, this startless lamps are used for rapid or instant start. In this type of lamps, the filament transformer replaces the starter. The two electrodes are heated with the help of secondary of this transformer. This preheating and the presence of full supply voltage across the electrodes cause the ionization of the medium.

For the satisfactory starting of the lamp, an earthed strip 'K' is used.

Advantages

The startless lamp has the following advantages.

- The life of the lamp is more.
- The instantaneous starting of the lamp.
- The flickering of the lamp and the false starting is completely eliminated.

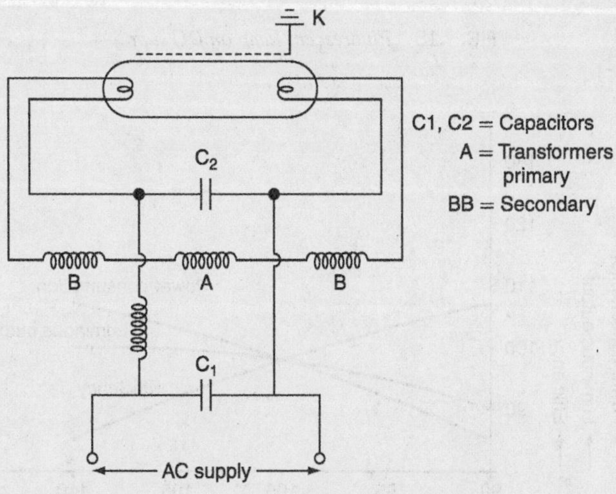

FIG. 7.14 *Startless fluorescent lamp*

7.9.4 Fluorescent lamp on DC supply

Fluorescent lamp fed from AC supply mainly suffers from the stroboscopic effect. This effect can be completely avoided if the lamp is fed from DC supply, with some modification of the circuitry. Figure 7.15 shows the fluorescent lamp operating on DC supply.

Additionally, a reversible switch is provided, which ensures the polarity reversal of the supply, every time the tube is switched on. This reversal of supply terminals is necessary to prevent the blackening of a single electrode, this is due to when the lamp starts working, the positive end of the tube gets blackened due to the migration of ionized mercury vapor to the negative end.

In order to work the fluorescent tube on DC, a resistance is connected in series with the choke, which results in the increased power consumption and lesser efficiency.

Performance curves of fluorescent lamp

The performance curves of fluorescent lamp are shown in Fig. 7.16.

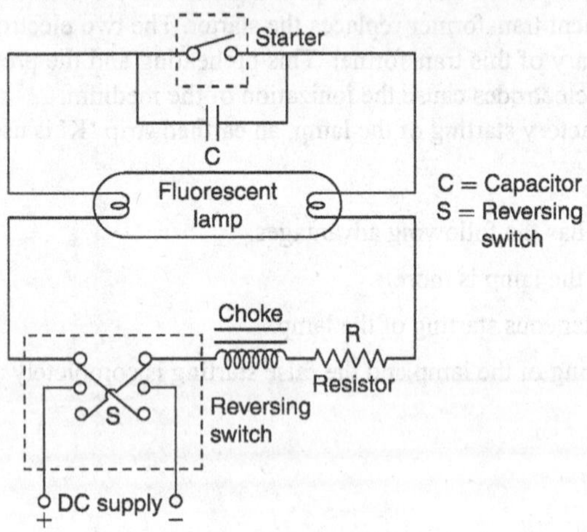

FIG. 7.15 *Fluorescent lamp on DC supply*

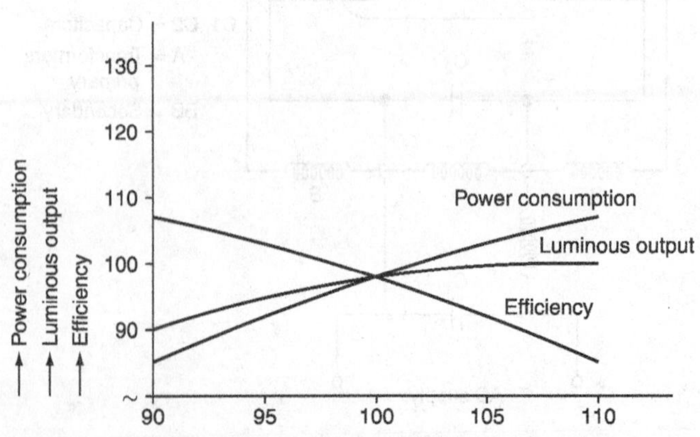

FIG. 7.16 *Performance curves of fluorescent lamp*

It is observed that the effect of the variation of the voltage on the fluorescent lamp is less when compared to the incandescent lamps. However, their life and performance are adversely affected by both low and high voltages. These lamps show better performance at an operating temperature of about 20–25°C. If the lamp is operating at high voltages, its luminous output increases but its efficiency decreases.

7.10 COMPARISON BETWEEN TUNGSTEN FILAMENT LAMPS AND FLUORESCENT LAMPS

Table 7.2 gives the comparison between tungsten filament lamps and fluorescent lamps.

TABLE 7.2 *Comparison between tungsten filament and fluorescent lamps*

Incandescent lamp	*Fluorescent lamp*
1. Initial cost is less.	1. Initial cost is more.
2. Fluctuation in supply voltage has less effect on light output, as the variations in voltage are absorbed in choke.	2. Fluctuations in supply voltage has comparatively more effect on the light output.
3. It radiates the light; the color of which resembles the natural light.	3. It does not give light close to the natural light.
4. It works on AC as well as DC.	4. Change of supply needs additional equipment.
5. The luminous efficiency of the lamp is high that is about 8–40 lumens/W.	5. The luminous efficiency is poor, which is about 8–10 lumen/W.
6. Different color lights can be obtained by using different colored glasses.	6. Different color lights can be obtained by using different composition of fluorescent powder.
7. Brightness of the lamp is more.	7. Brightness of the lamp is less.
8. The reduction in light output of the lamp is comparatively high, with the time.	8. The reduction in light output of the lamp is comparatively low, with the lamp.
9. The working temperature is about 2,000°C.	9. The working temperature is about 50°C.
10. The normal working life is 1,000 hr.	10. The normal working life is 5,000–7,500 hr.
11. No stroboscopic effect.	11. Stroboscopic effect is present.
12. These lamps are widely used for domestic, industrial, and street lighting.	12. They find wide application in domestic, industrial, and floodlighting.
13. The luminous efficiency increases with the increase in the voltage of the lamp.	13. The luminous efficiency increase with the increase in voltage and the increase in the length of tube.

7.11 BASIC PRINCIPLES OF LIGHT CONTROL

When light strikes the surface of an object, based on the properties of that surface, some portion of the light is reflected, some portion is transmitted through the medium of the surface, and the remaining is absorbed.

The method of light control is used to change the direction of light through large angle. There are four light control methods. They are:

1. reflection,
2. refraction,
3. riffusion, and
4. absorption.

7.11.1 Reflection

The light falling on the surface, whole of the light will not absorbed or transmitted through the surface, but some of the light is reflected back, at an angle equals to the angle of incidence. The ratio of reflected light energy to the incident light energy is known as reflection factor. The two basic types of reflection are:

(i) mirror or specular reflection and

(ii) diffuse reflection.

Specular reflection

When whole of the light falling on a smooth surfaces will be reflected back at an angle equal to the angle of incidence. Such a reflection is known as *specular reflection.* With such reflection, observer will be able to see the light source but not the illuminated surface. Most of the surfaces causing the specular reflection are silvered mirrors, highly polished metal surfaces. Specular reflection is shown in Fig. 7.17.

A surface that is almost free from reflection is called a matt surface.

Diffuse reflection

When the light ray falling on any surface, it is scattered in all directions irrespective of the angle of incidence. Such type of reflector is known as *diffuse reflection* and is shown in Fig. 7.18. Most of the surfaces causing the diffuse reflection are rough or matt surfaces such as blotting paper, frosted glass, plaster, etc.

In this reflection, observer will be able to see the illuminated surface but not the light source.

7.11.2 Refraction

When a beam of light passes through two different mediums having different densities, the light ray will be reflected. This phenomenon is known as *refraction.*

Figure 7.19 shows the refraction of light ray from dense medium to rare medium where μ_1 and μ_2 are the refractive indices of two medium, θ is the angle of incidence, and α is the angle of reflection.

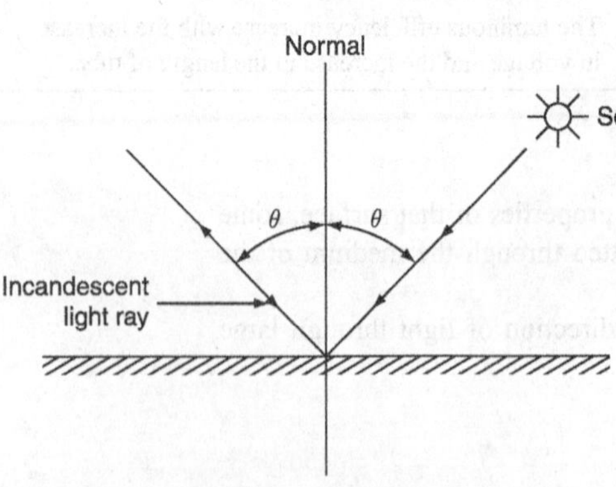

FIG. 7.17 *Specular reflection*

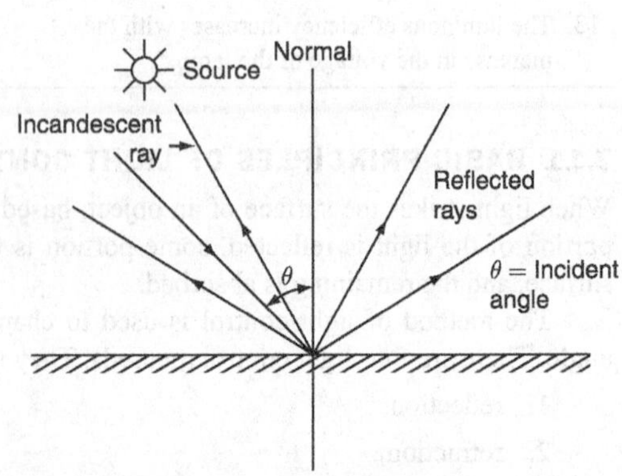

FIG. 7.18 *Diffuse reflection*

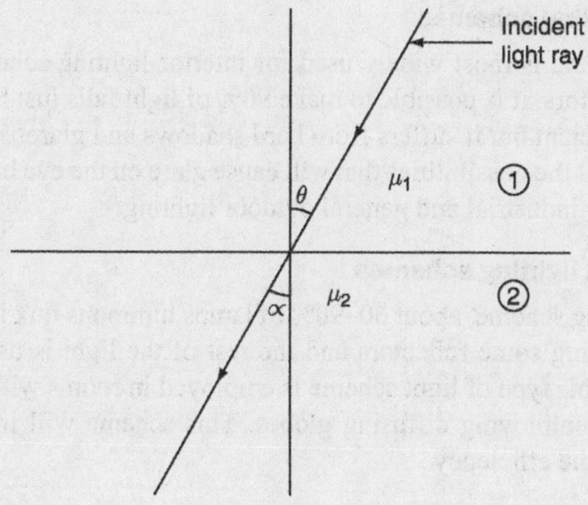

FIG. 7.19 *Refraction*

The angle of light ray with normal is comparatively less in dense medium than in rare medium.

7.11.3 Diffusion

When a ray of light falling on a surface is reflected in all possible directions, so that such surface appears luminous from all possible directions. This can be achieved with a diffusing glass screen introduced between the observer and the light source. The normally employed diffusing glasses are opal glass and frosted glass. Both are ordinary glasses, but frosted glass is an ordinary glass coated with crystalline substance.

Although frosted glass is cheaper than opal glass, the disadvantage of frosted glass is, it collects more dust particles and it is difficult to clean.

7.11.4 Absorption

In some of the cases, whole of the light emitted by tungsten filament lamp will be excessive, so that it is necessary to avoid that the amount of unwanted wavelengths without interference. This can be achieved by using a special bluish colored glass for the filament lamp to absorb the unwanted radiation.

7.12 TYPES OF LIGHTING SCHEMES

Usually, with the reflector and some special diffusing screens, it is possible to control the distribution of light emitted from lamps up to some extent. A good lighting scheme results in an attractive and commanding presence of objects and enhances the architectural style of the interior of a building. Depending upon the requirements and the way of light reaching the surface, lighting schemes are classified as follows:

(i) direct lighting,

(ii) semidirect lighting,

(iii) indirect lighting,

(iv) semi-indirect lighting, and

(v) general lighting.

7.12.1 Direct lighting schemes

Direct lighting scheme is most widely used for interior lighting scheme. In this scheme, by using deep reflectors, it is possible to make 90% of light falls just below the lamp. This scheme is more efficient but it suffers from hard shadows and glare. Hence, while designing such schemes, all the possibilities that will cause glare on the eye have to be eliminated. It is mainly used for industrial and general outdoor lighting.

7.12.2 Semidirect lighting schemes

In semidirect lighting scheme, about 60–90% of lamps luminous flux is made to fall downward directly by using some reflectors and the rest of the light is used to illuminate the walls and ceiling. This type of light scheme is employed in rooms with high ceiling. Glare can be avoided by employing diffusing globes. This scheme will improve not only the brightness but also the efficiency.

7.12.3 Indirect lighting schemes

In this lighting scheme, 90% of total light is thrown upwards to the ceiling. In such scheme, the ceiling acts as the lighting source and glare is reduced to minimum.

This system provides shadowless illumination, which is very useful for drawing offices and in workshops where large machines and other difficulties would cause trouble some shadows if direct lighting schemes were used.

7.12.4 Semi-indirect lighting schemes

In semi-indirect lighting scheme, about 60–90% of light from the lamp is thrown upwards to the ceiling and the remaining luminous flux reaches the working surface. Glare will be completely eliminated with such type of lighting scheme. This scheme is widely preferred for indoor lighting decoration purpose.

7.12.5 General lighting scheme

This scheme of lighting use diffusing glasses to produce the equal illumination in all directions. Mounting height of the source should be much above eye level to avoid glare. Lamp fittings of various lighting schemes are shown in Fig. 7.20.

7.13 DESIGN OF LIGHTING SCHEMES

The lighting scheme should be such that:

- It should be able to provide sufficient illumination.
- It should be able to provide the uniform distribution of light throughout the working plane.
- It should be able to produce the light of suitable color.
- It should be able to avoid glare and hard shadows as much as possible.

While designing a lighting scheme, the following factors should be taken into consideration.

(i) Illumination level.

(ii) The size of the room.

(iii) The mounting height and the space of fitting.

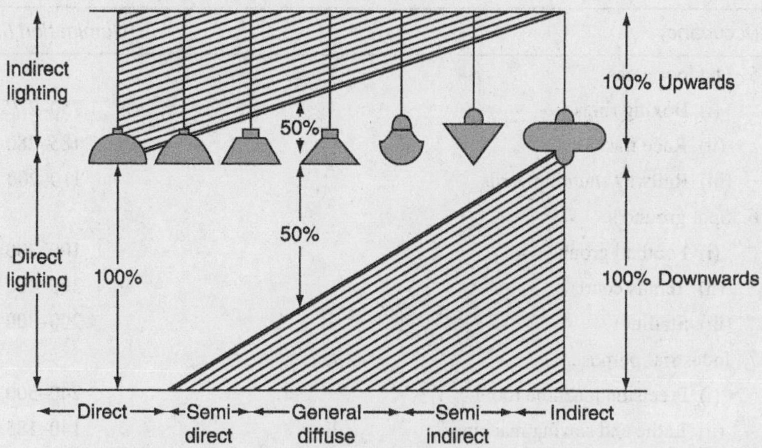

FIG. 7.20 *Lighting schemes*

7.13.1 Illumination level

The intensity of illumination required on the surface is depending on the type of work being done. For each type of work, there is a range of brightness that causes minimum fatigue and gives maximum output in terms of quality and quantity. Moving objects and the objects that are seen for longer duration require more illumination than those for stationary object and casual work.

The recommended levels of illumination for different aspects are given below.

Occupancy	*Illumination (lux)*
1. Covered areas:	
(i) Proofreading	95–185
(ii) Drawing an exhibition	55–95
(iii) Museums	35–55
(iv) Bedrooms and waiting room	18–32
(v) Hospital, railway yards, and platforms	5–10
2. Hotels and restaurants	
(i) Reception, dining room, and bedroom	150–200
(ii) Accounts and writing desks	300–400
3. Power station	
(i) Boiler house, turbine stage, transformer, and switch gear chamber	100–150
(ii) Control room	200–300
4. Canteens	100–200

(Continued)

Occupancy	Illumination (lux)
5. Outdoor areas	
(i) Boxing rings	1,750–2,750
(ii) Race tracks	185–280
(iii) Railway shunting yards	110–200
6. Spot ground's	
(i) Football ground	100–200
(ii) Tennis court	300–400
(iii) Stadium	200–300
7. Industrial purpose	
(i) Precision machine room	240–500
(ii) Lathe and sewing machine	140–185
(iii) General lighting factory	18–35
8. Schools and colleges	
(i) Laboratories, library, lecture hall, and workshop	200–300
(ii) Drawing rooms	400–500
(iii) Waiting rooms and stair	100–150

7.13.2 Size of the room

The luminous flux emitted from the source will not be completely utilized at the workplace. A portion of flux will be lost in the lamp fitting, some other will be absorbed, and the rest of it is reflected. This absorption and reflection are depending upon the size and color of the walls and ceiling. Illumination in any room depends upon the reflected light from the walls and ceiling. White color walls and ceiling reflect more light as compared to colored ones.

7.13.3 Mounting height and space of fittings

In general lighting, the illumination at any point should not vary throughout the room. So that, the lamp fittings for general lighting should be in such a way that the illumination received from each fitting overlaps with the other. In order to provide adequate illumination over the working plane, the distance of a light source from the wall should be half of the distance between the two adjacent lamps and also the distance between the source fitting or the spacing should not exceed more than 1.5 times the mounting height.

$$\therefore \frac{\text{space}}{\text{height}} \leq 1.5.$$

7.14 STREET LIGHTING

Street lighting not only requires for shopping centers, promenades, etc. but also necessary for the following.

- In order to make the street more attractive, so that obstructions on the road clearly visible to the drivers of vehicles.
- To increase the community value of the street.
- To clear the traffic easily in order to promote safety and convenience.

The basic principles employed for the street lighting are given below.

(i) Diffusion principle.

(ii) The specular reflection principle.

7.14.1 Diffusion principle

In this method, light is directed downwards from the lamp by the suitably designed reflectors. The design of these reflectors are in such a way that they may reflect total light over the road surface uniformly as much as possible. The reflectors are made to have a cutoff between 30° and 45°, so that the filament of the lamp is not visible expect just below the source, which results in eliminating glare. Illumination at any point on the road surface is calculated by applying inverse square low or point-by-point method.

7.14.2 Specular reflection principle

The specular reflection principle enables a motorist to see an object about 30 m ahead. In this case, the reflectors are curved upwards, so that the light is thrown on the road at a very large angle of incidence. This can be explained with the help of Fig. 7.21. An object resides over the road at 'P' in between the lamps S_1, S_2, and S_3 and the observer at 'Q'.

Thus, the object will appear immediately against the bright road surface due to the lamps at a longer distance. This method of lighting is only suitable for straight sections along the road. In this method, it is observed that the objects on the roadway can be seen by a smaller expenditure of power than by the diffusion method of lighting.

7.14.3 Illumination level, mounting height, and the types of lamps for street lighting

Normally, illumination required depends upon the class of street lighting installation. The illumination required for different areas of street lighting are given in Table 7.3.

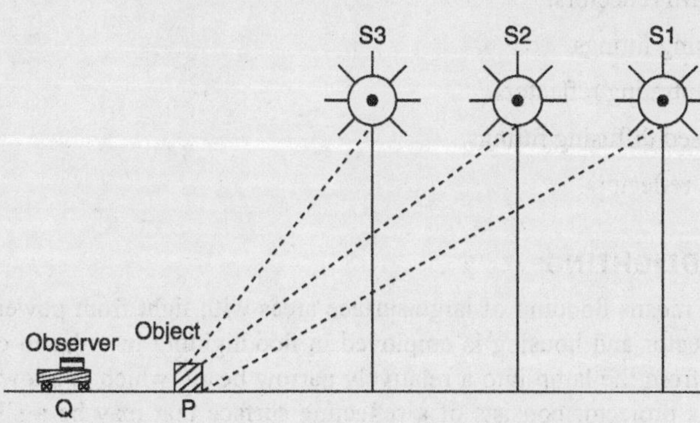

FIG. 7.21 *Specular reflection for street lighting*

TABLE 7.3 *Illumination required for different areas of street lighting*

	Area	*Illumination (lumen/m²)*
1.	Road junctions and important shopping centers.	30
2.	Poorly lighted sub-urban streets.	4
3.	Average well-lighted street.	8–15

Mercury vapor and sodium vapor discharge lamps are preferable for street lighting since the overall cost of the installation of discharge lamps are less than the filament lamps and also the less power consumption for a given amount of power output. Normal spacing for the standard lamps is 50 m with a mounting height of 8 m. Lamp posts should be fixed at the junctions of roads.

7.15 FACTORY LIGHTING

Industry or factory lighting must satisfy the following aspects.

1. The quality of work is to be improved.
2. Accidents must be reduced.
3. The productivity of labor should be increased.

The above requirements can be met by the factory lighting only when the lighting scheme provides:

(i) Adequate illumination on the working plane.
(ii) Minimum glare.
(iii) Clean and effective source fitting.
(iv) Uniform distribution of light over the working plane.

The lamps used for factory lighting are fitted with specially designed reflectors and they can be easily cleaned. The requirements of most of the installations of industrialized area can be met by the following lamp fitting.

- Industrial lighting fittings.
- Standard reflectors.
- Diffusing fittings.
- Concentrating reflectors.
- Enclosed diffusing fittings.
- Angle reflectors.

7.16 FLOODLIGHTING

Floodlighting means flooding of large surface areas with light from powerful projectors. A special reflector and housing is employed in floodlighting in order to concentrate the light emitted from the lamp into a relatively narrow beam, which is known as floodlight projector. This projector consists of a reflecting surface that may be a silvered glass or chromium plate or stainless steel. The efficiency of silvered glass and polished metal are 85–90% and 70%, respectively. Usually metal reflectors are robust; therefore, they can be preferred. An important application of illumination engineering is the floodlighting of large and open areas. It is necessary to employ floodlighting to serve one or more of the following purposes.

7.16.1 Esthetic floodlighting

They are used for enhancing the beauty of monuments, ancient buildings, and churches by floodlighting.

7.16.2 Industrial and commercial floodlighting

They are used for illuminating sports arenas, railway yards, quarries, car parks, etc.

7.16.3 Advertising

They are used for illuminating showcases and advertisement boards and for the decoration of houses, etc.

The projectors of floodlighting schemes are classified according to the light beam spread are discussed below.

Narrow beam projectors

Light beam with such a projectors spreads between 12° and 25°. They can be employed for a distance of 70 m.

Medium angle projectors

Projectors with beam spread between 25° and 40°. These are employed for a distance of 30–70 m.

Wide angle projectors

Projectors with beam spread between 40° and 90°. They can be employed for a distance of 30 m or below.

Economically, the wide angle projectors with high wattage lamps and narrow beam projectors with low wattage lamps are used.

7.16.4 Floodlighting calculations

While calculating the number of projectors required for floodlightings, it is necessary to know the level of illumination required; it is depending on the type of building and the purpose of floodlighting. And also the type of projector and the selection of projector depend upon the beam size as well as the light output.

∴ The total number of projectors required:

$$N = \frac{A \times E \times \text{depreciation factor} \times \text{waste} - \text{light factor}}{\text{utilization factor} \times \text{wattage of each lamp} \times \text{luminous efficiency of the lamp}},$$

where N is the number of projectors, A is the area of surface to be illuminated in square meter, and E is the illumination level required in lumen.

7.17 METHODS OF LIGHTING CALCULATIONS

There are so many methods have been employed for lighting calculation, some of those methods are as follows.

(i) Watts-per-square-meter method.

(ii) Lumen or light flux method

(iii) Point-to-point method

7.17.1 Watts-per-square-meter method

This method is more adoptive for rough calculation and checking also. According to the illumination required, this method makes an allowance of watt per square meter of area to be illuminated.

7.17.2 Lumen or light flux method

Lumen method is applicable for the cases in which all the sources produce uniform illumination over the working plane or an average value is required.

Total lumens received on working plane = No. of lamps × wattage of each lamp × efficiency of each lamp × coefficient of utilization.

7.17.3 Point-to-point or inverse square law method

This method is used to calculate the illumination at any particular point due to several number of sources whose candle powers are known values.

In general, illumination can be calculated by using the empirical formula:

$$N = \frac{E \times A}{\phi \times \text{UF} \times \text{MF}},$$

where N is the number of fitting required, E is the illumination required in lux, A is the working area in square meter, ϕ is the luminous flux produced per lamp in lumen, UF is the utilization factor, and MF is the maintenance factor.

Example 7.1: A room 20 × 10 m is illuminated by 60 W incandescent lamps of lumen output of 1,600 lumens. The average illumination required at the workplace is 300 lux. Calculate the number of lamps required to be fitted in the room. Assume utilization and depreciation factors as 0.5 and 1, respectively.

Solution:

The area of the room (A) = 20 × 10 m
= 200 m^2.

Total illumination required (E) = 300 lux.

The wattage of each lamp = 60 W

The luminous output of the lamp (ϕ) = 1,600 lumens

UF = 0.5, DF = 1.

$\therefore$ Maintenance factor, $\text{MF} = \frac{1}{DF} = \frac{1}{1} = 1.$

$\therefore$ The number of lamps required:

$$N = \frac{F \times A}{\phi \times \text{UF} \times \text{MF}}$$

$$= \frac{300 \times 200}{1{,}600 \times 1 \times 0.5} = 7.5 \text{ lamps.}$$

Example 7.2: The front of a building 35 × 18 m is illuminated by 15 lamps; the wattage of each lamp is 80 W. The lamps are arranged so that uniform illumination on the surface is obtained. Assuming a luminous efficiency of 20 lumens/W, the coefficient of utilization is 0.8, the waste light factor is 1.25, DF = 0.9. Determine the illumination on the surface.

Solution:

Area = (A) = 35 × 18 = 630 m^2.

The number of lamps, N = 15.

Luminous efficiency, η = 20 lumens/W.

UF = 0.8, DF = 0.9.

Waste light factor = 1.25, $E = ?$

$$\therefore N = \frac{A \times E \times \text{DF} \times \text{waste light factor}}{\text{UF} \times \eta \times \text{wattage of each lamp}}$$

$$15 = \frac{630 \times E \times 1.25 \times 0.9}{0.8 \times 20 \times 80}$$

$$= 0.554\, E.$$

$\therefore$ E = 27.07 lux (or) lumens/m^2.

Example 7.3: A room of size 10 × 4 m is to be illuminated by ten 150-W lamps. The MSCP of each lamp is 300. Assuming a depreciation factor of 0.8 and a utilization factor of 0.5. Find the average illumination produced on the floor.

Solution:

The area of the room (A) = 10 × 4 = 40 m^2.
The total luminous flux emitted by ten lamps (ϕ)
= 10 × 150 × 4π = 18, 849.5 lumens.
The total luminous flux reaching the working plane

$$= \frac{\phi \times \text{utilization factor}}{\text{depreciation factor}}$$

$$= \frac{18{,}849.5 \times 0.5}{0.8} = 11{,}780.97 \text{ lumens.}$$

The illumination on the working plane

$$E = \frac{\text{lumens on the working plane}}{\text{total area to be illuminated}}$$

$$= \frac{11{,}780.97}{40} = 294.52 \text{ lux.}$$

Example 7.4: The front of a building 25 × 12 m is illuminated by 20 1,200-W lamps arranged so that uniform illumination on the surface is obtained. Assuming a luminous efficiency of 30 lumens/W and a coefficient of utilization of 0.75. Determine the illumination on the surface. Assume DF = 1.3 and waste light factor 1.2.

Solution:

Area to be illuminated = 25 × 12 = 300 m^2.
The total lumens given out by 20 lamps is:
ϕ = number of lamps × wattage of each lamp × efficiency of each lamp
= 20 × 30 × 1,200 = 720,000 lumens.
The total lumens reaching the surface to be illuminated

$$= \frac{\phi \times UF}{DF \times \text{waste light factor}}.$$

$$= \frac{7,20,000 \times 0.75}{1.3 \times 1.2}$$

= 3,46,153.84 lumens.

The illumination on the surface:

$$E = \frac{3,46,153.84}{300} = 1,153.84 \text{ lux.}$$

Example 7.5: An illumination of 40 lux is to be produced on the floor of a room 16 × 12 m. 15 lamps are required to produce this illumination in the room; 40% of the emitted light falls on the floor. Determine the power of the lamp in candela. Assume maintenance factor as unity.

Solution:

Given data:

$E = 40$ lux

$A = 16 \times 12 = 192$ m^2

Number of lamps, $N = 15$

UF = 0.4, MF = 1

$$N = \frac{E \times A}{\phi \times \text{UF} \times \text{MF}}$$

$$15 = \frac{40 \times 192}{\phi \times 0.4 \times 1}$$

$\therefore \phi = 1,280$ lux.

So, the lumen output of the lamp in candela $= \frac{1,280}{4\pi} = 101.85$ cd.

Example 7.6: A drawing, with an area of 18 × 12 m, is to be illuminated with an average illumination of about 150 lux. The lamps are to be fitted at 6 m height. Find out the number and size of incandescent lamps required for an efficiency of 20 lumens/W. UF = 0.6, MF = 0.75.

Solution:

Given data:

$\eta = 120$ lumens/W

$E = 150$ lux

$A = 18 \times 12 = 216$ m^2

UF = 0.6

MF = 0.75

The total gross lumens required $\phi = \frac{E \times A}{\text{UF} \times \text{MF}}$.

$$= \frac{150 \times 216}{0.6 \times 0.75} = 72,000 \text{ lumens.}$$

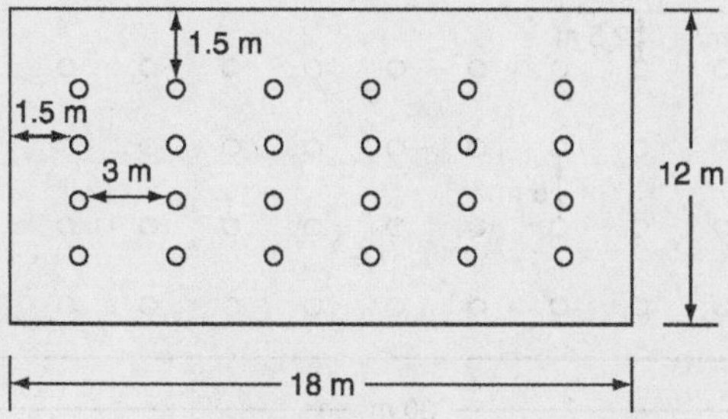

FIG. P.7.1 *Lamp arrangement*

$$\text{The total wattage required} = \frac{72,000}{\eta}$$

$$= \frac{72,000}{20} = 3,600 \text{ W.}$$

Let, if 24 lamps are arranged to illuminate the desired area. For space to height ratio unity, i.e., 6 lamps are taken along the length with a space of $18/6 \text{ m} = 3 \text{ m}$, and 4 lamps are along the width giving a space of $12/4 = 3$ m.

$$\therefore \text{The wattage of each lamp} = \frac{3,600}{24} = 150 \text{ W.}$$

The arrangement of 24 lamps in a hall of 18 × 12 m is shown in Fig. P.7.1

Example 7.7: A hall of 30 × 20 m area with a ceiling height of 6 m is to be provided with a general illumination of 200 lumens/m², taking a coefficient of utilization of 0.6 and depreciation factor of 1.6. Determine the number of fluorescent tubes required, their spacing, mounting height, and total wattage. Take luminous efficiency of fluorescent tube as 25 lumens/W for 300-W tube.

Solution:

Given data:

Area of hall (A) = 30 × 20 m = 600 m²

$E = 200$ lumens/m²

CU = 0.6

DF = 1.6

The wattage of fluorescent tube = 300 W

Efficiency $\eta = 25$ lumens/W

$$\therefore \text{Gross lumens required, } \phi = \frac{A \times E \times \text{DF}}{\text{UF}}$$

$$= \frac{600 \times 200 \times 1.6}{0.6} = 320,000 \text{ lux.}$$

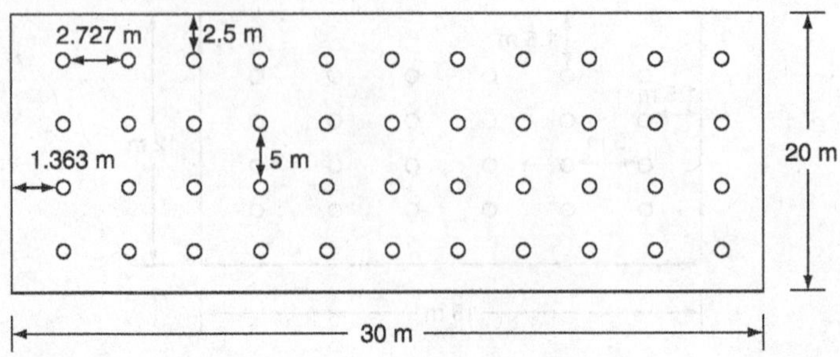

FIG. P.7.2 *Lamp arrangement*

$$\text{The total wattage required} = \frac{\phi}{\eta} = \frac{320,000}{25}.$$

$$\text{The number of tubes required} = \frac{\text{total wattage required}}{\text{wattage of each tube}}$$

$$= \frac{12,800}{300}$$

$$= 42.666 \cong 44.$$

Let us arrange 44 lamps in a 30 × 30 m hall, by taking 11 lamps along the length with spacing 30/11 = 2.727 m and 4 lamps along the width with spacing 20/4 = 5 m. Here the space to height ratio with this arrangement is, 2.727/5 = 0.545. Disposition of lamps is shown in Fig. P.7.2.

Example 7.8: A hall 40-m long and 16-m wide is to be illuminated and illumination required is 70-m candles. Five types of lamps having lumen outputs, as given below are available.

Watts:	50	100	150	200	250
Lumens:	1,500	1,830	2,500	3,200	4,000

Taking a depreciation factor of 1.5 and a utilization coefficient of 0.7, calculate the number of lamps required in each case to produce required illumination. Out of above five types of lamps, select most suitable type and design, a suitable scheme, and make a sketch showing location of lamps. Assume a suitable mounting height and calculate space to height ratio of lamps.

Solution:

Given data:

Area (A) = 30 × 12 = 360 m^2

DF = 1.5

CU = 0.7

E = 50-m candle

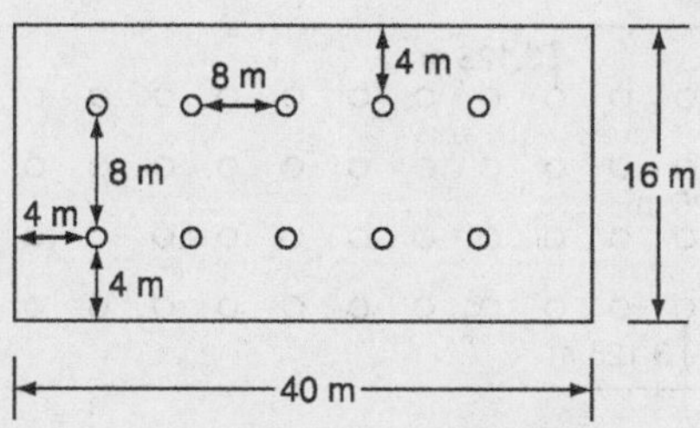

FIG. P.7.3 *Lamp arrangement*

Total gross lumens required:

$$\phi = \frac{A \times E \times \text{DF}}{\text{UF}}$$

$$= \frac{360 \times 50 \times 1.5}{0.7} = 38{,}572.42 \text{ lumens.}$$

(i) If 50-W lamps are used, the number of lamps required $= \frac{38{,}571.42}{1{,}500} = 25.7 \cong 26.$

(ii) If 100-W lamps are used, the number of lamps required $= \frac{38{,}571.42}{1830} = 7.416 \cong 8.$

(iii) If 150-W lamps are used, the number of lamps required $= \frac{38{,}571.42}{2{,}500} = 15.42 \cong 16.$

(iv) If 200-W lamps are used, the number of lamps required $= \frac{38.571.42}{3{,}200} = 12.05 \cong 14.$

(v) If 250-W lamps are used, the number of lamps required $= \frac{38{,}571.42}{4{,}000} = 9.642 \cong 10.$

Suitable type of lamp fitting will be 250-W lamps for a hall of 40 × 16 m.

Here, 10 lamps are arranged in two rows, each row having 5 lamps. By taking 5 lamps along the length with spacing $40/5 = 8$ m and 2 lamps along width side with spacing $16/2 = 8\,\text{m}$, i.e., space to height ratio $= 8/8 = 1$.

The disposition of lamps is shown in Fig. P.7.3.

Among the other lamps, some of wattage lamps require more number of lamp fittings and some other lamps will be few in requirement giving space–height ratio much more than required.

Example 7.9: An illumination on the working plane of 100 lux is required in a room 45 × 25 m in size. The lamps are required to be hung 3 m above the plane. Assuming a suitable space–height ratio, a utilization factor of 0.8, a lamp efficiency of 18 lumens/W, and a candle power depreciation of 30%, estimate the number, rating, and disposition of lamps.

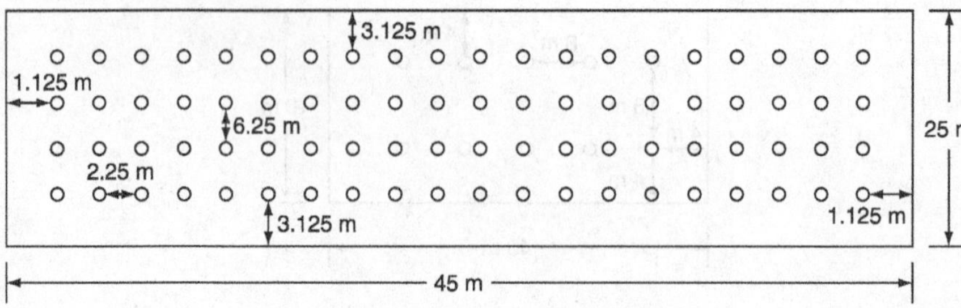

FIG. P.7.4 *Lamp arrangement*

Solution:

Given data:

Efficiency (η) = 18 lumens/W.

Area to be illuminated (A) = 45 × 25

= 1,125 m^2.

Illumination required (E) = 100 lux

UF = 0.8.

MF = 1 – candle power depreciation

= 1 – 0.3 = 0.7.

Total Gross lumens required:

$$\phi = \frac{A \times E}{\text{UF} \times \text{MF}} = \frac{1{,}125 \times 100}{0.8 \times 0.7}$$

= 200,892.857 lumens.

$$\text{Total wattage required} = \frac{\phi}{\eta} = \frac{200{,}892.857}{18}$$

= 11,160.714 W.

If 80 lamps are arranged to illuminate the desired area in 4 rows each row having 20 lamps

$$\text{Spacing in length wise} = \frac{45}{20} = 2.25.$$

$$\text{Spacing in width wise} = \frac{25}{4} = 6.25.$$

$$\therefore \text{Space–height ratio} = \frac{2.25}{6.25} = 0.36.$$

The disposition of lamps is shown in Fig. P.7.4.

Example 7.10: The front of a building 50 × 20 m^2 is desired to be illuminated by floodlighting projections placed at a distance of 25 m from the wall. The average illumination required is 40 lux. Estimate the number and size of the projectors required. Assume that waste light factor is 1.2, depreciation factor is 1.4, and coefficient of utilization is 0.35 (Fig. P.7.5).

Wattage:	250	500	1,000
Lumens:	4,000	6,000	12,000

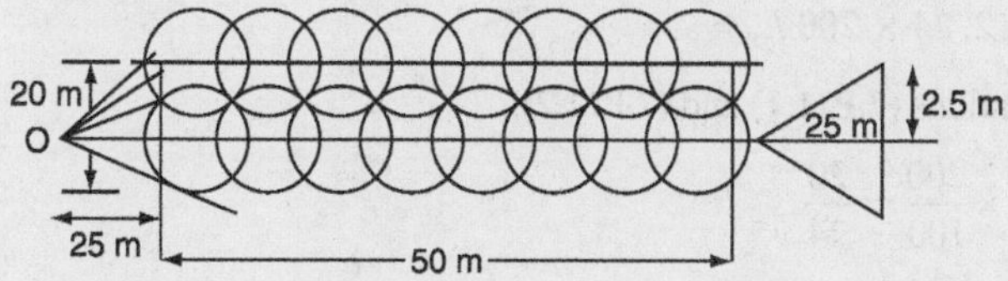

FIG. P.7.5 *Lamp arrangement*

Solution:

Given data:

Area $(A) = 50 \times 20 = 1{,}000\ \text{m}^2$

$E = 40$ lux.

Waste light factor $(W) = 1.2$

DF $= 1.4$

UF $= 0.35.$

The gross lumens emitted by the lamps (or) projectors

$$= \frac{A \times E \times W \times \text{DF}}{\text{UF}}$$

$$= \frac{1{,}000 \times 40 \times 1.2 \times 1.4}{0.35} = 192{,}000 \text{ lumens.}$$

If the wattage of each lamp is 1,000 W, then the number of projectors required is:

$$= \frac{\phi}{\text{wattage of each lamp} \times \eta}$$

$$= \frac{192{,}000}{1{,}000 \times 12}\left[\text{for 1,000-W lamp, } \eta = \frac{12{,}000}{1{,}000} \text{ lumens/W} = 12 \text{ lumens/W}\right]$$

$= 16$ lamps.

The angle of spread $(\theta) = \tan^{-1}\left(\frac{2.5}{25}\right) = 11°.$

Hence, 16 projectors of 1,000 W with beam angle of 11° will be required.

Example 7.11: A 100-V lamp develops 10 CP and a lamp of the same material and worked at the same efficiency develops 24 CP on 200 V. Compare diameter and length of the filaments prove the relationships used.

Solution:

Given that the two lamps are having the same efficiency so that their operating temperatures must be same.

Let d_1 and d_2 are the diameters of filaments, l_1 and l_2 are their lengths, and I_1 and I_2 are currents flowing through them.

Let us assume, input power taken by the two lamps is proportional to their output.

$\therefore$ For lamp 1, $10 \times 100\, I_1$. (7.P11.1)

For lamp 2, $24 \times 200\, I_2$. (7.P11.2)

From the Equations (7.P11.1) and (7.P11.2):

$$\frac{I_1}{I_2} = \frac{10}{24} \times \frac{200}{100} = \frac{20}{24}.$$

But their operating temperatures are same:

$$\therefore \frac{d_1}{d_2} = \left[\frac{I_1}{I_2}\right]^{2/3} = \left(\frac{20}{24}\right)^{2/3} = 0.885.$$

Since, output $\propto l_1\, d_1$:

$$\therefore \frac{10}{24} = \frac{l_1 d_1}{l_2 d_2}$$

$$\therefore \frac{l_1}{l_2} = \frac{10}{24} \times \frac{d_2}{d_1} = \frac{10}{24} \times \frac{1}{0.885}$$

$$\therefore \frac{l_1}{l_2} = 0.4705.$$

Example 7.12: A lamp of 50 W operates at 220 V and power factor 0.8. Its power factor is to be corrected to be unity. Determine the capacitance required for the condenser.

Solution:

Given data:

Voltage (V) = 220 V.

Wattage of lamp (P) = 50 W.

$\cos \phi_1 = 0.8.$

$\cos \phi_2 = 1.$

The current drawn by the lamp:

$$\mathrm{I} = \frac{P}{V \times \cos\phi} = \frac{50}{220 \times 0.8}$$

$$= 0.284 \text{ Amp.}$$

The reactive volt–amperes drawn by the lamp:

$$Q_1 = VI \sin\phi = 220 \times 0.284 \times 0.6$$

$$= 37.488 \text{ VAR.}$$

The corrected power factor $\cos\phi_2 = 1 \Rightarrow$ the reactive volt–amperes drawn by the lamp $Q_2 = VI \sin \phi_2 = 0$

The reactive volt–amperes to be compensated by the condenser is:

$$2\pi f C V^2 = 37.488$$

$$\therefore C = \frac{37.488}{2\pi f \times V^2} = \frac{37.488}{2\pi \times 50 \times (220)^2}$$

$$= 2.465\ \mu\text{F}.$$

Example 7.13: An incandescent lamp has a filament of 0.005-cm diameter and 60-cm length. It is required to construct another lamp of same type to work at double the supply voltage and given half the candle power. Assuming that the new lamp operates at same brilliancy; determine the dimensions of the filament.

Solution:

Given data:

The diameter of the lamp (1), $d_1 = 0.005$ cm.

The length of the filament of the lamp (1), $l_1 = 60$ cm

and $V_2 = 2\,V_1$.

Let the candle powers of the two lamps as I_1 and I_2 candle.

$$\therefore \frac{I_2}{I_1} = \frac{1}{2}$$

and dimensions of the filament of the lamp 2 as, l_2 and d_2. But we know that:

$$I_1 = l_1 d_1 \quad \text{and} \quad I_2 = l_2 d_2$$

$$\therefore \frac{I_2}{I_1} = \frac{l_2 d_2}{l_1 d_1} = \frac{1}{2}$$

and the power inputs of the two lamps are proportional to their outputs.

$$\therefore I_1 \,\alpha\, V_1\, l_1 \quad \text{and} \quad I_2 = V_2\, l_2$$

$$\frac{I_2}{I_1} = \frac{V_2 i_2}{V_1 i_1} = \left(\frac{V_2}{V_1}\right) \times \left(\frac{i_2}{i_1}\right)$$

$$\therefore \frac{i_2}{i_1} = \frac{I_2}{I_1} \times \frac{V_1}{V_2} = \frac{1}{2} \times \frac{1}{2} = \frac{1}{4}.$$

But, the current carrying capacity of the filament is depending upon, its diameter:

$$\therefore i \,\alpha\, (d)^{3/2}$$

$$\therefore \frac{i_2}{i_1} = \left(\frac{d_2}{d_1}\right)^{3/2} = \left(\frac{1}{4}\right)$$

$$\therefore \frac{d_2}{d_1} = (1/4)^{2/3} = 0.3968$$

$$d_2 = 0.3968 \times 0.005 = 0.001984 \text{ cm}$$

$$\text{and} \quad l_2 = \frac{1}{2} \times l_1 \times \frac{d_1}{d_2}$$

$$l_2 = \frac{1}{2} \times 60 \times \frac{1}{0.3968}$$

$$l_2 = 75.60 \text{ cm.}$$

Example 7.14: A 40-candle power and 300-V metal filament lamp has a measured candle power of 86.5 candle at 340 V and 60 candle at 320 V. Calculate the following:

(i) The constant for the lamp in expression $C = aV^b$, where C = candle power and V = voltage.

(ii) The change of candle power per volt at 300 V.

(iii) The percentage variation of candle power due to a voltage variation of 6% from the normal value.

Solution:

Given that:

At 300 V, the CP of the lamp = 40 cd.

At 320 V, the CP of the lamp = 60 cd.

And at 340 V, the CP of the lamp = 86.5 cd.

(i) The current for the lamp:

$C = a \cdot V^b$

$$60 = a \cdot (320)^b \qquad (7.P14.1)$$

$$86.5 = a \cdot (340)^b. \qquad (7.P14.2)$$

By dividing (7.P14.1) by (7.P14.2), we get:

$$\frac{60}{86.5} = \frac{(320)^b}{(340)^b}$$

$$0.6936 = (0.94117)^b$$

$$\ln (0.6936) = b \ln (0.94117)$$

$$b = \frac{\ln(0.6936)}{\ln(0.94117)} = \frac{-0.3658}{-0.0606}$$

$\therefore b = 6.036.$

By substituting value of b in Equation (7.P14.1), we get:

$60 = a\,(320)^{6.036}$

$\therefore a = 4.54 \times 10^{-14}.$

Hence, $C = a \cdot (V)^b$

$= 4.54 \times 10^{-4} \times (V)^{6.036}.$

(ii) The change of candle power per volt at 300 V

$C = a\,(V)^b = 4.54 \times 10^{-14} . (V)^{6.036}$

$$\frac{dc}{dv} = 4.54 \times 10^{-4} \times 6.036 \times V^{5.036}$$

$$\left(\frac{dc}{dv}\right)v = 300 \text{ V} = 4.54 \times 10^{-14} \times 6.036 \times (300)^{5.036}$$

$$= 0.8176.$$

(iii) The percentage variation of candle power

We know that:

$C \propto (V)^{6.036}.$

If voltage increases by 6% then:

$C_1 = (V_1)^{6.036}.$

$C_2 = (V_1 + 0.06\, V_1)^{6.036}.$

From Equations (7.P14.1) and (7.P14.2):

$$\frac{C_1}{C_2} = \left(\frac{V_1}{1.06\,V_1}\right)^{6.036}$$

$$= (1.06)^{6.036} = 1.4267.$$

∴ Percentage change in candle power:

$$= \frac{C_2 - C_1}{C_1} \times 100$$

$$= (1.4267 - 1) \times 100$$

$$= 42.67\%.$$

If voltage falls by 6%, then:

$$\frac{C_1}{C_2} = (0.94)^{6.036} = 0.688.$$

∴ Percentage change in candle power:

$$= \frac{C_2 - C_1}{C_1} \times 100 = (0.688 - 1) \times 100$$

$$= 31.2\%.$$

Example 7.15: A room 40 × 24 m is illuminated by indirect lighting. An average illumination of 50 lux is required to illuminate the working plane. Eighty-watt filament lamps having luminous efficiency of 16 lumens/W are to be used. The coefficient of utilization is 0.75 and depreciation factor is 0.85.

Calculate the following:

(i) Gross lumens required.

(ii) Power required for illumination.

(iii) Number of lamps.

(iv) Find the saving in power if instead of 80-W filament lamps, 30-W fluorescent tubes are used having efficiency of 40 lumens/W. Also find the number of tube lights required.

Solution:

Given data:

The area of the room (A) = 40 × 24
= 960 m^2.

Illumination (E) = 50 lux.

The wattage of the filament lamp = 80 W.

The efficiency of the filament lamp = 16 lumens/W.

CU = 0.7.

DF = 0.85.

The wattage of the fluorescent tube = 30 W.

The efficiency of the fluorescent tube = 40 lumens/W.

(i) Gross lumens required:

$$\phi = \frac{E \times A}{\text{CU} \times \text{DF}} = \frac{50 \times 960}{0.7 \times 0.85} = 80{,}672.268 \text{ lumens.}$$

(ii) Power required for illumination if 80-W lamps of 16 lumens/W efficiency are used:

$$= \frac{\text{Gross lumen}}{\eta} = \frac{80{,}672.268}{16} = 50.42.016 \text{ W.}$$

(iii) The number of lamps required:

$$= \frac{\text{power required}}{\text{wattage of each lamp}}$$

$$= \frac{5{,}042.016}{80} = 63.025 \cong 64 \text{ lamps.}$$

(iv) If 30-W tubes are used with efficiency 50 lumens/W, then the power input required is:

$$= \frac{\text{gross lumens}}{\text{efficiency}} = \frac{80{,}672.268}{50}$$

$$= 1{,}613.445 \text{ W.}$$

The number of tubes required:

$$= \frac{\text{total wattage}}{\text{wattage of each tube}}$$

$$= \frac{1613.445}{30} = 53.78 \text{ tubes.}$$

Saving in power when 30-W fluorescent tubes are used instead of 80-W lamps is:

$$= 5{,}042.016 - 1{,}613.445$$
$$= 3{,}428.571 \text{ W.}$$

KEY NOTES

- Types of sources of illumination are:
  - (i) Electric arc lamps.
  - (ii) Incandescent lamps.
  - (iii) Fluorescent lamps.
  - (iv) Gaseous discharge lamps.
- Arc lamps, in which light radiated out when electric current is made to flow through two electrodes in contact with each other and are separated by some distance apart.
- The various arc lamps are:
  - (i) Carbon arc lamp.
  - (ii) Flame arc lamp.
  - (iii) Metal arc lamp.
- Incandescent lamp emits radiation, which falls in the visible region of wavelength when its filament heated to high temperature. Incandescent lamps are also known as temperature radiators.
- Aging effect means that the light output of an incandescent lamp decreases, as the lamp ages.

- The alternating nature of supply goes through zero twice per a cycle, which causes the interruption of fluorescent lamp twice for each cycle. This effect is known as stroboscopic effect.
- Lighting schemes are:
  - (i) Direct lighting.
  - (ii) Semidirect lighting.
  - (iii) Indirect lighting.
  - (iv) Semi-indirect lighting.
  - (v) General lighting.

(a) Floodlighting means the flooding of large surface areas with the light from powerful projectors.

(b) Methods employed for the lighting calculations are:

(i) Watts per square meter method.

(ii) Lumen or light flux method.

(iii) Point-to-point method.

SHORT QUESTIONS AND ANSWERS

1. Give the principle of electric arc lamp.

   Arc lamp working is based on the principle that ionization of air present between the two electrodes produces an arc and provides intense light.

2. Write the principle of electric incandescent lamp.

   Operating principle of incandescent lamp is when the filaments of these lamps are heated to high temperature, emit light that falls in the visible region of wavelength.

3. List out the properties should be possessed by the filament material.

   The material used for the filament of incandescent lamp must have:

   (i) High melting point.

   (ii) Low temperature coefficient.

   (iii) Highly resistive nature

   (iv) Sufficient mechanical strength to withstand vibrations

4. Mention any two reasons why tungsten is preferred to carbon as filament material.

   Tungsten is preferred to carbon due to the following reasons.

   (i) Working temperature of tungsten is 2,500–3,000°C.

   (ii) Its resistance at working temperature is about 12–15 times the cold resistance.

   (iii) It has positive temperature coefficient of resistance of 0.0045.

   (iv) Its resistivity is 5.6–12.5 $\mu\Omega$-cm.

5. What is meant by aging effect?

   The light output of an incandescent lamp decreases, as the lamp ages is known as aging effect. This is mainly due to:

   (i) Less current drawn by the filament causes the reduction of lamp efficiency.

   (ii) The evaporation of filament blackens the inner side of the bulb.

6. What is stroboscopic effect?

   The alternating nature of the supply goes through zero twice per a cycle. At the usual alternating current supply frequency of 50 Hz, the discharge lamp will be extinguished 100 times per second. Due to the persistence of vision, the human eyes do not recognize that extinguish. If such light falls on any moving object, because of extinguishing nature of lamp, the object appearing to be either running slow or fast or running in reverse direction; sometimes, it may appear as stationary. This effect is known as '*stroboscopic effect*'.

7. Define refraction.

   When a beam of light passes through two different mediums having different densities, the light ray will be reflected. This phenomenon is known as *refraction*.

8. Define diffuse reflection.

   When a light ray falling on any surface, it is scattered in all the directions irrespective of the angle of incidence. Such type of reflector is known as *diffuse reflection*.

9. Define specular reflection.

   When whole of the light falling on a smooth surface, it will be reflected back at an angle equal to the angle of incidence. Such a reflection is known as *specular reflection*.

10. List the types of lamp fitting schemes.

Depending upon the requirements and the way of light reaching the surface, the lighting schemes are classified as follows:

(i) Direct lighting.

(ii) Semidirect lighting.

(iii) Indirect lighting.

(iv) Semi-indirect lighting.

(v) General lighting.

11. State the requirements of good lighting.

The lighting scheme should be in such a way that, it has:

(i) To provide sufficient illumination.

(ii) To provide uniform distribution of light throughout the working plane.

(iii) To produce the light of suitable color.

(iv) To avoid glare and hard shadows as much as possible.

12. What is meant by floodlighting?

Floodlighting means flooding of large surface areas with light from powerful projectors.

13. What is the empirical formula for calculating the number of lamps required for illumination?

In general, illumination can be calculated by using the empirical formula:

$$N = \frac{E \times A}{\phi \times \text{UF} \times \text{MF}}.$$

14. List out the methods of lighting calculations.

The Methods employed for lighting calculations are:

(i) Watts per square meter method.

(ii) Lumen or light flux method.

(iii) Point-to-point method:

15. What are the causes of aging effect of incandescent lamps?

Aging effect causes mainly due to the following reasons:

(i) Less current drawn by the filament causes the reduction of lamp efficiency.

(ii) The evaporation of filament blackens the inner side of the bulb.

16. How does the efficiency of a filament lamp increase with the increase in operating voltage?

The efficiency of a lamp increases with the increase in the operating voltage owing to increase in the temperature and is proportional to the square of the operating voltage.

17. Why sodium vapor discharge lamps are not used for general lighting?

Sodium vapor discharge lamps are not used for general lighting because they have got the drawback of color discrimination.

MULTIPLE-CHOICE QUESTIONS

1. Carbon arc lamps are commonly used in

(a) Cinema projectors.

(b) Domestic lighting.

(c) Factory lightning.

(d) Street lighting.

2. Which of the following filament material has the lowest melting point?

(a) Carbon.

(b) Tungsten.

(c) Osmium.

(d) Tantalum.

3. The melting point of tungsten is:

(a) 3,400°C.

(b) 2,800°C.

(c) 2,600°C.

(d) 2,400°C.

4. The vacuum inside an incandescent lamp is of the order of:

(a) 10^{-2} mmHg.

(b) 10^{-4} mmHg.

(c) 10^{-6} mmHg.

(d) 10^{-8} mmHg.

5. The rate of evaporation of tungsten filament in a lamp depends upon:
   (a) Exhaust tube diameter.
   (b) Glass shell diameter.
   (c) Vapor pressure inside.
   (d) None of the above.
6. Heat from the lamp filament is transmitted to the surrounding mainly through:
   (a) Circulation.
   (b) Conduction.
   (c) Convection.
   (d) Radiation.
7. Which of the following material is most commonly used for the filaments in incandescent lamps?
   (a) Tungsten.
   (b) Osmium.
   (c) Tantalum.
   (d) Silver.
8. A 0-W lamp consumes power of:
   (a) 0 W.
   (b) 5–10 W.
   (c) About 15 W.
   (d) About 25 W.
9. The output of a tungsten filament depends on:
   (a) The size of the shell.
   (b) The size of the lamp.
   (c) The temperature of filament.
   (d) All of the above.
10. The filament lamps normally operate at a power factor of:
    (a) 0.5 lagging.
    (b) 0.8 lagging.
    (c) Unity.
    (d) 0.85 lagging.
11. The expected life of an incandescent lamp is:
    (a) 100 hr.
    (b) 200 hr.
    (c) 500 hr.
    (d) 1,000 hr.
12. Most affected parameter of a filament lamp due to variation in operating voltage is:
    (a) Life.
    (b) Light output.
    (c) Luminous efficiency.
    (d) Wattage.
13. Which gas is sometimes used in filament lamps?
    (a) Nitrogen.
    (b) Carbon dioxide.
    (c) Argon.
    (d) Krypton.
14. Magnesium vapor in a filament lamp gives:
    (a) Green color light.
    (b) Pink color light.
    (c) Blue color light.
    (d) White color light.
15. The gas used in a gas-filled filament lamp is:
    (a) Helium.
    (b) Oxygen.
    (c) Nitrogen.
    (d) Ozone.
16. In the electric discharge lamps, the light is produced by:
    (a) The magnetic effect of current.
    (b) The heating effect of current.
    (c) Cathode ray emission.
    (d) The ionization in a gas or vapor.
17. In electric discharge lamps, for stabilization of arc:
    (a) A choke is connected in series with the supply.
    (b) A variable resistance is connected in series with the circuit.
    (c) A condenser is connected across the supply.
    (d) Any of the above.
18. Halogen lamps are useful for the illumination of:
    (a) Airports.
    (b) Parks and large gardens.
    (c) Playing fields.
    (d) All of the above.

19. Halogen lamps have the advantage(s) of:
    (a) Reduced dimensions of the lamp.
    (b) Better color rendition and longer life (about 2,000 hr)
    (c) High operating temperature with increased luminous efficiency.
    (d) All of the above.
20. Sodium vapor lamp needs an ionization voltage of about:
    (a) 5 V.
    (b) 20 V.
    (c) 50 V.
    (d) 100 V.
21. The ignition voltage for a sodium lamp is about:
    (a) 100–125 V.
    (b) 200–240 V.
    (c) 300–400 V.
    (d) 400–600 V.
22. When a sodium vapor lamp is switched on, initially the color is:
    (a) Red.
    (b) Pink.
    (c) Yellow.
    (d) Blue.
23. The color of the light given out by a sodium vapor discharge lamp is:
    (a) Pink.
    (b) Bluish green.
    (c) Yellow.
    (d) Blue.
24. An auto-transformer used with a sodium vapor lamp should have:
    (a) High step-up ratio.
    (b) High step-down ratio.
    (c) High leakage reactance.
    (d) High resistance.
25. The capacitor is used in auto-transformer circuit of a sodium vapor lamp in order to:
    (a) Regulate discharge voltage.
    (b) Improve the circuit power factor.
    (c) Control lamp illumination level.
    (d) Protect the lamp against over voltage.
26. The average life of a sodium vapor lamp is:
    (a) About 3,000 hr.
    (b) About 2,000 hr.
    (c) About 1,000 hr.
    (d) About 500 hr.
27. The luminous efficiency of a sodium vapor lamp is:
    (a) 40–50 lumens/W.
    (b) 50–100 lumens/W.
    (c) 120–200 lumens/W.
    (d) 10–12 lumens/W.
28. Sodium vapor lamps are used for the illumination of:
    (a) Streets and highways.
    (b) Rail yards and storage yards.
    (c) Parks.
    (d) All of the above.
29. A mercury vapor lamp gives light of:
    (a) Pink color.
    (b) Yellow color.
    (c) Greenish–blue color.
    (d) Red color.
30. The luminous efficiency of the high-pressure mercury vapor lamps ranges from:
    (a) 30–40 lumens/W.
    (b) 60–100 lumens/W.
    (c) 100–150 lumens/W.
    (d) 250 lumens/W.
31. Neon lamps:
    (a) Are of the size of the ordinary incandescent lamps.
    (b) Have the power consumption of the order of 5 W.
    (c) Are used as the indicator lamps, the night lamps, and for the determination of the polarity of DC mains.
    (d) All of the above.
32. Neon tubes are widely used for:
    (a) Advertising.
    (b) Indoor lighting.
    (c) Road signaling.
    (d) Airport lighting.

33. The vapor discharge tube used for the domestic light has:
    (a) One filament.
    (b) Two filaments.
    (c) Four filaments.
    (d) No filament.
34. The fluorescent tube is coated from inside with a thin layer of fluorescent material in the form of powder in order to:
    (a) Absorb invisible ultraviolet rays and radiate visible rays.
    (b) Improve the appearance.
    (c) Reduce glare.
    (d) Improve life.
35. A stabilizing choke is connected in the fluorescent tube circuit so as to:
    (a) Reduce the flicker.
    (b) Act as a ballast in operating conditions and provide a voltage impulse for starting.
    (c) Act as a starter.
    (d) Avoid radio interference.
36. A capacitor is connected across the fluorescent tube circuit in order to:
    (a) Eliminate the noise.
    (b) Limit the current.
    (c) Improve the power factor of the tube circuit.
    (d) All of the above.
37. In a fluorescent tube, a ballast resistance is connected in series with the choke:
    (a) When the tube is operated on DC supply.
    (b) When tube is operated on AC supply.
    (c) To reduce radio interference.
    (d) To reduce stroboscopic effects.
38. For the operation of the fluorescent tube on DC supply, the additional device incorporated in the tube circuit is a:
    (a) Transformer.
    (b) Resistor.
    (c) Inductor.
    (d) None of the above.
39. The light of fluorescent tube appears cooler than that from an incandescent lamp. This is due that fact that:
    (a) The tube consumes less power.
    (b) The surface area of the tube is larger than that of the incandescent lamp.
    (c) Tungsten is not used in the tube.
    (d) None of the above.
40. The flicker effect of the fluorescent lamps is more pronounced at:
    (a) Lower voltage.
    (b) Higher voltage.
    (c) Higher frequencies.
    (d) Lower frequencies.
41. Standard wattage of a 1-m fluorescent tube is:
    (a) 25 W.
    (b) 65 W.
    (c) 80 W.
    (d) 100 W.
42. The radio interference from a fluorescent tube can be reduced by:
    (a) Eliminating choke.
    (b) Connecting a small capacitor across starter terminals.
    (c) Putting tow tubes in parallel.
    (d) Any of the above.
43. In a mercury vapor lamp light, red objects appear black. This is on account of:
    (a) Color mixing.
    (b) High wavelengths of red color.
    (c) The absence of red light from the lamp radiations.
    (d) The absorption of red light by the lamp.
44. Blinking of a fluorescent tube may be on account of:
    (a) Low circuit voltage.
    (b) Loose contact.
    (c) Defective starter.
    (d) Any of the above.
45. Luminous efficiency of a fluorescent tube is about
    (a) 10 lumens/W.
    (b) 20 lumens/W.

(c) 40 lumens/W.

(d) 60 lumens/W.

46. The normal life of a fluorescent tube is about:

(a) 1,000 hr.

(b) 2,000 hr.

(c) 7,500 hr.

(d) 10,000 hr.

47. The bulb that takes the lowest power is:

(a) Neon bulb.

(b) Torch bulb.

(c) GLS bulb.

(d) Night bulb.

48. The lamp used in cinema projector is:

(a) Carbon arc lamp.

(b) Tungsten filament lamp.

(c) Fluorescent lamp.

(d) Sodium vapor lamp.

49. The direct lighting scheme is most efficient but is liable to cause:

(a) Glare.

(b) Hard shadow.

(c) Monotony.

(d) Both (a) and (b).

50. Floodlighting is not used for:

(a) Industrial purposes.

(b) Advertising purposes.

(c) Esthetic purposes.

(d) Any of the above.

51. Total flux required in any lighting scheme depends inversely on:

(a) Surface area.

(b) Space-height ratio.

(c) Illumination.

(d) The coefficient of utilization.

52. Desired illumination level on the working plane depends upon:

(a) The age group of the observers.

(b) Whether the object is stationary or moving.

(c) The size of the object to be seen and its distance from the observer.

(d) All of the above.

53. Which of the following will need the highest level of illumination?

(a) Living room.

(b) Kitchen.

(c) Proofreading.

(d) Hospital wards.

54. Which of the following will need the lowest level of illumination?

(a) Workshop.

(b) Displays.

(c) Railway platform.

(d) Garage.

55. The illumination level in houses is in the range of:

(a) 20–50 lux.

(b) 100–200 lux.

(c) 300–500 lux.

(d) 700 lux.

56. The illumination level required for important traffic routes carrying fast traffic is about:

(a) 30 lux.

(b) 100 lux.

(c) 200 lux.

(d) 5 lux.

57. Illumination due to moon light is about:

(a) 0.03 lumen/m^2.

(b) 0.3 lumen/m^2.

(c) 30–50 lumens/m^2.

(d) 300–500 lumens/m^2.

58. The luminous flux reaching the working plane least depends on:

(a) The proportion of the room.

(b) The lumen output of the lamps.

(c) The color of the working plane surface.

(d) The reflectance of the internal surfaces.

59. The depreciation factor depends upon:

(a) The ageing of the lamp(s).

(b) The cleaning schedules of lamps.

(c) The type of work carried out at the premises.

(d) All of the above.

60. The glass that transmits the maximum light is:

(a) Clear glass.

(b) Milk glass.

(c) Serrated glass.

(d) Opalescent glass.

61. The heat from light source is particularly of importance in:

(a) The designing for air-conditioning.

(b) The designing for illumination level.

(c) The designing for floor space utilization.

(d) All of the above.

62. The dimming systems for the lights are used in:

(a) Auditoriums.

(b) Theaters.

(c) Ball room.

(d) All of the above.

63. The optical instrument used for the comparison of candle powers of different sources is called the:

(a) Bunsen meter.

(b) Photometer.

(c) Candle meter.

(d) Ratio meter.

64. The photometer used for comparing the two sources giving light of different colors is:

(a) Bunsen photometer.

(b) Grease spot photometer.

(c) Guilds flicker photometer.

(d) Lummer–Brodhun photometer.

65. The photometer that utilizes a Lambert's cosine law for its operation is:

(a) Macbeth illumino-meter.

(b) Trotter illumination photometer.

(c) Guilds flicker photometer.

(d) Lummer–Brodhun photometer.

66. The optical instrument used for measurement of mean spherical candle power of a lamp is:

(a) Lummer–Brodhun photometer.

(b) Guilds flicker photometer.

(c) Integrating sphere.

(d) Any of the above.

REVIEW QUESTIONS

1. Describe various sources of illumination.
2. List out the properties should be possessed by a good filament material.
3. Describe the construction and working principle of a fluorescent lamp.
4. With neat sketch explain the working principle of an incandescent lamp.
5. Explain the working principle of the discharge lamps.
6. Describe the construction and working principle of a sodium vapor lamp.
7. What are the types of high-pressure mercury vapor lamps.
8. Describe the construction and working principle of MA type lamp.
9. Explain the working principle of MAT type lamp.
10. Compare the performance of MA type lamp with MB type lamp.
11. What is stroboscopic effect? How it can be prevented in fluorescent lamp.
12. Compare the performance of tungsten filament lamps and fluorescent lamps.
13. Discuss the performance of fluorescent lamp on DC supply.
14. Describe the basic principles of the control of the light.
15. Discuss the various designs of lighting schemes.
16. Explain the following principles of the light control.

    (i) Reflection.

    (ii) Refraction.

    (iii) Diffusion.

    (iv) Absorption.

EXERCISE PROBLEMS

1. A room of size 15× 6 m is to be illuminated by ten 400-W lamps. The MSCP of each lamp is 250. Assume depreciation factor 0.75 and utilization factor 0.7. Find the average illumination produced on the floor.

2. The front of a building 40 × 10 m is illuminated by 20 200-W lamps arranged so that uniform illumination on the surface is obtained. Assuming a luminous efficiency of 40 lumens/W and a coefficient of the utilization of 0.8. Determine the illumination on the surface. Assume DF = 1.5 and waste light factor 1.3.

3. A drawing is to be illuminated with an average illumination of about 400 lux; the area of the drawing is about 15 × 10 m. The lamps are to be fitted at 6-m height. Find out the number and size of incandescent lamps required for an efficiency of 45 lumens/W. UF = 0.9, MF = 0.6.

4. An illumination on the working plane of 300 lux is required in a room of 25 × 25 m in size. The lamps are required to be hung 7 m above the plane. Assuming a suitable space to height ratio, a utilization factor of 0.6, a lamp efficiency of 30 lumens/W, and a candle power depreciation of 20%; estimate the number, rating, and disposition of lamps.

5. An illumination of 60 lux is to be produced on the floor of a room of 20 × 12 m. Ten in size lamps are required to produces this illumination in the room; its 60% of the emitted light falls on the floor. Determine the power of the lamp in candela. Assume maintenance factor as unity.

ANSWERS

1. a	18. d	35. b	52. c
2. c	19. d	36. c	53. c
3. a	20. a	37. a	54. b
4. b	21. d	38. b	55. a
5. c	22. b	39. c	56. b
6. d	23. c	40. d	57. c
7. a	24. c	41. a	58. d
8. b	25. b	42. b	59. a
9. c	26. a	43. c	60. a
10. c	27. a	44. d	61. d
11. d	28. d	45. c	62. b
12. a	29. c	46. b	63. c
13. c	30. a	47. a	64. a
14. a	31. d	48. d	65. b
15. c	32. a	49. a	66. c
16. d	33. b	50. b	
17. a	34. a	51. d	

Electric Drives

OBJECTIVES

After reading this chapter, you should be able to:

- describe the function of an electric drive
- study the various characteristics of an electric motor
- know the temperature raise and the cooling of an electric motor
- analyze the load equalization process

8.1 INTRODUCTION

Motor control is required in large number of industrial and domestic applications such as transportation systems, rolling mills, paper machines, textile mills, machine tools, fans, pumps, robots, and washing machines. Systems employed for motion control are called *drives* and may employ any of the prime movers. Drives employing electric motors are known as electric drives.

Nowadays, in electric power stations generating large amounts of electric energy for agriculture, industry, domestic needs, and electrified traction facilities and in driving all kinds of working machines, electric motor is essential, which is the predominant type of drive so the term electric drive being applied to it.

Electric drive becomes more popular because of its simplicity, reliability, cleanliness, easiness, and smooth control. Both AC and DC motors are used as electric drives; however, the AC system is preferred because:

- It is cheaper.
- It can be easily transmitted with low-line losses.
- It can be easy to maintain the voltage at consumer premises within prescribed limits.
- It is possible to increase or decrease the voltage without appreciable loss of power.

In spite of the advantages of AC motor, sometimes DC motor is used because:

- In some processes, such as electrochemical and battery charging, DC is the only type of power that is suitable.
- The speed control of DC motors is easy rather than AC; thus, for variable speed applications such as lift and Ward Leonard system, the DC motors are preferred.
- DC series motor is suited for traction work because of high starting torque.

8.2 BLOCK DIAGRAM OF ELECTRIC DRIVE

8.2.1 Source

1-ϕ and 3-ϕ, 50-Hz AC supplies are readily available in most locations. Very low power drives are generally fed from 1-ϕ source; however, the high power drives are powered from 3-ϕ source; some of the drives are powered from a battery (Fig. 8.1).

Ex: Fork lifts trucks and milk vans.

8.2.2 Power modulator

Power modulator performs the following functions:

- It modulates flow of power from the source to the motor is impart speed–torque characteristics required by the load.
- It regulates source and motor currents within permissible values, such as starting, braking, and speed reversal conditions.
- Selects the mode of operation of motor, i.e., motoring or braking.
- Converts source energy in the form suitable to the motor.

8.2.3 Electrical motors

Motors commonly used in electric drives are DC motors, induction motors, synchronous motors, blushless DC motors, stepper motors, and switched reluctance motors, etc. In olden days, induction and synchronous motors were employed mainly for constant speed drives but not for variable speed drives, because of poor efficiency and are too expensive. But in nowadays, AC motors employed in variable speed drives due to the development of semiconductors employing SCRs, power transistors, IGBTs, and GTOs.

8.2.4 Load

It is usually a machinery, such as fans, pumps, robots, and washing machines, designed to perform a given task, usually load requirements, can be specified in terms of speed and torque demands.

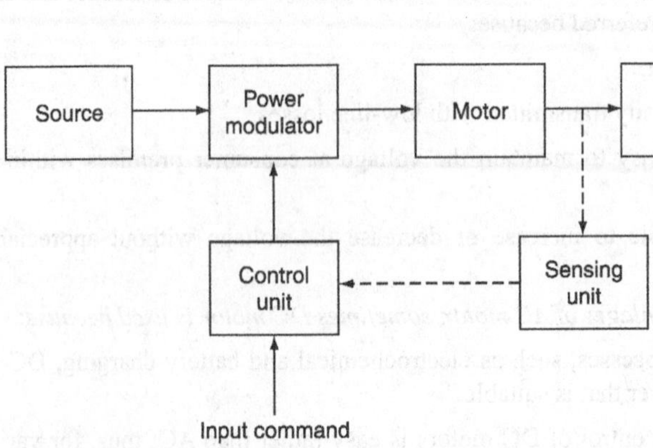

FIG. 8.1 *Block diagram of electric drive*

8.2.5 Control unit

Control unit controls the function of power modulator. The nature of control unit for a particular drive depends on the type of power modulator used. When semiconductor converters are used, the control unit will consists of firing circuits. Microprocessors also used when sophisticated control is required.

8.2.6 Sensing unit

Sensing unit consists of speed sensor or current sensor. The sensing of speed is required for the implementation of closed loop speed control schemes. Speed is usually sensed using tachometers coupled to the motor shaft. Current sensing is required for the implementation of current limit control.

Advantages of electric drives

There are a number of inherent advantages that the electric drive possesses over the other forms of conventional drives are:

- They have comparatively long life than the mechanical drive.
- It is cleaner, as there are no flue gases, etc.
- It is more economical.
- They have flexible control characteristics.
- There is no need to store fuel or transportation.
- It requires less maintenance.
- Do not pollute environment.
- It is the reliable source of drive.
- The electrical energy can be easily transmitted by using transmission lines over long distances.
- Available in wide range of torque, speed, and power.
- High efficiency.
- Electric braking system is much superior and economical.
- Smooth speed control is easy.
- They can be started instantly and can immediately be fully loaded.
- They can operate in all the quadrants of speed torque plane.
- Being compactness, they require less space.
- They can be controlled remotely.

Disadvantages of electric drives

The two inherit disadvantages of the electric drive system are:

- The non-availability of drive on the failure of electrical power supply.
- It cannot be employed in distant places where electric power supply is not available.

8.3 TYPES OF ELECTRIC DRIVES

Depending on the type of equipment used to run the electric motors in industrial purpose, they may be classified into three types. They are:

(i) Group drives.

(ii) Individual drives.

(iii) Multi-motor drives.

8.3.1 Group drives

Electric drive that is used to drive one or more than two machines from line shaft through belts and pulleys is known as *group drive.* It is also sometimes called the *line shaft drive.* This drive is economical in the consideration of the cost of motor and control gear. A single motor of large capacity cost is less than the total cost of a number of small motors of the same total capacity. In switch over from non-electric drive to electric drive, the simplest way is to replace the engine by means of motor and retaining the rest of power transmission system.

Advantages

- The cost of installation is less. For example, if the power requirement of each machine is 10 HP and there are five machines in the group, then the cost of five motors will be more than one 50-HP motor.
- If it is operated at rated load, the efficiency and power factor of large group drive motor will be high.
- The maintenance cost of single large capacity motor is less than number of small capacity motors.
- It is used for the processes where the stoppage of one operation necessitates the stoppages of sequence of operations as in case of textile mills.
- It has overload capacity.

Disadvantage

Even though group drive has above advantages, it suffers from the following disadvantages.

- If there is any fault in the main motor, all the machines connected to the motor will fail to operate; thereby, paralyzing a part of industry until the fault is removed.
- It is not possible to install any machine at a distant place.
- The possibility of the installation of additional machines in an existing industry is limited.
- The level of noise produced at the work site is quite large.
- The speed control of different machines using belts and pulleys is difficult.
- The flexibility of layout is lost due to line shaft, belts, and pulleys.

8.3.2 Individual drive

In individual drive, a single electric motor is used to drive one individual machine. Such a drive is very common in most of the industries.

Advantages

- It is more clean and safety.
- Machines can be located at convenient places.
- If there is a fault in one motor, the output and operation of the other motors will not be effected.
- The continuity in the production of the industry is ensured to a higher degree.
- Individual drive is preferred for new factories, as it causes some saving in the cost.

Disadvantage

- Initial cost will be high.
- Power loss is high.

8.3.3 Multi-motor drive

In multi-motor drives, several separate motors are provided for operating different parts of the same machine.

Ex: In traveling cranes, three motors are used for hoisting, long travel, and cross-travel motions. Multi-motor drive is used in complicated metal cutting machine tools, rolling mills, paper making machines, etc.

8.4 CHOICE OF MOTORS

The selection of the driving motor for a given service depends upon the conditions under which it has to operate. Due to the universal adoption of electric drive, it has become necessary for the manufacturer to manufacture motors of various designs according to the suitability and the use in various designs according to the suitability and the use in various classes of industry. This has resulted into numerous types of motors. For this reason, the selection of motor itself has become an important and tedious process. The conditions under which an electric motor has to operate and the type of load it has to handle, determine its selection.

While selecting a motor, the following factors must be taken into consideration:

1. Cost:
   (i) initial cost and
   (ii) running cost.
2. Electric characteristics:
   (i) starting characteristics,
   (ii) running characteristics,
   (iii) speed control characteristics, and
   (iv) braking characteristics.
3. Mechanical characteristics:
   (i) type enclosure and bearings,
   (ii) arrangement for the transmission of power,
   (iii) noise, and
   (iv) cooling.

4. Size and vetting of motors:
   (i) requirements for continuous, Intermittent, or variable load cycle and
   (ii) overload capacity.
5. Type of drive:
   (i) the drive is for one or more machines and
   (ii) the type of transmission through gears, belts, etc.
6. Nature of supply.

From the above, it is seen that a large number of factors are to be considered in making the choice of an electric motor for a given drive. The motor selected must fulfill all the necessary load requirements and at the same time, it should not be very costly if it has to be a commercial success. The factors motioned above will be individually discussed in the following sections to bring home to the reader the importance of each. While making the final choice of the motor, a satisfactory compromise may have to be made in some cases on account of the conflicting requirements.

8.5 CHARACTERISTICS OF DC MOTOR

The performance and, therefore, suitability of a DC motor are determined from its characteristics. The important characteristics of DC motor are:

(i) *Torque vs. armature current characteristics (T vs. I_a):*
This characteristic curve gives relation between torque developed in the armature (T) and armature current (I_a). This is also known as *electrical characteristic.*

(ii) *Speed vs. armature current characteristics (N vs. I_a):*
This characteristic curve gives relation between speed (N) and armature current (I_a). This is also known as *speed characteristics.*

(iii) *Output (HP) vs. armature current characteristics (HP vs. I_a):*
The horse power of the motor is dependent on the shaft torque, so its characteristics follows *shaft torque characteristic.*

(iv) *Speed vs. characteristics (N vs. T):*
This characteristic gives relation between speed (N) and torque (T) developed in the armature. This curve may be derived from the two characteristics mentioned in characteristics (i) and (ii) above.

Characteristics (i), (ii), and (iii) are called *starting characteristics*, and (iv) is known as *running characteristics.*

While discussing motor characteristics, the following relations should always be kept in mind.

$$T \propto \phi I_a \quad \text{and} \quad N \propto \frac{E_b}{\phi},$$

where T_a is the torque developed in the armature in N-m, I_a is the armature current in ampere, E_b is the back emf in volts, and ϕ is the flux in weber.

8.5.1 Characteristics of shunt motor

The field winding connected across the armature terminals called as *shunt motor* as shown in Fig. 8.2. Rated voltage is applied across the field and armature terminals.

Starting characteristics

The study of starting characteristics of a motor is essential to know the starting torque necessary to accelerate the motor from standstill position is also to require to overcome the static friction and the standstill load or, to provide load torque.

Torque vs. armature current (T Vs I_a)

In the expression for the torque of a DC motor, torque is directly proportional to the product of flux per pole (ϕ) and armature current (I_a):

$$\therefore T \propto \phi I_a \quad \textbf{(8.1)}$$

Since, in case of a DC shunt motor, the flux per pole (ϕ) is considered to be constant.

$$\therefore T \propto I_a.$$

So, the torque is proportional to armature current and is practically a straight line passing through the origin as shown if Fig. 8.3.

To generate high starting torque, this type of motor requires a large value of armature current at starting. This may damage the motor, hence DC shunt motors can develop moderate starting torque and hence suitable for such applications where starting torque requirement is moderate.

Speed vs. armature current (N Vs I_a)

In shunt motor, the applied voltage '*V*' is kept constant, the field current will remain constant, and hence the flux will have maximum value on no load due to the armature reaction; if load on the motor increases, the flux will be slightly decrease. By neglecting the armature reaction, the flux is almost constant.

From the speed equation of DC shunt motor:

$$N \propto \frac{E_b}{\phi},$$

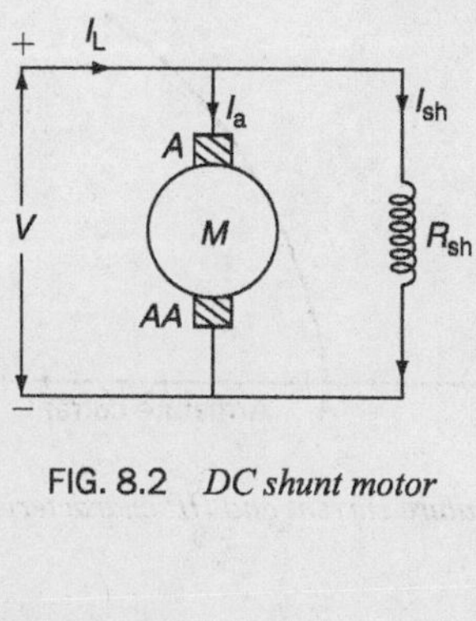

FIG. 8.2 *DC shunt motor*

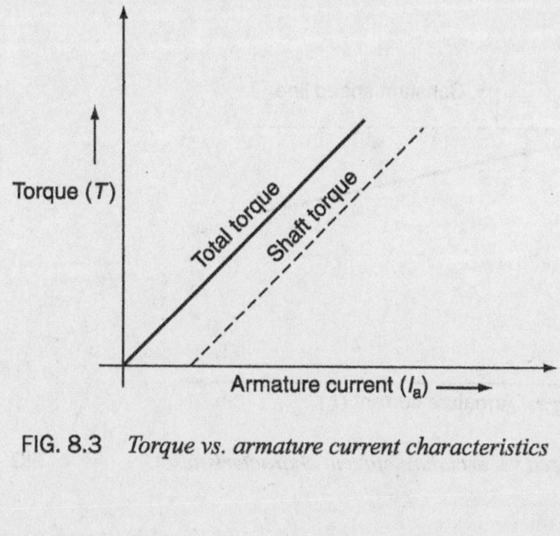

FIG. 8.3 *Torque vs. armature current characteristics*

where $E_b = V - I_a R_a$

$$\therefore N \propto \frac{V - I_a R_a}{\phi}.$$

Since, for DC shunt motor, the flux per pole is considered to be constant.

$$\therefore N \propto V - I_a R_a. \quad \textbf{(8.2)}$$

So, as the load on the motor increases, the armature current increases and hence $I_a R_a$ drop also increases. For constant supply, the voltage ($V - I_a R_a$) decreases and hence the speed reduces. Hence, as armature current increases, the speed of the DC motor decreases. The variation of speed with armature current is shown in Fig. 8.4.

Output vs. armature current

The output of the motor is dependent on the shaft torque. If the armature current increases, the output of the motor gradually increases. The variation of output with the armature current is shown in Fig. 8.5.

Running characteristics

Speed–torque characteristics (N vs. T)

These characteristics can be derived from its staring characteristics of (i) and (ii). During the steady-state operation of the motor, the voltage equation of the armature circuit is given by:

$$V = E_b + I_a R_a, \quad \textbf{(8.3)}$$

where V is the applied voltage, E_b is the back emf of motor, I_a is the armature current, and R_a is the armature resistance.

The back emf of motor can be expressed as:

$$E_b \propto \phi N$$

$$\therefore E_b = K \phi N,$$

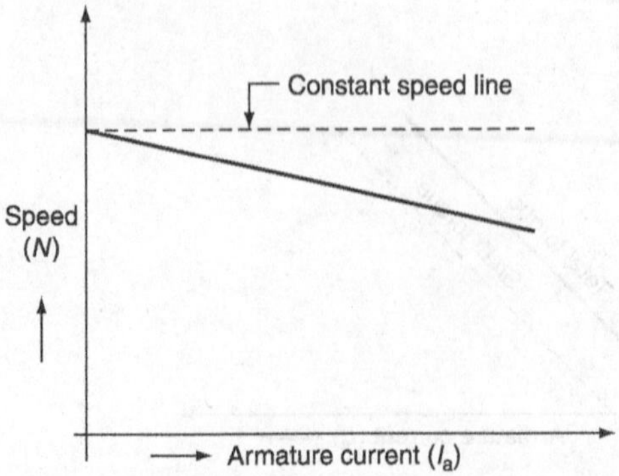

FIG. 8.4 *Speed vs. armature current characteristics*

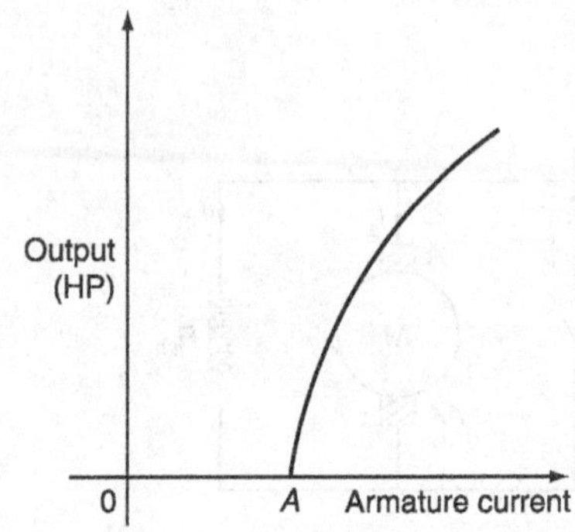

FIG. 8.5 *Armature current and HP characteristics*

where K is the constant, $N = \dfrac{E_b}{K\phi}$.

Substituting E_b from Equation (8.3) in above equation:

$$\text{Speed, } N = \frac{V - I_a R_a}{K\phi}. \quad \textbf{(8.4)}$$

The torque of the motor is directly proportional to product of flux and armature current.

$$\therefore T \propto \phi I_a$$

$$= K \phi I_a \quad \textbf{(8.5)}$$

$$I_a = \frac{T}{K\phi} \quad \textbf{(8.6)}$$

Substitute Equation (8.6) in Equation (8.4), we get:

$$N = \frac{V}{K\phi} - \frac{T}{(K\phi)^2} \times R_a. \quad \textbf{(8.7)}$$

Since, the shunt motor flux is constant, the speed of the motor is:

$$N = \frac{V}{K_1} - \frac{T}{K_1^2} R_a, \quad \textbf{(8.8)}$$

where $K_1 = K\phi$.

When V and R_a are kept constant, the speed torque characteristic is a *straight line*.

If the load on the motor increases, thus the torque increases and hence the speed of the motor decreases. The characteristic curve can be drawn from the Equation (8.8) and is shown in Fig. 8.6.

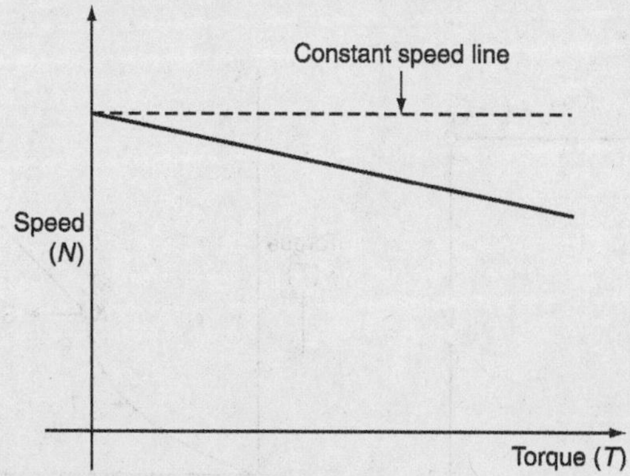

FIG. 8.6 *Speed and torque characteristics*

8.5.2 Characteristics of DC series motor

In case of series motor, the field windings are connected in series with armature terminals as shown in Fig. 8.7. Since, the field winding is connected in series with the armature winding, the load current (I_L) is equals to the armature current (I_a) or the series field current (I_{se}).

$$\therefore I_L = I_a = I_{se}.$$

Starting characteristics

Torque vs. armature current (T Vs I_a)

In case of DC motors, torque is directly proportional to the product of flux per pole (ϕ) and armature current (I_a).

$$\therefore T \propto \phi I_a.$$

Up to the saturation point, the flux is proportional to the field current and hence the armature current:

i.e., $\phi \propto I_{se} \propto I_a.$

Therefore, the torque is proportional to the square of the armature current.

$$\therefore T \propto I_a^2 \, (\because I_a = I_{se}). \quad \textbf{(8.9)}$$

Hence, the curve drawn in Fig. 8.8; the torque and the armature currents are parabolas, up to saturation point. After saturation, the flux (ϕ) is almost independent of the excitation current and so the torque is proportional to the armature current, i.e., $T \propto I_a$. Hence the characteristics become a straight line. The variation of torque with the armature current is shown in Fig. 8.8.

Speed vs. armature current

From the speed equation of DC series motor, the speed is directly proportional to the back emf and is inversely proportional to flux:

i.e., $N \propto \dfrac{E_b}{\phi},$

where $E_b = V - I_a R_{se}.$

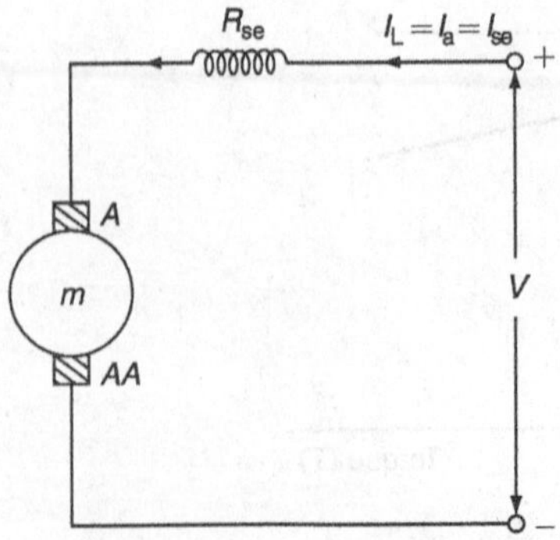

FIG. 8.7 *DC series motor*

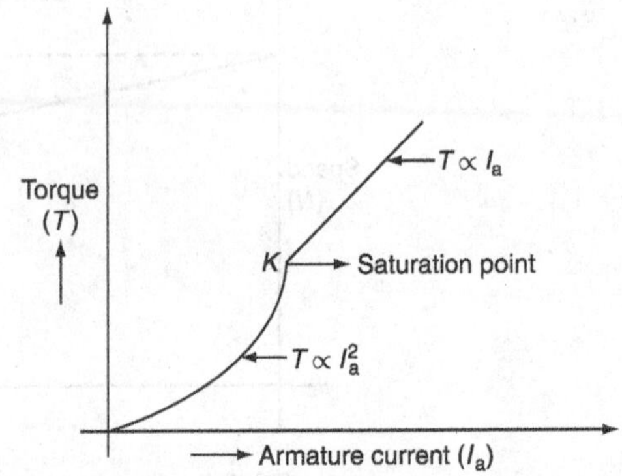

FIG. 8.8 *Torque and armature current*

When the armature current increases, the voltage drop in the armature resistance and the field resistance increases, but under the normal conditions, the voltage drop is small and it is negligible. Hence, $V = E_b$ and it is constant:

$$\therefore N \propto \frac{1}{\phi} \propto \frac{1}{I_a}$$

$$\therefore N \propto \frac{1}{I_a}. \tag{8.10}$$

This relation shows the variation of speed with the armature current and it will be a *rectangular hyperbola*, which is shown in Fig. 8.9.

Running characteristics

Speed–torque characteristics

These characteristics can be derived its starting characteristics. It is also known as *mechanical characteristic.*

In case of series motors:

$$T \propto \phi I_a \propto I_a^2$$

$$\text{and } N \propto \frac{1}{I_a}.$$

As the torque of a DC machine is directly proportional to armature current and flux, the speed will be inversely proportional to the square root of the torque, i.e., from the above two relations:

$$N \propto \frac{1}{\sqrt{T}}. \tag{8.11}$$

But at higher loads, the flux becomes saturated and the torque will be proportional to armature current, so the speed can be represented as:

$$N \propto \frac{1}{T}. \tag{8.12}$$

The speed–torque characteristics of a DC series motor is shown in Fig. 8.10.

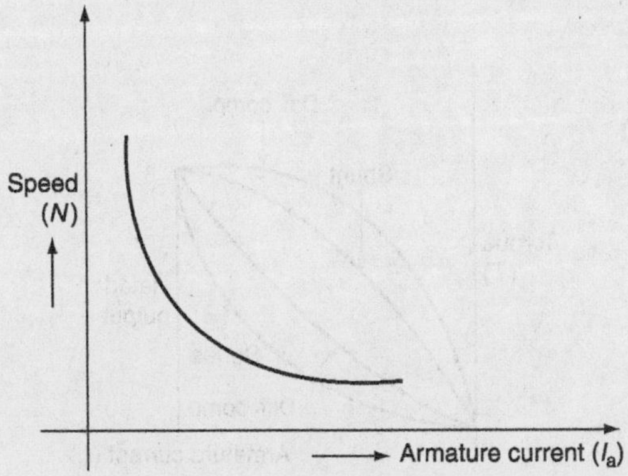

FIG. 8.9 *Speed and armature current*

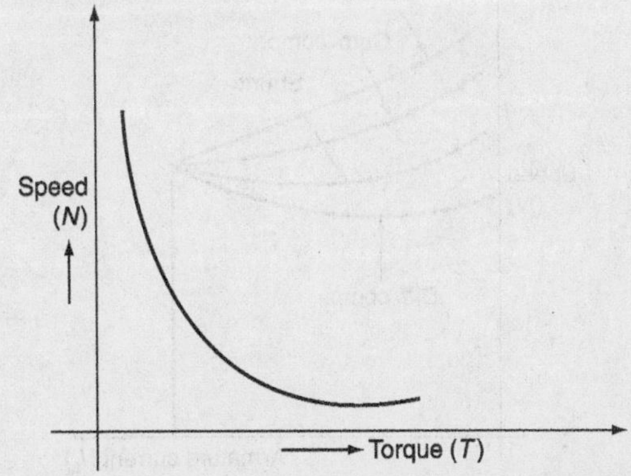

FIG. 8.10 *Speed–torque characteristics*

Hence, the series motors are best suited for services where the motor is directly coupled to the load such as whose speed falls with the increase in load torque.

8.5.3 Characteristics of DC compound wound motors

Compound motors have both series. If the series field excitation aids with the shunt excitation, then the motor is said to be *cumulatively compounded.* If the series field opposes the shunt field excitation, it is known as *differential compound motor.*

The characteristics of such motors lie in between shunt and series motors.

Cumulative-compound motor

Since, the series field aids with the shunt field winding, the flux is increased, as load is applied to the motor, and due to this reason, the motor speed slightly decreases. Such machines are used where series characteristics are required. Due to the shunt field, the winding speed will not become excessively high, but due to the series field winding, it will be able to take heavy loads.

Compound wound motors have the greatest application with loads that require high starting torques or pulsating load.

Differential-compound motors

In this motor, the series field opposes the shunt field and the flux is decreased, as load is applied to the motor. This results in the motor speed that is almost constant or even increasing with increase in load.

The speed–armature current and the torque–armature current characteristics of both the cumulative and the differential compound motors are shown in Figs. 8.11 and 8.12.

8.6 THREE-PHASE INDUCTION MOTOR

Three-phase induction motors are simple in design, rugged in construction with the absence of commentator, and reliable in service. Besides this, they have low initial cost, simple maintenance, easy operation, and simple control gear for starting and speed control.

The speed–torque characteristics of the induction motor are quite important in the selection of the induction motor drive. These characteristics can be effectively determined by means of the equivalent circuit of the induction motor. The simplified equivalent circuit of induction motor is shown in Fig. 8.13.

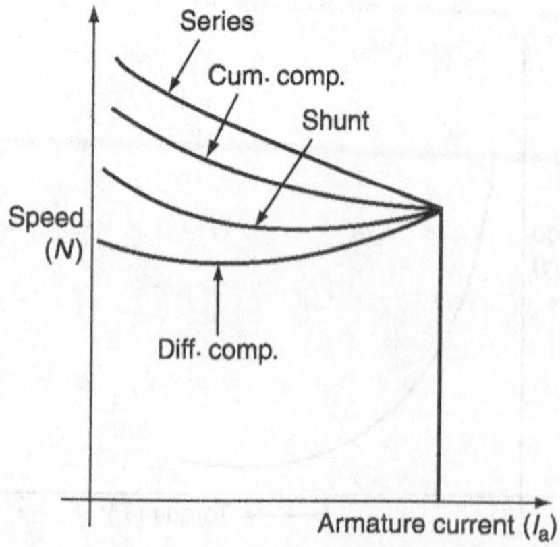

FIG. 8.11 *Speed and armature current characteristics*

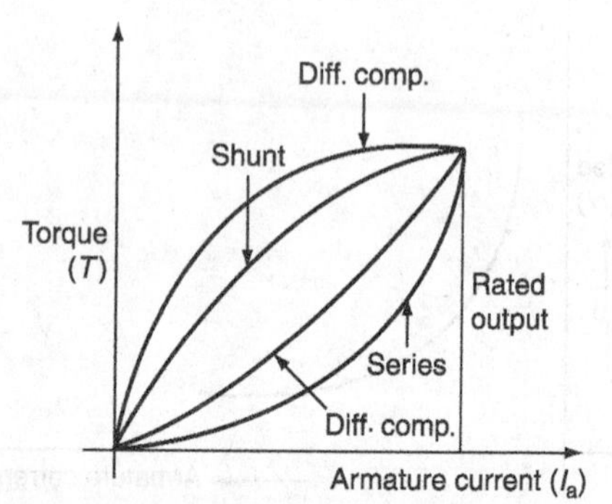

FIG. 8.12 *Torque and armature current characteristics*

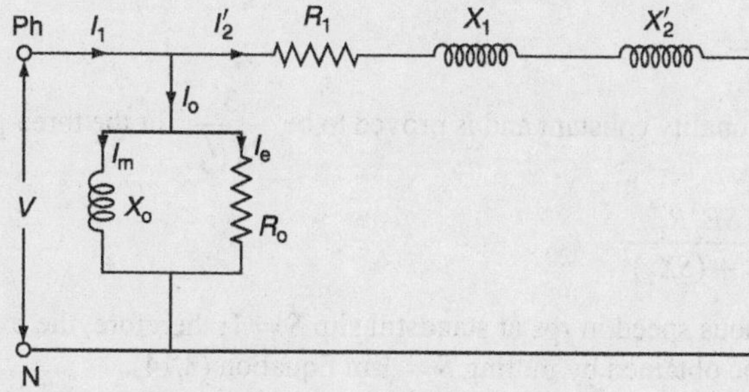

FIG. 8.13 *Equivalent circuit of induction motor*

In Fig. 8.13, V is the applied voltage per phase, R_1 and X_1 are the stator resistance and leakage reactance per phase, R_2' and X_2' are the rotor resistance and leakage reactance per phase, R_0 and X_0 are the resistance and reactance per phase of the magnetizing branch, and I_2' is the rotor current per phase.

From the equivalent circuit of induction motor, as shown in Fig. 8.13, the rotor current referred to the stator is given by:

$$I_2' = \frac{V}{\sqrt{\left(R_1 + \frac{R_2'}{S}\right)^2 + (X_1 + X_2')^2}}.$$

If the induction motor is rotating at slip is then:

Induced emf of rotor $= SE_2$.

Rotor resistance $= R_2$.

Rotor reactance $= SX_2$.

Rotor current/phase, $I_2 = \frac{SE_2}{\sqrt{R_2^2 + (SX_2)^2}}$.

8.6.1 Torque equation

The torque produced in the induction motor is mainly depends on the magnitude of rotor current, the power factor of the rotor circuit, and the part of rotating magnetic field that interacts with the rotor.

$$\therefore T \propto E_2 I_2 \cos \phi_2. \tag{8.13}$$

Substituting the values of I_2 and $\cos \phi_2$ in Equation (8.13):

$$T \propto E_2 \times \frac{SE_2}{\sqrt{R_2^2 + (SX_2)^2}} \times \frac{R_2}{\sqrt{R^2 + (SX_2)^2}}$$

$$\therefore\ T \propto \frac{SE_2^2 R_2}{R_2^2 + (SX_2)^2}$$

$$T = \frac{KSE_2^2R_2}{R_2^2 + (SX_2)^2}, \tag{8.14}$$

where 'K' is proportionality constant and is proved to be $\frac{3}{2\pi N_s}$ for the three-phase induction motor.

$$\therefore T = \frac{3}{2\pi N_s} \frac{SE_2^2R_2}{R_2^2 + (SX_2)^2}, \tag{8.15}$$

where N_s is synchronous speed in rps at standstill slip $S = 1$; therefore, the expression for starting torque may be obtained by putting $S = 1$ in Equation (8.14).

$$\therefore T_{st} = \frac{KE_2^2R_2}{R_2^2 + X_2^2}. \tag{8.16}$$

Condition for maximum torque

The torque developed by the motor under running condition mainly depends on slip at which motor is running.

Therefore, the torque will be maximum when:

$$\frac{dT}{dS} = 0; \qquad \text{where } T = \frac{KSE_2^2R_2}{R_2^2 + (SX_2)^2}.$$

By differentiating torque w.r.t. 'S', we get:

$$\frac{dT}{dS} = \frac{(KSE_2^2R_2)\frac{d}{dS}(R_2^2 + S^2X_2^2) - (R_2^2 + S^2X_2^2)\frac{d}{dS}(KSE_2^2R_2)}{(R_2^2 + S^2X_2^2)^2}$$

$$\therefore KS_mE_2^2(2S_mX_2^2) - (R_2^2 + S_m{}^2X_2^2) = 0$$

$$2S_m{}^2X_2^2 - (R_2^2 + S_m{}^2X_2^2) = 0$$

$$R_2 = S_mX_2$$

$$S_m = \frac{R_2}{X_2} \tag{8.17}$$

Equation (8.17) reveals that the slip 'S_m' at which maximum torque will be developed by the induction motor.

From Eq. (8.14), the maximum torque corresponding to slip $S_m = R_2/X_2$ is given by:

$$T_{max} = \frac{K \times \frac{R_2}{X_2} \times R_2 \times E_2^2}{R_2^2 + \frac{R_2^2}{X_2^2} \times X_2^2}$$

$$= \frac{KE_2^2}{2X_2}. \tag{8.18}$$

8.6.2 Torque ratios

The performance of motor is estimated in terms of the ratios of different torques such as full-load, starting, and maximum torques.

Ratio of full-load torque to maximum torque

Let,

S_f = full-load slip of the motor
S_m = slip corresponding to maximum torque $= \dfrac{R_2}{X_2}$.

According to the torque, the equation of motor is:

$$\text{Full-load torque } T_{\text{f.l}} \propto \frac{S_f E_2^2 R_2}{R_2^2 + (S_f X_2)^2}.$$

$$\text{Maximum torque } T_m \propto \frac{S_m E_2^2 R_2}{R_2^2 + (S_m X_2)^2}.$$

$$\therefore \frac{T_{\text{F.L}}}{T_m} = \left[\frac{S_f E_2^2 R_2}{R_2^2 + (S_f X_2)^2}\right]\left(\frac{R_2^2 + (S_m X_2)^2}{S_m E_2^2 R_2}\right)$$

$$= \frac{S_f}{S_m}\left[\frac{\left(\dfrac{R_2}{X_2}\right)^2 + {S_m}^2}{\left(\dfrac{R_2}{X_2}\right)^2 + {S_f}^2}\right].$$

We know that $\dfrac{R_2}{X_2} = S_m$

$$\therefore \frac{T_{\text{F.L}}}{T_m} = \frac{S_f}{S_m}\left[\frac{2{S_m}^2}{{S_f}^2 + {S_m}^2}\right]$$

$$= \frac{2 S_f S_m}{{S_f}^2 + {S_m}^2}.$$

Ratio of starting torque to maximum torque

From Equations (8.16) and (8.18):

$$\frac{T_{st}}{T_m} = \frac{K E_2^2 R_2}{R_2^2 + X_2^2} \times \frac{2 X_2}{K E_2^2}$$

$$= \frac{2(R_2/X_2)}{(R_2/X_2)^2 + 1}$$

$$= \frac{2 S_m}{{S_m}^2 + 1}.$$

8.6.3 Torque–speed and torque–slip characteristics

The torque–speed and torque–slip characteristics are shown in Fig. 8.14 (a) and (b). According to the torque equation of motor:

$$T \propto \frac{SE_2^2 R_2}{R_2^2 + (SX_2)^2}.$$

But for constant supply voltage, E_2 is also constant:

$$\text{i.e., } T \propto \frac{SR_2}{R_2^2 + (SX_2)^2}.$$

From the above expression, it is evident that, when torque is zero, slip $S = 0$ in low-slip region, slip is very very small, so that (SX_2) is so small compared to R_2; hence, it can be neglected.

$$T \propto \frac{SR_2}{R_2^2} \propto S.$$

Therefore, torque T is proportional to slip 'S' if rotor resistance R_2 is constant. That is speeds nearer to synchronous speeds, the torque–speed, and torque–slip curves are approximately straight lines.

In high-slip region, the slip value approaches to unity. Here, it can be assumed that R_2^2 is very very small as compared to $(SX_2)^2$; hence, it can be neglected.

$$T \propto \frac{SR_2}{S^2 X_2^2}.$$

When slip increases, the torque increases to its maximum value when $S = R_2/X_2$. The maximum torque is also known as pullout or breakdown torque. Beyond this, if slip further increases torque is inversely proportional to slip if R_2 and X_2 are constant.

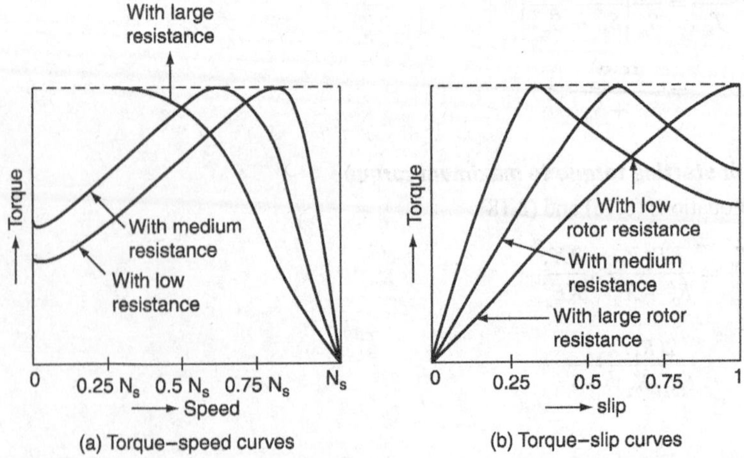

FIG. 8.14 *(a) Torque-speed characteristics and (b) torque–slip characteristics*

This means that the torque–speed and the torque–slip curves are approximately straight lines. Figure 8.14 (a) and (b) shows the torque speed and the torque–slip curves for the different values of rotor resistance.

Example 8.1: A 3-ϕ induction motor has a ratio of maximum torque to full-load torque as 2:1. Determine the ratio of actual starting torque to full-load torque for $Y-\Delta$ starting. Given $R_2 = 0.2\ \Omega$ and $X_2 = 2\Omega$.

Solution:

Given data:

$$\frac{T_m}{T_f} = 2.$$

$$R_2 = 0.2\Omega.$$

$$X_2 = 2\Omega.$$

The slip at maximum torque is $S_m = \frac{R_2}{X_2} = \frac{0.2}{2} = 0.1.$

We know that $\frac{T_f}{T_m} = \frac{2S_f S_m}{S_f^2 + S_m^2}$

$$\frac{1}{2} = \frac{2S_f \times 0.1}{S_f^2 + (0.1)^2}.$$

$$S_f^2 + (0.1)^2 = 0.4\ S_f$$

$$S_f^2 - 0.4S_f + 0.01 = 0$$

$$S_f = \frac{0.4 \pm \sqrt{(0.4)^2 - 4\times1\times0.01}}{2\times1}$$

$$= \frac{0.4 \pm 0.346}{2}$$

$$= 0.054 \qquad \text{(taking small value)}.$$

$\therefore$ Full-load current per phase:

$$I_f = \frac{S_f E_2}{\sqrt{R_2^2 + (S_f X_2)^2}}.$$

Short-circuit rotor current per phase:

$$I_{sc} = \frac{E_2}{\sqrt{R_2^2 + X_2^2}}$$

$$\therefore \frac{I_{sc}}{I_f} = \frac{\sqrt{R_2^2 + (S_f X_2)^2}}{S_f\left(\sqrt{R_2^2 + X_2^2}\right)}$$

$$= \frac{\sqrt{(0.2)^2 + (0.054\times 2)^2}}{0.054\times\sqrt{(0.2)^2 + 2^2}} = \frac{0.227}{0.1085} = 2.091.$$

∴ Starting torque with star–delta starter:

$$T_{st} = \frac{1}{3}\, T_f \left(\frac{I_{sc}}{I_f}\right)^2 \times S_f$$

$$T_{st} = \frac{1}{3}\times T_f (2.091)^2 \times 0.054$$

$$\frac{T_{st}}{T_f} = 0.078.$$

Example 8.2: The supply voltage to a cage rotor motor is 70% instead of 100%. Determine the reduction in starting torque and starting current.

Solution:

Let 'I_{sc}' be the starting current with normal voltage.

The starting current with 70% of supply voltage = 0.7 I_{sc}.

The reduction in the starting current $= \frac{I_{sc} - 0.7 I_{sc}}{I_{sc}} \times 100 = 30\%.$

The starting torque with normal $= T_f \left(\frac{I_{sc}}{I_f}\right)^2 S_f$

$$= T_f \left(\frac{0.7 I_{sc}}{I_f}\right)^2 S_f$$

= 49 times the starting torque,

where T_f, I_f, S_f, and I_{sc} are the full-load torque, full-load current, full-load slip, and short-circuit current, respectively.

Reduction in starting torque = (1 − 49) × 100
= 51%.

Example 8.3: Determine the ratio of actual starting torque to full-load torque for star–delta starting. If a 3-ϕ induction motor has a ratio of maximum torque to full-load torque as 3:1 and the resistance and the reactance are 0.4 Ω and 5 Ω, respectively.

Solution:

The ratio of maximum torque to full-load torque:

$$\frac{T_{max}}{T_f} = \frac{R_2^2 + S_f^2 X_2^2}{2S_f R_2 X_2}$$

$$3 = \frac{(0.4)^2 + (5)^2 S_f^2}{2S_f \times 0.4\times 5}$$

$$12S_f = 0.16 + 25S_f^2$$

$$25S_f^2 - 12S_f + 0.16 = 0$$

$S_f = 0.013$, neglecting higher values.

$$\text{Full-load rotor current/phase} = \frac{S_f E_2}{\sqrt{R_2^2 + S_f^2 X_2^2}} = I_f.$$

$$\text{Short-circuit rotor current/phase, } I_{sc} = \frac{E_2}{\sqrt{R_2^2 + X_2^2}}$$

$$\frac{I_{sc}}{I_f} = \frac{\sqrt{R_2^2 + S_f^2 X_2^2}}{S_f\sqrt{R_2^2 + X_2^2}}$$

$$= \frac{\sqrt{(0.4)^2 + (0.013)^2 \times (5)^2}}{(0.013)\sqrt{(0.4)^2 + (5)^2}}$$

$$= 6.23.$$

Starting torque with star–delta starter:

$$T_{st} = \frac{1}{3}T_f\left(\frac{I_{sc}}{I_f}\right)^2 S_f$$

$$= \frac{1}{3} \times T_f \times (6.23)^2 \times 0.013$$

$$= 0.168\ T_{f^*}$$

Example 8.4: A 50-kVA, 400-V, 3-ϕ, and 50-Hz squires cage induction motor has full-load slip of 6%. Its standstill impedance is 0.866 Ω/phase. It is started using a tapped autotransformer. Calculate the tap position and the ratio of starting torque to full load. The maximum allowable supply current at the time of starting is 100 A.

Solution:

$$\text{Full-load current } I_f = \frac{\text{output in kVA} \times 1{,}000}{\sqrt{3} \times \text{line voltage}}$$

$$= \frac{50 \times 1000}{\sqrt{3} \times 400}$$

$$= 72.16\text{ A.}$$

$$\text{Short-circuit current } I_{sc} = \frac{\text{supply voltage}}{\text{standstill impedence/phase}}$$

$$= \frac{400\sqrt{3}}{0.866}$$

$$= 266.67\text{ A.}$$

Supply current at starting $I_{st} = 100$ A

The tap position of autotransformer, $K = \frac{I_{st}}{I_{sc}}$

$$= \sqrt{\frac{100}{266.7}}$$

$$= 61.24\%.$$

Full-load slip $S = 0.06$.

The ratio of starting torque to full-load torque:

$$\frac{T_{st}}{T_f} = K^2 \left(\frac{I_{sc}}{I_f}\right)^2 \times S_f$$

$$= (0.6124)^2 \left(\frac{266.7}{72.16}\right)^2 \times 0.06$$

$$= 0.306.$$

Example 8.5: The rotor of four-pole and 50-Hz slip ring induction motor has a resistance of 0.25 Ω per phase and runs at 1,440 rpm at full load. Determine the external resistance per phase that must be added to lower the speed to 1,300 rpm.

Solution:

The synchronous speed of the motor $N_S = \frac{120f}{P}$

$$= \frac{120 \times 50}{4}$$

$$= 1{,}500 \text{ rpm}.$$

Full-load slip $S = \frac{N_S - N}{N_S} \times 100 = \frac{1500 - 1440}{1500} \times 100 = 4\%.$

Let 'R' be the resistance added in the rotor circuit. Then:

Motor speed $N_2 = 1{,}200$ rpm.

$$\text{Slip } S_2 = \frac{1{,}500 - 1{,}300}{1{,}500} = 0.13.$$

$$\text{Slip } S = \frac{3I_2^2 R_2}{\text{input power to rotor}}.$$

For constant power input to rotor and rotor current:

$$S \propto R_2.$$

So that:

$$\frac{S_2}{S_2} = \frac{P_2 + R}{P_2}$$

$$\frac{0.13}{0.04} = \frac{0.25 + R}{0.25}$$

$$R = 0.5625\ \Omega.$$

Example 8.6: Determine the new value of stator current if a 3-ϕ, 440-V and 1,200-rpm slip ring induction motor is operating with 3% slip and taking a stator current of 50-A speed of the motor is reduced at constant torque to 600 rpm using stator voltage control.

Solution:

Slip at the reduced speed:

$$S^1 = \frac{N_S - N^1}{N_S} = \frac{1{,}200 - 600}{1{,}200}$$

$$= 0.5.$$

Torque developed by the induction motor $T \propto SV^2$ for the constant torque:

$$V \propto \sqrt{\frac{1}{S}}$$

$$V^1 = V\sqrt{\frac{S}{S^1}} = 440 \times \sqrt{\frac{0.03}{0.5}}$$

$$= 107.77 \text{ V.}$$

Stator current $I_1 \propto SV$.
The new stator current:

$$I_1^1 = I_1 \times \frac{S^1V^1}{SV}$$

$$= \frac{50 \times 0.5 \times 107.77}{0.03 \times 440}$$

$$= 204.1 \text{ A.}$$

Example 8.7: A 9.5-kW, 240-V, three-phase, star-connected, 50-Hz, and four-pole squirrel cage induction has its full-load internal torque at a slip of 0.05. The parameters of the motor are:

$R_1 = 0.4\Omega$/phase, $R_2 = 0.3\Omega$/phase

$X_1 = X_2 = 0.5\Omega$/phase, $X_m = 16\Omega$/phase.

Assume that the shunt branch is connected across the supply terminals. Determine (a) maximum internal torque at rated voltage and frequency, (b) slip at maximum torque, and (c) internal starting torque at rated, voltage, and frequency.

Solution:

Phase voltage, $V = \frac{240}{\sqrt{3}} = 138.56$ V.

At maximum torque:

Maximum slip $S_{max} = \frac{R_2}{X_2} = \frac{0.3}{0.5} = 0.6.$

At maximum slip, the equivalent impedance of the motor is:

$$Z = \left(R_1 + \frac{R_2}{S}\right) + j(X_1 + X_2)$$

$$= \left(0.4+\frac{0.3}{0.6}\right)+j(0.5+0.5)$$

$$= 0.9 + 1j = 1.3456148\ \Omega.$$

Rotor current per phase, $I_2 = \frac{E_2}{Z} = \frac{138.56}{1.345\angle 48}$

$= 103$ A.

Rotor copper losses $= 3I_2^2R_2$

$= 3 \times (103)^2 \times 0.3$

$= 9,548.1$ W.

The power input to rotor $P_2 = \frac{\text{rotor copper loss}}{S_{max}}$

$= \frac{9548.1}{0.6}$

$= 1,5913.5$ synchronous W.

Synchronous speed $N_S = \frac{120f}{P} = \frac{120 \times 50}{4} = 1,500$ rpm.

Maximum torque $T = \frac{9.55 \times P_2}{N_S} = \frac{9.55 \times 15,913.5}{1,500}$

$= 101.31$ N-m.

At standstill:

At standstill, the slip $S = 1.0$.

Equivalent motor impedance, $Z = (R_1 + R_2) + j(X_1 + X_2)$

$= (0.3 + 0.4) + j\,(0.5 + 0.5)$

$= 0.7 + j1$

$= 1.22\ \angle 55\Omega$.

Rotor current $I_2 = \frac{E_2}{Z} = \frac{138.56}{1.22} = 113.57$ A.

Power input to rotor P_2 = total rotor copper losses

$= 3 \times (113.57)^2 \times (0.3)$

$= 11,608.33$ synchronous W.

Starting torque $T_{st} = \frac{9.55 \times P_2}{N_S}$

$= \frac{9.55 \times 11,608.33}{1,500}$

$= 73.9$ N-m.

Example 8.8: A 30-HP, six-pole, 50-Hz, and three-phase induction motor has stator/rotor phase voltage ratio of 7/5. The stator and rotor impedances per phase are $(0.35 + j0.65)\ \Omega$ and $(0.15 + j0.65)\ \Omega$, respectively. Find the starting torque exerted by the motor when an external resistance of 1.5 Ω is inserted in each phase; the motor being started directly on the 440-V supply system. Assume *Y/Y* connection.

Solution:

Supply voltage per phase $V = \frac{440}{\sqrt{3}} = 254$ V.

Rotor to stator phase voltage ration $K = 5/7 = 0.714$.

Equivalent resistance of motor as referred to rotor:

$$R_{02} = R_2 + K_1^2 R_1$$
$$= (0.15) + (0.714)^2 (0.35)$$
$$= 0.328\ \Omega.$$

Similarly, the equivalent reactance referred to rotor:

$$X_{02} = X_2 + K^2 X_1$$
$$= 0.65 + (0.714)^2 (0.65)$$
$$= 0.98\ \Omega.$$

When the external resistance is inserted then, the equivalent motor impedance referred to rotor is:

$$Z = \sqrt{(R_{02} + 1.5)^2 + X_{02}^2}$$
$$= \sqrt{(328 + 1.5)^2 + (0.98)^2}$$
$$= 2\ \Omega.$$

At standstill, the induced emf in the rotor:

$$E_2 = V_1 \times K$$
$$= 254.714$$
$$= 181.356 \text{ V}.$$

Rotor current $R_2 = \frac{E_2}{Z} = \frac{181.356}{2}$.

The rotor copper losses $= 3I_2^2 R_2$

$$= 3 \times 90.67^2 \times (0.15)$$
$$= 3{,}699.47 \text{ W}.$$

At standstill, rotor power input:

$$P_2 = 3{,}699.47 \text{ W} \qquad (\therefore \text{ slip } S = 1).$$

Synchronous speed $N_S = \frac{120f}{P} = \frac{120 \times 50}{6} = 1{,}000$ rpm.

$$\text{Starting torque } T_{st} = \frac{9.55 \times 3699.47}{1{,}000}$$

$$= 35.32 \text{ N-m.}$$

Example 8.9: For a three-phase induction motor, maximum torque is thrice the full-load torque and starting torque is 1.9 times the full-load torque. In order to get a full-load slip of 6%, calculate the percentage reduction in rotor circuit resistance neglect stator impedance.

Solution:

The ratio of starting torque to maximum torque is given by:

$$\frac{T_{st}}{T_m} = \frac{2}{\frac{S_m}{1} + \frac{1}{S_m}}$$

$$\frac{1.9T_{fl}}{3T_{fl}} = \frac{2}{\frac{S_{ml}}{1} + \frac{1}{S_{ml}}}$$

$$0.64 = \frac{2}{S_{ml} + \frac{1}{S_{ml}}}$$

$$S_{ml}^2 - 3.125S_{ml} + 1 = 0$$

$S_{ml} = 0.362$ neglecting higher values.

$$\text{Maximum slip } S_{ml} = \frac{R_2}{X_2}$$

$$\frac{R_2}{X_2} = 0.362$$

$$R_2 = 0.362\, X_2.$$

For a full-load slip of 0.06, the ratio of full-load torque to maximum torque is given by:

$$\frac{T_f}{T_m} = \frac{2}{\frac{S_{m2}}{0.06} + \frac{0.06}{S_{m2}}}$$

$$\frac{1}{3} = \frac{2}{\frac{S_{m2}}{0.06} + \frac{0.06}{S_{m2}}}$$

$$S_{m2}^2 - 0.36S_{m2} + 0.0036 = 0$$

$$S_{m2} = 0.35$$
$$= \frac{R_2^1}{X_2}$$
$$0.35 = \frac{R_2^1}{X_2}$$
$$R_2^1 = 0.35X_2.$$

∴ The reduction in rotor circuit resistance $= 0.362\,X_2 - 0.35\,X_2$

$$= 0.012\,X_2.$$

∴ The percentage reduction in rotor circuit resistance $= \frac{0.362X_2 - 0.35X_2}{0.362X_2} \times 100$

$$= \frac{0.012X_2}{0.362X_2} \times 100$$
$$= 3.315\%.$$

Example 8.10: The rotor of a three-phase induction motor has 0.05-Ω resistance per phase and 0.3-standstill reactance per phase. What external resistance is required in the rotor circuit in order to get half of the maximum torque at starting? Neglect stator impedance by what percentage will this external resistance change the current and power factor at starting?

Solution:

The ratio of starting torque to the maximum torque is given by:

$$\frac{T_{st}}{T_m} = \frac{1/2\,T_m}{T_m} = \frac{2}{\frac{S_m}{1} + \frac{1}{S_m}}$$

$$S_m^2 - 4S_m + 1 = 0$$

$S_m = 0.27$ neglecting higher values.

We know that:

$$S_m = \frac{R_2^1}{X_2}$$
$$R_2^1 = 0.27 \times 0.3$$
$$= 0.081\ \Omega.$$

The external resistance inserted in the rotor circuit $= R_2^1 - R_2$

$$= 0.081 - 0.05$$
$$= 0.031\ \Omega.$$

Without external resistance:

$$\text{Starting current } I_{st} = \frac{E_2}{\sqrt{R_2^2 + X_2^2}}$$

$$= 3.28\ E_2.$$

$$\text{Power factor} = \frac{R_2}{\sqrt{R_2^2 + X_2^2}}$$

$$= \frac{0.05}{\sqrt{(0.05)^2 + (0.3)^2}}$$

$$= 0.1643.$$

With external resistance:

$$\text{Starting current } I_{st} = \frac{E_2}{\sqrt{(0.08)^2 + (.3)^2}}$$

$$= 3.218 E_2 \text{ A.}$$

$$\text{Power factor } \cos\phi = \frac{0.081}{\sqrt{(0.081)^2 + (0.3)^2}} = 0.26.$$

Percentage reduction in the starting current:

$$= \frac{3.28E_2 - 3.218E_2}{3.28E_2} \times 100$$

$$= 1.89\%.$$

Percentage improvement in the power factor:

$$= \frac{0.26 - 0.1643}{0.1643} \times 100$$

$$= 58.24\%.$$

8.7 SPEED CONTROL OF DC MOTORS

In practical applications, a motor may be required to perform a number of desirable jobs conforming different load conditions and speed requirements. The availability of DC motors to adjustment of their operating speed over wide ranges and by a variety of methods is one of the important reasons for the strong competitive position of DC machinery in the industrial applications.

The natures of speed control required by different industrial drives are:

- Some drives require a continuously variable speed over the range from zero to full speed, such drives are known as *variable-speed drives.*
- Some drives require only two to three fixed speeds over a region, such drives are known as *multi-speed drives.*
- In some cases, speed is needed for adjusting or setting up the work on driven machine only for a few revolutions per minute. Such a speed is known as *creeping speed.*

For example, crane or hoist requires same torque at all speeds, while a fan or centrifugal pump requires a torque proportional to the square of the speed. For most of the drives, however, a control of speed within ±25% of the normal speed is required.

The speed and torque of a DC motor can be expressed by the following relationships.

$$N \propto \left(\frac{V - I_a R_a}{\phi} \right) \quad \textbf{(8.19)}$$

$$T \propto \phi I_a,$$

where V is the terminal voltage in volts, I_a is th armature current in ampere, R_a is the armature resistance in ohm, ϕ is the flux per pole in *wb*, N is the speed of DC motor in rpm, and T is the torque in N-m.

Therefore, the speed of DC motors can be regulated by varying ϕ, R, or V. The speed of DC motors can be controlled by the following methods:

1. Field control or flux control method.
2. Armature control method.
3. Applied voltage control.

8.7.1 Speed control of DC shunt motors

Speed of DC shunt motor can be controlled by varying the flux, armature resistance, and applied voltage to the armature terminals.

Various methods of controlling the speed of the shunt motor is given as follows.

Field control method

The speed adjustment of the DC shunt motors by field control may be obtained by one of the following methods.

(i) Field rehostatic control method.
(ii) Reluctance control method.
(iii) Field voltage control.

Field-rehostatic control method

In this method, speed control is obtained by controlling the field current or flux by means of a variable resistance inserted in series with the shunt filed winding. The external resistance (R_e) connected in series with the field winding is shown as shunt field regulator. The method of regulating the speed by varying the flux or field current in the shunt field winding is known as flux control method. Circuit diagram illustrating the speed control of a shunt motor is shown in Fig. 8.15.

The variation of external resistance 'R_e' in the filed reduces the field current and hence the flux 'ϕ' also reduces. The reduction in flux will also results in an increase in the speed. For DC shunt motor, speed is inversely proportional to field flux (ϕ). Since in this method of speed control, flux can be only reduced. Consequently, the motor runs at a speed higher than the normal speed. For this reason, this method of speed control is used to give motor speeds above normal or to correct for a fall in speed due to load.

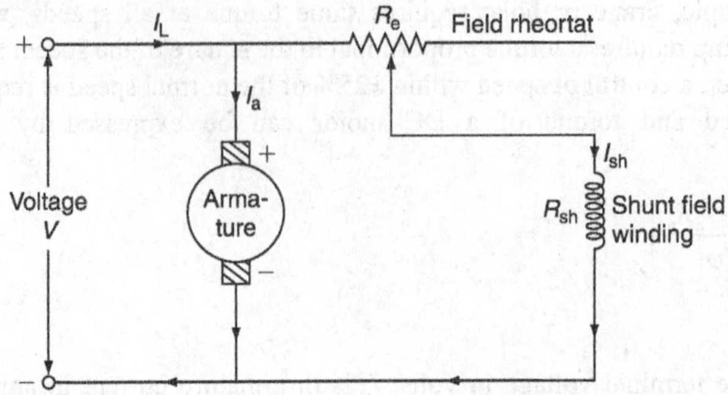

FIG. 8.15 *Field–rehostatic control of shunt motor*

Reluctance control

In this method of speed control, the motor must be constructed with special mechanical features so that the reluctance of the magnetic circuit can be changed, which makes the motor more expensive. Hence, the variable reluctance type of motor is seldom used.

Field voltage control

This method requires a variable voltage for the field circuit; such a variable supply can be obtained by means of an adjustable electronic rectifier.

Armature control method of DC shunt motor

The speed adjustment of the DC shunt motors by armature control may be obtained by one of the following methods.

(i) Armature rehostatic control method.

(ii) Armature diverter method or potential devider method.

Armature rheostat control method

In armature or rehostatic control method of speed, a variable rehostatic or resistance connected in series with the armature is known as controller resistance. The circuit diagram of the armature control method is shown in Fig. 8.16.

The speed is directly proportional to the voltage applied across the armature. Voltage across the armature can be controlled by changing resistance connected in series with it. As the controller resistance is increased, the potential difference across the armature is decreased thereby decreasing the armature speed. There is a particular load current at which the speed would be zero is called stating current. The main disadvantage of this method is speed up to zero is not possible, as it requires large rehostat in series with the armature that is practically impossible.

Armature diverter method or potential devider method

The main disadvantage of the above method can be overcome by connecting a rheostat in a potential devider arrangement as shown in Fig. 8.17.

When the variable rehostat is at minimum position, the voltage across the armature is zero. If rehostat is moved toward maximum position, the voltage across the armature increases then speed also increases. The variation of speed with the armature voltage is shown in Fig. 8.18.

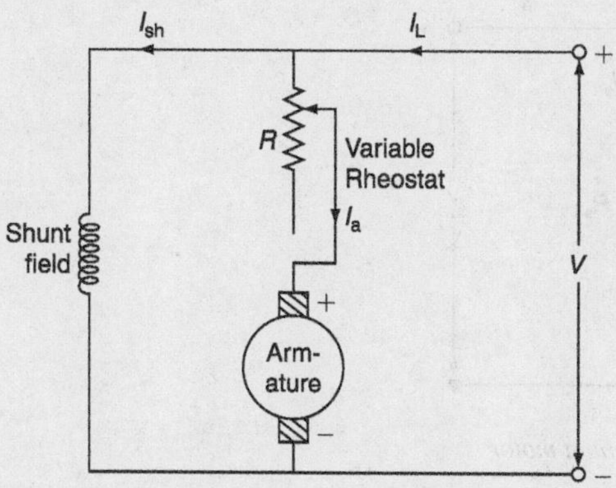

FIG. 8.16 *Armature rheostatic control of shunt motor*

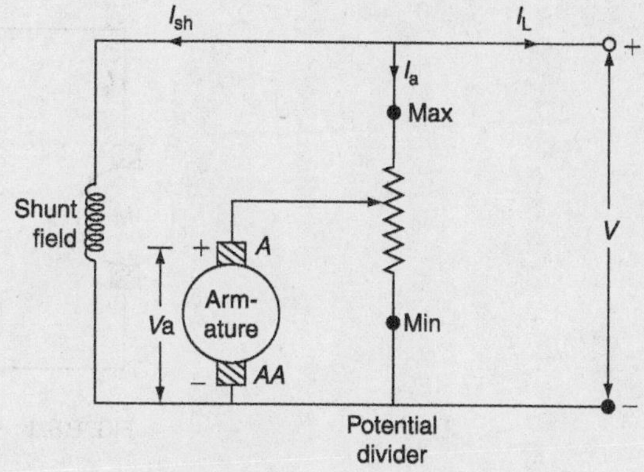

FIG. 8.17 *Potential divider method of shunt motor*

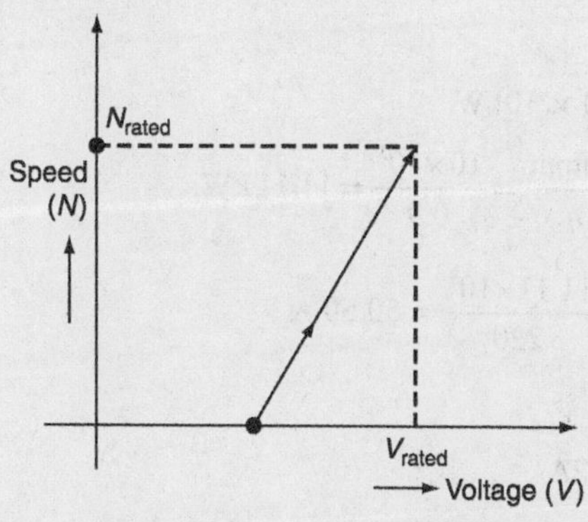

FIG. 8.18 *Speed–voltage characteristics*

Example 8.11: A DC shunt motor rated at 220 V, 15 kW, and 1,500 rpm has a full-load efficiency of 90%. Its field and armature resistances are 110 Ω and 0.25 Ω, respectively. Determine the value of the resistance to be inserted in series with the armature and the power lost in the armature circuit to reduce the speed to 1,000 rpm when:

(i) The load torque is independent of the speed.

(ii) The load torque is directly proportional to the square of the speed.

Solution:

Given data:

$V = 200$ V

$P = 15{,}000$ W

$N_1 = 1{,}500$ rpm

$N_2 = 1{,}000$ rpm

$R_{sh} = 110\ \Omega$

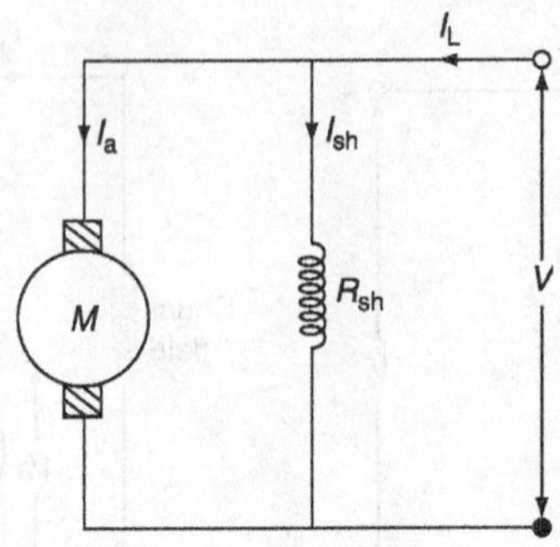

FIG. P.8.1 *DC shunt motor*

$R_a = 0.25\ \Omega$
$\eta = 0.9.$

(i) Motor output = 10×10^3 W.

$$\text{Motor input} = \frac{\text{Output}}{\eta} = \frac{10 \times 10^3}{0.9} = 11.11\ \text{kW}.$$

$$\text{Line current } I_L = \frac{11.11 \times 10^3}{220} = 50.50\ \text{A}.$$

From Fig. P.8.1:

$$I_{sh} = \frac{V}{R_{sh}} = \frac{220}{110} = 2\text{A}$$

$$I_L = I_a + I_{sh}$$

$$I_a = I_L - I_{sh} = 50.50 - 2$$
$$= 48.50\ \text{A}.$$

$$\text{Back emf, } E_{b1} = V - I_a R_a$$
$$= 220 - 48.50 \times 0.25$$
$$= 207.87\ \text{V}.$$

Now, back emf corresponding to 1,000 rpm will be:

$$\frac{E_{b1}}{E_{b2}} = \frac{N_1}{N_2} \qquad (\because \text{ for shunt motor } E_b \propto N)$$

$$E_{b2} = E_{b1} \times \frac{N_2}{N_1}$$

$$= 207.87 \times \frac{1,000}{1,500}$$

$$= 138.58\ \text{V}.$$

But, $E_{b2} = V - I_a (R_a + R_{Ext})$

$$138.58 = 220 - 50.50\,(0.25 + R_{Ext})$$

$$0.25 + R_{Ext} = \frac{220 - 138.58}{50.50} = 1.612$$

$$R_{Ext} = 1.612 - 0.25 = 1.362\ \Omega.$$

$\therefore$ The power loss in the armature circuit $= I_a^2 (R_a + R_{ext})$

$$= (50.50)^2 \times 1.612$$

$$= 4.11 \text{ kW}.$$

(ii) Given $T \propto N$: (i)

But for shunt motor:

$T \propto I_a$ (ϕ is constant). (ii)

From Equations (i) and (ii):

$I_a \propto N.$ (iii)

I_a corresponding to 15,000 rpm is 50.50 A, then I_a' corresponding to 1,000 rpm is:

$$I_a = I_a^1 \times \frac{N_2}{N_1} = 50.50 \times \frac{1{,}000}{1{,}500}$$

$$= 33.66 \text{ A}.$$

Back emf, $E_b = V - I_a^1 (R_a + R_{Ext})$

$$138.58 = 220 - 33.66\,(0.25 + R_{Ext})$$

$$R_{Ext} = \frac{200 - 138.58}{33.66} - 0.25$$

$$= 2.168\ \Omega.$$

The power loss in the armature $= I_a^2 R$

$$= (33.66)^2 \times (0.25 + 2.168)$$

$$= 2.739 \text{ kW}.$$

(iii) Given $T \propto N^2$:

i.e., for shunt motor, $T \propto I_a \propto N^2$

$I_a \propto N^2.$

The armature current corresponding to 1,000 rpm is:

$$I_a^1 = 50.50 \times \left(\frac{1{,}000}{1{,}500}\right)^2$$

$$= 22.44 \text{ A}.$$

Back emf, $E_b = V - I_a^1 (R_a + R_{Ext})$

$$138.58 = 220 - 22.44 (0.25 + R_{Ext})$$

$$R_{Ext} = \frac{220 - 138.58}{22.44} - 0.25$$

$$= 3.378\ \Omega.$$

$\therefore$ The power loss in the armature $= I_a^2 (R_a + R_{Ext})$

$$= (22.44)^2 \times (0.25 + 3.378)$$

$$= 1.827 \text{ kW}.$$

Example 8.12: The armature and the field resistances of a 260-V DC shunt motor are 0.25 Ω and 160 Ω, respectively. When driving a load of constant torque at 500 rpm, the armature current is 20 A. If it is desired to raise the speed from 500 to 1,000 rpm, what resistance should be inserted in the field circuit? Assume that the magnetic circuit is unsaturated.

Solution:

Given data:

$V = 200$ V

$R_a = 0.25\ \Omega$

$R_{sh} = 160\ \Omega$

$I_a = 20$ A

$N_1 = 500$ rpm

$N_2 = 1{,}000$ rpm.

We know that, for shunt motor:

$$E_b \propto N\phi$$

$$\therefore N \propto \frac{E_b}{\phi}$$

$$\text{i.e., } \frac{N_1}{N_2} = \frac{E_{b_1}}{E_{b_2}} \times \frac{\phi_1}{\phi_2} \qquad \text{(i)}$$

$$\therefore E_{b_1} = V - I_a R_a$$

$$= 200 - 20 \times 0.25 = 195 \text{ V}.$$

Given that magnetic circuit is unsaturated and torque remains constant:

i.e., $\phi \propto I_{sh}$ and $T \propto \phi I_a$.

From the two reaction:

$$\phi_1 I_{a1} = \phi_2 I_{a2}$$

and $I_{sh_1} I_{a1} = I_{sh2} I_{a2}$

$$\therefore \; I_{a2} = I_{a1} \times \frac{I_{sh1}}{I_{sh_2}}. \qquad \text{(ii)}$$

Let $R_{sh2} = R_{sh1} + R_{Ext}$

$$I_{sh1} = \frac{V}{R_{sh1}} = \frac{200}{160} = 1.25$$

$$I_{sh2} = \frac{V}{R_{sh2}} = \frac{200}{R_{sh2}}.$$

Now, by substituting I_{sh1} and I_{sh2} in Equation (ii), we get:

$$I_{a2} = 20 \times \frac{1.25}{200 / R_{sh2}}$$

$$= 0.125 \times R_{sh2}. \qquad \text{(iii)}$$

But $E_{b2} = V - I_{a2} R_a$

$$E_{b2} = 200 - (0.125 \times R_{sh2}) R_a$$

$$= 200 - (0.125 \times 0.25) \times R_{sh2}$$

$$= 200 - 0.03125 \, R_{sh2}.$$

By substituting E_{b1} and E_{b2} in Equation (i):

We get $\dfrac{N_1}{N_2} = \dfrac{E_{b1}}{E_{b2}} \times \dfrac{\phi_2}{\phi_1}$

$$\frac{N_1}{N_2} = \frac{E_{b1}}{E_{b2}} \times \frac{I_{sh1}}{I_{sh2}} \qquad (\therefore \phi \propto I_{sh})$$

$$\frac{500}{1{,}000} = \left(\frac{195}{200 - 0.03125 \, R_{sh2}} \right) \times \frac{(200 / R_{sh2})}{1.25}$$

$$0.5 = \frac{31{,}200}{R_{sh2}(200 - 0.03125 R_{sh2})}$$

$$200 R_{sh2} - 0.03125 R_{sh2}^2 = 62{,}400.$$

$$0.03125 R_{sh2}^2 - 200 R_{sh2} + 62{,}400 = 0.$$

$$\therefore R_{sh2} = \frac{200 \pm \sqrt{(200)^2 - 4 \times 0.03125 \times 62{,}400}}{2 \times 0.03125}$$

$$= \frac{200 \pm 179.44}{0.0625}$$

$$= \frac{20.56}{0.0625} = 328.96\Omega \qquad \text{(neglecting positive sign)}$$

$$\therefore R_{sh2} = 328.96\ \Omega$$

$$\text{i.e., } R_{sh1} + R_{Ext} = 328.96\ \Omega$$

$$\therefore R_{Ext} = 328.96 - 160 = 168.96\ \Omega.$$

Example 8.13: A 220-V DC shunt motor, having an armature resistance of 0.5 Ω, draws from the main current of 30 A on half-full load. The speed is to be increased to twice half-full-load speed. If the torque of the motor is of constant magnitude, determine the percentage change in flux required.

Solution:

Given data:

$V = 220$ V

$R_a = 0.5\ \Omega$

$I_{a1} = 30$ A.

Given that speed (N_2) at full load is twice the speed at half-full load

$$\text{i.e., } \frac{N_2}{N_1} = 2$$

$$\text{back emf, } E_{b1} = V - I_{a1}R_a$$

$$= 220 - 30 \times 0.5$$

$$= 205\ \text{V.}$$

$$E_{b2} = V - I_{a2}R_a = 220 - I_{a2} \times 0.5.$$

We know that , for shunt motor:

$$E \propto N\phi$$

$$\text{i.e., } \frac{N_2}{N_1} = \frac{E_{b2}}{E_{b1}} \times \frac{\phi_1}{\phi_2}$$

$$2 = \frac{220 - 0.5I_{a2}}{200} \times \frac{\phi_1}{\phi_2}$$

$$\frac{\phi_1}{\phi_2} = \frac{220 - 0.5I_{a2}}{410}. \qquad \text{(i)}$$

Given that the torque remains constant so that:

$$\phi_1 I_{a1} = \phi_2 I_{a2}$$

$$I_{a2} = \frac{\phi_1}{\phi_2} \times I_{a1}$$

$$= \frac{\phi_1}{\phi_2} \times 30. \qquad \text{(ii)}$$

Subsisting Equation (ii) in Equation (i):

$$\frac{\phi_2}{\phi_1} = \frac{220 - \left(0.5 \times 30 \times \dfrac{\phi_1}{\phi_2}\right)}{410}.$$

Now, let $\frac{\phi_1}{\phi_2} = K.$

$$\therefore K = \frac{220 - \dfrac{15}{K}}{410}$$

$$410K = 220 - \frac{15}{K}$$

$$410K^2 = 220K - 15$$

$$410K^2 - 220K + 15 = 0$$

$$K = \frac{200 \pm \sqrt{(220)^2 - 4 \times 410 \times 15}}{2 \times 410}$$

$$= \frac{220 \pm 154.272}{820}$$

$$= \frac{220 + 154.272}{820} \quad \text{(neglecting negative sign)}$$

$$K = 0.45642$$

i.e., $\frac{\phi_2}{\phi_1} = 0.45642$

$$\therefore \phi_2 = 0.456421$$

$\therefore$ The percentage change in flux $= \frac{\phi_1 - \phi_2}{\phi_1} \times 100$

$$= \frac{\phi_1 - 0.45641}{\phi_1} \times 100$$

$$= 54.35\%.$$

Example 8.14: A 250-V shunt motor develops a total torque of 150 N-m and takes 20 A at 800 rpm. The armature and shunt field resistances are 0.1Ω and 200 Ω, respectively. If the speed is to be increased to 1,200 rpm, determine the percentage reduction of the field and the additional resistance to be inserted in the field circuit. The total torque developed at 1,200 rpm is 100 N-m. Neglect armature reaction and assume that magnetization characteristics is a straight line.

Solution:
Given data:

Voltage $V = 250$ V

$$R_a = 0.1\Omega$$
$$R_{sh1} = 200\ \Omega$$
$$T_1 = 1{,}500 \text{ N-m at } 800 \text{ rpm}$$
$$T_2 = 100 \text{ N-m at } 1{,}200 \text{ rpm.}$$
$$I_{L1} = 20\text{A}$$
$$I_{sh1} = \frac{V}{R_{sh1}} = \frac{250}{200} = 1.25\text{ A}$$
$$I_{L1} = I_{a1} + I_{sh1}$$
$$I_{a1} = I_{L1} - I_{sh1}$$
$$I_{a1} = 20 - 1.25 = 18.75\text{ A}$$
$$\therefore E_{b1} = V - I_{a1} R_a$$
$$= 250 - 18.75 \times 0.1 = 248.125\text{ V.}$$

Let 'ϕ_1' be the flux at 800 rpm and 'ϕ_2' be the flux at 1,200 rpm. We know that for the shunt motor:

$$T \propto \phi I_a$$
$$\therefore \frac{T_2}{T_1} = \frac{\phi_2}{\phi_1} \times \frac{I_{a2}}{I_{a1}}.$$

Let $\frac{\phi_2}{\phi_1} = K,$

then $\frac{T_2}{T_1} = K\frac{I_{a2}}{I_{a1}}$

$$I_{a2} = \frac{T_2}{T_1} \times \frac{I_{a2}}{K}$$
$$= \frac{100}{150} \times \frac{18.75}{K}. \quad \text{(i)}$$

And also:

$$\frac{N_2}{N_1} = \frac{E_{b2}}{E_{b1}} \times \frac{\phi_1}{\phi_2}$$

$$\frac{E_{b2}}{E_{b1}} = \frac{N_2}{N_1} \times \frac{\phi_1}{\phi_2}$$

$$= \frac{1,200}{800} \times \mathrm{K}$$

$$\therefore E_{b2} = 1.5\, K\, E_{b1} \qquad \text{(ii)}$$

But $E_{b2} = V - I_{a2} R_a$. (iii)

Substitute Equations (i) and (ii) in Equation (iii):

$$1.5K\, E_{b1} = 250 - \left(\frac{100}{150}\right) \times \frac{18.75}{K} R_a$$

$$(1.5K) \times 248.125\mathrm{K} = 250 - \frac{12.5}{K} R_a$$

$$372.187K = 250 - \frac{12.5}{K} \times 0.1$$

$$372.187K = 250 - \frac{1.25}{K}$$

$$372.187K^2 - 250K + 1.25 = 0$$

$$\therefore K = \frac{250 \pm \sqrt{(250)^2 - 4 \times 372.2187 \times 1.25}}{2 \times 372.187}$$

$$K = \frac{250 \pm 246.25}{744.374}$$

$$K = \frac{250 + 246.25}{744.374}$$

$$K = 0.667.$$

$$\therefore \phi_2 = 0.667\, \phi_1.$$

$$\text{The percentage reduction in flux} = \frac{\phi_1 - \phi_2}{\phi_1} \times 100$$

$$= \frac{\phi_1 - 0.667\phi_1}{\phi_1} \times 100$$

$$= 33.33\%.$$

Given that for shunt motor magnetization characteristic is a straight line, so that:

$$\phi \propto I_{sh}$$

$$\text{i.e., } \frac{\phi_2}{\phi_1} = \frac{I_{sh2}}{I_{sh1}} = K$$

$$I_{sh2} = 0.667\, I_{sh1}$$

$$= 0.667 \times 1.25$$

$$= 0.83375 \text{ A.}$$

Let 'R_{ext}' be the additional resistance to be inserted in the field circuit, then:

$$I_{sh2} = \frac{V}{R_{sh1} + R_{ext}}$$

$$0.83375 = \frac{250}{200 + R_{ext}}$$

$$200 + R_{ext} = \frac{250}{0.83375} = 299.85$$

$$\therefore R_{ext} = 299.85 - 200$$

$$\therefore R_{ext} = 99.85\ \Omega.$$

Additional resistance to be added in the field circuit = 99.85 Ω.

Example 8.15: The speed of a 15-HP (metric) 400-V DC shunt motor is to be reduced by 25% by the use of a controller. The field current is 2.5 A and the armature resistance is 0.5 Ω. Calculate the resistance of the controller, if the torque remains constant and the efficiency is 82%

Solution:

Rating of motor = 15 HP

$V = 400$ volts.

$R_a = 0.5\ \Omega$

$I_{sh} = 2.5$ A

$\eta = 82\% = 0.82.$

The speed of a DC motor can be controlled (reduced) by adding resistance in series with armature.

∴ Let the speed of the motor, $N_1 = N$ rpm.

After adding resistance, $N_2 = 75\%$ of N

$$N_2 = 0.75\, N.$$

$$\text{Motor input} = \frac{\text{output}}{\eta} = \frac{15 \times 735.5}{0.82}$$

$$= 13{,}454.26 \text{ W.}$$

Input $= V \cdot I_L = 13{,}454.26$.

$$I_L = \frac{13{,}454.26}{400} = 33.63 \text{ A.}$$

From Fig. P.8.1:

$$I_L = I_a + I_{sh}$$

$$I_{a1} = I_L - I_{sh}$$

$$= 33.63 - 2.5$$

$$= 31.31 \text{ A.}$$

Given that torque is constant:

$$\therefore T \propto \phi I_a$$

$$\therefore \phi_1 I_{a1} = \phi_2 I_{a2}.$$

For the shunt motor flux 'ϕ' is constant.

$$\therefore I_{a1} = I_{a2} = 31.13 \text{ A.}$$

We know that:

$$N \propto E_b/\phi$$

$$\Rightarrow N \propto E_b$$

$$\frac{N_1}{N_2} = \frac{E_{b1}}{E_{b2}}$$

$$\frac{N}{0.75N} = \frac{V - I_{a1}R_a}{V - I_{a2}(R_a + R_{ext})}$$

$$1.33\,(400 - 31.13 \times (0.5 + R_{ext})) = 400 - 31.13 \times 0.5$$

$$400 - 31.3(0.5 + R_{ext}) = 289.04$$

$$3.544 = 0.5 + R_{ext}$$

$$R_{ext} = 3.544 - 0.5 = 3.04\ \Omega$$

$$R_{ext} = 3.04\ \Omega.$$

Example 8.16: A 200-V shunt motor has an armature resistance of 0.5 Ω it takes a current of 16 A on full load and runs at 600 rpm. If a resistance of 0.5 Ω is placed in the armature circuit, find the ratio of the stalling torque to the full-load torque.

Solution:

Given data:

$V = 200$ volts.

$R_a = 0.5\ \Omega$.

$I_f = I_a = 16$ A.

$N = 600$ rpm.

$R_{ext} = 0.5\ \Omega$.

Total full-load current = 16 A.

$$\text{Total stalling current} = \frac{V}{R_a + R_{ext}} = \frac{200}{0.5 + 0.5} = 200 \text{ A.}$$

∴ For shunt motor 'ϕ' is constant, so that:

$T \propto I_a$

$$\therefore \frac{\text{Stalling torque}}{\text{Full-load torque}} = \frac{\text{stalling current}}{\text{full-load current}} = \frac{200}{16} = 12.5.$$

Example 8.17: A 100-HP and 500-rpm DC shunt motor is driving a grinding mill through gears. The moment of inertia of the mill is 1,265 kg-m². If the current taken by the motor must not to exceed twice full-load current during starting, estimate the minimum timetaken to run the mill up to full speed.

Solution:

Given data:

Motor rating (P) = 100 HP.

Motor output power = 100 × 735.5 W

= 73,550 W.

The speed of motor (N) = 500 rpm.

The moment of inertia (J) = 1,265 kg-m².

$$\text{Motor output } P = \frac{2\pi N T_{F.L}}{60}.$$

$$\therefore \text{Full-load torque } T_{FL} = \frac{P \times 60}{2\pi N}$$

$$= \frac{73{,}550 \times 60}{2\pi \times 500} = 1{,}404.70 \text{ N-m}$$

$$= \frac{1{,}404.70}{9.81} = 143.19 \text{ kg-m.}$$

Given that motor takes twice the [$\therefore$ 1 kg = 9.81N] full-load current; hence, it exerts twice the full-load torque.

$= 2 \times T_{FL}$.

$\therefore$ Accelerating torque = 2 × 143.19

= 286.38 kg-m.

$$\text{Angular acceleration } \alpha = \frac{T_{FL} \times g}{1,265} = 2.223 \text{ rad/sec}^2.$$

We know that:
Angular speed (ω) = angular acceleration × time

$$\omega = \alpha \times t.$$

$$\therefore t = \frac{\omega}{\alpha} = \frac{2\pi N}{\alpha \times 60} = \frac{2\pi \times 500}{60 \times 2.23}$$

= 23.55 s.

8.7.2 Speed control of DC series motor

The speed control of DC series motor can be obtained by changing the series field current, flux, or voltage applied across the armature. The methods of the speed control of the series motor are:

(i) Field control method.

(ii) Armature control method.

Field control method

In the series motor, the variation of flux can be brought about by diverting the current flowing through the series field winding by any one of the following methods.

Field diverter's method

In this method, the series field winding is shunted by a variable resistor 'R' known as series field divertor. Any desired amount of current can be passed through the divertor by adjusting its resistance. Hence, the flux can be controlled, i.e., decreased, and consequently the speed of the motor is increased.

The arrangement of field diverter and the speed–armature current characteristics with change in resistance 'R' is shown in Figs. 8.19 (a) and (b).

Armature diverter method

In this method, the armature of the motor is shunted with an external variable resistance (R) as shown in Fig. 8.20 is known as armature diverter.

For a given constant load torque, if armature current is reduced due to armature divertor then flux (ϕ) must increase ($\because T \propto I_a$). So that, the motor reacts by drawing more current from the supply. So, the current through field winding increase, so the flux increases and the speed of the motor reduces.

This method of speed control is used to have the speed below the normal value.

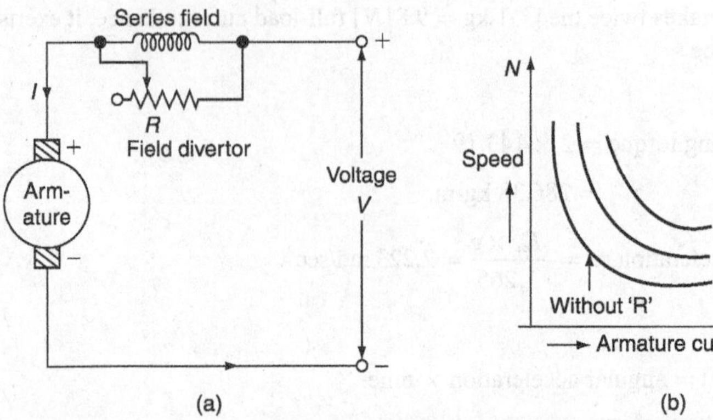

FIG. 8.19 *(a) Field diverter method of speed control and (b) Speed–current characteristics*

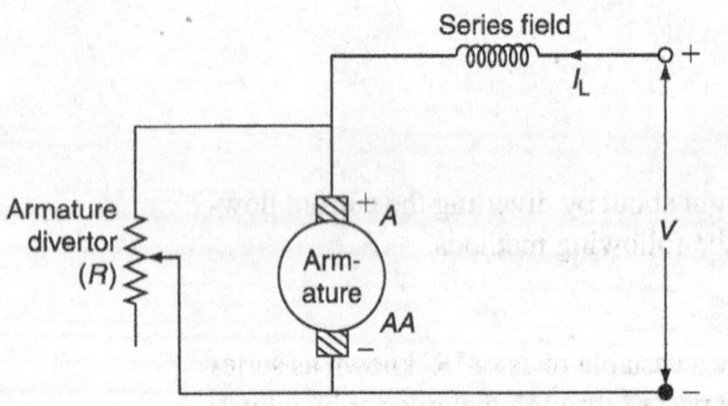

FIG. 8.20 *Armature diverter method of speed control*

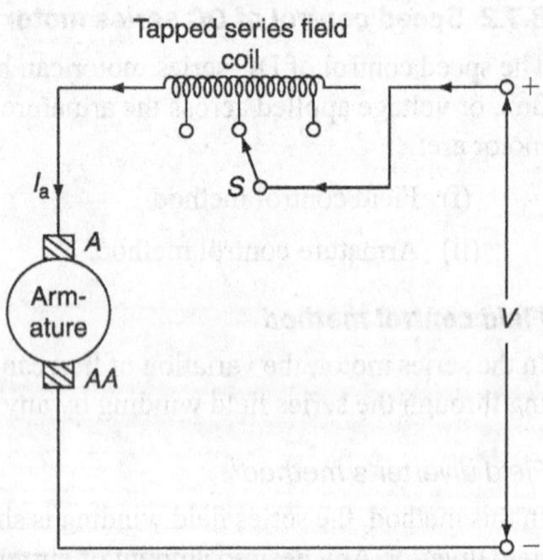

FIG. 8.21 *Tapped field speed control*

Tapped filed method

In this method, the flux change is achieved by providing a number of tapings from the field winding, which are brought out side as shown in Fig. 8.21.

As shown in Fig. 8.21, the selector switch 'SW' is provided to select number of turns. So, the net mmf will change. This will cause the change in the speed of DC series motor.

This method is used in electric traction.

Series–parallel connection of field coils

In this method of speed control, several speeds can be obtained by grouping the several field coils as shown in Figs. 8.22 (a) and (b). This method is used generally in case of fan motors.

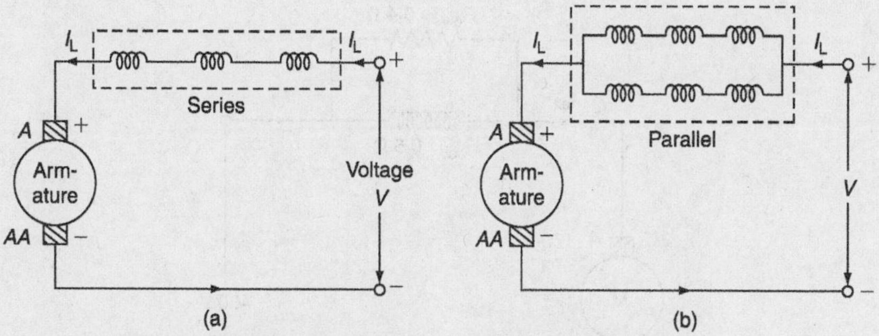

FIG. 8.22 *Series–parallel connection of field winding*

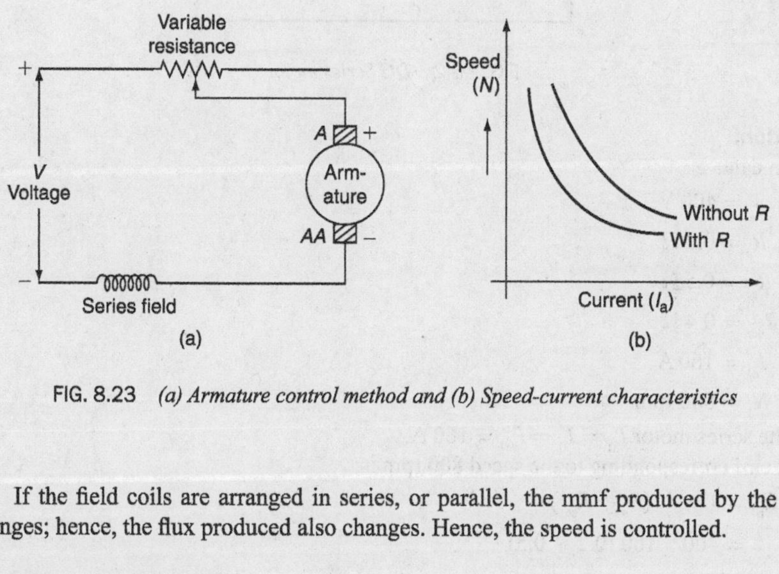

FIG. 8.23 *(a) Armature control method and (b) Speed-current characteristics*

If the field coils are arranged in series, or parallel, the mmf produced by the coils changes; hence, the flux produced also changes. Hence, the speed is controlled.

Armature control method

Armature resistance control method is the most common method employed for DC series motor. The arrangement and speed–current characteristics of series motor is shown in Figs. 8.23 (a) and (b).

By increasing the resistance in series with the armature, voltage drop across this resistance occurs. So that, the voltage applied across the armature terminals can be decreased. As the speed is directly proportional to the voltage across the armature, the speed reduces.

Example 8.18: A 400-V series motor has an armature resistance of 0.2 Ω and a series field resistance of 0.5 Ω. It takes a current of 160 A at a speed of 800 rpm. Find the speed of the motor if a diverter of resistance 0.4 Ω is connected across the field, the load torque being kept constant.

Neglect armature reaction and assume that flux is proportional to the current.

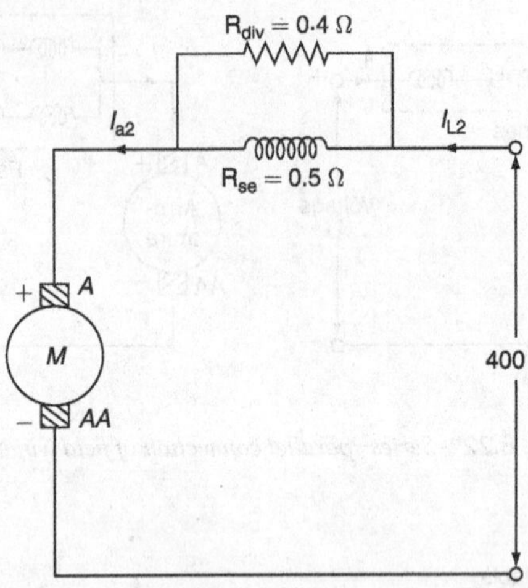

FIG. P.8.2 *DC Series motor*

Solution:

Given data:

$V = 400$ V

$R_a = 0.2\ \Omega$

$R_{se} = 0.5\ \Omega$

$R_{div} = 0.4\ \Omega$

$I_{a1} = 160$ A

$N_1 = 800$ rpm.

For the series motor $I_{a1} = I_{l1} = I_{se1} = 160$ A.

Back emf corresponding to the speed 800 rpm is:

$$E_{b1} = V - I_{a1}(R_a + R_{se1})$$
$$= 400 - 160\,(0.2 + 0.5)$$
$$= 288 \text{ V.}$$

Let, when a diverter of resistance 0.4 Ω is connected across field winding current flowing through the armature be I_{a2}.

Given that the torque remains constant, then:

$$\phi_1 I_{a1} = \phi_2 I_{a2}.$$

But for the series motor $\phi \,\alpha\, I_{se}$:

$$\therefore I_{a1}^2 = \phi_2 I_{a2}.$$

Now, from the Fig. P.8.2, the current flowing through the diverter is:

$$I_{se2} = I_{a2} \times \frac{R_{div}}{R_{div} + R_{se}}$$

$$= I_{a2} \times \frac{0.4}{0.4 + 0.5} = 0.44 I_{a2}.$$

But $\phi_2 \,\alpha\, I_{se2}$

$$\therefore I_{a1}^{\,2} = 0.44\, I_{a2}^{\,2}$$

$$I_{a2}^{\,2} = \frac{I_{a1}^{\,2}}{0.44}$$

$$I_{a2} = \frac{I_{a1}}{0.44} = \frac{160}{\sqrt{0.44}} = 241.20 \text{ A.}$$

And $I_{se2} = 0.44\, I_{a2} = 0.44 \times 241.20$

$$= 106.1319 \text{ A.}$$

Now, back emf, $E_{b2} = V - I_{a2}R_a - I_{se2}R_{se}$

$$= 400 - 241.20 \times 0.2 - 106.1319 \times 0.5$$

$$= 400 - 48.24 - 53.06$$

$$= 298.7 \text{ V.}$$

We know that:

$$N \propto \frac{E_b}{\phi}$$

$$\therefore \frac{N_1}{N_2} = \frac{E_{b1}}{E_{b2}} \times \frac{\phi_2}{\phi_1}$$

$$N_2 = N_1 \times \frac{E_{b1}}{E_{b2}} \times \frac{\phi_1}{\phi_2}$$

$$\therefore N_2 = N_1 \times \frac{E_{b1}}{E_{b2}} \times \frac{I_{se1}}{I_{se2}}$$

$$= 800 \times \frac{298.7}{288} \times \frac{160}{106.1319}$$

$$= 1{,}250.85 \text{ rpm.}$$

Example 8.19: A 220-V and 10-HP (metric) shunt motor has field and armature resistances as of 120 Ω and 0.25 Ω respectively. Calculate the resistance to be inserted in the armature circuit to reduce the speed to 700 rpm from 950 rpm, if the full-load efficiency is 80% and the torque varied as the square of the speed.

Solution:

Given data:

$V = 220$ V

Motor rating = 10 HP

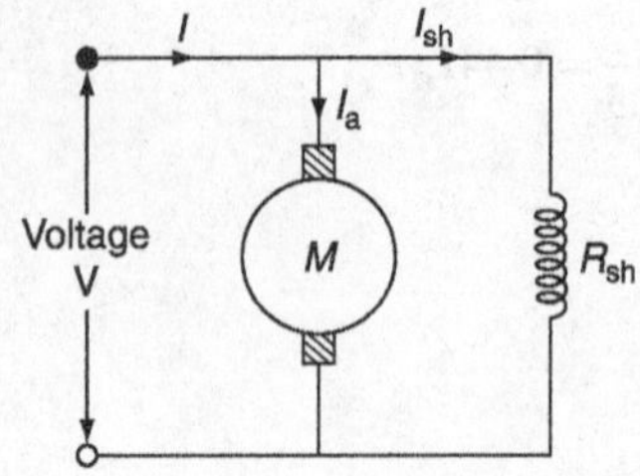

FIG. P.8.3 *DC shunt motor*

$R_{sh} = 120\ \Omega$

$R_a = 0.25\ \Omega$

$N_1 = 950$ rpm

$N_2 = 700$ rpm

$\eta = 80\% = 0.8.$

Motor output power $= (P_0) = 10$ HP

$$= 10 \times 735.5 \quad [\because 1\text{HP} = 735.5\,\text{W}]$$

$$= 7{,}355 \text{ W}.$$

$$\text{Motor input power } (P_i) = \frac{P_0}{\eta} = \frac{7{,}355}{0.8}$$

$$= 9{,}193.75$$

$$\cong 9.194 \text{ W}.$$

We know that, motor electric input $= VI = P_i$

$$\therefore 9{,}194 = 220 \times I$$

$$I = 41.78 \text{ A}.$$

We know that $T \propto \phi I_a$.
For shunt motor 'ϕ' is constant.

$$\text{Hence, } \frac{T_1}{T_2} = \frac{I_{a1}}{I_{a2}}. \tag{8.19.1}$$

Given that $T \propto N^2$

$$\therefore \frac{T_1}{T_2} = \frac{N_1^2}{N_2^2}. \tag{8.19.2}$$

From Equations (8.19.1) and (8.19.2):

$$\frac{N_1^2}{N_2^2} = \frac{I_{a1}}{I_{a2}}. \tag{8.19.3}$$

From Fig. P.8.3,

$$I_{a1} = I_{L1} - I_{sh1}$$

$$= 41.78 - \frac{V}{R_{sh}}$$

$$= 41.78 - \frac{220}{120}$$

$$= 39.94 \text{ A.}$$

From Equation (8.19.3):

$$I_{a2} = \frac{N_1^2}{N_2^2} \times I_{a1}$$

$$= \frac{700^2}{950^2} \times 39.94 = 21.68 \text{ A.}$$

We know that:

$$N \propto \frac{E_b}{\phi}.$$

For the shunt motor $N \propto E_b$:

$$\frac{N_1}{N_2} = \frac{E_{b1}}{E_{b2}}. \tag{8.19.4}$$

From the data:

$$E_{b1} = V - I_{a1}R_a \qquad [\text{for motor, } V = E_b + I_a R_a]$$

$$= 220 - 39.94 \times 0.25$$

$$= 210 \text{ V.}$$

And $E_{b2} = V - I_{a2}(R_a + R_{ext})$

$$= 220 - 21.68\,(0.25 + R_{ext}).$$

Substitute E_{b1} and E_{b2} in Equation (8.19.4):

$$\therefore \frac{950}{700} = \frac{210}{220 - 21.68\,(0.25 + R_{ext})}$$

$$220 - 21.68(0.25 + R_{ext}) = \frac{210 \times 700}{950} = 154.73$$

$$0.25 + R_{ext} = \frac{-154.73 + 220}{21.68} = 3.01$$

$$R_{ext} = 3.01 - 0.25 = 2.76\ \Omega$$

$$\therefore R_{ext} = 2.76\ \Omega.$$

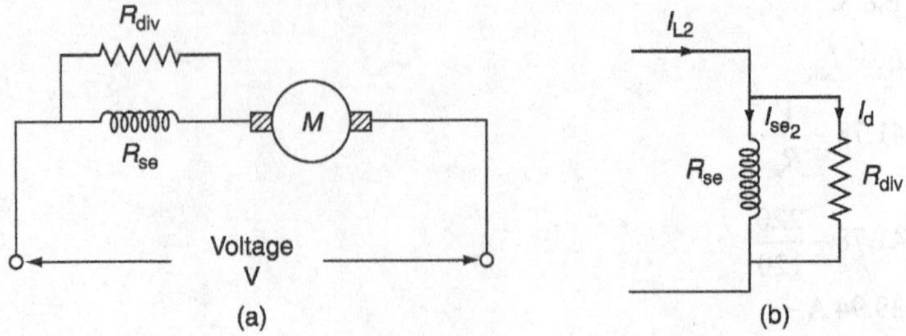

FIG. P.8.4 *DC series motor*

Example 8.20: A DC series motor drives a load, the torque of which varies as the square of the speed. The motor takes a current of 30 A, when the speed is 600 rpm. Determine the speed and current when the field winding is shunted by a diverter; the resistance of which is 1.5 times that of the field winding. The losses may be neglected.

Solution:

Given data (Fig. P.8.4):

$$I = I_{a1} = I_{se1} = I_{L1} = 30 \text{ A}$$

$$N_1 = 600 \text{ rpm}$$

$$R_{div} = 1.5\, R_{se}.$$

After connecting the diverter:

$$\text{Total resistance} = R_{div} + R_{se}.$$

$$\text{Line current} = I_{L2} = I_{a2}.$$

$$\text{Speed} = N_2$$

$$I_{se2} = I_{L2} \times \frac{R_{div}}{R_{se} + R_{div}}$$

$$= I_{L2} \times \frac{1.5 \times R_{se}}{R_{se} + 1.5 R_{se}} = 0.6\, I_{L2}.$$

We know that:

$$N \propto \frac{E_b}{\phi}, \quad \phi \propto I_{se}.$$

Since the losses are negligible $E_b = V =$ constant:

$$\therefore N \propto \frac{1}{\phi}$$

$$\frac{N_2}{N_1} = \frac{\phi_1}{\phi_2} = \frac{I_{se1}}{I_{se2}}$$

$$= \frac{30}{0.6\,I_{L2}} = \frac{50}{I_{L2}}. \qquad \text{(i)}$$

We know that:

$$T \propto \phi I_a, \quad \phi \propto I_{se}$$

$$\frac{T_1}{T_2} = \frac{\phi_1}{\phi_2} \times \frac{I_{a1}}{I_{a2}} = \frac{I_{a1}}{I_{a2}} \times \frac{I_{se1}}{I_{se2}}$$

$$= \frac{30 \times 30}{I_{L2} \times 0.6 I_{L2}} = \frac{1,500}{I_{L2}^{\ 2}}. \qquad \text{(ii)}$$

Given that $T \propto N^2$:

$$\frac{N_2}{N_1} = \left(\frac{N_1}{N_2}\right)^2$$

$$\frac{1500}{I_{L2}^2} = \left(\frac{I_{L2}}{50}\right)^2 \qquad [\therefore \textit{from Equations} \text{ (i) and (ii)}]$$

$$I_{L2}^4 = 1,500 \times 50^2$$

$$I_{L2} = 44 \text{ A}.$$

Substitute I_{L2} in Equation (i):

$$\frac{N_2}{N_1} = \frac{50}{I_{L2}}$$

$$N_2 \frac{50}{44} \times 600 = 681.7 \text{ rpm}$$

$$\therefore N_2 = 681.7 \text{ rpm}.$$

Example 8.21: A 500-V DC series motor runs at 500 rpm and takes 60 A; the resistances of the field and the armature are 0.3 and 0.2 Ω, respectively. Calculate the value of the resistance to be shunted with series field winding in order that the speed may be increased to, 600 rpm, if the torque were to remain constant. Saturation may be neglected.

Solution:

Given that:

$V = 500$ V

$N_1 = 500$ rpm

$I_{a1} = 60$ A

$R_a = 0.2\ \Omega$

$R_{se} = 0.3\ \Omega$

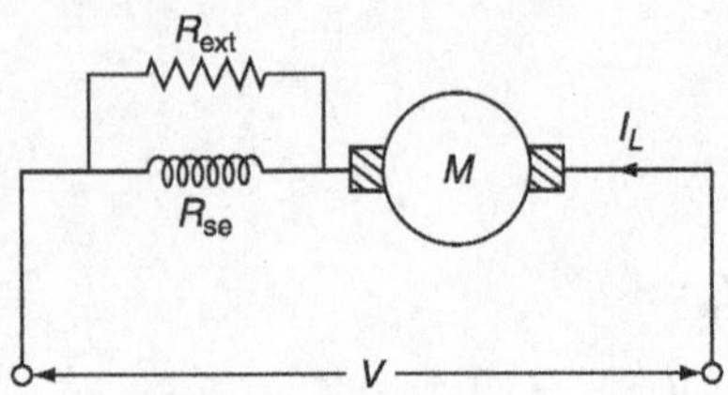

FIG. P.8.5 *DC series motor*

$N_2 = 600$ rpm

$I_{L1} = I_{a1} = I_{se1} = 60$ A.

After connecting resistance across field winding, let I_{a2} be the armature current (Fig. P.8.5).

$$\therefore I_{se2} = I_{a2} \times \frac{R_{ext}}{R_{ext} + 0.3}. \quad \text{(i)}$$

Given that the load torque is constant:

$T_1 = T_2$.

We know that:

$$T \propto \phi I_a \quad \text{and} \quad N \propto E_b/\phi$$

$$\therefore I_{a1}\, \phi_1 = I_{a2}\, \phi_2. \quad \text{(ii)}$$

For series motor $\phi \propto I_{se}$:

$$\therefore I_{a1}\, I_{se1} = I_{a2}\, I_{se2}$$

$$I_{a1}^2 = I_{a2} \times I_{se2}$$

$$I_{a1}^2 = I_{a2} \times \left[I_{a2} \times \frac{R_{ext}}{R_{ext} + 0.3} \right]$$

$$60^2 = I_{a2}^2 \left[\frac{R_{ext}}{R_{ext} + 0.3} \right]. \quad \text{(iii)}$$

And from the circuit:

$$N \propto E_b/\phi$$

$$\frac{N_1}{N_2} = \frac{E_{b1}}{E_{b2}} \times \frac{\phi_2}{\phi_1}$$

$$\text{or} \quad \frac{N_2}{N_1} = \frac{E_{b2}}{E_{b1}} \times \frac{\phi_1}{\phi_2}$$

$$\frac{600}{500} = \frac{V - I_{a2}R_{a2}}{V - I_{a1}R_{a1}} \times \frac{\phi_1}{\phi_2}$$

$$\frac{600}{500} = \frac{500 - I_{a2}\left[0.2 + \dfrac{R_{ext} \times 0.3}{R_{ext} + 0.3}\right]}{500 - 60(0.2 + 0.3)} \times \frac{I_{se1}}{I_{se2}}$$

$$\frac{600 \times 470}{500 \times 60} = \frac{500 - I_{a2}\left(0.2 + \dfrac{R_{ext} \times 0.3}{R_{ext} + 0.3}\right)}{I_{a2} \times \dfrac{R_{ext}}{R_{ext} + 0.3}}$$

$$9.4 = \frac{500 - I_{a2}\left(0.2 + \dfrac{R_{ext} \times 0.3}{R_{ext} + 0.3}\right)}{I_{a2}\left(\dfrac{R_{ext}}{R_{ext} + 0.3}\right)}. \qquad \text{(iv)}$$

From Equation (iii):

$$\frac{60^2}{I_{a2}} = I_{a2}\left[\frac{R_{ext}}{R_{ext} + 0.3}\right]. \qquad \text{(v)}$$

Substitute Equation (v) in Equation (iv):

$$9.4 = \frac{500 - 0.2I_{a2} - 0.3 \times \dfrac{60^2}{I_{a2}}}{\dfrac{60^2}{I_{a2}}}$$

$$\frac{60^2 \times 9.4}{I_{a2}} = 500 - 0.2I_{a2} - \frac{0.3 \times 60^2}{I_{a2}}$$

$$500I_{a2} - 0.2I_{a2}^2 - 1{,}080 - 33{,}840 = 0$$

$$0.2I_{a2}^2 - 500I_{a2} + 34{,}920 = 0$$

$$I_{a2} = \frac{+500 \pm \sqrt{500^2 - 4 \times 0.2 \times 34{,}920}}{2 \times 0.2}$$

$$I_{a2} = \frac{500 \pm 471.23}{0.4} = 71.9 \text{ A}.$$

Substitute I_{a2} in Equation (iii):

$$60^2 = I_{a2}^2 \times \frac{R_{ext}}{R_{ext} + 0.3}$$

$$60^2 = 71.9^2 \times \frac{R_{ext}}{R_{ext} + 0.3}$$

$$R_{ext} + 0.3 = 1.436\ R_{ext}$$

$$0.436\ R_{ext} = 0.3$$

$$R_{ext} = \frac{0.3}{0.436} = 0.6878\ \Omega$$

$$\therefore\ R_{ext} = 0.6878\ \Omega.$$

Example 8.22: A 440-V series motor takes a line current of 60 A and runs at a speed of 750 rpm. What resistance should be connected in series with the armature to reduce the speed to 500 rpm. The load torque at this new speed is 75% of its previous value. The resistance of the armature and the series field are 0.05 Ω and 0.015 Ω, respectively. Assume that flux is proportional to load.

Solution:
Given data:

$V = 440$ V

$I_L = 60$ A

$N_1 = 750$ rpm

$N_2 = 500$ rpm

Torque at 500 rpm $= T_1$

Torque at 750 rpm$= T_2 = 0.75T_1$

$R_a = 0.05\ \Omega$

$R_{se} = 0.015\ \Omega.$

We know that:

$$T \propto \phi I_a \quad \text{and} \quad \phi \propto I_{se}$$

$$T_1 \propto \phi_1\ I_{a1} \propto I_{a1}^2$$

$$T_2 \propto \phi_2\ I_{a2} \propto I_{a2}^2$$

$$\therefore \frac{T_1}{T_2} = \frac{I_{a1}^2}{I_{a2}^2}$$

$$I_{a2}^2 = I_{a1}^2 \times \frac{T_2}{T_1}$$

$$= (60)^2 \times \frac{0.75\ T_1}{T_1}$$

$= 2,700$

$\therefore I_{a2} = 51.96\text{ A}$

$\therefore E_{b1} = V - I_{a1}(R_a + R_{se})$

$= 440 - 60\ (0.05+0.015)$

$= 436.1\text{ V}.$

$E_{b2} = V - I_{a2}(R_a + R_{se} + R_{ext})$

$= 440 - 51.96\ (0.05 + 0.015 + R_{ext})$

$= 440 - 51.96\ (0.065 + R_{ext}).$ (i)

But $E_b \propto N\phi$

$$\frac{E_{b1}}{E_{b2}} = \frac{N_1}{N_2} \times \frac{\phi_1}{\phi_2}$$

$$\frac{E_{b1}}{E_{b2}} = \frac{N_1}{N_2} \times \frac{I_{a1}}{I_{a2}}$$

$$E_{b2} = \frac{N_2}{N_1} \times \frac{I_{a2}}{I_{a1}} \times E_{b1}$$

$$= \frac{1}{\frac{750}{500}} \times \frac{1}{\frac{60}{51.96}} \times 436.1 \qquad \text{(ii)}$$

$= 251.77\text{ V}.$

From Equations (i) and (ii):

$\therefore\ 251.77 = 440 - 51.96\ (0.065 + R_{ext})$

$0.065 + R_{ext} = 3.622$

$R_{ext} = 3.55\ \Omega.$

Example 8.23: A series motor with series field and armature resistance of 0.06 Ω and 0.02 Ω, respectively, is connected across 440-V mains. The armature takes 60 A and its speed is 850 rpm. Determine its speed when it takes 85 A from this very and the excitation is increased by 20%.

Solution:

Given data (Fig. P.8.6):

$V = 440\text{ V}$

$N_1 = 850\text{ rpm}$

$I_{a1} = 60\text{ A}$

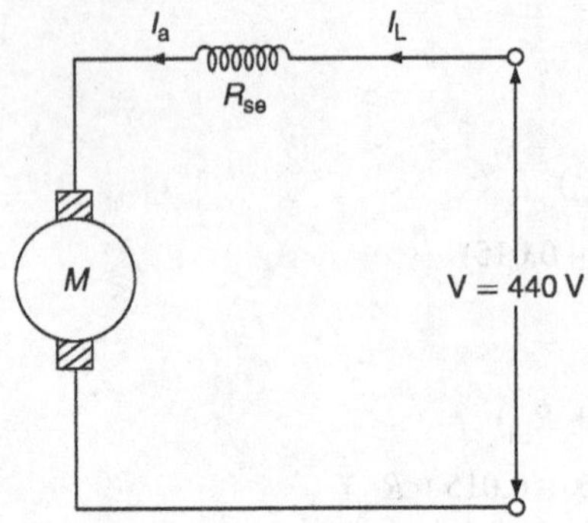

FIG. P.8.6 *DC series motor*

$I_{a2} = 85$ A

$R_a = 0.02\ \Omega$

$R_{se} = 0.06\ \Omega$

$\phi_2 = 1.15\ \phi_1.$

From the speed equation:

$$\frac{E_{b2}}{E_{b1}} = \frac{N_2}{N_1} \times \frac{\phi_1}{\phi_2}$$

$$E_{b1} = 440 - 60 \times (0.02 + 0.06)$$
$$= 435.2 \text{ V}$$

$$E_{b2} = 440 - 85\ (0.8)$$
$$= 372 \text{ V}$$

$$\therefore \frac{N_2}{N_1} = \frac{E_{b2}}{E_{b1}} \times \frac{\phi_2}{\phi_1}.$$

$$= \frac{372}{435.2} \times \frac{1.15\phi_1}{\phi_1}$$
$$= 0.98$$

$$\therefore N_2 = 0.98 \times 850 = 833 \text{ rpm.}$$

Example 8.24: A six-pole and 200-V DC series motor taking 45 A and runs at 800 rpm, with all the coils connected in series. Find the current taken by the motor if the coils are rearranged and their sets are put in series of two in parallel. The torque is proportional to the cube of speed. Assume unsaturated magnetic circuit and neglect losses.

Solution:

Given data (Fig. P.8.7):

$P = 6$ poles

$V = 200$ V

$I_1 = 45$

$N = 800$ rpm.

Case (i): If all the coils are connected in series, then, $\phi_1 \propto I_1$:

$$E_{b1} \propto \phi_1 N_1$$

$$E_{b1} \propto I_1 N_1 \quad \text{(i)}$$

$$T_1 \propto \phi_1 I_1.$$

$$T_1 \propto I_1^2.$$

Given that $T \propto N^3$: (ii)

$$\therefore T_1 \propto I_1^2 \propto N_1^3$$

$$I_1 \propto N_1^{3/2}. \quad \text{(iii)}$$

Case (ii):

$$\phi_2 \propto I_2/3$$

$$\therefore E_{b2} \propto \phi_2 N_2$$

$$E_{b2} \propto (I_2/3)\, N_2 \quad \text{(iv)}$$

And $\quad T_2 \propto \phi_2 I_2 \propto \frac{I_2^2}{3}.$ (v)

From Equation (ii) and (v):

$$T_2 \propto \frac{I_2^2}{3} \propto N^3$$

$$I_2 \propto \sqrt{3} N^{3/2}. \quad \text{(vi)}$$

From Equation (iii) and (vi):

$$\frac{I_1}{I_2} = \frac{N_1^{3/2}}{\sqrt{3} N_2^{3/2}}.$$

If losses are neglected $E_{b1} = E_{b2}$:

$$N_1 I_1 = \frac{N_2 I_2}{3}$$

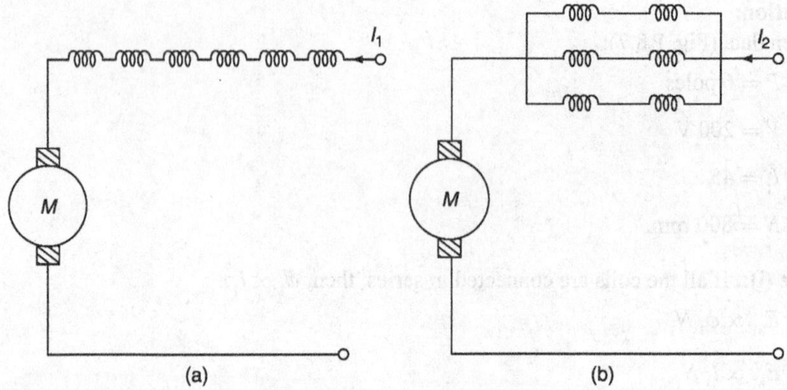

FIG. P.8.7 *DC series motor*

$$\therefore \frac{I_1}{I_2} = \frac{N_2}{3N_1}$$

$$\therefore \frac{N_2}{3N_1} = \frac{N_1^{3/2}}{\sqrt{3}N_2^{3/2}}$$

$$N_2^{5/2} = \sqrt{3}\ \ N_1^{5/2}$$

$$\therefore N_2 = (\sqrt{3})^{2/5} N_1$$

$$= 1.245\, N_1$$

$$= 1.245 \times 800 = 996.5 \text{ rpm.}$$

$$\text{Now } \frac{I_1}{I_2} = \frac{N_2}{3N_1}$$

$$I_2 = I_1 \times \frac{3N_1}{N_2}$$

$$= 45 \times 3 \times \frac{800}{996.5}$$

$$= 108.37 \text{ A.}$$

8.7.3 Ward–Leonard method of speed control

The speed control of DC motor accomplished by means of an adjustable voltage generator is called the Ward–Leonard system. If it is desired to have wide and very sensitive speed control, then this system is more generally used. The system is as shown in Fig. 8.24.

In Fig. 8.24, *R* is the potential devider, M_1 is the main motor whose speed is to be controlled, *G* is the separately excited generator that feeds the armature of the motor M_1, M_2 is the driving motor that drive generator and main motor, and *S* is a double-throw switch.

As shown in Fig. 8.24, M_1 is the main motor whose speed control is required. The field winding of this motor is permanently connected to DC supply and armature is fed from variable voltage so that the motor can run at any desired speed. To provide this variable, the voltage motor generator set is used, in which the generator is directly coupled to a constant speed motor. The field circuit of this generator is separately excited from the available DC supply through a reversing switch and a potential divider '*R*' so that its excitation can be varied from zero to maximum in both the directions. Thus, the generator output voltage can be varied from zero to maximum value. The polarity of generating voltage will be reversed with the help of reversing switch; thus, the change of the direction of the motor M_1 can be achieved.

This system is commonly employed for elevators, hoists, and main drive in steel mills, as this method can give unlimited speed control in either direction. Since the generator voltage can be varied gradually from zero, no extra starting equipment is required to start up the main motor smoothly. The important feature of the Ward–Leonard system is its regenerative action. The modified Ward–Leonard is called Ward–Leonard–Ilgner system in which a flywheel is used in addition to motor-generator set, whose function is to reduce fluctuations in the power demand from the supply circuit. When the main motor M_1 becomes suddenly overloaded, the driving motor M_2 slows down, thus allowing the inertia of the flywheel to supply a part of the overload. However, when the load is suddenly thrown of the main motor M_1, then M_2 speeds up thereby again stores energy in the flywheel.

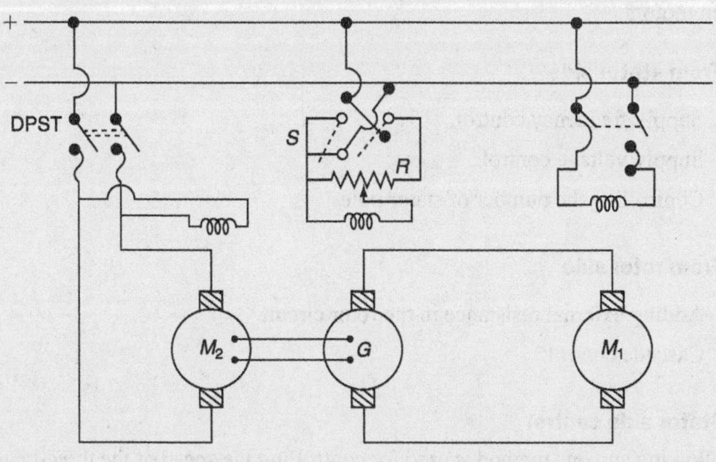

FIG. 8.24 *Ward–Leonard speed control system*

Advantages of Ward–Leonard system

- A wide range of speed from standstill to high speed in either direction.
- Starting without any extra starting equipment.
- Extremely good speed regulation at any speed.

Disadvantages

- High capital cost due to the motor generator set.
- The efficiency of this method is not so high.

8.8 SPEED CONTROL OF INDUCTION MOTORS

A three-phase induction motor is practically a constant-speed motor as the DC shunt motor. The speed control of DC shunt motor can be achieved easily, but it is difficult to achieve the smooth speed control of the induction motor because the performance of the induction motor in terms of its power factor, efficiency, etc. gets adversely effected.

We know that for the induction motor:

The speed of motor $N = N_s(1-S)$. **(8.20)**

And, the torque $T \propto \dfrac{SE_2^2R_2}{R_2^2+(SX_2)^2}$. **(8.21)**

From the above two relations:

The speed of the induction motor can be changed either by changing its synchronous speed (N_s) or by changing the slip and also the parameters R_2 and E_2 are changed then to keep torque constant for constant load condition, slip will change, then its speed gets effected.

Thus, the following methods are used for controlling the speed of the three-phase induction motors.

8.8.1 From stator side

1. Supply frequency control.
2. Supply voltage control.
3. Controlling the number of stator poles.

8.8.2 From rotor side

1. Adding external resistance in the rotor circuit.
2. Cascade control.

8.8.3 Stator side control

Thus, following any one method is used for controlling the speed of the three-phase induction motors on stator side.

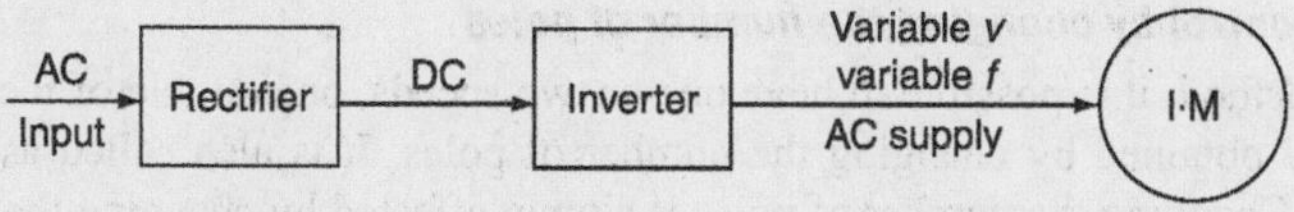

FIG. 8.25 *Electronic circuit for variable frequency*

Speed control by varying the supply frequency

This method is impractical for most applications because the frequency of the supply system must remain fixed. The synchronous speed is given by:

$$N_s = \frac{120f}{P}. \quad \textbf{(8.22)}$$

Thus, by controlling the supply frequency, the synchronous speed can be controlled over a wide range that gives the smooth speed control of the induction motor. Hence, in this method, variable voltage and frequency is achieved by using converter and inverter circuit as shown in Fig. 8.25.

Rectifier converts normal AC supply to constant DC voltage. This DC supply is then given to inverter that converts constant DC to variable AC voltage and frequency.

Supply voltage control

This is a slip-control method with constant frequency variable supply voltage. In this method, the voltage applied to the stator is varied.

We know that:

$$T \propto \frac{SE_2^2 R_2}{R_2{}^2 + (SX_2)^2}.$$

But, at standstill, rotor-induced emf depends on the supply voltage.

i.e., $E_2 \propto V$.

In the operating region of an induction motor or for low-slip region $(SX_2) << R_2$.

So that $T = \frac{SE_2^2}{R_2}$.

Rotor resistance is constant; therefore:

$$T \propto SE_2^2 \propto SV^2. \quad \textbf{(8.23)}$$

From the above relation, if the supply voltage 'V' is reduced below the rated value torque developed by the induction motor reduce. But, so as to maintain the torque constant for constant load, it is necessary to increase the slip thereby decreasing the speed of induction motor.

This method of speed control is simple, low initial cost, and has low maintenance cost, but it has limited use because, the operation at voltage is restricted by magnetic saturation and also large change in voltage is required for relatively for small change in speed.

Speed control by changing the number of poles

In this method, it is possible to have one or two speeds, one double of the other which is generally obtained by changing the number of poles. It is also called as pole-changing method. Changing the number of poles is simply affected by changing the connections of stator winding with the help of simple switches. Due to this number of stator poles gets changed, in the ratio 2:1. Hence, either of the two speeds can be selected.

Consider the single phase of a certain three-phase winding when the supply is across the two terminals and the third is kept open, as shown in Fig. 8.26

Let the conductors which are carrying current in upward direction from South Pole, while the conductors which carry current in downward direction from north polarity. The distribution of current is as shown in Fig. 8.26 due to these eight poles get formed.

Now, the two terminals 1 and 2 which the supply was given earlier are joined together and supply is given to the common point of the first two terminals and the third terminal, on observing the direction of current, it will be found that total eight poles are changed to four poles only as shown in Fig. 8.27; so that, the speed now will be double of the previous value.

8.8.4 Control on rotor side

The following method is used for controlling the speed of three-phase induction motors on rotor side.

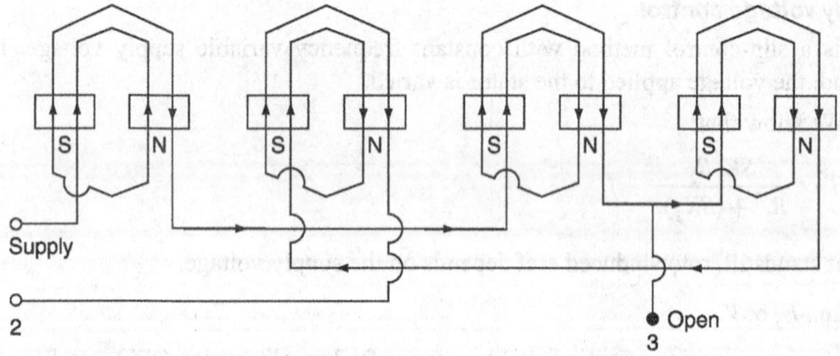

FIG. 8.26 *Eight-pole winding*

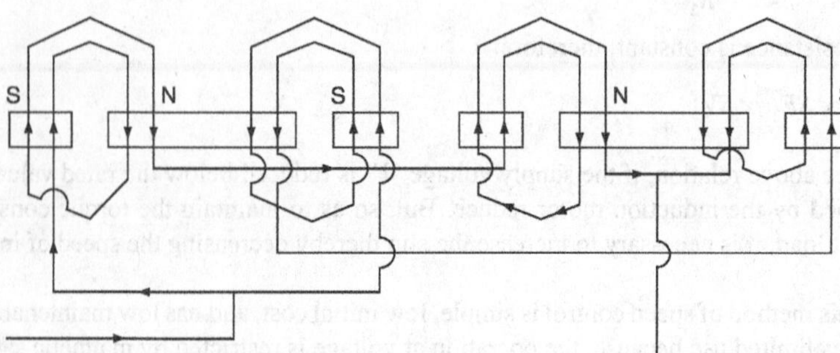

FIG. 8.27 *Four-pole winding*

Cascade control

Multiple speeds are derived and motors are sometimes operated in tandem or cascade. If two motors are to be mechanically coupled together, one of the machines must be phase-wound motor while the other can be a squirrel-cage motor. The first is connected to the mains in the usual way, while that of the second stator is fed from the rotor winding of the first, as shown in Fig. 8.28.

When two motors are operated in tandem, they may be running in the same direction, or the phase rotation of one motor may be reversed, thus tending to make it in reverse direction. In both the cases, the set will run after it is started, but in the later case, no starting torque is developed so that this connection is rarely used.

If P_1 and P_2 be the number poles of both the machines, then the synchronous speed of the set is depending on total number of poles $P_1 + P_2$ in the first case and $P_1 - P_2$ in the second. If the number of poles of the two motors is not equal; four speeds possible: two for tandem operation and one for each motor separately.

Let 'P_1' be the poles of main motor and 'P_2' be the poles of the auxiliary motor.

If 'S' is the slip, the actual rotating speed of the motor is:

$$N_1 = (1 - S)\, N_s$$

$$= (1-S)\left(\frac{120f}{P_1}\right) = \frac{120}{P_1}(f - S \times f). \quad \textbf{(8.24)}$$

But, for the induction motor, the frequency of the rotor current is 'S' times of supply frequency.

$$\text{Frequency } f_r = Sf. \quad \textbf{(8.25)}$$

Let, f_{r1} be the frequency of the rotor current of the main motor and the frequency of the rotor current of the auxiliary motor is f_{r2} then:

$$\text{The speed of the main motor } N_1 = N_1 = \frac{120}{P_1}(f - f_{r1}). \quad \textbf{(8.26)}$$

$$\text{The speed of the auxiliary motor } N_2 = N_2 = \frac{120}{P_2}(f_{r1} - f_{r2}).$$

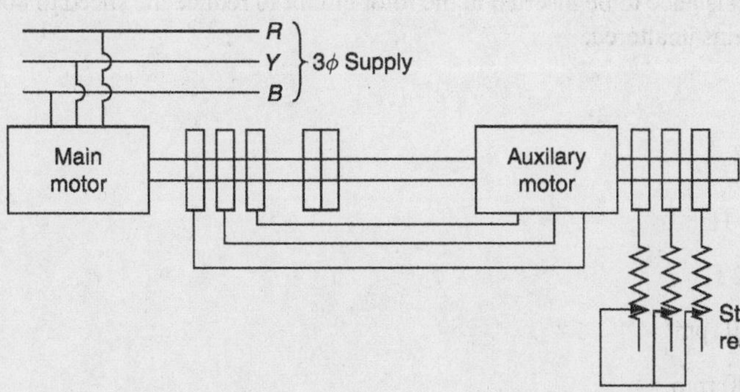

FIG. 8.28 *Cascade control of induction motor*

As f_{r1} is so small, so f_{r2} will be very small; so that, it can be neglected.

$$\therefore N_2 = \frac{120}{P_2}(f_{r1}). \quad \textbf{(8.27)}$$

Since, the two motors are coupled together:

$$N_1 = N_2$$

$$\frac{120}{P_1}(f - f_{r1}) = \frac{120}{P_2} f_{r1}$$

$$(f - f_{r1})P_2 = P_1 f_{r1}$$

$$f P_2 = f_{r1}(P_1 + P_2)$$

$$\therefore f_{r1} = \frac{f P_2}{P_1 + P_2}.$$

Substituting f_{r1} from above equation in Equation (8.27), we get:

$$N_2 = \frac{120}{P_2} \times \frac{f P_2}{P_1 + P_2}$$

$$= \frac{120 f}{P_1 + P_2}. \quad \textbf{(8.28)}$$

Equation (8.28) relation shows that the speed of the set is that of a single machine having the number of poles equal to the sum of the numbers of poles of the two machines. Hence, the set can give four different speeds. If it is required to have the speeds above the normal, the torque of the second motor is reversed by simply changing two of the leads of the second. This is known as differential cascading.

Example 8.25: A six-pole and 50-Hz slip ring induction motor with a rotor resistance per phase of 0.2 Ω and a stand-still reactance of 1.0 Ω per phase runs at 960 rpm at full load. Calculate the resistance to be inserted in the rotor circuit to reduce the speed to 800 rpm, if the torque remains unaltered.

Solution:

Given data:

$$P = 6$$

$$f = 50 \text{ Hz}$$

$$R/\text{ph} = 0.2\ \Omega$$

$$N_1 = 960 \text{ rpm}$$

$$N_2 = 800 \text{ rpm}.$$

$$\text{Synchronous speed } N_s = \frac{120f}{P} = \frac{120 \times 50}{6} = 1{,}000 \text{ rpm.}$$

The slip of the motor when N running at 960 rpm is:

$$S = \frac{N_s - N_r}{N_s} = \frac{1{,}000 - 960}{1{,}000} = 0.04.$$

Let the motor input $= P$.
Full-load current $= I_2$.

$\therefore$ Rotor copper loss $= I_2^2 R = I_2^2 \times 0.4.$

For the induction motor:
Rotor copper loss $= S \times$ Rotor input.

$$I_2^2 \times 0.4 = 0.04 \times P. \qquad \text{(i)}$$

Slip at $N_2 = 800$ rpm is:

$$= \frac{1{,}000 - 800}{1{,}000} = 0.2.$$

Let new rotor resistance be R^1 in rotor circuit:

New copper loss $= I_2^2 R^1$

$$\frac{I_2^2 R^1}{P} = 0.2 \qquad \text{(ii)}$$

$$\frac{I_2^2 R^1 \times 0.04}{I_2^2 \times 0.4} = 0.2$$

$$R^1 = \frac{0.2 \times 0.4}{0.04} = 2\Omega.$$

The external resistance to be added in the rotor circuit is:

$$R_{ext} = R^1 - R$$
$$= 2 - 0.4 = 1.6\,\Omega.$$

Example 8.26: The rotor resistance and the reactance at stand-still condition of a 3-ϕ, six-pole, and 440-V induction motor are. 0.2 Ω and 1.0 Ω, respectively, per phase. Calculate the starting current, and when the speed is 960 rpm, the frequency of the supply is 50 Hz.

Solution:

Rotor resistance per phase $= 0.2\,\Omega$.
Rotor reactance per phase $= 1.0\,\Omega$.

$$\text{Synchronous speed } N_s = \frac{120f}{P} = \frac{120 \times 50}{6} = 1{,}000 \text{ rpm.}$$

The slip of the induction motor $S = \frac{N_s - N_r}{N_s} = \frac{1,000-960}{1,000} = 0.04.$

At the time of starting $S = 1$.

$$\therefore \text{Rotor current per phase } (I_2) = \frac{V_{\text{Ph}}}{R_2^2 + X_2^2}$$

$$= \frac{440/\sqrt{3}}{\sqrt{(0.2)^2 + 1^2}} = 249.05 \text{ A.}$$

At a speed of 960 rpm, the rotor resistance per phase is:

$$R_2^1 = \frac{R_2}{S} = \frac{0.2}{0.04} = 5\ \Omega.$$

$$\therefore \text{The rotor current per phase, } I_2^1 = \frac{440/\sqrt{3}}{\sqrt{5^2 + 1^2}}$$

$$= \frac{254.034}{\sqrt{26}} = 49.82 \text{ A.}$$

Example 8.27: A 10-HP, four-pole, 50-Hz, and 220-V induction motor is connected in cascade with another 15-HP, six-pole, 50-Hz, and 220-V motor. What are the possible speeds obtained with this combination and what is the maximum load in HP which may be delivered without overloading either machine? Also determine the ratio of the mechanical power outputs of the two machines at this load?

Solution:

The synchronous speed of the six-pole machine $N_{S1} = \frac{120f}{P_1}$

$$= 120 \times 50/6$$

$$= 1,000 \text{ rpm.}$$

The synchronous speed of the four-pole machine $N_{S2} = \frac{120f}{P_2}$

$$= \frac{120 \times 50}{4}$$

$$= 1,500 \text{ rpm.}$$

The speed of the I machine when it is connected to II machine:

$$N_{SC1} = \frac{120f}{P_1 + P_2} = \frac{120 \times 50}{6+4} = 600 \text{ rpm.}$$

Similarly, the speed of the second machine is:

$$N_{SC2} = \frac{120f}{P_1 - P_2} = \frac{120 \times 50}{6-4} = 3,000 \text{ rpm.}$$

In cascade connection, the rating of the main motor should be such that it covers the power output of the auxiliary motor also. So that, the maximum load which can be delivered is 15 HP.

The ratio of the mechanical power outputs of the main and the auxiliary induction motors are:

$$P_1 : P_2 = 6 : 4 = 1 : 5 : 1.$$

Example 8.28: The open circuit voltage across the slip rings of a 100-HP induction motor is 280 volts at standstill. What resistance in rotor circuit will reduce its full-load speed by 20%. The full-load slip is 3% with no additional rotor resistance. Assume rotor to be star-connected. And full-load sip $S_1 = 0.03$.

Solution:

The mechanical power developed by the rotor:

$$P_{mech} = 100 \times 735.5$$

$$= 73{,}550 \text{ W}.$$

The standstill induced emf per phase in rotor:

$$E_2 = \frac{280}{\sqrt{3}}$$

$$= 161.65 \text{ V}.$$

The rotor current per phase $I_2 = \dfrac{S_1 E_2}{R_2}$ (X_2 is neglected)

$$= \frac{0.03 \times 161.65}{R_2}$$

$$= \frac{4.85}{R_2}.$$

The mechanical power developed by the rotor is:

$$= \frac{\text{Total rotor copper loss}}{S} \times (1 - S)$$

$$73{,}550 = \frac{3 I_2^2 R_2}{0.03} \times (1 - 0.03)$$

$$= 97 I_2^2 R_2.$$

Substituting $I_2 = \dfrac{4.05}{R_2}$ value in the above expression.

$$= 97 \left(\frac{4.05}{R_2} \right)^2 R_2$$

$$73{,}550 = \frac{2{,}281.68}{R_2}$$

$R_2 = 0.03\ \Omega.$

The new speed $N_2 = N_S\,(1 - 0.03)\,(1 - 0.2)$

$$N_2 = 0.776\,N_S.$$

$$\text{Slip } S_2 = \frac{N_S - N_2}{N_S} = \frac{N_S - 0.776N_S}{N_S}$$

$S_2 = 0.224.$

The load torque is assumed to be constant.

$S \propto$ rotor resistance.

$$\frac{S_2}{S_1} = \frac{R_2 + R}{R_2}$$

$$\frac{.224}{0.03} = \frac{0.03 + R}{0.03}$$

$R = 0.193\ \Omega.$

Example 8.29: A eight-pole, 50-Hz, and 3-ϕ induction motor is running at 4% slip when delivering full-load torque. It has a standstill rotor resistance of 0.3 Ω and a reactance of 0.8 Ω per phase. Calculate the speed of the motor if an additional resistance of 0.3 Ω per phase is inserted in the rotor circuit. The full-load torque remains constant.

Solution:

$$\text{The synchronous speed of the motor } N_S = \frac{120f}{P}$$

$$= \frac{120 \times 50}{8}$$

$$= 750 \text{ rpm.}$$

Full-load slip $S_1 = 0.04.$

$$\text{The motor torque } T = \frac{KSR_2E_2^2}{R_2^2 + S^2X_2^2}.$$

At full load, the new slip is S_2 then:

$$T_1 = \frac{KS_1R_2E_2^2}{R_2^2 + S_1^2X_2^2}$$

$$= \frac{K \times 0.04 \times 0.3 \times E_2^2}{(0.3)^2 + (0.04 \times 0.8)^2}$$

$$= KE_2^2(0.1318).$$

$$T_2 = \frac{KS_2(R_2 + R)E_2^2}{(R_2 + R)^2 + (S_2X_2)^2}$$

$$= \frac{KS_2(0.3 + 0.3)E_2^2}{(0.3 + 0.3)^2 + (0.8S_2)^2}$$

$$= \frac{0.6KS_2E_2^2}{0.36 + 0.64S_2^2}.$$

The two torques are remains same i.e., $T_1 = T_2$:

$$0.1318 = \frac{0.6S_2}{0.36 + 0.64S_2^2}$$

$$0.0474 + 0.084S_2^2 = 0.6S_2$$

$$0.084S_2^2 - 0.6S_2 + 0.0474 = 0$$

$S_2 = 0.079$ neglecting higher values.

∴ The speed of motor $N_2 = N_S(1 - S)$

$$= 750\,(1 - 0.079)$$

$$= 690.75 \text{ rpm.}$$

Example 8.30: A four-pole induction motor and six-pole induction motor are connected in cumulative cascade at 50-Hz supply. The frequency in the secondary circuit of the six-pole motor is observed to be 1.0 Hz. Calculate the slip of each machine and the combined speed of the test.

Solution:
The cascaded speed of the set, $N_0 = \frac{120f}{P_1 + P_2} = \frac{120 \times 50}{4 + 6} = 600$ rpm

$$N_r = 600 \text{ rpm.}$$

The frequency in the secondary circuit of the six-pole motor:

$$f'' = S''f$$

$$S'' = \frac{f''}{f} = \frac{1.0}{50}.$$

$$\text{Actual speed } N = N_r(1 - S'')$$

$$= 600\,(1 - 0.02)$$

$$= 588 \text{ rpm.}$$

The synchronous speed of the four-pole induction motor:

$$N_S = \frac{120 \times f}{P}$$

$$= \frac{120 \times 50}{4}$$

$$= 1{,}500 \text{ rpm.}$$

The slip for four-pole induction motor $S = \frac{N_S - N}{N_S} = \frac{1500 - 588}{1500}$

$$= 0.608$$

$$= 60.8\%.$$

The impressed frequency on the second motor:

$$f^1 = S_f$$

$$= 0.608 \times 50$$

$$= 30.4 \text{ Hz.}$$

At $f^1 = 30.4$, the synchronous speed of the six-pole induction motor is:

$$N_S^1 = \frac{120 \times f^1}{P_2}$$

$$= \frac{120 \times 30.4}{6}$$

$$= 608 \text{ rpm.}$$

The slip for the six-pole induction motor $S = \frac{N_S^1 - N}{N_S^1}$

$$= \frac{608 - 588}{608}$$

$$= 0.033$$

$$= 3.3\%.$$

Example 8.31: The rotor of a six-pole, 50-Hz, and 3-ϕ induction motor has a resistance of 0.3 Ω per phase and sums at 960 rpm. If the load torque remains unchanged, calculate the additional rotor resistance that will reduce the speed by 20%.

Solution:

The synchronous speed of the motor, $N_S = \frac{120f}{P}$

$$= \frac{120 \times 50}{6}$$

$$= 1{,}000 \text{ rpm.}$$

Full-load speed = 960 rpm.

$$\text{Full-load slip } S_1 = \frac{N_S - N_1}{N_S} = \frac{1{,}000 - 960}{1{,}000}$$

$$= 0.04.$$

$$\text{New speed } N_2 = N_1 (1 - 0.2)$$

$$= 960 \times 0.8$$

$$= 768 \text{ rpm.}$$

$$\text{New slip } S_2 = \frac{N_S - N_2}{N_S}$$

$$= \frac{1{,}000 - 768}{1{,}000}$$

$$= 0.232.$$

For the constant load torque:

$$S \propto R_2$$

$$\frac{S_2}{S_1} = \frac{R_2 + R}{R_2}$$

$$\frac{0.232}{0.04} = \frac{0.3 + R}{0.3}$$

$$R = 1.44\ \Omega.$$

Example 8.32: A cascade it consists of two motors A and B with four and six poles, respectively. The motor is connected to a 50-Hz supply. Find (i) the speed of the set and (ii) the electric power transferred to motor B when the input to motor A is 30-kW neglect losses.

Solution:

The synchronous speed of the test, $N = \frac{120 \times f}{P_A + P_B} = \frac{120 \times 50}{4+6}$

$$= 600 \text{ rpm.}$$

The power output of motor $B = P \times \frac{P_B}{P_A + P_B}$

$$= 30 \times \frac{6}{4+6}$$

$$= 18 \text{ kW.}$$

∴ The outputs of the two motors are proportional to the number of their poles.

8.9 RATING OF MOTOR

The selection of motor for particular drive application based on the size of motor depends upon the following two factors:

(i) Maximum temperature raise for a given load.

(ii) Maximum torque required.

The size of motor and its rating are mainly dependent upon the raise in temperature. The temperature raise in turn depends upon the type of insulation used.

8.9.1 Temperature raise of motor

The various losses takes place in any motor will be converted into heat. The heat thus produced will increase the temperature of various parts of the motor. The increase in temperature is mainly dependent on the following two factors:

(i) Amount of heat developed internally at uniform rate.

(ii) The amount of heat dissipated from the surface of the motor.

In fact, the continuous rating of a machine is that rating for which the final temperature raise is equal to or just below the permissible value of the temperature raise for the insulating material used in protection of motor windings. When the machine is overloaded for such a long time that its final temperature raise exceeds the permissible limit, it is likely to be damaged. Sometimes, it will results immediate breakdown of insulating material which will cause a sudden short circuit in the motor, which may also lead to a fire. Since temperature raise is one of the chief features in fixing the size of motor. The temperature raise will be high in the beginning and will decrease gradually with the passage of time and finally the temperature of the motor attains a steady-state value. At this point, the heat produced and dissipated will be equal.

The above circumstances make the heating calculations very complex and practically impossible unless certain assumptions are made as:

(i) Heat developed, i.e., losses remains constant during temperature raise.

(ii) The heat dissipation is directly proportional to the difference in the temperature of motor and cooling medium, i.e., Newton's law of cooling hold's good.

(iii) The temperature of cooling medium remains unchanged.

(iv) The motor is assumed to be a homogeneous mass having the same and uniform temperature in all parts. It implies high thermal conductivity.

(v) For the determination of an expression for the temperature raise of an electrical machine after time 't' seconds from the instance of switching it on.

Let P is the electrical power converted into heat (W or J/sec), M is the mass of active parts of motor (kg), S is the specific heat of material (J/kg/°C), O is the temperature raise above the cooling medium or ambient temperature (°C), A is the surface area of cooling, (m^2), θ_f is the final temperature raise with constant load (°C), and λ is the coefficient of cooling or the rate of heat dissipation (W/m^2/°C raise).

Now, let us assume that the machine attains a temperature raise of θ°C above ambient temperature after 't' seconds of switching on the machine and further raise of temperature by dθ in very small time 'dt' seconds.

The rate at which the loss takes place or the heat is absorbed by the motor $= MS\dfrac{d\theta}{dt}$ J/sec.

The rate at which heat is dissipated $= A\theta\lambda$ J/sec.

But, the rate at which the electrical power converted into heat = the rate at which the heat is absorbed + the rate at which the heat dissipated by the motor.

$$P = MS\frac{d\theta}{dt} + A\lambda\theta \quad \textbf{(8.29)}$$

$$P - A\theta\lambda = MS\frac{d\theta}{dt}$$

$$dt = \frac{MSd\theta}{P - A\theta\lambda}. \quad \textbf{(8.30)}$$

Integrating the Equation (8.30):

$$\int dt = \int \frac{MS}{P - A\theta\lambda} d\theta$$

$$t = MS\log_e(P - A\theta\lambda) \times \left(\frac{-1}{A\lambda}\right) + K, \quad \textbf{(8.31)}$$

where K is the integration constant.

Initially, at time $t = 0$ sec, temperature raise $\theta = 0$°C.

By substituting $t = 0$ and $\theta = 0$ in Equation (8.31), we get the integration constant (K):

$$\text{i.e., } 0 = \frac{-MS}{A\lambda}\log_e(P - 0) + K$$

$$\text{or} \quad K = \frac{MS}{A\lambda}\log_e P.$$

Substituting the value of 'K' in Equation (8.31), we get:

$$t = \frac{-MS}{A\lambda}\log_e(P - A\lambda\theta) + \frac{MS}{A\lambda}\log_e P \quad \textbf{(8.32)}$$

$$= \frac{-MS}{A\lambda}[\log_e(P - A\theta\lambda) - \log_e P]$$

$$= \frac{-MS}{A\lambda}\log_e\left[\frac{P - A\theta\lambda}{P}\right]$$

$$\therefore \frac{-A\lambda t}{MS} = \log_e\left[\frac{P - A\theta\lambda}{P}\right].$$

By applying exponential on both side, we get:

$$e^{\frac{-A\lambda t}{MS}} = 1 - \frac{A\theta\lambda}{P} \qquad [\because \log_e^e = 1]$$

$$\frac{A\theta\lambda}{P} = 1 - e^{\frac{-A\lambda t}{MS}}$$

$$\theta = \frac{P}{A\lambda}\left(1 - e^{\frac{-A\lambda t}{MS}}\right). \qquad \textbf{(8.33)}$$

When 't' is infinity, 'θ' approaches to its final steady-state temperature 'θ_f'. So, by substituting $t = \infty$ and $\theta = \theta_f$ in Equation (8.33), we get:

$$\theta_f = \frac{P}{A\lambda}[1 - e^{-\infty}]$$

$$= \frac{P}{A\lambda}[1 - 0] = \frac{P}{A\lambda}. \qquad \textbf{(8.34)}$$

Substituting $\theta_f = \frac{P}{A\lambda}$ in Equation (8.33), we get:

$$\theta = \theta_f\left[1 - e^{\frac{-A\lambda}{MS}t}\right]$$

$$= \theta_f\left[1 - e^{\frac{-t}{T_h}}\right], \qquad \textbf{(8.35)}$$

where 'T_h' $= \frac{MS}{A\lambda}$ is known as heating time constant of motor.

The above relation is the equation of temperature rise with time. The temperature raise time curve or heating curve is exponential in nature as shown in Fig. 8.29.

From the equation of temperature raise:

$$\theta = \theta_f\left[1 - e^{\frac{-t}{T_h}}\right].$$

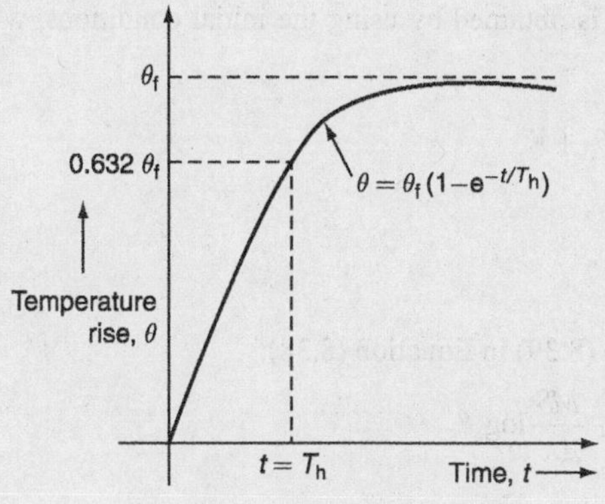

FIG. 8.29 *Heating curve*

At $t = T_h$, $\theta = \theta_f[1 - e^{-1}]$

$\therefore\ \theta = 0.632\ \theta_f$.

Thus, *heating time constant* can be defined as follows:

The heating time constant is the time taken by the machine to attain 63.2% of its final steady temperature raise (θ_f).

The heating time constant of the conventional electrical machines is usually within the range of 0.5–3 for 4 h.

8.9.2 Cooling of motor

Let us assume, if the supply to the motor is switched off, after attaining the final steady temperature raise of 'θ_f', the motor starts cooling. When the machine is switched off, no heat is produced, therefore:

Heat absorbed + heat dissipated = 0

$$\therefore MS\frac{d\theta}{dt} + A\lambda'\theta = 0, \qquad (8.36)$$

where λ = heat dissipation during cooling of motor.

$$MSd\,\theta + A\lambda'\,\theta \cdot dt = 0$$

$$dt = -\frac{MS}{A\lambda'}d\theta. \qquad (8.37)$$

Integrating the Equation (8.37):

$$\int dt = \frac{-MS}{A\lambda^1}\int d\theta$$

$$t = \frac{-MS}{A\lambda^1}\log_e\theta + K^1, \qquad (8.38)$$

where K^1 is the integration constant.

The value of K^1 is obtained by using the initial conditions, when $t = 0$ and $\theta = \theta_f$, we get:

$$0 = \frac{-MS}{A\lambda^1}\log_e\theta_f + K^1$$

$$K^1 = \frac{MS}{A\lambda^1}\log_e\theta_f. \tag{8.39}$$

Substituting Equation (8.39) in Equation (8.38):

$$t = \frac{-MS}{A\lambda^1}\log_e\theta + \frac{MS}{A\lambda^1}\log_e\theta_f$$

$$= \frac{-MS}{A\lambda^1}[\log_e\theta - \log_e\theta_f]$$

$$= \frac{-MS}{A\lambda^1}\log_e\left(\frac{\theta}{\theta_f}\right)$$

$$\therefore \frac{-A\lambda^1 t}{MS} = \log_e\left(\frac{\theta}{\theta_f}\right).$$

Applying exponentials on both side λ:

$$e^{\frac{-A\lambda^1}{MS}t} = \log_e e\left(\frac{\theta}{\theta_f}\right) \qquad [\because \log_e e^x = x]$$

$$\frac{\theta}{\theta_f} = e^{\frac{-A\lambda^1}{MS}t}$$

$$= \theta_f e^{\frac{-A\lambda^1}{MS}t}$$

$$= \theta_f e^{\frac{-t}{T_c}}, \tag{8.40}$$

where $T_c = \frac{MS}{A\lambda^1}$ is know as cooling time constant.

The above relation is the equation of cooling of motor. The cooling curve is exponentially decaying in nature as shown in Fig. 8.30.

From the cooling equation, at time $t = T_c$:

We have $\theta = \theta_f(e^{-1})$

$\therefore \theta = 0.368\theta_f$.

Thus, we can define the *cooling time constant* as:

The cooling time constant is defined as the time required cooling the machine down to 36.8% of the initial temperature raise above the ambient temperature.

The heating and cooling curves follows an exponential law. Heating time constant and cooling time constant may be different for the same machine and also the cooling

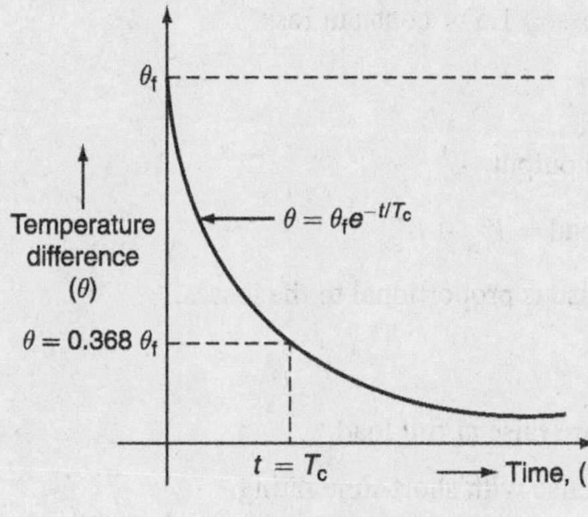

FIG. 8.30 *Cooling curve*

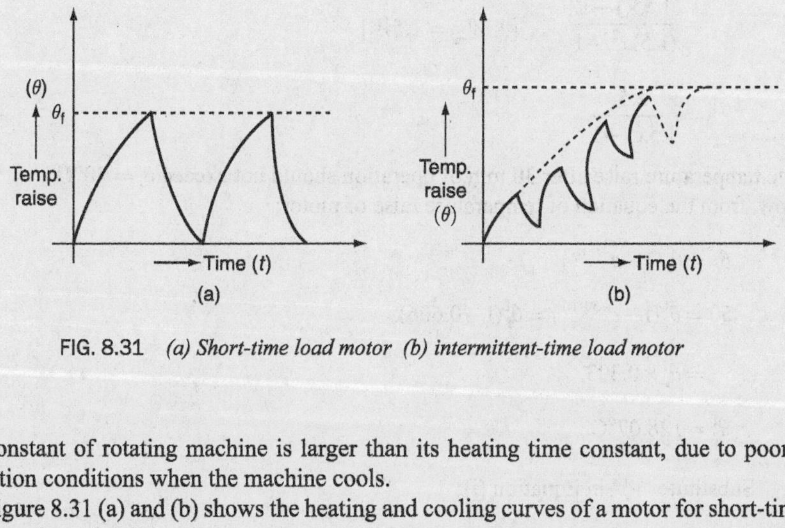

FIG. 8.31 *(a) Short-time load motor (b) intermittent-time load motor*

time constant of rotating machine is larger than its heating time constant, due to poorer ventilation conditions when the machine cools.

Figure 8.31 (a) and (b) shows the heating and cooling curves of a motor for short-time and intermittent loads.

Example 8.33: An induction motor has a final steady-state temperature raise of 50°C when running at its rated output. Calculate its half-hour rating for the same temperature raise if the copper losses at the rated output are 1.5 times its constant losses. The heating time constant is 60 min.

Solution:

Given data:

Final steady temperature (θ_f) = 50°C.

Time constant (τ_h) = 60 min.

$$\text{Rating}(t) = \frac{1}{2}\,\text{hour} = 30\,\text{min}.$$

And, the copper loss = 1.5 × constant loss

i.e., $W_{cu} = 1.5 \times W_i$

let '*P*' be the rated output,

Total loss at full load = $W_{cu} + W_i$.

But, the temperature raise is proportional to the losses.

$\therefore \theta \propto W_{loss}$.

Let, θ_f be the temperature raise at full load.

θ_f^1 be the temperature raise with short-time rating.

$$\therefore \frac{\theta_f}{\theta_f^1} = \frac{W_{cu} + W_i}{x^2 W_{cu} + W_i}$$

$$= \frac{1.5 \times 1 + 1}{(1.5)x^2 + 1} \qquad [\because W_{cu} = 1.5 W_i]$$

$$= \frac{2.5}{1.5x^2 + 1}. \qquad \text{(i)}$$

The temperature raise after 30 min of operation should not exceed $\theta_f = 50°C$. Now, from the equation of temperature raise of motor:

$$\theta_f = \theta_f^1 (1 - e^{-t/\tau_h})$$

$$50 = \theta_f^1 (1 - e^{-30/60}) = \theta_f^1 (1 - 0.606)$$

$$= \theta_f^1 \times 0.393.$$

$$\therefore \theta_f^1 = 128.07°C.$$

Substitute 'θ_f^1' in Equation (i):

$$\frac{\theta_f}{\theta_f^1} = \frac{2.5}{1.5x^2 + 1}$$

$$\theta_f^1 = \theta_f \times \left(\frac{1.5x^2 + 1}{2.5} \right)$$

$$1.5x^2 + 1 = \frac{\theta_f^1}{\theta_f} \times 2.5$$

$$1.5x^2 = \frac{127.07}{50} \times 2.5$$

$$1.5x^2 = 6.3537$$

$$x^2 = 4.235$$

$$\therefore x = 2.058.$$

$\therefore$ Hence, the half-hour rating of machine is 2.058 times its continuous rating.

Example 8.34: A 10-kW motor has a heating time constant and cooling time constant of 45 and 70 min, respectively. The final temperature attained is 60°C. Find the temperature of motor after 45 min full-load run and then switched of for 30 min.

Solution:

Given data:

$$\tau_h = 45\text{min}$$

$$\tau_c = 70 \text{ min}$$

$$\theta_f = 60°\text{C}$$

$$t = 45 \text{ min.}$$

We know that:

$$\theta = \theta_f\left(1 - e^{-t/\tau_h}\right)$$

$$= 60\left(1 - e^{-45/45}\right)$$

$$= 60 \times 0.632 = 37.927°\text{C.}$$

When the motor is switched off for 30 min, the temperature is:

$$\theta = \theta_f \; e^{-t/\tau_c}$$

$$= 37.927\, e^{-30/70}$$

$$= 37.927 \times 0.6514 = 24.707 \cong 25°\text{C.}$$

Example 8.35: The heating time constant of a 80-kW motor is 60 min. The temperature raise is 65°C when runs continuously on full load. Find the half-hour rating of motor for the same temperature raise. Assume that the losses are proportional to the square of the load and the motor cools to ambient temperature between each load cycle.

Solution:

Let 'x' be the half-hour rating in kW.

$$\text{Losses at half-hour rating} = \left(\frac{x}{80}\right)^2 \times \text{losses at 80 kW.}$$

Let θ is the temperature raise at x kW and θ_f is the temperature raise at 80 kW. We know that the losses $\propto$ load^2 and temperature raise $\propto$ losses

$$\frac{\theta}{\theta_f} = \left(\frac{x}{80}\right)^2$$

$$\therefore \theta = \theta_f \times \left(\frac{x}{80}\right)^2$$

$$\therefore \theta = 65 \times \left(\frac{x}{80}\right)^2.$$

Now, $65 = \theta\left(1 - e^{-t/\tau_h}\right)$

$$= 65\left(\frac{x^2}{80}\right)\left(1 - e^{-30/60}\right).$$

$$0 = \frac{x^2}{80}\left(1 - e^{-1/2}\right)$$

$$6{,}400 = x^2\,(1 - e^{-0.5}) = x^2\,(0.393)$$

$$x = \sqrt{\frac{6400}{0.393}}$$

$$= 127.5 \text{ kW}.$$

Example 8.36: The heating time constant and final steady temperature of a motor on continuous running is 60 min and 40°C. Find out the temperature (i) after 25 min at this load, (ii) after 45 min at this load, (iii) if the temperature raise at half-hour rating is 40°C, find the maximum steady temperature, (iv) what will be the time required to increase the temperature from 25°C to 40°C at one-and-half-hour rating.

Solution:

Given data:

$\theta_f = 40°\text{C}$

$t = 25$ min

$\tau_h = 60$ min.

(i) We know that:

$$\theta = \theta_f\left(1 - e^{-t/\tau_h}\right)$$

$$= 60\left(1 - e^{-25/60}\right)$$

$$= 60 \times 0.340 = 20.44°\text{C}.$$

(ii) For 45 min at the same load:

$$\theta = \theta_f\left(1 - e^{-t/\tau_h}\right)$$

$$= 60\left(1 - e^{-45/60}\right)$$

$$= 31.658°\text{C}.$$

(iii) If the temperature raise is 40°C after half an hour, the maximum temperature:

$$\therefore \theta_{\mathrm{f}} = \frac{\theta}{\left(1-e^{-t/\tau_{\mathrm{h}}}\right)} = \frac{40}{1-e^{-30/60}}$$

$$= \frac{40}{1-e^{-1/2}} = 101.65^{\circ}\,\mathrm{C}.$$

(iv) Given, time taken to attain temperature raise of 40°C is one-and-half hour. Then, the maximum temperature θ_{f} is 101.65°C.

Let 't' be the taken in min needed to raise the temperature from 25°C to 40°C.

$$\theta = \theta_{\mathrm{f}}\left(1-e^{-t/\tau_{\mathrm{h}}}\right)$$

$$25 = 40\left(1-e^{-t/60}\right)$$

$$0.625 = \left(1-e^{-t/60}\right)$$

$$e^{-t/60} = 0.375$$

$$-t/60 = \ln(0.375) = -0.98$$

$$\therefore t = 60 \times 0.98 = 58.84^{\circ}\mathrm{C}.$$

Thus, the temperature will increase from 25°C to 40°C in time, $t^{1} = 90 - 58.84$

$$= 31.15 \text{ min.}$$

Example 8.37: The heating time constant of a motor is 90 min with 1-hr rating as 200 W. The maximum efficiency of motor occurs at 80% of full load. Determine the continuous rating of the motor.

Solution:

Given that, the maximum efficiency occurs at 80% of full load. Therefore, at 80% of full load, the copper loss is equal to the iron loss.

Let iron loss = copper loss = W_{C} W.

Copper loss at 80% of full load = W_{C}.

$$\text{Copper loss at full load} = \left(\frac{1}{0.8}\right)^{2} \times W_{\mathrm{C}}.$$

$$\text{Losses at full load} = W_{\mathrm{C}} + \left(\frac{1}{0.8}\right)^{2} \times W_{\mathrm{C}}$$

$$= W_{\mathrm{C}}\left(1+\left(\frac{1}{0.8}\right)^{2}\right)$$

$$= 2.5625\, W_{\mathrm{C}}.$$

$$\text{Losses at load of 200 W} = W_C + \left[\frac{200}{0.8 + \text{full load}}\right]^2 \times W_C.$$

$$\theta_f = \text{Total loss on full load.}$$

$$\theta_f^1 = \text{Total loss on 30 min rating.}$$

$$\frac{\theta_f^1}{\theta_f} = \frac{1}{1-e^{-t/\tau_h}} = \frac{\text{total loss on 30 min rating}}{\text{total loss on full load}}$$

$$\frac{1}{1-e^{-60/90}} = \frac{W_C + \left[\frac{200}{0.8 \times \text{full load}}\right]^2 \times W_C}{2.5625\, W_C}$$

$$2.055 = \frac{1 + \left(\frac{250}{\text{full load}}\right)^2}{2.5625}$$

$$5.265 = 1 + \left(\frac{250}{\text{full load}}\right)^2$$

$$\frac{250}{\text{full load}} = 4.265$$

$$\therefore \text{Full load} = \frac{250}{2.065} = 121.04 \text{ W}.$$

8.10 TYPES OF LOADS

While selecting a motor, in addition to the information of load–speed–torque characteristics, the variation of load torque, losses, and temperature raise with time is also needed. In case the load and torque verses time variation is periodic and repetitive, such one cycle of variation of load with time is known as load or *duty cycle*. The various types of loads that occur in industrial practice can be classified depending upon their variation with time and duty cycle, which can be specified by the load diagram.

Figure 8.32 shows the typical duty cycle or load cycle which will give the variation of load with time and also the type of load.

8.10.1 Classification of loads with respect to time

The loads are classified with respect to time as follows.

Continuous and constant loads

The loads on the motor operate for a long time under the same conditions.

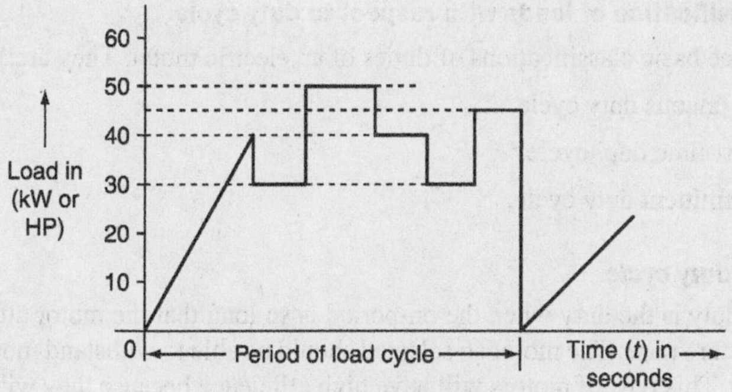

FIG. 8.32 *Duty cycle or load cycle*

Ex: fan, compressors, conveyors, centrifugal pumps, etc.

Continuous and variable loads

The load on the motor operates repetitively for a longer duration but varies continuously over a period.

Ex: metal cutting lathes, hoist winches, conveyors, etc.

Pulsating loads

The load on the motor which can be viewed as constant torque superimposed by pulsations.

Ex: tile looms, reciprocating pumps, certain type of loads with crankshaft, frame saws, etc.

Impact loads

The load on the motor having regular and repetitive load peaks or pulses, i.e., load increases to a maximum level suddenly.

Ex: rolling mills, shearing machines, etc.

Short-time intermittent loads

The load on the motor occurs periodically in identically duty cycle, each duty cycle having a period of application of load and rest.

Ex: Roller trains, cranes, hoisting mechanisms, etc.

Short-time loads

The load on the motor occurs periodically remains constant for short time and then remains idle or off for longer time.

Ex: servomotors, motor–generator sets, used for charging batteries, drilling machines, etc.

8.10.2 Classification of loads with respect to duty cycle

There are three basic classifications of duties of an electric motor. They are:

1. Continuous duty cycle.
2. Short-time duty cycle.
3. Intermittent duty cycle.

Continuous duty cycle

Continuous duty is the duty when the on-period is so long that the motor attains a steady-state temperature raise. The motor so selected should be able to withstand momentary over-load capacity. This type of motors will have high efficiency because they will be operating almost at its full load and also have good power factor.

There are mainly two types of continuous duty cycle. They are:

(i) Continuous duty at constant load cycle.

(ii) Continuous duty at variable load cycle.

In continuous duty with constant load cycle, the load torque remains constant for a sufficiently longer period. The variation of torque against time for continuous duty is shown in Fig. 8.33.

Ex: Conveyors, compressors, fan, etc. in which continuous duty at constant load occurs.

In continuous duty with variable load cycle, the load on the motor is not constant, but it has several phases in one cycle. The variation of load against time for variable load cycle is shown in Fig. 8.34. The selection of motor for this type of duty involves thermal calculation, which is a difficult task. The motors operating for such type of duties will have poor efficiency and also poor power factor.

The selection of motor for this type of duty may be based on average power or average current method.

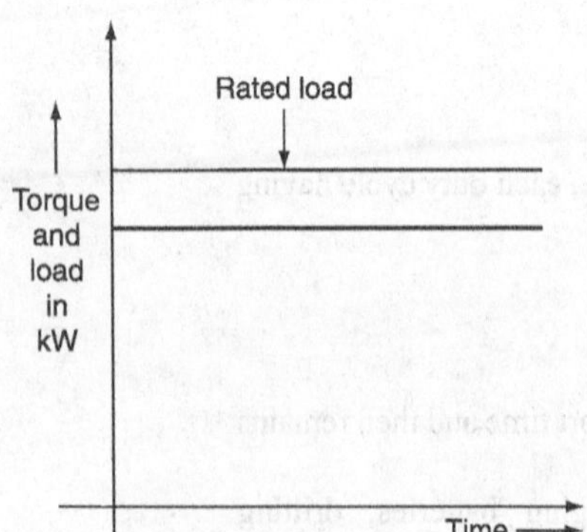

FIG. 8.33 *Continuous duty with constant load*

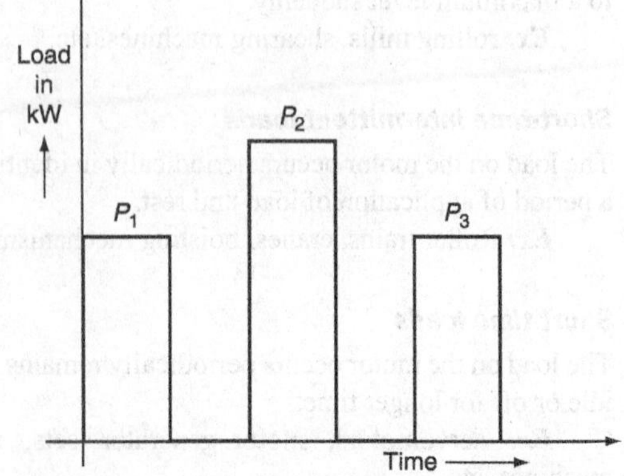

FIG. 8.34 *Continuous duty with variable load*

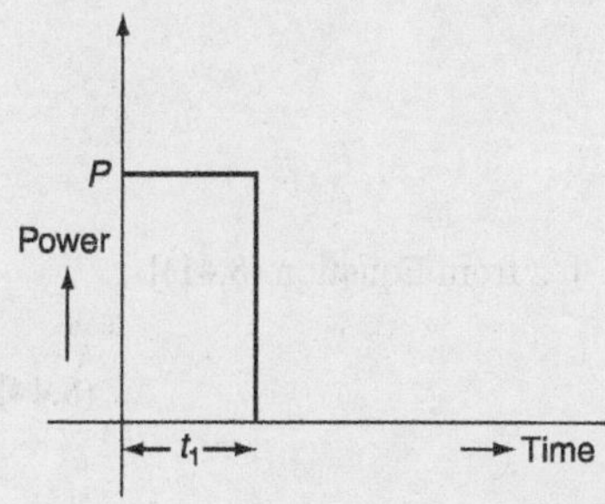

FIG. 8.35 *Load cycle for short time duty*

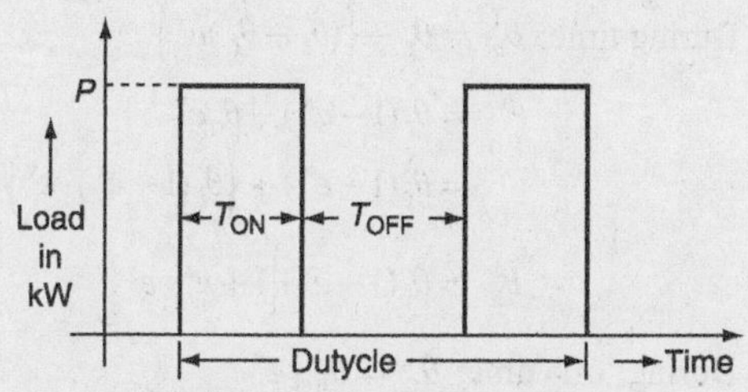

FIG. 8.36 *Load cycle for intermittent duty*

Short-time duty

In this type of duty, the load occurs on the motor during a small interval and the remains idle for long time to re-establish the equality of temperature with the cooling medium. The variation of the load against time for short-time duty is shown in Fig. 8.35.

Usually, such type of short-time duty occurs in bridges, lock gates, and some other household appliances such as mixies.

Intermittent duty

The duty in which load on the motor varies periodically in a sequence of identical cycles shown in Fig. 8.36, in which motor is loaded for sometimes 't_{on}' and shut off for a period of 't_{off}'.

Motor heats during 'on' period 't_{on}' and cools down during 'off' period 't_{off}'. The ratio of 't_{on}' to $(t_{on} + t_{off})$ is known as *duty ratio*.

$$\text{Duty ratio} = \frac{t_{on}}{(t_{on} + t_{off})}.$$

Maximum temperature attained with intermittent loading can be obtained by using the temperature raise and cooling equations of motor, and is given as follows.

Let $\theta_h, \theta_h^{\,1}, \theta_h^{\,2} \cdots \theta_h^{\,n-1}$ be the temperature raise and $\theta_C, \theta_C^{\,1}, \theta_C^{\,2} \cdots \theta_C^{n-1}$ be the fall in temperature for '*n*' times intermittency.

Let t_1 be the duration of heating in second, t_2 be the duration of cooling in second, τ_n be the heating time constant in second, τ_C be the cooling time constant in second, and θ_f be the maximum permissible temperature raise of motor.

During on time: $\theta_h = \theta_f \left(1 - e^{(-t_1/\tau_n)}\right)$

$$\theta_h = \theta_f (1 - e^x),$$

$$\text{where } x = \frac{-t_1}{\tau_n}. \quad \textbf{(8.41)}$$

$$\text{During off time } \theta_C = \theta_h e^{-t_2/\tau_C} = \theta_h e^y, \quad \textbf{(8.42)}$$

$$\text{where } y = \frac{-t_2}{\tau_C}$$

Substituting Equation (8.41) in Equation (8.42):

$$\text{We get } \theta_C = \theta_f (1 - e^x) e^y. \quad \textbf{(8.43)}$$

Similarly, for the next intermittent loading:

During time: $\theta_h^1 = \theta_f - \{(\theta_f - \theta_C)e^x\}$

$$= \theta_f(1-e^x) + \theta_C e^x$$

$$= \theta_f(1-e^x) + \{\theta_f(1-e^x)\cdot e^y\}\cdot e^x \qquad [\because \text{from Equation (8.41)}]$$

$$\therefore \theta_h^1 = \theta_f(1-e^x)[1+e^x\cdot e^y]. \qquad \textbf{(8.44)}$$

During 'off' time: $\theta_C^1 = \theta_h^1 \cdot e^y$

$$\theta_C^1 = \{\theta_f(1-e^x)(1+e^x\cdot e^y)\}\cdot e^y. \qquad \textbf{(8.45)}$$

Similarly, for the next 'on' and 'off' periods:

During 'on' time: $\theta_h^2 = \theta_f - \{(\theta_f - \theta_C^1)e^x\}$

$$= \theta_f(1-e^x) + \theta_C^1 e^x$$

$$= \theta_f(1-e^x) + \{\theta_f(1-e^x)(1+e^x\cdot e^y)\}e^y\cdot e^x$$

[∴ form equation (7.47)]

$$= \theta_f(1-e^x)[1+(e^x\cdot e^y + e^{2x}\cdot e^{2y})]$$

$$= \theta_f(1-e^x)[1+(e^x\cdot e^y + e^{2x}\cdot e^{2y})] \qquad \textbf{(8.46)}$$

During 'off' time: $\theta_C^2 = \theta_h^2 e^y$

$$= \theta\cdot(1-e^x)(1+e^x\cdot e^y + e^{2x}\cdot e^{2y})\cdot e^y. \qquad \textbf{(8.47)}$$

Similarly, for '*n*' times intermittency:

$$\theta_h^{n-1} = \theta_f(1-e^x)[1+e^x e^y + e^{2x}\cdot e^{2y} + \cdots + e^{(n-1)x}\cdot e^{(n-1)y}]$$

$$= \theta_f(1-e^x)\left[\frac{1-e^{nx}\cdot e^{ny}}{1-e^x\cdot e^y}\right]. \qquad \textbf{(8.48)}$$

As $n \to \infty$ both e_{nx} and e_{ny} will be zero, as *x* and *y* are negative. If 'θ_m' be the maximum temperature with intermittent loading then:

$$\theta_m = \theta_f(1-e^x)\left[\frac{1-0}{1-e^x\cdot e^y}\right]$$

$$= \theta_f\left[\frac{1-0}{1-e^x\cdot e^y}\right].$$

By substituting *x* and *y* values in the above equations:

$$\therefore \theta_m = \theta_f\left[\frac{1-e^{-t_1/\tau_h}}{1-e^{[-t_1/\tau_h + t_2/\tau_C]}}\right]. \qquad \textbf{(8.49)}$$

8.11 RATING OF MOTOR

In cases, where the load fluctuates over a given cycle, as in rolling mills, etc., the raise of motor is determined accurately by finding the heating and cooling curves of motor, when working on given cycle. The various methods for determining the rating of motor for continuous duty and variable load are:

1. Equivalent current method.
2. Equivalent torque method.
3. Equivalent power method.

8.11.1 Equivalent current method

In this method, the actual current may be replaced by an equivalent current method (I_{eq}), which produces the same losses in the motor as the actual current.

$$I_{eq} = \sqrt{\frac{I_1^2 t_1 + I_2^2 t_2 + I_3^2 t_3 + \cdots + I_n^2 t_n}{t_1 + t_2 + \cdots + t_n}},$$

where $I_1, I_2, I_3, \ldots, I_n$ be the load currents within short intervals of $t_1, t_2, \ldots, t_n$ over a period of time 'T' seconds (Fig. 8.37).

8.11.2 Equivalent power method

In this method, if the load cycle is given in HP or kW verses time, then the motor rating can be directly found as follows (Fig. 8.38).

$$\text{Motor rating} = \sqrt{\frac{P_1^2 t_1 + P_2^2 t_2 + \cdots P_n^2 t_n}{t_1 + t_2 + \cdots + t_n}}.$$

Load changes uniformly; load cycle varies as shown in Fig. 8.39. The motor rating is given by:

$$\text{Motor rating} = \sqrt{\frac{1/3 P_1^2 t_1 + P_2^2 t_2 + P_0^2 t_3 + P_2^2 t_4 + 1/3 P_5^2 t_5}{t_1 + t_2 + t_3 + t_4 + t_5}}.$$

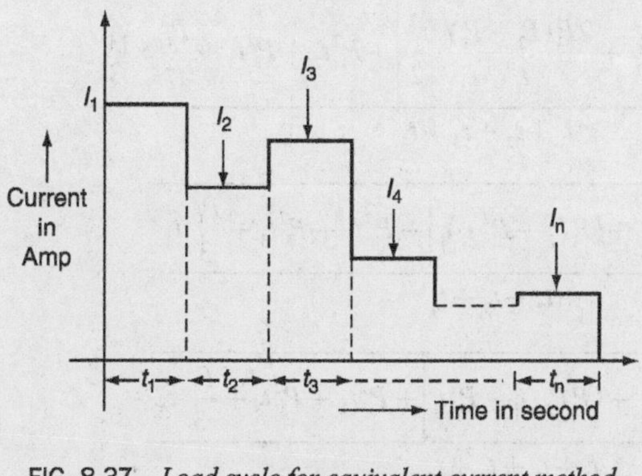

FIG. 8.37 *Load cycle for equivalent current method*

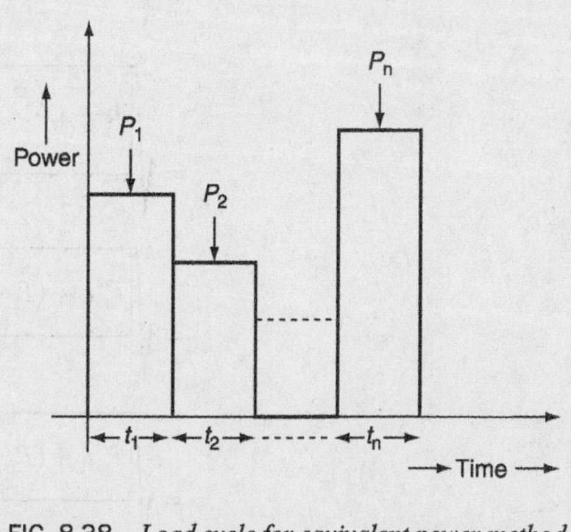

FIG. 8.38 *Load cycle for equivalent power method*

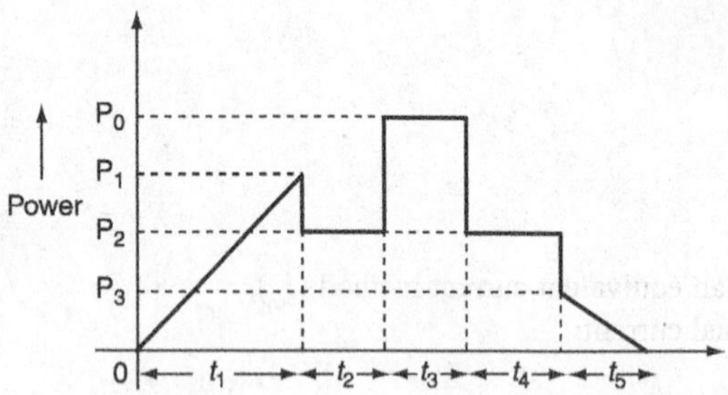

FIG. 8.39 *Load cycle for uniform load variation*

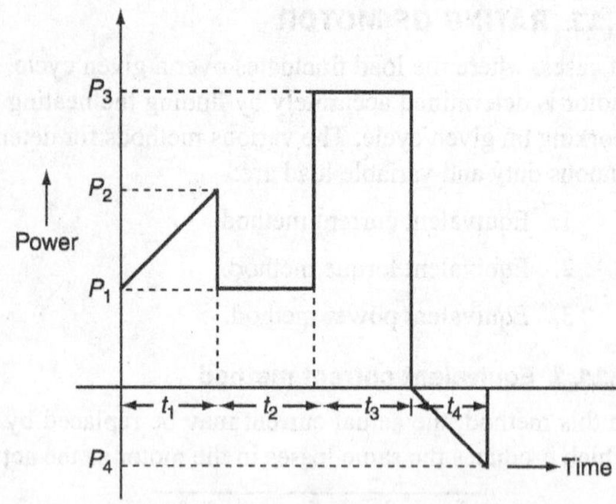

FIG. 8.40 *Load cycle for negative power*

Note: If the power, load, or torque changes uniformly, then $\int P^2 dt$ has to be taken for that period.

If the load curve consisting of negative power, i.e., power returned to the source, as shown in Fig. 8.40, the motor rating can be directly determined as follows.

$$\text{Motor rating} = \sqrt{\frac{\int_0^{t1}\left[P_1 + \frac{(P_2 - P_1)}{t_1}t\right]^2 dt + P_1^2 t_2 + P_3^2 t_3 + \int_0^{t4}\left[\left(\frac{P_4}{t_4} \times t\right)\right]^2 dt}{t_1 + t_2 + t_3 + t_4}}$$

(or)

$$= \sqrt{\frac{\int_0^{t1}\left[P_1^2 + \frac{(P_2 - P_1)^2}{t_1^2} \times t^2 + \frac{2P_1(P_2 - P_1)}{t_1} \times t\right] dt + P_1^2 t_2 + P_3^2 t_3 + \frac{P_4^2}{t_4^2} \times \frac{t^3}{3}\Big|_0^{t4}}{t_1 + t_2 + t_3 + t_4}}$$

$$= \sqrt{\frac{\left[P_1^2 t_1 + \frac{(P_2 - P_1)^2}{t_1^2}\frac{t_1^3}{3} + \frac{2P_1(P_2 - P_1)}{t_1}\cdot\frac{t_1^2}{2}\right] + P_1^2 t_2 + P_3^2 t_3 + \frac{P_4^2}{t_4^2} \times \frac{t_4^3}{3}}{t_1 + t_2 + t_3 + t_4}}$$

$$= \sqrt{\frac{\left[P_1^2 t_1 + (P_2 - P_1)^2 \frac{t_1}{3} + (P_1 P_2 - P_1^2)\cdot t_1\right] + P_1^2 t_2 + P_3^2 t_3 + \frac{P_4^2}{3} t_4}{t_1 + t_2 + t_3 + t_4}}$$

$$= \sqrt{\frac{\left[P_1^2 t_1 + (P_2 - P_1)^2 \frac{t_1}{3} + (P_1 P_2 \cdot t_1 - P_1^2 t_1\right] + P_1^2 t_2 + P_3^2 t_3 + \frac{P_4^2}{3} t_4}{t_1 + t_2 + t_3 + t_4}}$$

$$= \sqrt{\frac{\left(P_1^2 + P_2^2 - P_1P_2\right)\frac{t_1}{3} + P_1P_2t_1 + P_1^2t_2 + P_3^2t_3 + \frac{P_4^2}{3}t_4}{t_1 + t_2 + t_3 + t_4}}$$

$$= \sqrt{\frac{\left(P_1^2 + P_2^2\right)\frac{t_1}{3} + P_1P_2\left(1 - 2/3\right)t_1 + P_1^2t_2 + P_3^2t_3 + \frac{P_4^2}{3}t_4}{t_1 + t_2 + t_3 + t_4}}$$

$$= \sqrt{\frac{\left(P_1^2 + P_2^2\right)^2\frac{t_1}{3} + P_1P_2\left(t_1/3\right) + P_1^2t_2 + P_3^2t_3 + \frac{P_4^2}{3}t_4}{t_1 + t_2 + t_3 + t_4}}$$

$$= \sqrt{\frac{\left(\left(P_1^2 + P_2^2\right)^2 + P_1P_2\right)\frac{t_1}{3} + P_1^2t_2 + P_3^2t_3 + \frac{P_4^2}{3}t_4}{t_1 + t_2 + t_3 + t_4}}$$

$$= \sqrt{\frac{\left(P_1^2 + P_1P_2 + P_2^2\right)\frac{t_1}{3} + P_1^2t_2 + P_3^2t_3 + \frac{P_4^2}{3}t_4}{t_1 + t_2 + t_3 + t_4}}.$$

8.11.3 Equivalent torque method

This method is used to compute the motor heating rating effect, for short time and intermittent loads where the torque is varying as shown in Fig. 8.41.

In Fig. 8.41, T_1, T_2, T_3, T_4, and T_5 be the load torques develop during the periods t_1, t_2, t_3, t_4, and t_5 seconds now the equivalent torque can be calculated by considering time for one complete cycle and RMS value of load torques at different times.

$$\therefore \text{Equivalent torque } (T) = \sqrt{\frac{T_1t_1^2 + T_2t_2^2 + T_3t_3^2 + T_4t_4^2 + T_5t_5^2}{t_1 + t_2 + t_3 + t_4 + t_5}}.$$

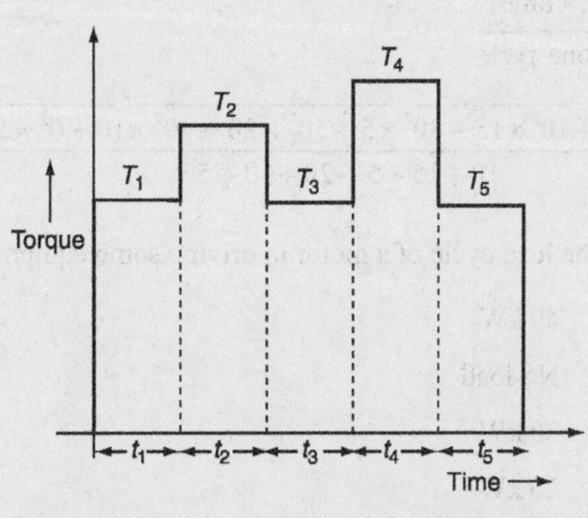

FIG. 8.41 *Load cycle for equivalent torque method*

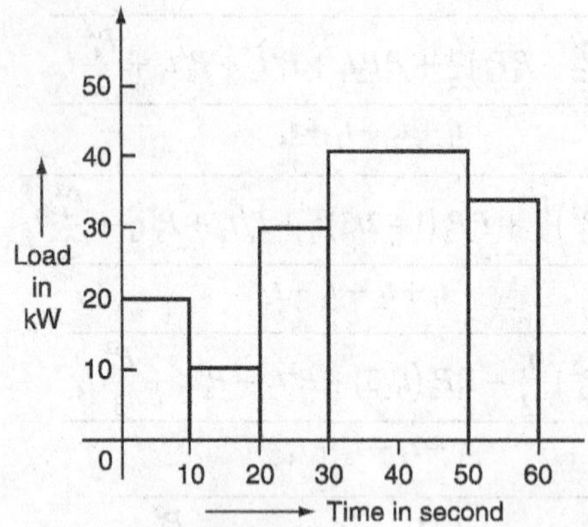

FIG. P.8.8 *Load cycle*

Example 8.38: A motor operates continuously on the following load cycle.

20 kW for 10 sec,

10 kW for 15 sec,

30 kW for 5 sec,

50 kW for 20 sec,

40 kW for 10 sec,

and idle for 5 sec.

Draw the load diagram and find the size of the motor required.

Solution:

The rating of the motor = RMS value of the load (Fig. P.8.8).

$$= \sqrt{\frac{\Sigma(\text{kW}^2 \times \text{time})}{\text{time for one cycle}}}$$

$$= \sqrt{\frac{20^2 \times 10 + 10^2 \times 15 + 30^2 \times 5 + 50^2 \times 20 + 40^2 \times 10 + 0^2 \times 5}{10 + 15 + 5 + 20 + 10 + 5}} = 34.19 \text{ kW}.$$

Example 8.39: The load cycle of a motor in driving some equipment is as follows.

0–3 min	40 kW
3–7 min	No-load
7–12 min	30 kW
12–15 min	20 kW
15–18 min	50 kW.

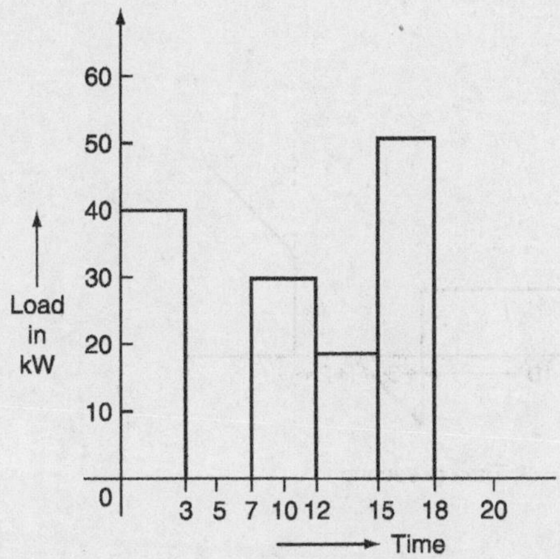

FIG. P.8.9 *Load cycle*

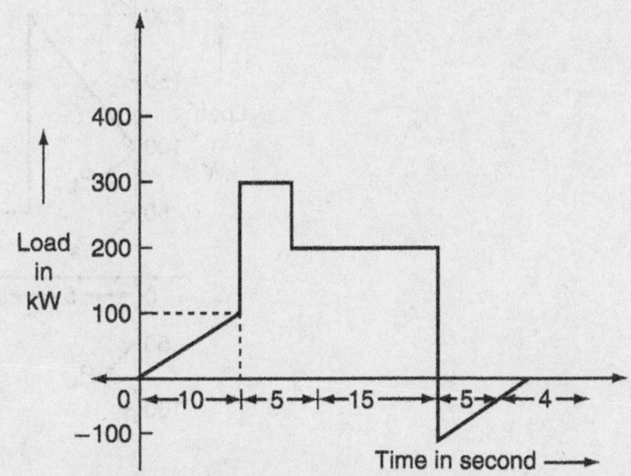

FIG. P.8.10 *Load cycle*

The load repeated indefinitely. Draw the load cycle and suggest suitable continuous rating of the motor.

Solution:
From Fig. P.8.9,

$$\text{Motor rating} = \sqrt{\frac{(40)^2\times3+0^2\times4+30^2\times5+20^2\times3+50^2\times3}{3+4+5+3+3}}$$

$$= 31.622\text{ kW} \cong 32\text{ kW}.$$

Example 8.40: A motor has to perform the following load cycle:

Load raising uniformly from 0 to 100 kW in 10 s.

Constant load 300 kW for 5 sec.

Constant load 200 kW for 15 sec.

Regenerative braking power returned falling uniform from 50 to 0 kW in 5 s. Decking period 4 s, motor stationary. Draw the load cycle and suggest a suitable continuous rated motor.

Solution:
From Fig. P.8.10,

$$\text{Motor rating} = \sqrt{\frac{1/3\times100^2\times10+300^2\times5+200^2\times15+\frac{1}{3}\times(-50)^2\times5+0^2\times4}{10+5+15+5+4}}$$

$$= \sqrt{\frac{1{,}087{,}500}{34}} = 178.84\text{ kW} \cong 179\text{ kW}.$$

Example 8.41: A motor has the following load cycle.

Load raising uniformly from 100 to 200 kW in 5 s.

Continuous load 50 kW for 10 s regenerative braking kW returned to the supply 50 kW to 0 kW for 3 s and idle for 2 s.

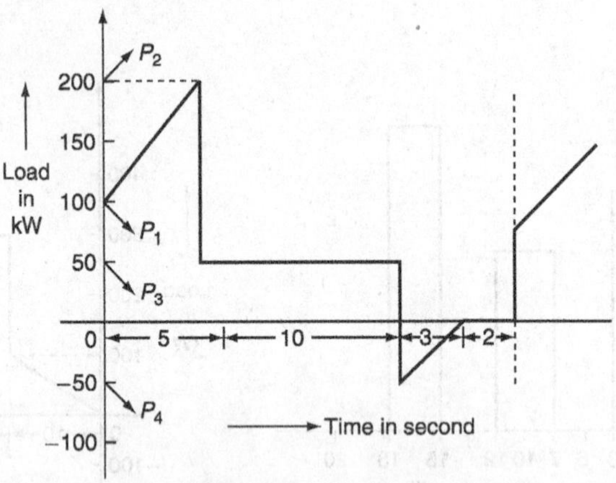

FIG. P.8.11 *Load cycle*

Draw the load diagram neatly for one cycle. Find the size of continuously rated motor for the above duty. The load cycle is repeated indefinitely.

Solution:

$$(\text{Motor rating})^2 = \left[\frac{\int_0^{t1}\left[P_1+\frac{(P_2-P_1)}{t_1}\cdot t\right]^2 dt + P_3^2 t_2 + \frac{1}{3}P_4^2 t_3 + P_5^2 \times t_4}{T}\right].$$

From the load curve (Fig. P.8.11),

$P_1 = 100$ kW, $t_1 = 5$ s

$P_2 = 200$ kW, $t_2 = 10$ s

$P_3 = 50$ kW, $t_3 = 3$ s

$P_4 = -50$ kW, $t_4 = 2$ s

$P_5 = 0$ kW.

$$(\text{Motor rating})^2 = \frac{\int_0^5 \left(100+\frac{(200-100)}{5}t\right)^2 dt + 50^2\times 10 + \frac{1}{3}(-50)^2\times 3 + 0^2\times 2}{5+10+3+2+0}$$

$$= \frac{\int_0^5 (100+20t)dt + 25{,}000 + 2{,}500}{20}$$

$$= \frac{\int_0^5 (100)^2 + 400t^2 + 4{,}000t)dt + 25{,}000 + 2{,}500}{20}$$

$$= \frac{\left[100^2 t + \frac{400t^3}{3} + 4{,}000\frac{t^2}{2}\right]_0^5 + 27{,}500}{20}$$

$$= \frac{100^2 \times 5 + 400 \times \frac{5^3}{3} + 4{,}000 \times \frac{5^2}{2} + 27{,}500}{20}$$

$$= \frac{5 \times 10^4 + 16666.67 + 50{,}000 + 27500}{20}$$

$$= 7{,}208.3.$$

$$\therefore \text{ Motor rating } = \sqrt{7208.33} = 84.90 \text{ kW} \cong 85 \text{ kW}.$$

Alternative method:

$$(\text{Motor rating})^2 = \frac{1/3\left(P_1^2 + P_1P_2 + P_2^2\right)t_1 + P_3^2 t_2 + \frac{1}{3}P_4^2 t_3}{T}$$

$$= \frac{1/3\left(100^2 + 100 \times 200 + 200^2\right) \times 5 + 50^2 \times 10 + 1/3 \times (-50)^2 \times 3}{20}$$

$$= 5{,}541.66.$$

$$\therefore \text{ Motor rating} = \sqrt{5541.66} = 74 \text{ kW}.$$

8.12 LOAD EQUALIZATION

The load fluctuations take place in many of the industrial drives such as rolling mills, planning machines presses, and reciprocating pumps, where the load on the motor varies widely within a span of few seconds. The sudden and peak load requires very large current from the supply results high voltage drop in the system or alternately would require very large size of cables. It is very essential to smooth out fluctuating load is known as '*load equalization*'. The load equalization involves the storage of energy during the off-peak period and gives out during the peak load period.

Load equalization process is commonly achieved by means of a flywheel. A flywheel is nothing but a big wheel that is mounted on the same shaft of motor, if the speed of the motor is not to be reversed or a heavy rotating body that acts as a reservoir for absorbing and redistributing stored energy is also known as flywheel.

8.12.1 Function of flywheel

To operate the flywheel efficiently, the driving motor should have drooping speed characteristics. The various models of flywheel are shown in Fig. 8.42 (a) and (b). During the lightload, the acceleration of the flywheel is increased and it stores the kinetic energy and at the time of peak load, the flywheel slows down and the stored kinetic energy is given out to the load; so that, the demand of the load from the motor or supply is reduced.

It is necessary that the motor used for load equalization should have drooping characteristics. The flywheel is not used with motors having constant speed for example

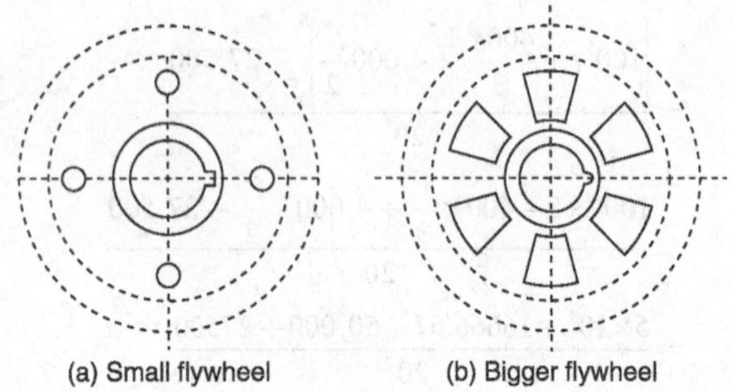

FIG. 8.42 *Flywheel*

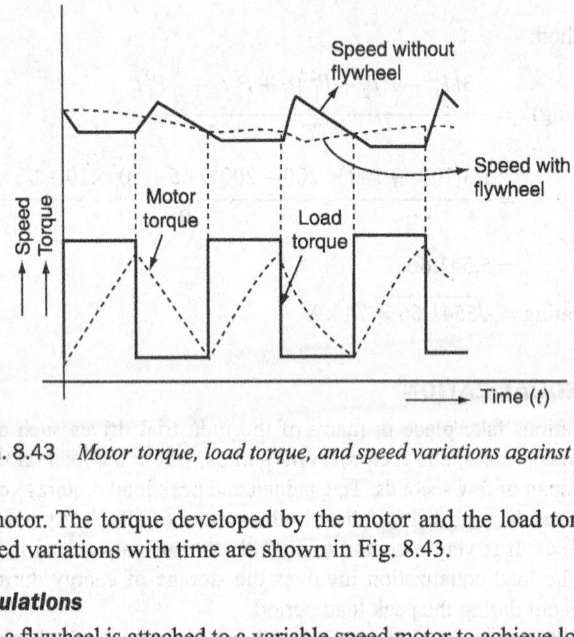

FIG. 8.43 *Motor torque, load torque, and speed variations against time*

synchronous motor. The torque developed by the motor and the load torque required as well as the speed variations with time are shown in Fig. 8.43.

Flywheel calculations

Let us consider a flywheel is attached to a variable speed motor to achieve load equalization.

Let T_L be the load torque (assumed constant during particular interval) in N-m. T_M is the motor torque in N-m, T_F is the flywheel torque in N-m, T_0 is the no-load torque in N-m, ω_0 is the motor speed on no-load in rad/sec, ω is the motor speed at any instant in rad/sec, and J is the moment of inertia of flywheel in kg-m².

$$S = (\omega_0 - \omega) = \text{motor slip.}$$

Case (i): Let us consider that the load on the motor is increasing; during this period, the flywheel will decelerate and impart its stored kinetic energy to the load. The torque required to be supplied by the motor:

$$T_M = T_L - T_F \qquad \textbf{(8.50)}$$

The kinetic energy given by the flywheel when its speed reduced from ω_0 to ω is:

$$\begin{aligned}\text{KE} &= \frac{1}{2}J\left(\omega_0{}^2 - \omega^2\right)\\ &= \frac{1}{2}J\left(\omega_0 + \omega\right)\left(\omega_0 - \omega\right)\\ &= J\left(\frac{\omega_0 + \omega}{2}\right)\left(\omega_0 - \omega\right).\end{aligned} \quad \textbf{(8.51)}$$

Let $\left(\dfrac{\omega_0 + \omega}{2}\right) = \omega$ (mean speed)

$\omega_0 - \omega = S$ (Slip).

Then, Equation (8.51) becomes:

$$\text{KE} = J\omega S. \quad \textbf{(8.52)}$$

The power given out by the flywheel = the rate of change of the energy given up by the flywheel.

$$\begin{aligned}&= \frac{\text{d}}{\text{d}t}(J\omega S)\\ &= J\omega\frac{\text{d}S}{\text{d}t}.\end{aligned} \quad \textbf{(8.53)}$$

$$\begin{aligned}\text{The flywheel torque } (T_\text{F}) &= \frac{\text{power given out by flywheel}}{\omega}\\ &= \frac{J\omega\left(\dfrac{\text{d}S}{\text{d}t}\right)}{\omega}\\ &= J\frac{\text{d}S}{\text{d}t}.\end{aligned} \quad \textbf{(8.54)}$$

By substituting Equation (8.54) in Equation (8.50), we get:

$$\begin{aligned}T_\text{M} &= T_\text{L} - T_\text{F}\\ &= T_\text{L} - J\frac{\text{d}S}{\text{d}t}.\end{aligned} \quad \textbf{(8.55)}$$

If the slip, i.e., drop in speed limited to 10%, then the slip is proportional to the motor torque:

i.e., $S \propto T_\text{M}$

$\therefore S = KT_\text{M}.$

Then, $T_\text{M} = T_\text{L} - J\dfrac{\text{d}(KT_\text{M})}{\text{d}t}$

$$T_M = T_L - JK\frac{dT_M}{dt}$$

$$T_L - T_M = JK\frac{dT_M}{dt}$$

$$\frac{dT_M}{T_L - T_M} = \frac{dt}{JK}. \tag{8.56}$$

Integrating the Equation (8.56):

$$\int\frac{dT_M}{T_L - T_M} = \int\frac{dt}{JK}$$

$$-\log_e(T_L - T_M) = \frac{t}{JK} + C, \tag{8.57}$$

where C is proportionality constant.

At time $t = 0$, the motor torque will be equals to the no-load torque:

$$\text{i.e., at } t = 0, \quad T_M = T_0. \tag{8.58}$$

The value of 'C' can be determined by using the initial conditions. Substituting Equation (8.58) in Equation (8.57):

$$-Log_e(T_L - T_0) = \frac{0}{JK} + C \tag{8.59}$$

$$\therefore C = -\log_e(T_L - T_0).$$

Substituting 'C' value in Equation (8.57):

$$\therefore -\text{Log}_e(T_L - T_M) = \frac{t}{JK} - \text{Log}_e(T_L - T_0)$$

$$-\text{Log}_e(T_L - T_0) + \log_e(T_L - T_M) = \frac{-t}{JK}$$

$$\text{Log}_e\left[\frac{T_L - T_M}{T_L - T_0}\right] = \frac{-t}{JK}.$$

Applying exponentials on both sides:

$$\left[\frac{T_L - T_M}{T_L - T_0}\right] = e^{-t/JK}$$

$$T_L - T_M = (T_L - T_0)e^{-t/JK}. \tag{8.60}$$

Case (ii): Now consider that the load is totally removed or decreasing, the motor starts accelerating and so the KE is stored by the flywheel.

Hence, the flywheel regains its normal speed; therefore, the slip decreases, i.e., $\frac{dS}{dt}$ is negative.

Now, motor torque will be:

$$T_M = T_0 + T_F. \quad \textbf{(8.61)}$$

But,

$$T_F = -J\frac{dS}{dt}. \quad \textbf{(8.62)}$$

Substitute Equation (8.62) in Equation (8.61):

$$\therefore T_M = T_0 - J\frac{dS}{dt}. \quad \textbf{(8.63)}$$

We know that $S \propto T_M$:

$$S = KT_M$$

$$\therefore T_M = T_0 - JK\frac{dT_M}{dt}$$

$$-JK\frac{dT_M}{dt} = T_M - T_0$$

$$\frac{dT_M}{T_M - T_0} = -\frac{dt}{JK}.$$

Integrating on both sides:

$$\int \frac{dT_M}{T_M - T_0} = -\int \frac{dt}{JK}$$

$$\text{Log}_e(T_M - T_0) = \frac{-t}{JK} + C_2, \quad \textbf{(8.64)}$$

where 'C_2' is integration constant.

The value of constant can be obtained by substituting the initial conditions in Equation (8.64).

At $t = 0$; $T_M = T_M^1$ (motor torque when load is decreased)

$$\therefore \text{Log}\left(T_M^1 - T_0\right) = \frac{0}{JK} + C_2$$

$$\therefore C_2 = \text{Log}_e\left(T_M^1 - T_0\right).$$

By substituting 'C_2' in Equation (8.64), we get:

$$\text{Log}_e(T_M - T_0) = \frac{-t}{JK} + \text{Log}_e\left(T_M^1 - T_0\right)$$

$$\text{Log}_e(T_M - T_0) - \text{Log}_e\left(T_M^1 - T_0\right) = \frac{-t}{JK}$$

$$\text{Log}_e\left(\frac{T_M - T_0}{T_M^1 - T_0}\right) = \frac{-t}{JK}.$$

Applying exponentials on both sides:

$$\frac{T_M - T_0}{T_M^1 - T_0} = e^{-t/JK}$$

$$\therefore T_M - T_0 = \left(T_M^1 - T_0\right)\left(e^{-t/JK}\right)$$

$$\therefore T_M = T_0 + \left(T_M^1 - T_0\right)e^{-t/JK}. \quad \textbf{(8.65)}$$

Example 8.42: A 15-HP, three-phase, eight-pole, and 50-Hz induction motor provided with a flywheel has to supply a load torque of 600 N-m for 10 s followed by a no-load during which the flywheel regains the full speed. The full-load slip of the motor is 4% and the torque–speed curve may be assumed linear over the working range. Find the moment of inertia of the flywheel if the motor torque is not to exceed twice the full-load torque.

Solution:

Given data:

$P_0 = 15$ HP

$= 15 \times 735.5 = 11.03$ kW.

No. of poles $P = 8$

$f = 50$ Hz

$S_f = 0.04$

$t = 10$ sec

$T_L = 600$ N-m

$T_M = 2.\ TFL$

$T_0 = 0.$

Now, synchronous speed $N_s = \frac{120f}{P}$

$= \frac{120 \times 50}{8} = 750$ rpm.

Full-load torque $T_{FL} = \frac{60 \times P_0}{2\pi N_{FL}}$

$N_{FL} = N_s(1 - S_f) = 750\ (1 - 0.04)$

$= 720$ rpm.

$$T_{FL} = \frac{60 \times 11.03 \times 10^3}{2\pi \times 720} = 146.39 \text{ N-m.}$$

$\therefore T_M = 2T_{FL} = 2 \times 146.39 = 292.78$ N-m.

Slip speed $= S_f \times N_s = 0.04 \times 750$

$$= 30 \text{ rpm}$$

$$= \frac{30 \times 2\pi}{60} = 3.14 \text{ rad/s.}$$

And, $K = \frac{S}{T_{FL}} = \frac{3.14}{146.39} = 0.0214$

$$\therefore T_M = T_L - (T_L - T_0)e^{-t/JK}$$

$$-t/JK = \ln\left(\frac{T_L - T_M}{T_L - T_0}\right)$$

$$-t/JK = \ln\left[\frac{600 - 292.78}{600}\right] = 0.669$$

$$J = \frac{t}{0.669 \times K} = \frac{10}{0.669 \times 0.0914} = 698.49 \text{ kg-m}^2.$$

Example 8.43: A motor fitted with a flywheel that supplies a load of torque 800 N-m for 5 s. During no-load period, the flywheel regains its original speed. The motor torque is required to be limited to 600 N-m. The no-load speed of the motor is 650 rpm and its full-load speed slip is 10%; determine the moment of inertia of the flywheel.

Solution:

Given data:

$T_L = 800$ N-m

$T_0 = 0$

$T_M = 600$ N-m

$N = 600$ rpm

$t = 5$ sec

$S_f = 10\% = 0.1.$

$\therefore$ Slip speed $= S_f \times N$

$$= 0.1 \times 650 = 65 \text{ rpm}$$

$$= \frac{65 \times 2\pi}{60} = 6.806 \text{ rad/s.}$$

We know that:

$$S \propto T_M$$

$$S = KT_M$$

$$\therefore \quad K = \frac{S}{T_M} = \frac{6.806}{600} = 0.01134.$$

Now:

$$T_M = T_L - (T_L - T_0)e^{-t/KJ}$$

$$e^{-t/JK} = \frac{T_L - T_M}{T_L - T_0}$$

$$e^{-t/JK} = \frac{800-600}{800-0} = 0.25$$

$$-t/JK = \ln(0.25) = \frac{5}{1.386 \times 0.01134}$$

$$J = 318.12 \text{ kg-m}^2.$$

Example 8.44: A 3-ϕ, 25-kW, six-pole, and 940-rpm induction motor has a constant load torque of 350 N-m and at wide intervals additional torque of 1,100 N-m for 8 s.

Calculate:

(i) The moment of the inertia of the flywheel used for the load equalization, if motor torque is not to exceed twice the rated torque.

(ii) Time taken after the removal of additional load, before the motor torque becomes 500 N-m.

Solution:

Given data:

$$P_0 = 25 \text{ kW}$$

$$P = 6$$

$$N = 940 \text{ rpm}$$

$$T_L = 350 + 1{,}100 = 1{,}450 \text{ N-m}$$

$$T_M = 2\, T_{FL}$$

$$T_{FL} = \frac{P_0}{(2\pi N/60)} = \frac{25 \times 10^3}{(2\pi \times 940/60)}$$

$$= 253.9 \text{ N-m}$$

$$T_M = 2T_{FL} = 2 \times 253.9 = 507.94 \text{ N-m}$$

$$N_s = \frac{120f}{P} = \frac{120 \times 50}{6} = 1{,}000 \text{ rpm.}$$

Full-load slip = 1,000 − 940 = 60 rpm = 0.06.

But, $S \propto T_{FL}$

$$S = KT_{FL}.$$

$$\text{Slip speed} = \frac{2\pi \times 60}{60} = 6.283$$

$$K = \frac{S}{T_{FL}} = \frac{6.283}{253.9} = 0.024$$

$$T_M = T_L - (T_L - T_0)e^{-t/JK}$$

$$-t/JK = \ln\left[\frac{T_L - T_M}{T_L - T_0}\right]$$

$$-t/JK = \ln\left[\frac{1450 - 507.94}{1450 - 0}\right]$$

$$-t/JK = \ln(0.649) = -0.431$$

$$J = \frac{t}{0.431 \times K} = \frac{8}{0.431 \times 0.024}$$

$$J = 773.39 \text{ kg-m}^2.$$

(ii) Time taken after removal of additional load:

$$T_M = T_0 + \left(T_M^1 - T_0\right)e^{-t/JK}.$$

Now, $T_M = 500$ N-m

$$T_M^1 = 507.94 \text{ N-m}$$

$$T_0 = 350 \text{ N-m}$$

$$500 = 350 + (507.94 - 350)e^{-t/JK}$$

$$0.997 = e^{-t/JK}$$

$$t = 0.05 \times JK$$

$$t = 0.05 \times 773.39 \times 0.024$$

$$t = 0.9409 \text{ s.}$$

KEY NOTES

- The functions of *power modulators* are:
  - It modulates the flow of power from the source to the motor is impart speed–torque characteristics required by the load.
  - It regulates source and motor currents within permissible values, such as starting, braking, and speed reversal conditions.
  - Selects the mode of the operation of motor, i.e., motoring or braking.
  - Converts source energy in the form suitable to the motor.
- The varies types of *electric drives* are:
  - (i) Group drives.
  - (ii) Individual drives.
  - (iii) Multi-motor drives.
- Electric drive is used to driven one or more than two machines from line shaft through belts and pulleys, is known as group drive.
- A single electric motor is used to drive one individual machine is known as individual drive
- *Multi-motor drives* several separate motors are provided for operating different parts of the same machine.
- The *choice of motors* the following factors must be taken into consideration:
  - (i) Cost.
  - (ii) Electric characteristics.
  - (iii) Mechanical characteristics.
  - (iv) Size and vetting of motors.
  - (v) Type of drive.
- *Starting characteristics* are:
  - (i) Torque vs. armature current (T vs I_a).
  - (ii) Speed vs. armature current (N vs I_a).
  - (iii) Output vs. armature current.
- *Maximum torque* corresponding to slip is: $S_m = R_2/X_2$.
- The ratio of the full-load torque to the maximum torque is:

$$\frac{T_{FL}}{T_m} = \frac{S_F}{S_m}\left[\frac{2S_m^2}{S_f^2 + S_m^2}\right].$$

- The ratio of the starting torque to the maximum torque is:

$$\frac{T_{st}}{T_m} = \frac{2S_m}{S_m^2 + 1}.$$

- The speed control of DC motors are:
  - i. Field control or flux control method.
  - ii. Armature control method.
  - iii. Applied voltage control.
- The speed controls of the three-phase induction motor on stator side are:
  - i. Supply frequency control.
  - ii. Supply voltage control.
  - iii. Controlling the number of stator poles.
- The speed controls of the three-phase induction motor on rotor side are:
  - i. Adding external resistance in the rotor circuit.
  - ii. Cascade control.

SHORT QUESTIONS AND ANSWERS

1. Mention any two advantages of electric drives.

   There are a number of inherent advantages that the electric drive possesses over the other forms of conventional drives are:

   - They have comparatively long life as compared to mechanical drive.
   - It is cleaner, as there are no flue gases, etc.
   - It is more economical.
   - They have flexible control characteristics.

2. What are the types of electric drives?

   Depending on the type of equipment used to run the electric motors in industrial purpose may be classified into three types such as:

   (i) Group drives.

(ii) Individual drives.

(iii) Multi-motor drives.

3. Define group drive.

Electric drive used to drive one or more than two machines from line shaft through belts and pulleys, is known as group drive.

4. What are the advantages of group drive?

- If it is operated at the rated load, the efficiency and the power factor of the large group drive motor will be high.
- The maintenance cost of a single large capacity motor is less than a number of small capacity motors.
- It is used for the processes where the stoppage of one operation necessitates the stoppages of the sequence of the operations as in case of the textile mills.
- It has overload capacity.

5. Define individual drive.

In individual drive, a single electric motor is used to drive one individual machine is know as individual drive.

6. What are the speed control methods of DC motors?

(i) Field control or flux control method.

(ii) Armature control method.

(iii) Applied voltage control.

7. What is the advantages of Ward–Leonard speed control system?

(i) Accurate speed control.

(ii) Bi-directional speed control is possible.

8. Give the expression for temperature raise of an electric motor.

$$\theta = \theta_f \left[1 - e^{-\frac{-t}{T_n}}\right],$$

where 'T_n' $= \dfrac{MS}{A\lambda}$ is known as the heating time constant of the motor.

9. Define heating time constant.

The heating time constant is the time taken by the machine to attain 63.2% of its final steady temperature rise (θ_f).

10. Define cooling time constant.

The cooling time constant is defined as the time required cooling the machine down to 36.8% of the initial temperature raise above ambient temperature.

11. What is meant by duty cycle?

In case the load and torque verses time variation is periodic and repetitive, such one cycle of the variation of the load with the time is known as load or duty cycle.

12. Give a couple of examples for impact loads.

Rolling mills and shearing machines.

13. What is a flywheel?

A flywheel is nothing but a big wheel that is mounted on the same shaft of motor; if the speed of the motor is not to be reversed or a heavy rotating body that acts as a reservoir for absorbing and redistributing stored energy is also known as flywheel.

14. What is meant by load equalization?

The sudden and peak load require very large current from the supply results high voltage drop in the system or alternately would require very large size of cables. It is very essential to smooth out fluctuating load is known as 'load equalization'.

15. What are the various methods that are used to determine the rating and size of electric motor?

The various methods that are used for determining the rating of the motor for the continuous duty and the variable load are:

(i) Equivalent current method.

(ii) Equivalent torque method.

(iii) Equivalent power method.

MULTIPLE-CHOICE QUESTIONS

1. The main drawback of electric drive is that:

(a) It is cumbersome drive.

(b) The initial as well as the maintenance costs are costlier.

(c) The electrical power supply failure makes the drive standstill.

(d) All of the above.

2. An existing workshop is to be changed over from an engine drive to an electric drive. The type of drive to be adopted is:
   (a) Individual drive.
   (b) Group drive.
   (c) Multi-motor drive.
   (d) Any of the above
3. Group drive has become obsolete nowadays because of:
   (a) High noise level, untidy look, the loss of the flexibility of layout, the cumbersome speed control of individual machine, less safe operation, low operation efficiency.
   (b) The stoppage of all operations on the break-down of large single motor.
   (c) The high initial cost of the motor and the control gear.
   (d) Both (a) and (b).
4. Which of the following alternatives will be cheaper in initial cost?
   (a) One motor of 100 kW.
   (b) Four motors of 25 kW each.
   (c) Five motors of 20 kW each.
   (d) 10 motors of 10 kW each.
5. The type of drive used for a paper mill requiring a constant speed operation and the flexibility of control is:
   (a) Group drive.
   (b) Multi-motor drive.
   (c) Individual or multi-motor drive.
   (d) Individual drive.
6. In individual drive:
   (a) Each operator has complete control of each machine.
   (b) No-load losses are eliminated.
   (c) The initial cost is less than that of the group drive.
   (d) Both (a) and (b).
7. The selection of an electric motor is governed by:
   (a) The nature of load to be handled.
   (b) Environmental conditions.
   (c) The nature of electric supply available.
   (d) All of the above.
8. DC motors are not so widely used as AC ones because:
   (a) DC motors are usually more expensive than AC motors for similar operating conditions.
   (b) Additional equipment is required for converting the existing AC supply into a DC supply.
   (c) DC motors have commutators that subject to trouble resulting from sparking, brush wear, arc over, and the presence of moisture and destructive fumes in the surrounding air.
   (d) All of the above.
9. The least significant electrical characteristic in the selection of the electric motor for a flour mill is:
   (a) Starting characteristics.
   (b) Braking.
   (c) Running characteristics.
   (d) Efficiency.
10. The least significant feature while selecting a motor for centrifugal pump is:
    (a) Speed control.
    (b) Power rating of motor.
    (c) Operating speed.
    (d) Starting characteristics.
11. The load, for which the motor always starts on the load is:
    (a) Fan motor.
    (b) Conveyor motor.
    (c) Flour mill motor.
    (d) All of the above.
12. The load cycle for a motor driving a power press will be:
    (a) Continuous.
    (b) Variable.
    (c) Intermittent and variable.
    (d) Continuous but periodical.
13. The application(s) that need(s) frequent starting and stopping is/are:
    (a) Paper mills.
    (b) Lifts and hoists.
    (c) Blowers and fans.
    (d) Grinding mills.

14. The machine having heavy fluctuation of load is:
    (a) Lathe.
    (b) Planer.
    (c) Punching machine.
    (d) Printing machine.
15. Variable speed operation is preferred in:
    (a) Water pump.
    (b) Ceiling fan.
    (c) Exhaust fan.
    (d) Refrigerator.
16. The application in which the motor is to start with high acceleration is:
    (a) Lifts and hoists.
    (b) Centrifugal pump.
    (c) Oil expeller.
    (d) Flour mill.
17. While selecting a motor for an air-conditioner, the feature of utmost importance is:
    (a) The type of enclosure.
    (b) The type of bearing.
    (c) Noise.
    (d) Power transmission arrangement.
18. Belt conveyors offer:
    (a) High starting torque.
    (b) Medium starting torque.
    (c) Low starting torque.
    (d) Zero starting torque.
19. The characteristics(s) of the drive for crane hoisting and lowering is/are:
    (a) Fast speed control.
    (b) Smooth movement.
    (c) Precise control.
    (d) All of the above.
20. In case of centrifugal pumps, the starting torque is usually:
    (a) Less than running torque.
    (b) Same as running torque.
    (c) Slightly higher than running torque.
    (d) Double of running torque.
21. The starting torque of a DC motor is independent of:
    (a) Armature current.
    (b) Flux.
    (c) Speed.
    (d) All of the above.
22. The diameter of the rotor shaft for an electric motor depends upon:
    (a) Speed only.
    (b) Power output only.
    (c) Speed and power output.
    (d) Speed, power output, and power factor.
23. In series motor, the method used for controlling the flux per pole is/are:
    (a) Diverter field control.
    (b) Series-parallel control.
    (c) Tapped field control.
    (d) All of the above.
24. The DC series motors are very suitable for the heavy duty applications such as electric railways and rolling mills because of:
    (a) Low initial as well as maintenance cost.
    (b) High starting torque.
    (c) Possibility of speed control.
    (d) Nearly constant speed.
25. The motor having wider range of speed control is:
    (a) DC shunt motor.
    (b) Squirrel cage induction motor.
    (c) Synchronous motor.
    (d) Double squirrel cage induction motor.
26. DC shunt motor has:
    (a) High starting torque and it is suitable for heavy duty applications.
    (b) Almost constant speed.
    (c) Torque varying nearly as the square of the current.
    (d) Dangerously high speed at no load.
27. For smooth and precise speed control over a wide range, preferable motor is:
    (a) DC motor.
    (b) Squirrel cage induction motor.

(c) Synchronous motor.

(d) Single-phase induction motor.

28. For quick speed reversal, the motor preferred is:

(a) DC motor.

(b) Squirrel cage induction motor.

(c) Slip-ring induction motor.

(d) Synchronous motor.

29. Ward–Leonard controlled DC drives are usually used for:

(a) Light duty excavators.

(b) Medium duty excavators.

(c) Heavy duty excavators.

(d) All of the above.

30. A slip-ring induction motor is preferred over squirrel cage induction motor for the following main characteristics:

(a) Low starting current.

(b) High starting torque.

(c) Speed control over limited range.

(d) All of the above.

31. A slip-ring induction motor is preferred over the squirrel cage induction motor for the following main consideration(s):

(a) Low windage losses.

(b) High starting torque.

(c) Slow speed operation.

(d) Both (a) and (c)

32. The squirrel cage induction motors with high slip and the slip-ring induction motors develop maximum torque at standstill are used for:

(a) Elevators.

(b) Machine tools.

(c) Presses and punches.

(d) All of the above.

33. Which of the following motors has series characteristics?

(a) Capacitor start induction motor.

(b) Repulsion motor.

(c) Shaded-pole motor.

(d) Reluctance motor.

34. Which of the following motors has the least range of speed control?

(a) DC shunt motor.

(b) Wound rotor induction motor.

(c) Schrage motor.

(d) Synchronous motor.

35. The speed can be controlled by injecting emf in the rotor circuit in case of:

(a) Squirrel-cage induction motor.

(b) Wound rotor induction motor.

(c) Synchronous motor.

(d) Schrage motor.

36. The squirrel cage induction motor cannot be started by using:

(a) Resistance in rotor circuit.

(b) Resistance in stator circuit.

(c) Autotransformer starting.

(d) Star-delta starter.

37. In start-delta starting:

(a) Applied voltage across motor terminals is reduced.

(b) Starting current is reduced.

(c) Operation speed is reduced.

(d) Both (a) and (b).

38. The least expensive drive is:

(a) Belt drive.

(b) Rope drive.

(c) Chain drive.

(d) Gear drive.

39. 15-min rated motors are suitable for:

(a) Light-duty cranes.

(b) Medium-duty cranes.

(c) Heavy-duty cranes.

(d) All of the above.

40. In plugging of DC motors:

(a) Connection to armature are reversed.

(b) Connection to field are reversed.

(c) Connection to both armature and field are reversed.

(d) Connections to supply are reversed.

41. During the rheostatic braking of a DC motor:
    (a) Armature is reverse connected.
    (b) The direction of the field current is reversed.
    (c) The field is disconnected from the supply.
    (d) It is operated as a DC generator.
42. The rheostatic braking may be applied to an induction motor provided:
    (a) It is a squirrel cage type.
    (b) It Is a wound type.
    (c) Separate DC source for field excitation is available.
    (d) Variable external resistance is available.
43. During the regenerative braking:
    (a) The motors are disconnected from the supply.
    (b) The motors are reverse connected to the supply.
    (c) The motors are operated as generators.
    (d) The motors are made to come to standstill.
44. The regenerative braking:
    (a) Can be easily applied to the DC shunt motors.
    (b) Can be easily applied to the DC series motors.
    (c) Can be used for stopping the motor.
    (d) Cannot be used when the load on motor has overhauling characteristics.
45. Net energy saved during the regenerative braking of an electric train:
    (a) Is independent of the train weight.
    (b) Decreases with the reduction in the train sped due to braking.
    (c) Increase with the increase in the specific resistance.
    (d) Increases with the increase in the down gradient.
46. The motors used along with the flywheels for fluctuating loads are:
    (a) DC shunt motors.
    (b) DC cumulative compound motors and three-phase induction motors.
    (c) Synchronous motors.
    (d) All of the above.
47. Load equalization is desirable in the case of:
    (a) Very large refrigeration and air-conditioning plants.
    (b) Rolling mills, electric hammers, presses, and reciprocating pumps.
    (c) Lathes, wood-working machines, paper-making machines, shapers, and slotters.
    (d) Traveling cranes and lifts.
48. The motor preferred for the traction work is:
    (a) DC shunt motor.
    (b) DC series motor.
    (c) Plan-squirrel cage induction motor.
    (d) Synchronous motor.
49. The motors best suited for the rolling mills are:
    (a) DC shunt motors.
    (b) Plain-squirrel cage induction motors.
    (c) Synchronous motors.
    (d) Any of the above.
50. Motor preferred for kiln drive is usually:
    (a) Wound rotor induction motor.
    (b) Cascaded controlled AC motor.
    (c) Ward–Leonard controlled DC shunt motor.
    (d) Any of the above.
51. Centrifugal pumps are usually driven by:
    (a) DC series motors.
    (b) DC shunt motors.
    (c) Plain squirrel cage induction motors.
    (d) Any of the above.
52. The motor used in mines is:
    (a) Flame proof squirrel cage induction or wound rotor motor.
    (b) DC series motor.
    (c) DC shunt motor.
    (d) Any of the above.
53. The motor used in punches, presses, and shears is:
    (a) DC series or shunt motors.
    (b) DC cumulative compound motor.
    (c) High slip squirrel cage or wound rotor induction motor.
    (d) Both (b) and (c).

REVIEW QUESTIONS

1. Compare and contrast slip ring and squirrel cage induction motor from application viewpoint.
2. Discuss the advantages and the disadvantages of the electrical drive over the other drives.
3. Compare group drives and individual drives.
4. Explain characteristics of DC shunt motor.
5. Explain characteristics of DC series motor.
6. What are the various speed–control methods of DC motors?
7. Derive an expression for temperature raise of an electric motor.
8. What is load equalization? How it achieved?
9. Explain how would you estimate the rating of motor for the intermittent duty cycle?
10. Explain how the maximum torque can be obtained at the time of stating of a three-phase slip ring induction motor.
11. Discuss the various factors that govern the size and the rating of a motor for particular service.
12. Discuss the various losses that occur in insulating materials and how they can be reduced?

EXERCISE PROBLEMS

1. A 3-ϕ induction motor has a ratio of maximum torque to full-load torque as 3:2. Determine the ratio of actual starting torque to full-load torque for $Y-\Delta$ starting. Given $R_2 = 0.32\ \Omega$ and $X_2 = 3\ \Omega$.
2. A 40-kVA, 440-V, 3-ϕ, and 50-Hz squires cage induction motor has full-load slip of 5%. Its standstill impedance is 0.7Ω/phase. It is started using a tapped autotransformer. Calculate the tap position and the ratio of starting torque to full load. It the maximum allowable supply current at the time of starting is 100 A.
3. The armature and the field resistances of a 230-V DC shunt motor are 0.35 and 150 Ω, respectively. When driving a load of constant torque at 600 rpm, the armature current is 25 A. If it is desired to raise the speed from 700 to 1,000 rpm, what resistance should be inserted in the field circuit? Assume that the magnetic circuit is unsaturated.
4. The speed of a 25-HP (metric) and 440-V DC shunt motor is to be reduced by 35% by the use of a controller. The field current is 3 A and the armature resistance is 0.4 Ω. Calculate the resistance of the controller, if the torque remains constant and the efficiency is 70%.
5. A 230-V, and 15-HP (metric) shunt motor has the field and the armature resistance as of 150 Ω. Calculate the resistance to be inserted in the armature circuit to reduce the speed to 600 rpm from 1,000 rpm, if the full-load efficiency is 85% and the torque varied as the square of the speed.
6. A 400-V series motor takes a line current of 65 A and runs at a speed of 800 rpm. What resistance should be connected in series with the armature to reduce the speed to 400 rpm. The load torque at this new speed is 70% of its previous value. The resistance of the armature and the series field are 0.04 and 0.02 Ω, respectively. Assume that flux is proportional to the load.
7. A six-pole, 50-Hz, and 3-ϕ induction motor is running at 3% slip when delivering full-load torque. It has the standstill rotor resistance of 0.25 Ω and the reactance of 0.7 Ω per phase. Calculate the speed of the motor if an additional resistance of 0.25 Ω per phase is inserted in the rotor circuit. The full-load torque remains constant.
8. A cascade is consists of two motors A and B with six and eight poles, respectively. The motor is connected to a 50-Hz supply. Find (i) the speed of the set and (ii) the electric power transferred to motor B when the input to motor A is 50-kW neglect losses.
9. A 15-kW motor has a heating time constant and cooling time constant are 50 and 80 min, respectively. The final temperature attained is 50°C. Find the temperature of the motor after 45 min full-load run and then switched off for 40 min.
10. A motor fitted with a flywheel supplies a load of torque 900 N-m for 6 s. During no-load period, the flywheel regains its original speed. The motor torque is required to be limited to 700 N-m. The no-load speed of the motor is 750 rpm and its full-load speed slip is 15%. Determine the moment of inertia of the flywheel.

ANSWERS

1. c	15. b	29. c	43. c
2. b	16. a	30. d	44. a
3. d	17. c	31. b	45. d
4. a	18. a	32. c	46. b
5. c	19. d	33. b	47. b
6. d	20 a	34. d	48. b
7. d	21. c	35. d	49. a
8. d	22. c	36. a	50. d
9. b	23. d	37. d	51. c
10. a	24. b	38. a	52. a
11. d	25. a	39. a	53. d
12. c	26. b	40. a	
13. b	27. a	41. d	
14. c	28. a	42. c	

CHAPTER 9

Electric Traction-I

OBJECTIVES

After reading this chapter, you should be able to:

- know about the various traction systems
- understand the various track electrifications
- analyze the different braking methods of traction motors

9.1 INTRODUCTION

The system that causes the propulsion of a vehicle in which that driving force or tractive force is obtained from various devices such as electric motors, steam engine drives, diesel engine dives, etc. is known as traction system.

Traction system may be broadly classified into two types. They are electric-traction systems, which use electrical energy, and non-electric traction system, which does not use electrical energy for the propulsion of vehicle.

9.1.1 Requirements of ideal traction system

Normally, no single traction system fulfills the requirements of ideal traction system, why because each traction system has its merits and suffers from its own demerits, in the fields of applications.

The requirements of ideal traction systems are:

- Ideal traction system should have the capability of developing high tractive effort in order to have rapid acceleration.
- The speed control of the traction motors should be easy.
- Vehicles should be able to run on any route, without interruption.
- Equipment required for traction system should be minimum with high efficiency.
- It must be free from smoke, ash, durt, etc.
- Regenerative braking should be possible and braking should be in such a way to cause minimum wear on the break shoe.
- Locomotive should be self-contained and it must be capable of withstanding over-loads.
- Interference to the communication lines should be eliminated while the locomotive running along the track.

9.1.2 Advantages and disadvantages of electric traction

Electric traction system has many advantages compared to non-electric traction systems. The following are the advantages of electric traction:

- Electric traction system is more clean and easy to handle.
- No need of storage of coal and water that in turn reduces the maintenance cost as well as the saving of high-grade coal.
- Electric energy drawn from the supply distribution system is sufficient to maintain the common necessities of locomotives such as fans and lights; therefore, there is no need of providing additional generators.
- The maintenance and running costs are comparatively low.
- The speed control of the electric motor is easy.
- Regenerative braking is possible so that the energy can be fed back to the supply system during the braking period.
- In electric traction system, in addition to the mechanical braking, electrical braking can also be used that reduces the wear on the brake shoes, wheels, etc.
- Electrically operated vehicles can withstand for overloads, as the system is capable of drawing more energy from the system.

In addition to the above advantages, the electric traction system suffers from the following drawbacks:

- Electric traction system involves high erection cost of power system.
- Interference causes to the communication lines due to the overhead distribution networks.
- The failure of power supply brings whole traction system to stand still.
- In an electric traction system, the electrically operated vehicles have to move only on the electrified routes.
- Additional equipment should be needed for the provision of regenerative braking, it will increase the overall cost of installation.

9.2 REVIEW OF EXISTING ELECTRIC TRACTION SYSTEM IN INDIA

In olden days, first traction system was introduced by Britain in 1890 (600-V DC track). Electrification system was employed for the first traction vehicle. This traction system was introduced in India in the year 1925 and the first traction system employed in India was from Bombay VT to Igatpuri and Pune, with 1,500-V DC supply. This DC supply can be obtained for traction from substations equipped with rotary converters. Development in the rectifiers leads to the replacement of rotary converters by mercury arc rectifiers. But nowadays further development in the technology of semiconductors, these mercury arc valves are replaced by solid-state semiconductors devices due to fast traction system was introduced on 3,000-V DC. Further development in research on traction system by French international railways was suggested that, based on relative merits and demerits, it is advantageous to prefer to AC rather than DC both financially and operationally.

Thus, Indian railways was introduced on 52-kV, 50-Hz single-phase AC system in 1957; this system of track electrification leads to the reduction of the cost of overhead, locomotive equipment, etc. Various systems employed for track electrification are shown in Table 9.1.

TABLE 9.1 *Track electrification systems*

S. no	*System*	*Voltage*	*Frequency*
1	DC system	600 V, 1,500 V, or 3,000 V	–
2	Single-phase AC system	15–25 kV is stepped down to 300–400 V	$\frac{162}{3}$ Hz and 25 Hz
3	Three-phase AC system	15–25 kV is stepped down to 3,300–3,600 V	$\frac{162}{3}$ Hz and 50 Hz

9.3 SYSTEM OF TRACTION

Traction system is normally classified into two types based on the type of energy given as input to drive the system and they are:

(i) Non-electric traction system

Traction system develops the necessary propelling torque, which do not involve the use of electrical energy at any stage to drive the traction vehicle known as electric traction system.

Ex: Direct steam engine drive and direct internal combustion engine drive.

(ii) Electric traction system

Traction system develops the necessary propelling torque, which involves the use of electrical energy at any stage to drive the traction vehicle, known as electric traction system.

Based upon the type of sources used to feed electric supply for traction system, electric traction may be classified into two groups:

1. Self-contained locomotives.
2. Electric vehicle fed from the distribution networks.

9.3.1 Self-contained locomotives

In this type, the locomotives or vehicles themselves having a capability of generating electrical energy for traction purpose. Examples for such type of locomotives are:

(i) Steam electric drive

In steam electric locomotives, the steam turbine is employed for driving a generator used to feed the electric motors. Such types of locomotives are not generally used for traction because of some mechanical difficulties and maintenance problems.

(ii) Diesel electric trains

A few locomotives employing diesel engine coupled to DC generator used to feed the electric motors producing necessary propelling torque. Diesel engine is a variable high-speed type that feeds the self- or separately excited DC generator. The excitation for generator can be supplied from any auxiliary devices and battery.

Generally, this type of traction system is suggested in the areas where coal and steam tractions are not available. The advantages and disadvantages of the diesel engine drive are given below:

Advantages

- As these are no overhead distribution system, initial cost is low.
- Easy speed control is possible.

- Power loss in speed control is very low.
- Time taken to bring the locomotive into service is less.
- In this system, high acceleration and braking retardation can be obtained compared to steam locomotives.
- The overall efficiency is high compared to steam locomotives.

Disadvantages

- The overloading capability of the diesel engine is less.
- The running and maintenance costs are high.
- The regenerative braking cannot be employed for the diesel engine drives.

9.3.2 Petrol electric traction

This system of traction is used in road vehicles such as heavy lorries and buses. These vehicles are capable of handling overloads. At the same time, this system provides fine and smooth control so that they can run along roads without any jerking.

9.3.3 Battery drives

In this drive, the locomotive consists of batteries used to supply power to DC motors employed for driving the vehicle. This type of drives can be preferred for frequently operated services such as local delivery goods traction in industrial works and mines, etc. This is due to the unreliability of supply source to feed the electric motors.

9.3.4 Electric vehicles fed from distribution network

Vehicles in electrical traction system that receives power from over head distribution network fed or substations with suitable spacing. Based on the available supply, these groups of vehicles are further subdivided into:

(i) System operating with DC supply. Ex: tramways, trolley buses, and railways.
(ii) System operating with AC supply. Ex: railways.

Systems operating with DC supply

In case if the available supply is DC, then the necessary propelling power can be obtained for the vehicles from DC system such as tram ways, trolley buses, and railways.

Tramways: Tramways are similar to the ordinary buses and cars but only the difference is they are able to run only along the track. Operating power supply for the tramways is 500-V DC tramways are fed from single overhead conductor acts as positive polarity that is fed at suitable points from either power station or substations and the track rail acts as return conductor.

The equipment used in tramways is similar to that used in railways but with small output not more than 40–50 kW. Usually, the tramways are provided with two driving axels to control the speed of the vehicles from either ends. The main drawback of tramways is they have to run along the guided routes only. Rehostatic and mechanical brakings can be applied to tramways. Mechanical brakes can be applied at low speeds for providing better saturation where electric braking is ineffective, during the normal service. The erection and maintenance costs of tramways are high since the cost of overhead distribution structure is costlier and sometimes, it may cause a source of danger to other road users.

Trolley buses: The main drawback of tramways is, running along the track is avoided in case of trolley buses. These are electrically operated vehicles, and are fed usually 600-V DC from two overhead conductors, by means of two collectors. Even though overhead distribution structure is costlier, the trolley buses are advantageous because, they eliminate the necessity of track in the roadways.

In case of trolley buses, rehostatic braking is employed, due to high adhesion between roads and rubber types. A DC compound motor is employed in trolley buses.

9.4 SYSTEM OF TRACK ELECTRIFICATION

Nowaday, based on the available supply, the track electrification system are categorized into.

(i) DC system.

(ii) Single-phase AC system.

(iii) Three-phase AC system.

(iv) Composite system.

9.4.1 DC system

In this system of traction, the electric motors employed for getting necessary propelling torque should be selected in such a way that they should be able to operate on DC supply. Examples for such vehicles operating based on DC system are tramways and trolley buses. Usually, DC series motors are preferred for tramways and trolley buses even though DC compound motors are available where regenerative braking is desired. The operating voltages of vehicles for DC track electrification system are 600, 750, 1,500, and 3,000 V. Direct current at 600–750 V is universally employed for tramways in the urban areas and for many suburban and main line railways, 1,500–3,000 V is used. In some cases, DC supply for traction motor can be obtained from substations equipped with rotary converters to convert AC power to DC. These substations receive AC power from 3-ϕ high-voltage line or single-phase overhead distribution network. The operating voltage for traction purpose can be justified by the spacing between stations and the type of traction motors available. Theses substations are usually automatic and remote controlled and they are so costlier since they involve rotary converting equipment. The DC system is preferred for suburban services and road transport where stops are frequent and distance between the stops is small.

9.4.2 Single-phase AC system

In this system of track electrification, usually AC series motors are used for getting the necessary propelling power. The distribution network employed for such traction systems is normally 15–25 kV at reduced frequency of 16$\frac{2}{3}$ Hz or 25 Hz. The main reason of operating at reduced frequencies is AC series motors that are more efficient and show better performance at low frequency. These high voltages are stepped down to suitable low voltage of 300–400 V by means of step-down transformer. Low frequency can be obtained from normal supply frequency with the help of frequency converter. Low-frequency operation of overhead transmission line reduces the line reactance and hence the voltage drops directly and single-phase AC system is mainly preferred for main line services where the cost of overhead structure is not much importance moreover rapid acceleration and retardation is not required for suburban services.

9.4.3 Three-phase AC system

In this system of track electrification, 3-ϕ induction motors are employed for getting the necessary propelling power. The operating voltage of induction motors is normally 3,000–3,600-V AC at either normal supply frequency or 16⅔-Hz frequency.

Usually 3-ϕ induction motors are preferable because they have simple and robust construction, high operating efficiency, provision of regenerative braking without placing any additional equipment, and better performance at both normal and seduced frequencies. In addition to the above advantages, the induction motors suffer from some drawbacks; they are low-starting torque, high-starting current, and the absence of speed control. The main disadvantage of such track electrification system is high cost of overhead distribution structure. This distribution system consists of two overhead wires and track rail for the third phase and receives power either directly from the generating station or through transformer substation.

Three-phase AC system is mainly adopted for the services where the output power required is high and regeneration of electrical energy is possible.

9.4.4 Composite system

As the above track electrification system have their own merits and demerits, 1-ϕ AC system is preferable in the view of distribution cost and distribution voltage can be stepped up to high voltage with the use of transformers, which reduces the transmission losses. Whereas in DC system, DC series motors have most desirable features and for 3-ϕ system, 3-ϕ induction motor has the advantage of automatic regenerative braking. So, it is necessary to combine the advantages of the DC/AC and 3-ϕ/1-ϕ systems. The above cause leads to the evolution of composite system.

Composite systems are of two types.

(i) Single-phase to DC system.

(ii) Single-phase to three-phase system or kando system.

Single-phase to DC system

In this system, the advantages of both 1-ϕ and DC systems are combined to get high voltage for distribution in order to reduce the losses that can be achieved with 1-ϕ distribution networks, and DC series motor is employed for producing the necessary propelling torque. Finally, 1-ϕ AC distribution network results minimum cost with high transmission efficiency and DC series motor is ideally suited for traction purpose. Normal operating voltage employed of distribution is 25 kV at normal frequency of 50 Hz. This track electrification is employed in India.

Single-phase to 3-ϕ system or kando system

In this system, 1-ϕ AC system is preferred for distribution network. Since single-phase overhead distribution system is cheap and 3-ϕ induction motors are employed as traction motor because of their simple, robust construction, and the provision of automatic regenerative braking.

The voltage used for the distribution network is about 15–25 kV at 50 Hz. This 1-ϕ supply is converted to 3-ϕ supply through the help of the phase converters and high voltage is stepped down transformers to feed the 3-ϕ induction motors. Frequency converters are also employed to get high-starting torque and to achieve better speed control with the variable supply frequency.

9.5 COMPARISON OF DC AND AC TRACTIONS

Table 9.2 gives a comparison between DC traction and AC traction.

TABLE 9.2 *Comparison between DC and AC tractions*

Factor	*DC traction*	*AC traction*
1. Cost	DC series motors are cheaper.	AC series motors are expensive.
2. Efficiency	It is more efficient.	Less efficient.
3. Maintenance	It requires less maintenance.	It requires more maintenance.
4. Acceleration	It is capable of giving high acceleration.	It is capable of giving less acceleration.
5. Speed control	The speed control of DC series motor is limited.	Wide range of speed control is possible.
6. Interference	DC system causes less interference with communication lines.	It will produce more interference with communication lines.
7. Regenerative braking	Regenerative braking is more efficient in DC series motor.	Regenerative braking is less efficient in AC series motor.
8. Overhead distribution	Overhead distribution.	Overhead distribution.
9. System	System is less costly in DC system.	System is costlier in AC system.
10. Torque	The torque developed by the DC series motor is less.	The starting and running torque developed by the AC series motor is more.
11. Substations	The number of substations required for a given track distance on DC traction is more.	The number of substations required in AC traction is less.

9.6 SPECIAL FEATURES OF TRACTION MOTORS

The general features of the electric motors used for traction purpose are:

1. Mechanical features.
2. Electrical features.

9.6.1 Mechanical features

1. A traction motor must be mechanically strong and robust and it should be capable of withstanding severe mechanical vibrations.
2. The traction motor should be completely enclosed type when placed beneath the locomotive to protect against dirt, dust, mud, etc.
3. In overall dimensions, the traction motor must have small diameter, to arrange easily beneath the motor coach.
4. A traction motor must have minimum weight so the weight of locomotive will decrease. Hence, the load carrying capability of the motor will increase.

9.6.2 Electrical features

High-starting torque

A traction motor must have high-starting torque, which is required to start the motor on load during the starting conditions in urban and suburban services.

Speed control

The speed control of the traction motor must be simple and easy. This is necessary for the frequent starting and stopping of the motor in traction purpose.

Dynamic and regenerative braking

Traction motors should be able to provide easy simple rehostatic and regenerative braking subjected to higher voltages so that system must have the capability of withstanding voltage fluctuations.

Temperature

The traction motor should have the capability of withstanding high temperatures during transient conditions.

Overload capacity

The traction motor should have the capability of handling excessecive overloads.

Parallel running

In traction work, more number of motors need to run in parallel to carry more load. Therefore, the traction motor should have such speed–torque and current–torque characteristics and those motors may share the total load almost equally.

Commutation

Traction motor should have the feature of better commutation, to avoid the sparking at the brushes and commutator segments.

9.7 TRACTION MOTORS

No single motor can have all the electrical operating features required for traction.

In earlier days, DC motor is suited for traction because of the high-starting torque and having the capability of handling overloads. In addition to the above characteristics, the speed control of the DC motor is very complicated through semiconductor switches. So that, the motor must be designed for high base speed initially by reducing the number of turns in the field winding. But this will decrease the torque developed per ampere at the time of staring. And regenerative braking is also complicated in DC series motor; so that, the separately excited motors can be preferred over the series motor because their speed control is possible through semi-controlled converters. And also dynamic and regenerative braking in separately excited DC motor is simple and efficient.

DC compound motors are also preferred for traction applications since it is having advantageous features than series and separately excited motors.

But nowadays squirrel cage induction and synchronous motors are widely used for traction because of the availability of reliable variable frequency semiconductor inverters.

The squirrel cage induction motor has several advantages over the DC motors. They are:

(i) Robust construction.

(ii) Highly reliable.

(iii) Low maintenance and low cost.

(iv) High efficiency.

Synchronous motor features lie in between the squirrel cage induction motor and the DC motor. The main advantages of the synchronous motor over the squirrel cage induction motor are:

(i) The synchronous motors can be operated at leading power by varying the field excitation.

(ii) Load commutated thyristor inverter is used in synchronous motors as compared to forced commutation thyristor inverter in squirrel cage induction motors.

Even though such forced commutation reduces the weight and volume of induction motor, the synchronous motor is less expensive.

9.7.1. DC series motor

From the construction and operating characteristics of the DC series motor, it is widely suitable for traction purpose. Following features of series motor make it suitable for traction.

(i) DC series motor is having high-starting torque and having the capability of handling overloads that is essential for traction drives.

(ii) These motors are having simple and robust construction.

(iii) The speed control of the series motor is easy by series parallel control.

(iv) Sparkless commutation is possible, because the increase in armature current increases the load torque and decreases the speed so that the emf induced in the coils undergoing commutation.

(v) Series motor flux is proportional to armature current and torque. But armature current is independent of voltage fluctuations. Hence, the motor is unaffected by the variations in supply voltage.

(vi) We know that:

$$N \alpha \frac{1}{\phi} \alpha \frac{1}{I_a} \quad \text{and} \quad T \propto \phi I_a.$$

But for series motor $\phi \propto I_a$

$$\therefore T \alpha I_a^2$$

$$\therefore N \alpha \frac{1}{I_a} \alpha \frac{1}{\sqrt{T}}.$$

But the power output of the motor is proportional to the product of torque and speed.

$\therefore$ Motor output $\alpha\, T\, N\, \alpha \sqrt{T}$.

That is motor input drawn from the source is proportional to the square root of the torque. Hence, the series motor is having self-retaining property.

(vii) If more than one motor are to be run in parallel, their speed–torque and current–torque characteristics must not have wide variation, which may result in the unequal wear of driving wheels.

9.7.2 DC shunt motor

From the characteristics of DC shunt motor, it is not suitable for traction purpose, due to the following reasons:

(i) DC shunt motor is a constant speed motor but for traction purpose, the speed of the motor should vary with service conditions.

(ii) In case of DC shunt motor, the power output is independent of speed and is proportional to torque. In case of DC series motor, the power output is proportional to $\sqrt{T}$. So that, for a given load torque, the shunt motor has to draw more power from the supply than series motor.

(iii) For shunt motor, the torque developed is proportional to armature current ($T \propto I_a$). So for a given load torque motor has to draw more current from the supply.

(iv) The flux developed by shunt motor is proportional to shunt field current and hence supply voltage. $\left[\because \phi_{sh} \propto I_{sh} \propto \frac{V}{R_{sh}}\right]$. But the torque developed is proportional to ϕ_{sh} and I_a. Hence, the torque developed by the shunt motor is affected by small variations in supply voltage.

(v) If two shunt motors are running in parallel, their speed–torque and speed–current characteristics must be flat and same. Otherwise, the currents drawn by the motor from the supply mains will be different and cause to unequal sharing of load.

Example 9.1: A DC series motor drives a load. The motor takes a current of 13 A and the speed is 620 rpm. The torque of the motor varies as the square of speed. The field winding is shunted by a diverter of the same resistance as that of the field winding, then determine the motor speed and current. Neglect all motor losses and assume that the magnetic circuit is unsaturated.

Solution:

Before connecting field diverter:

Speed, $N_1 = 620$ rpm.

Series field current, $I_{se1} = 13$ A.

The same current flows through the armature; so that,

$$I_1 = I_{se1} = I_{a1} = 13 \text{ A}.$$

After connecting field diverter, the field winding is shunted by the diverter of the same refinance; so that:

$$\text{Series field current} = I_{se2} = \frac{1}{2} I_2.$$

Since torque developed:

$$T \propto \phi I_a$$

$$\propto \phi I_1$$

$$\frac{T_2}{T_1} = \frac{T_2 I_2}{\phi_1 I_1} = \frac{\frac{1}{2} I_2^2}{2 I_1^2} \quad \text{(i)} \quad (\phi \propto I_{se} \text{ magnetic circuit is unsaturated}).$$

According to given data, the torque varies as the square of the speed.

$$\frac{T_2}{T_1} = \frac{N_2^2}{N_1^2}. \quad \text{(ii)}$$

From Equations (i) and (ii):

$$\frac{I_2^2}{2 I_1^2} = \frac{N_2^2}{N_1^2}$$

$$\frac{N_2}{N_1} = \frac{I_2}{\sqrt{2I_1}}. \quad \text{(iii)}$$

All the losses are neglected, and assume that the supply voltage is constant.

$$N \alpha \frac{1}{\phi}$$

$$\frac{N_2}{N_1} = \frac{\phi_1}{\phi_2}$$

$$= \frac{I_1}{\frac{1}{2}I_2}. \quad \text{(iv)}$$

From Equations (iii) and (iv):

$$\frac{I_2}{\sqrt{2I_1}} = \frac{2I_1}{I_2}$$

$$I_2^2 = 2\sqrt{2}\, I_1^2$$

$$= 2 \times \sqrt{2} \times (13)^2$$

$$= 478.004.$$

$$\therefore I_2 = 21.86 \text{ A}.$$

From Equation (iv):

$$\frac{N_2}{N_2} = \frac{2I_1}{I_2}$$

$$N_2 = \frac{2I_1}{I_2} \times N_1$$

$$= 2 \times \frac{13}{21.86} \times 620$$

$$= 737.42 \text{ rpm}.$$

Example 9.2: Two DC traction motors, each takes a current of 45 from 450 V mains and runs at the speed of 600 and 625 rpm, respectively. Each motor has an effective resistance of 0.4 Ω. Calculate the speed and voltage across each machine when mechanically coupled and electrically connected in series and taking a current of 45 A from 450 V mains the resistance of each motor being unchanged.

Solution:

The resistance of each motor = 0.4 Ω.

The speed of motor 1, $N_1 = 600$ rpm.

The speed of motor 2, $N_2 = 325$ rpm.

The line voltages across the two machines are V_1 and V_2

$$\therefore V_1 + V_2 = 450 \text{ V}. \quad \text{(i)}$$

The back emf of two motors, when they are connected across 450 V and 45 A, respectively:

$$E_1 = E_2 = 450\ (45 \times 0.4) \qquad [\text{since } E_2 = \text{V} - I_a R].$$

When the two motors are coaled mechanically and connected in series, the speed of each motor will be the same (say N rpm), the current will be equals to 45 A, and the sum of voltage drops across the motors be 450 V. In view of the above explanation, E_1^1 and E_2^1 can be given as:

$$E_1^1 = E_1 \times \frac{N}{N_1}$$

$$= 432 \times \frac{N}{600}.$$

Voltage across the motor 1 is:

$$V_1 = E_1^1 + I_a R$$

$$= 0.72N + 45 \times 0.4$$

$$= 0.72N + 18. \quad \text{(ii)}$$

The back emf of the motor 2 is:

$$E_1^1 = E_2 \times \frac{N}{N_2}$$

$$= 432 \times \frac{N}{625}.$$

$$E_2^1 = 0.6912\ N.$$

The voltage across the motor 2 is:

$$V_2 = E_2^1 + I_a R$$

$$= 0.6912N + 18. \quad \text{(iii)}$$

Substitute Equations (ii) and (iii) in Equation (i):

$$0.72N + 18 + 0.6912N + 18 = 450$$

$$N = 293.36 \text{ rpm.}$$

Substitute the value of 'N' in Equations (ii) and (iii):

$$V_1 = 0.72 \times 293.36 + 18$$

$$= 229.2193\ V.$$

$$V_2 = 0.6192 \times 293.36 + 18$$

$$= 220.77\ V.$$

Example 9.3: A 230-V, and 12-HP motor has shunt and armature resistance of 120 and 0.3 Ω, respectively. Calculate the resistance to be inserted in the armature circuit to reduce the speed by 20%, assuming the torque remains constant. The efficiency of the motor is 90%.

Solution:

Supply voltage = 230 V.

Motor output = 12 × 735.5

= 8,826 W.

$$\text{Motor input} = \frac{\text{output}}{\eta} = \frac{8,826}{0.9}$$

$$= 9,806.67 \text{ W.}$$

$$\text{Motor input} = VI$$

$$9,806.67 = 230 \times 2$$

$$\text{I} = 42.63 \text{ Ans.}$$

We know that $T \propto \phi I_a$:

$$\frac{T_1}{T_2} = \frac{I_{a1}}{I_{a2}}.$$

According to given data, the torque remain constant.

$$I_{a1} = I_{a2}$$

$$I_{sh} = \frac{V}{R_{sh}} = \frac{230}{120}$$

$$= 1.916 \text{ A.}$$

At rated speed:

$$I_{a1} = I - I_{sh}$$

$$= 42.63 - 1.916$$

$$= 40.713 \text{ A.}$$

We know that $N \alpha \frac{E_1}{\phi}$.

$$\frac{E_{b2}}{E_{b1}} = \frac{N_2}{N_1}$$

$$E_{b1} = V - I_{a_1} R_a$$

$$= 230 - (40.713 \times 0.3) = 217.786 \text{ V.}$$

$$\frac{E_{b1}}{E_{b2}} = \frac{N_2}{N_1}$$

$$E_{b2} = \frac{0.8 N_1}{N_1} \times E_{b1}$$

$$= 0.8 \times 217.786 = 174.22 \text{ V.}$$

Or, $V - I_{a2} R_t = 174.22$

$$230 - (40.713) R_t = 174.22$$

$$\therefore R_t = 1.37\ \Omega$$

$$R_t = R_a + \text{external resistance}$$

$$1.37 = 0.3 + \text{external resistance}$$

$$\therefore \text{External resistance} = 1.07\ \Omega.$$

Example 9.4: A 230-V, 10-HP, and DC shunt motor with $R_a = 0.2\ \Omega$ and $R_{sh} = 80\ \Omega$, runs at 1000 rpm on full load. The efficiency on the full load is 80%. If the speed is to be raised to 1200 rpm keeping load constant, determine extra resistance to be added in the field ckt. Assume 1 HP = 736 W.

Solution:

$V = 230$ V, HP rating = 10.

The method of speed control is flux control. Figure 9.4 (a) and (b) is showing two conditions.

$R_a = 0.2\ \Omega$ and $R_{sh} = 80\ \Omega$.

In first case, $I_{sh1} = \dfrac{V}{R_{sh}} = \dfrac{230}{80} = 2.875$ A

While output on full load = 10 HP = 10 × 73.6 = 7360 W.

Now, $$\%\eta = \frac{O/p}{I/p}\times 100$$

$$80 = \frac{7360}{\text{Input}}\times 100$$

$$\text{Input} = \frac{7360}{80}\times 100 = 9200 \text{ W.}$$

$$\text{But input} = V \times I_{L1}$$

$$9200 = 230 \times I_{L1}$$

$$I_{L1} = \frac{9200}{230} = 40 \text{ A}$$

$$I_{a1} = I_{L1} - I_{sh1} = 40 - 2.875 = 37.125 \text{ A.}$$

Use torque equation, $T \propto \phi I_a \propto I_{sh} I_a$:

$$\frac{T_1}{T_2} = \frac{I_{sh1}}{I_{sh2}} \times \frac{I_{a1}}{I_{a2}}.$$

But, as load is constant, $T_1 = T_2$

$$I_{a1} I_{sh1} = I_{a2} I_{sh2}$$

$$37.125 \times 2.875 = I_{a2} I_{sh2}$$

$$I_{a2} I_{sh2} = 106.734. \qquad \text{(i)}$$

Now, $E_{b1} = V - I_{a1} R_a = 230 - (37.125)(0.2)$

$$= 222.575 \text{ V.}$$

$$E_{b2} = V - I_{a2} R_a = 230 - 0.2\, I_{a2}.$$

Use speed equation:

$$N \propto \frac{E_b}{\phi} \propto \frac{E_b}{I_{sh}}$$

$$\frac{N_1}{N_2} = \frac{E_{b1}}{E_{b2}} \times \frac{I_{sh2}}{I_{sh1}}$$

$$\frac{1000}{1200} = \frac{222.575}{230 - 0.2I_{a2}} \times \frac{I_{sh2}}{2.875}.$$

Replacing I_{a2} by $\dfrac{106.734}{I_{sh2}}$ from Equation (i):

$$\frac{1000}{1200} = \frac{222.575}{230 - 0.2\left(\dfrac{106.734}{I_{sh2}}\right)} \times \frac{I_{sh2}}{2.875}$$

$$\frac{230 - 0.2\left(\dfrac{106.734}{I_{sh2}}\right)}{I_{sh2}} = \frac{222.575 \times 1200}{1000 \times 2.875} = 92.90$$

$$230I_{sh2} - 21.348 = 92.90\, I_{sh}^2$$

$$(I_{sh2})^2 - 2.476\, I_{sh2} + 0.229 = 0.$$

Solving the quadratic:

$$I_{sh2} = 2.379 \text{ A} \quad \text{or} \quad 0.096 \text{ A}.$$

It is always easy to achieve 2.379 A from 2.875 A, as less resistance is to be required in series with the field winding.

So, neglecting lower value of I_{sh2}, we get:

$$I_{sh2} = 2.379 \text{ A}.$$

Now, with R_x in series with field:

$$I_{sh2} = \frac{V}{R_{sh} + R_x}$$

$$2.379 = \frac{230}{80 + R_x}$$

$$\therefore R_x = 16.679\ \Omega.$$

Example 9.5: A series motor having a resistance of 0.8 Ω between its terminal drives. The torque of a fan is proportional to the square of the speed. At 220 V, its speed is 350 rpm and takes 12 A. The speed of the fan is to be raised to 400 rpm by supply voltage control. Estimate the supply voltage required.

Solution:

$R_a + R_{se} = 0.8\ \Omega$, $V_1 = 220$ V, $N_1 = 350$ rpm, $I_1 = I_{a1} = 12$ A

$N_2 = 400$ rpm.

Use the torque equation, $T \propto \phi I_a \propto I_a^2$ as $\phi \propto I_a$:

$$\frac{T_1}{T_2} = \left(\frac{I_{a1}}{I_{a2}}\right)^2. \qquad \text{(i)}$$

Also $T \propto N^2$ (given)

$$\frac{T_1}{T_2} = \left(\frac{N_1}{N_2}\right)^2. \tag{ii}$$

Equating Equations (i) & (ii):

$$\left(\frac{N_1}{N_2}\right)^2 = \left(\frac{I_{a1}}{I_{a2}}\right)^2$$

$$\left(\frac{350}{400}\right)^2 = \left(\frac{12}{I_{a2}}\right)^2$$

$$\therefore I_{a2} = 13.7\text{ A.}$$

Use the speed equation $N \propto \frac{E_b}{\phi} \propto \frac{E_b}{I_a}$

$$\frac{N_1}{N_2} = \frac{E_{b1}}{E_{b2}} \times \frac{I_{a2}}{I_{a1}}. \tag{iii}$$

Now, $E_{b1} = V_1 - I_{a1}(R_a + R_{se})$

$= 220 - 12(0.8) = 210.4$ V.

In second case, the voltage is to be changed from V_1 to V_2.

$\therefore E_{b2} = V_2 - I_{a2}(R_a + R_{se})$

$= V_2 - 13.7(0.8) = V_2 - 10.96.$

E_{b1} and E_{b2} are substituted in Eq (iii):

$$\frac{350}{400} = \frac{210.4}{V_2 - 10.96} \times \frac{13.7}{12}$$

$V_2 - 10.96 = 274.52$

$V_2 = 284.52$ V.

$\therefore$ This is the new supply voltage required to raise the speed from 350 rpm to 400 rpm.

Example 9.6: A 230-V DC shunt motor takes a current of 20 A on a certain load. The armature resistance is 0.8 Ω and the field circuit resistance is 250 Ω. Find the resistance to be inserted in series with the armature to have the speed is half if the load torque is constant.

Solution:

$I_{L1} = 20$ A.

$$I_{sh} = \frac{V}{R_{sh}} = \frac{230}{250} = 0.92\text{ A.}$$

$I_{a1} = I_{L1} - I_{sh} = 20 - 0.92 = 19.08.$

$E_{b1} = V - I_{a1}R_a = 230 - 19.08\,(0.8) = 214.736$ V.

$T \propto \phi I_a \propto I_a$ ($\because \phi$ is constant).

$$\frac{T_1}{T_2} = \frac{I_{a1}}{I_{a2}} = 1 \qquad \text{as torque is constant}$$

$$\therefore I_{a1} = I_{a2} = 19.08 \text{ A.}$$

R_x = external resistance in armature

$$E_{b2} = V - I_{a2}(R_a + R_x) = 230 - (19.08)(0.8 + R_x).$$

$$\text{Now,} \quad N \propto \frac{E_b}{\phi} \propto E_b \qquad (\because \phi \text{ is constant})$$

$$\therefore \frac{N_1}{N_2} = \frac{E_{b1}}{E_{b2}}$$

$$\frac{1}{0.5} = \frac{214.736}{230 - 19.08(0.8 + R_x)}$$

$$230 - 19.08\,(0.8 + R_x) = 214.736 \times 0.5 = 107.368$$

$$19.08\,(0.8 + R_x) = -\,107.368 + 230$$

$$= 122.632$$

$$0.8 + R_x = 6.43$$

$$R_x = 6.43 - 0.8$$

$$R_x = 5.62\ \Omega.$$

9.7.3 AC series motor

Practically, AC series motor is best suited for the traction purpose due to high-starting torque (Fig. 9.1). When DC series motor is fed from AC supply, it works but not satisfactorily due to some of the following reasons:

(i) If DC series motor is fed from AC supply, both the field and the armature currents reverse for every half cycle. Hence, unidirectional torque is developed at double frequency.

(ii) Alternating flux developed by the field winding causes excessive eddy current loss, which will cause the heating of the motor. Hence, the operating efficiency of the motor will decrease.

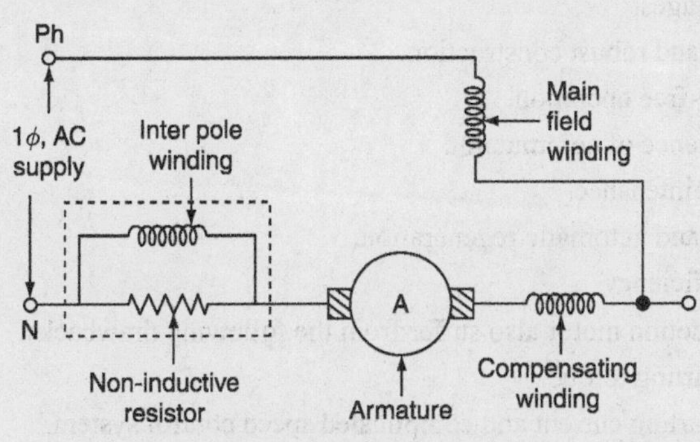

FIG. 9.1 *AC series motor*

(iii) Field winding inductance will result abnormal voltage drop and low power factor that leads to the poor performance of the motor.

(iv) Induced emf and currents flowing through the armature coils undergoing commutation will cause sparking at the brushes and commutator segments.

Hence, some modifications are necessary for the satisfactory operation of the DC series motor on the AC supply and they are as follows:

(i) In order to reduce the inductive reactance of the series field, the field winding of AC series motor must be designed for few turns.

(ii) The decrease in the number of turns of the field winding reduces the load torque, i.e., if field turns decrease, its mmf decrease and then flux, which will increase the speed, and hence the torque will decrease. But in order to maintain constant load torque, it is necessary to increase the armature turns proportionately.

(iii) If the armature turns increase, the inductive reactance of the armature would increase, which can be neutralized by providing the compensating winding.

(iv) Magnetic circuit of an AC series motor should be laminated to reduce eddy current losses.

(v) Series motor should be operating at low voltage because high voltage low current supply would require large number of turns to produce given flux.

(vi) Motor should be operating at low frequency, because inductive reactance is proportional to the frequency. So, at low frequency, the inductive reactance of the field winding decreases.

The operating characteristics of the AC series motor are similar to the DC series motor. Weight of an AC series motor is one and a half to two times that of a DC series motor. And operating voltage is limited to 300 V. They can be built up to the size of several hundred kW for traction work.

At the time of starting operation, the power factor is low; so that, for a given current, the torque developed by the AC motor is less compared to the DC motor. Thus, the AC series motor is not suitable for suburban services with frequent stops and preferred for main line service where high acceleration is not required.

9.7.4 Three-phase induction motor

The three-phase induction motors are generally preferred for traction purpose due to the following advantages.

1. Simple and robust construction.
2. Trouble-free operation.
3. The absence of commutator.
4. Less maintenance.
5. Simple and automatic regeneration.
6. High efficiency.

Three-phase induction motor also suffer from the following drawbacks.

1. Low-starting torque.
2. High-starting current and complicated speed control system.
3. It is difficult to employ three-phase induction motor for a multiple-unit system used for propelling a heavy train.

Three-phase induction motor draws less current when the motor is started at low frequencies. When a three-phase induction motor is used, the cost of overhead distribution system increases and it consists of two overhead conductors and track rail for the third phase to feed power to locomotive, which is a complicated overhead structure and if any person comes in contact with the third rail, it may cause danger to him or her.

This drawback can be overcome by employing kando system. In this system, 1-ϕ supply from the overhead distribution structure is converted to 3-ϕ supply by using phase converters and is fed to 3-ϕ induction motor.

The speed controller of induction motor becomes smooth and easy with the use of thyristorized inverter circuits to get variable frequency supply that can be used to control the speed of three-phase induction motor.

Nowadays, by overcoming the drawbacks of three-phase induction motor, it can be used for traction purpose.

9.7.5 Linear induction motor

It is a special type of induction motor that gives linear motion instead of rotational motion, as in the case of a conventional motor.

In case of linear induction motor, both the movement of field and the movement of the conductors are linear.

A linear induction motor consists of 3-ϕ distributed field winding placed in slots, and secondary is nothing but a conducting plate made up of either copper or aluminum as shown in Fig. 9.2.

The field system may be either single primary or double primary system. In single primary system, a ferro magnetic plate is placed on the other side of the copper plate; it is necessary to provide low reluctance path for the magnetic flux. When primary is excited by 3-ϕ AC supply, according to mutual induction, the induced currents are flowing through secondary and ferro magnetic plate. Now, the ferro magnetic plate energized and attracted toward the primary causes to unequal air gap between primary and secondary as shown in Fig. 9.2(a). This drawback can be overcome by double primary system as shown in Fig. 9.2(b). In this system, two primaries are placed on both the sides of secondary, which will be shorter in length compared to the other depending upon the use of the motor.

When the operating distance is large, the length of the primary is made shorter than the secondary because it is not economical to place very large 3-ϕ primary. Generally, the short secondary form of system is preferred for limited operating distance, as shown in Fig. 9.2(c).

When 3-ϕ primary winding is excited by giving 3-ϕ AC supply, magnetic field is developed rotating at linear synchronous speed, V_s.

The linear synchronous speed is given by:

$$V_s = 2\tau f \text{ m/s},$$

where τ is the pole pitch in m and f is the supply frequency in hertzs.

Note: here, the synchronous speed does not depend upon the number of poles but depends upon the pole pitch and the supply frequency.

(a) Short single primary.

(b) Short double primary.

(c) Short secondary.

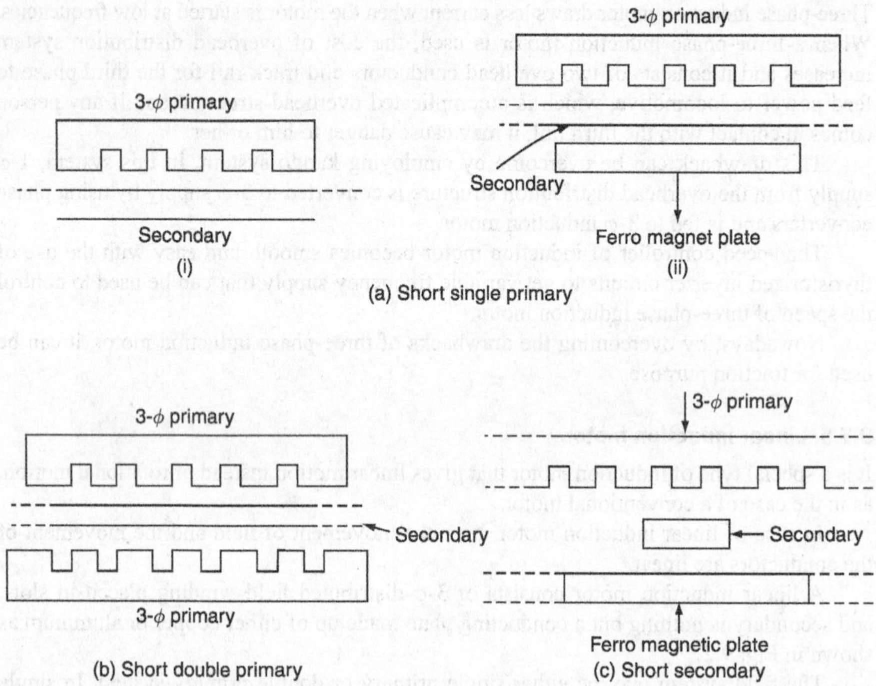

FIG. 9.2 *Linear induction motor*

The flux developed by the field winding pulls the rotor same as to the direction of the magnetic field linearly, which will reduce relative speed between field and rotor plate. If the speed of the rotor plate is equal to the magnetic field, then the field would be stationary when viewed from the rotor plate. If rotor plate is rotating at a speed more than linear synchronous, the direction of a force would be reversed, which causes regenerative braking.

The slip of the linear induction motor is given by:

$$s = \frac{V_s - V}{V_s},$$

where 'V' is the actual speed of the rotor plate.

The speed–torque (tractive effort) characteristics is shown in Fig. 9.3.

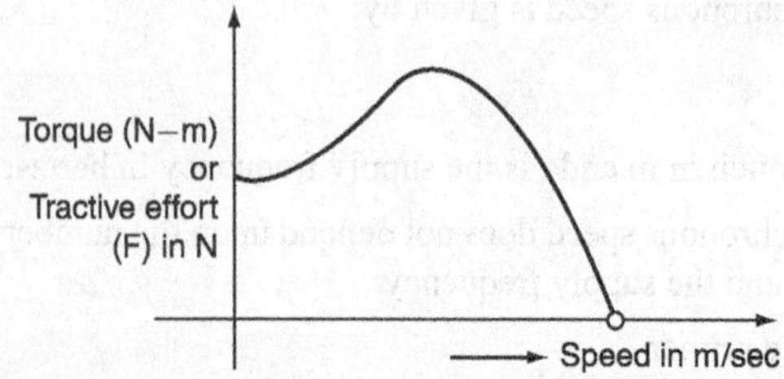

FIG. 9.3 *Torque–speed characteristics*

Therefore, force or tractive effort is given by:

$$F = \frac{P_2}{V_s},$$

where 'P_2' is the actual power supply to the rotor.

Advantages

(i) Simple in construction.

(ii) Low initial cost.

(iii) Maintenance cost is low.

(iv) Maximum speed is not limited due centrifugal forces.

(v) Better power to weight ratio.

Disadvantages

(i) High cost of providing collector system.

(ii) Poor efficiency and low power factor, due to high currents drawn by the motor because of large air gap.

Applications

Linear induction motor are generally used in:

- High-speed rail traction.
- Trolley cars and metallic belt conveyors.
- Electromagnetic pumps.

9.7.6 Synchronous motor

The synchronous motor is one type of AC motor working based upon the principle of magnetic lacking. It is a constant speed motor running from no-load to full load. The construction of the synchronous motor is similar to the AC generator; armature winding is excited by giving three-phase AC supply and field winding is excited by giving DC supply. The synchronous motor can be operated at leading and lagging power factors by varying field excitation.

The synchronous motor can be widely used various applications because of constant speed from no-load to full load.

- High efficiency.
- Low-initial cost.
- Power factor improvement of three-phase AC industrial circuits.

9.8 BRAKING

If at any time, it is required to stop an electric motor, then the electric supply must be disconnected from its terminals to bring the motor to rest. In this method, even though supply is cut off, the motor continue to rotate for long time due to inertia. In some cases, there is delay in bringing the other equipment. So that, it is necessary to bring the motor to rest quickly. The process of bringing the motor to rest within the pre-determined time is known as braking.

A good braking system must have the following features:

- Braking should be fast and reliable.
- The equipment to stop the motor should be in such a way that the kinetic energy of the rotating parts of the motor should be dissipated as soon as the brakes are applied.

Braking applied to bring the motor to rest position is of two types and they are:

(i) Electric braking.
(ii) Mechanical braking.

9.8.1 Electric braking

In this process of braking, the kinetic energy of the rotating parts of the motor is converted into electrical energy which in turn is dissipated as heat energy in a resistance or in sometimes, electrical energy is returned to the supply. Here, no energy is dissipated in brake shoes.

9.8.2 Mechanical braking

In this process of braking, the kinetic energy of the rotating parts is dissipated in the form of heat by the brake shoes of the brake lining that rubs on a wheel of vehicle or brake drum.

The advantages of the electric braking over the mechanical braking

- The electric braking is smooth, fast, and reliable.
- Higher speeds can be maintained; this is because the electric braking is quite fast. This leads to the higher capacity of the system.
- The electric braking is more economical; this is due to excessive wear on brake blocks or brake lining that results frequent and costly replacement in mechanical braking.
- Heat produced in the electric braking is less and not harmful but heat produced in the mechanical braking will cause the failure of brakes.
- In the electric braking, sometimes, it is possible to fed back electric energy during braking period to the supply system. This results in saving in the operating cost. This is not possible in case of mechanical braking.

Disadvantages

In addition to the above advantages, the electric braking suffers from the following disadvantages.

- During the braking period, the traction motor acts generator and electric brakes can almost stop the motor but it cannot hold stationary. Hence, it is necessary to employ mechanical braking in addition to electric braking.
- Traction motor has to work as a generator during braking period. So that, motor has to select in such a way that it should have suitable braking characteristics.
- The initial cost of the electric braking equipment is costlier.

9.9 TYPES OF ELECTRIC BRAKING

Electric braking can be applied to the traction vehicle, by any one of the following methods, namely:

(i) Plugging.
(ii) Rehostatic braking.
(iii) Regenerative braking.

9.9.1 Plugging

In this method of braking, the electric motor is reconnected to the supply in such a way that it has to develop a torque in opposite direction to the movement of the rotor. Now, the motor will decelerates until zero speed is zero and then accelerates in opposite direction. Immediately, it is necessary to disconnect the motor from the supply as soon as system comes to rest.

The main disadvantage of this method is that the kinetic energy of the rotating parts of the motor is wasted and an additional amount of energy from the supply is required to develop the torque in reverse direction, i.e., in this method, the motor should be connected to the supply during braking. This method can be applied to both DC and AC motors.

Plugging applied to DC motors

Pulling is nothing but reverse current braking. This method of braking can be applied to both DC shunt and DC series motors by reversing either the current through armature or the field winding in order to produce the torque in apposite direction, but not both. The connection diagrams for both DC shunt and DC series motors during normal and braking periods are given as follows.

The connection diagram for normal running conditions of both DC shunt and DC series motors are shown in Figs. 9.4 (a) and 9.5 (a). The back emf developed by the motor

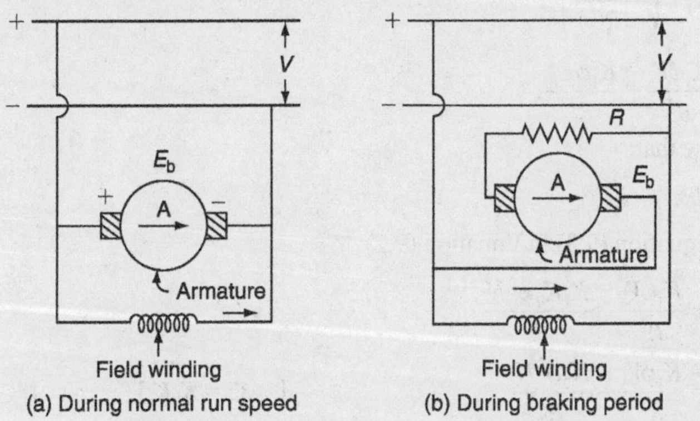

FIG. 9.4 *Plugging of DC shunt motor*

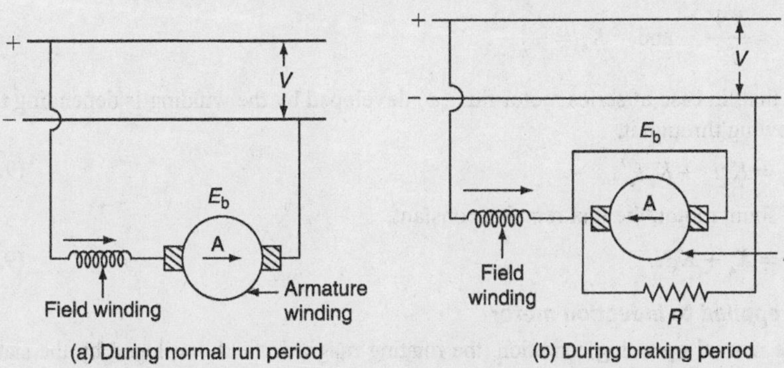

FIG. 9.5 *Plugging of DC series motor*

is equal in magnitude and same as to the direction of terminal or supply voltage. During the braking, the armatures of both shunt and series motors are reversed as shown in Fig. 9.4 (b) and Fig. 9.5 (b). Now, the back emf developed by the motor direction of terminal voltage. A high resistance '*R*' is connected in series with the armature to limit high-starting current during the braking period.

Current flowing through the armature during normal run condition:

$$I_1 = \frac{V - E_b}{R_a}, \tag{9.1}$$

where V is the supply voltage, E_b is the back emf, and R_a is the armature resistance.

Current flowing through the armature during braking period:

$$I_2 = \frac{V - (-E_b)}{R_a + R}$$

$$= \frac{V + E_b}{R_a + R} = \frac{V + E_b}{R'} \qquad [\because R' = R_a + R].$$

$\therefore$ Electric braking torque, $T_B \propto \phi\, I_2$.

$$T_B = K_1\, \phi\, I_2$$

$$= K_1\, \phi \left(\frac{V + E_b}{R'} \right)$$

$$= \frac{K_1 \phi V}{R'} + \frac{K_1 \phi E_b}{R'}. \tag{9.2}$$

But we know that:

$$E_b \propto N\phi. \tag{9.3}$$

Substitute Equation (9.3) in Equation (9.2):

$$\therefore T_B = \frac{K_1 \phi V}{R'} + \frac{K_1 K_2 \phi^2 N}{R'}$$

$$= \frac{K_1 \phi V}{R'} + \frac{K_3 \phi^2 N}{R'} \qquad [\because K_3 = K_1 K_2]$$

$$= K_4 \phi = K_5 \phi^2, \tag{9.4}$$

where $K_4 = \frac{K_1 V}{R'}$ and $K_4 = \frac{K_3 N}{R'}$.

We know that, in case of series motor flux (ϕ) developed by the winding is depending the current flowing through it.

$$\therefore T_B = K_6 I_a + K_7 I_a^{\,2}. \tag{9.5}$$

In case of shunt motor, the flux remains constant.

$$\therefore T_B = K_4 + K_5 N. \tag{9.6}$$

Plugging applied to induction motor

During the normal operating condition, the rotating magnetic field developed by the stator and the rotation of rotor are in the same direction. But during the braking period, plugging

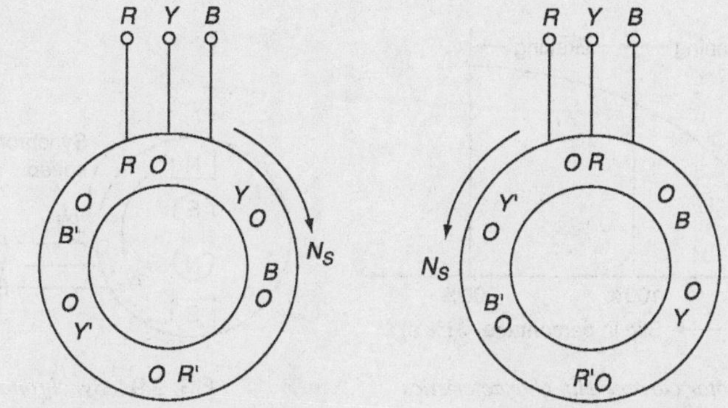

(a) During normal run condition (b) During braking period

FIG. 9.6 *Plugging applied to induction motor*

is applied to an induction motor by reversing any two phases of the three phases of stator winding in order to change the direction of the rotating magnetic field as shown in Fig. 9.6. So that, the rotating magnetic filed and the rotor will be rotating in opposite direction. So that, the relative speed between emf and rotor is nearly twice the synchronous speed $N_s - (-N_s) = 2N_s$.

$\therefore$ Slip during the braking period:

$$S = \frac{-N_s - N_s}{N_s} = \frac{-2N_s}{N_s} = -2.$$

But the voltage induced in the rotor (E_2) is proportional to the slip (S) $\times$ stator voltage (V):

$$\therefore E_2 \propto SV.$$

So, the rotor voltage during the braking period is twice the normal voltage. To avoid the damage of the rotor winding, it should be provided with additional insulation, to withstand the high induced voltage.

The rotation of the magnetic field in the reverse direction produce torque in reverse direction; thereby applying the brakes to the motor. The braking of induction motor can be analyzed by the torque–slip characteristics shown in Fig. 9.7.

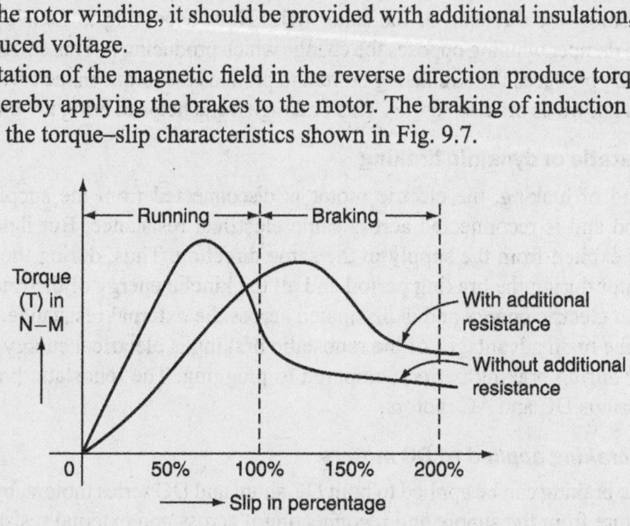

FIG. 9.7 *Torque–slip characteristics*

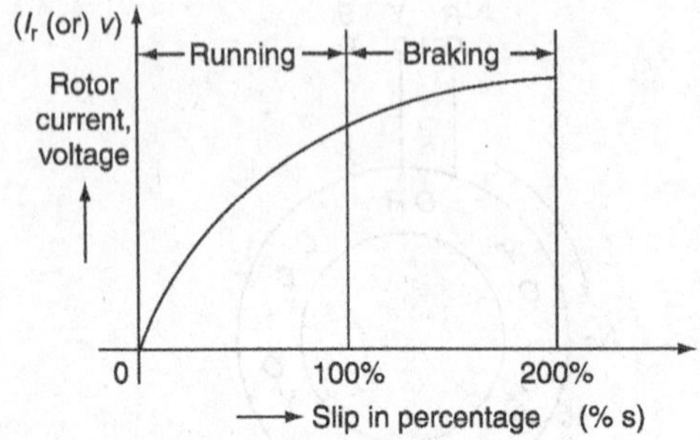

FIG. 9.8 *Rotor current–slip characteristics*

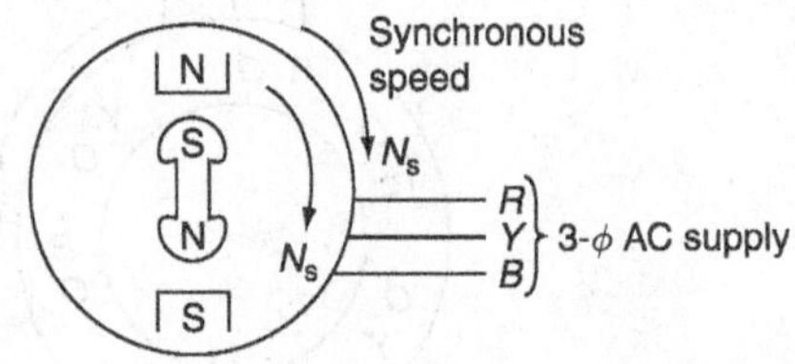

FIG. 9.9 *Synchronous motor*

$$\text{Rotor current during the braking period, } I_{2B} = \frac{SE_2}{\sqrt{R_2^2 + (SX^2)^2}}.$$

The characteristic curve for the rotor current and the rotor voltage with the variation of the slip is shown in Fig. 9.8.

Plugging applied to synchronous motor

Normally, the stator winding of the synchronous motor is fed with 3-ϕ AC supply to produce the rotating magnetic field that induces stator poles. And, the field winding is excited by giving the DC supply thereby inducing the rotor poles. At any instant, the stator poles gets locked with the rotor poles and the synchronous motor rotating at the synchronous speed. In this method of plugging applied to synchronous motor, simply it is not possible to produce the counter torque during the braking period by interchanging any two of three phases. This is due to the magnetic locking of stator and rotor poles (Fig. 9.9).

In order to develop the counter torque, the rotor of synchronous motor should be provided with damper winding. The EMF induced in the damper winding whenever there is any change, i.e., the reversal of the direction of the stator field. Now, according to Lenz's law, the emf induced in the damper winding opposes the change which producing it. This emf induced in the damper winding produces the circulating current to produce the torque in the reverse direction. This torque is known as braking torque. This braking torque helps to bring the motor to rest.

9.9.2 Rheostatic or dynamic braking

In this method of braking, the electric motor is disconnected from the supply during the braking period and is reconnected across same electrical resistance. But field winding is continuously excited from the supply in the same direction. Thus, during the starts working as generator during the braking period and all the kinetic energy of the rotating parts is converted into electric energy and is dissipated across the external resistance.

One of the main advantages of the rehostatic braking is electrical energy is not drawn by the motor during braking period compared to plugging. The rehostatic braking can be applied to various DC and AC motors.

Rehostatic braking applied to DC motors

The rheostatic braking can be applied to both DC shunt and DC series motors, by disconnecting the armature from the supply and reconnecting it across and external resistance. This is required to dissipate the kinetic energy of all rotating parts thereby brining the motor to rest.

DC shunts motor

Figure 9.10 shows the connection diagram of the DC shunt motor during both normal and braking conditions. In case of DC shunt motor, both armature and field windings are connected across the DC supply, as shown in Fig. 9.10 (a.)

During the braking period, the armature is disconnected from the supply and field winding is continuously excited by the supply in the same direction, as shown in Fig. 9.10 (b). The kinetic energy of all rotating parts is dissipated in the resistor 'R' now the machine starts working as generator. Now, braking developed is proportional to the product of the field and the armature currents. But the shunt motor flux remains constant, so the braking torque is proportional to armature current at low-speeds braking torque is less and in order to maintain constant braking torque, the armature is gradually disconnected. Hence, the armature current remains same thereby maintaining the uniform braking torque.

DC series motor

In this braking, which is applied to DC series motor, the armature is disconnected from the supply and is reconnected across an external resistance 'R' shown in Fig. 9.11 (a) and (b). But, simply, it is not possible to develop the retarding torque by the DC series motor after connecting armature across the resistance as DC shunt motor.

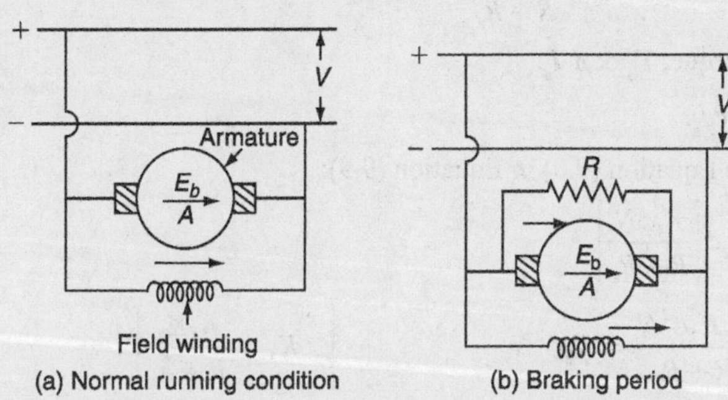

FIG. 9.10 *Rheostatic braking of DC shunt motor*

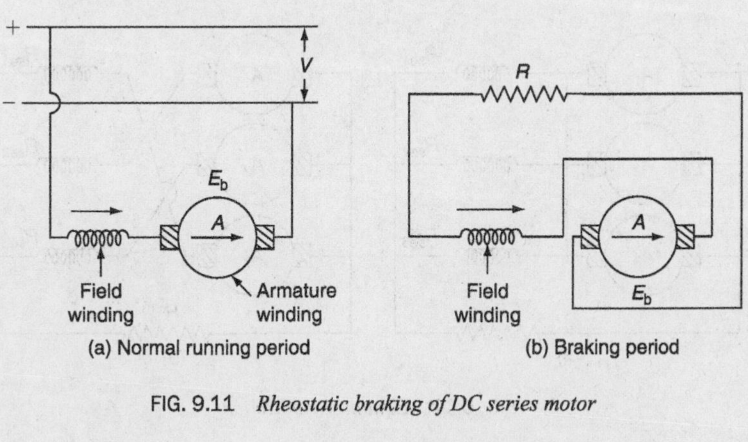

FIG. 9.11 *Rheostatic braking of DC series motor*

In case of DC series motor, both the field and armature windings are connected across the resistance after disconnecting the same from the supply; current directions of both the field and armatures are reversed. This results in the production of torque in same direction as before. So, in order to produce the braking torque only the direction of current in the armature has to be reversed. The connection diagram of DC series is shown in Fig. 9.11.

If more than one motor has to be used as in electric traction. All motors can be connected in equalizer connection as shown in Fig. 9.12. In this connection, one machine is excited by the armature current of another machine.

Braking torque

The current flowing through the armature during braking period:

$$I_a = \frac{E_b}{R+R_a}, \tag{9.7}$$

where E_b is the back emf developed, R is the external resistance, and R_a is the armature resistance.

And we know that, back emf $E_b \propto \phi N$

$$E_b = K_1\phi N.$$

$$\therefore \text{Braking current } I_a = \frac{K_1\phi N}{R+R_a}. \tag{9.8}$$

Braking torque, $T_B \propto \phi I_a$.

$$\therefore T_B = K_2\phi I_a. \tag{9.9}$$

Now, substitute Equation (9.8) in Equation (9.9):

$$\therefore T_B = K_2\phi\left[\frac{K_1\phi N}{R+R_a}\right]$$

$$= \frac{K_1K_2\phi^2 N}{R+R_a} = K_3\phi^2 N \qquad \left[\because K_3 = \frac{K_1K_2}{R+R_a}\right].$$

For shunt motor flux is practically constant:

$$\therefore T_B = K_5 I_a^2 N. \tag{9.10}$$

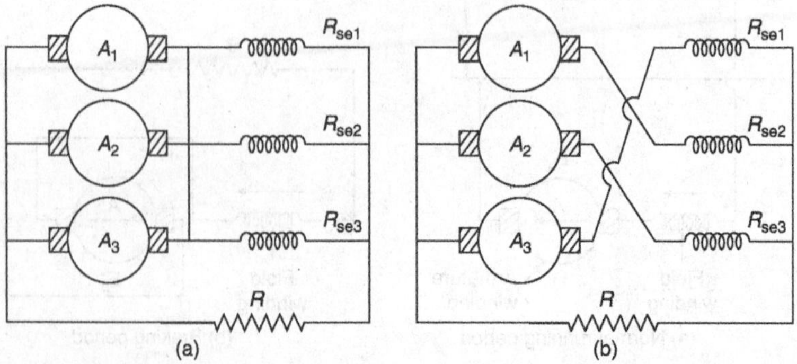

FIG. 9.12 *Equalizer connection*

Rehostatic braking applied to induction motor

In case of an induction motor, normally, under running condition, the stator is fed from AC supply. If rehostatic braking applied to an induction motor, stator must be disconnected from supply; therefore, there is no rmf, no rotor short-circuit current, and no retarding torque produced. To avoid the above difficulty, the stator must be excited by giving DC supply, to produce the constant air gap flux that is cut by the rotor conductors, which will induce currents in the short-circuited rotor. This rotor current will produce the required braking torque. This braking torque can be controlled either by controlling DC excitation or by varying rotor resistance. The various connections for giving to the stator winding are shown in Fig. 9.13.

9.9.3 Regenerative braking

Regenerative braking is the most efficient method of braking to stop the motor. In previous method of rehostatic braking, the kinetic energy of all rotating parts is wasted in external braking resistor and in case of plugging extra energy is drawn from the supply during braking period. But in this method of braking, no energy is drawn from the supply during the braking period and some of the energy is fed back to the supply system.

Regenerative braking can be applied to both DC and AC motors.

Regenerative braking applied to DC shunt motor

In case of DC shunt motor, energy can be fed back to the supply system whenever rotational emf is more than the supply voltage. During the braking period, the excitation and speed of DC shunt motor are suitably adjusted such that the rotational emf is more than the supply voltage ($E_b > V_t$). Since, back emf or rotational emf is directly proportional to the field flux and the speed of the rotation of the shaft of the machine.

Now, a motor acts as generator and the direction of current through armature is reversed $\left[\because I_a = \frac{V - E_b}{R_a}\right]$ so that, the torque developed by the armature is reversed.

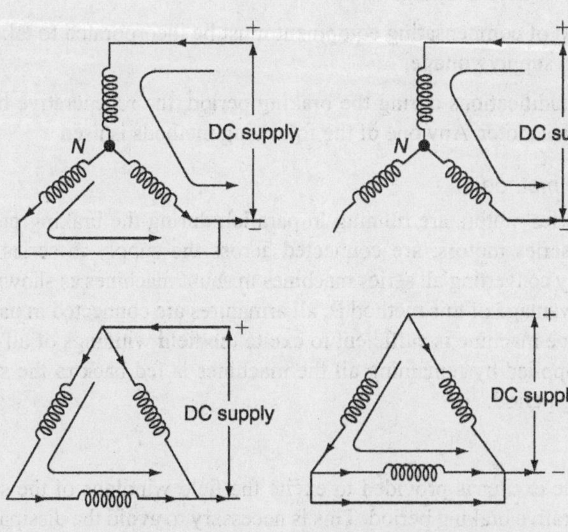

FIG. 9.13 *DC excitation to three-phase winding*

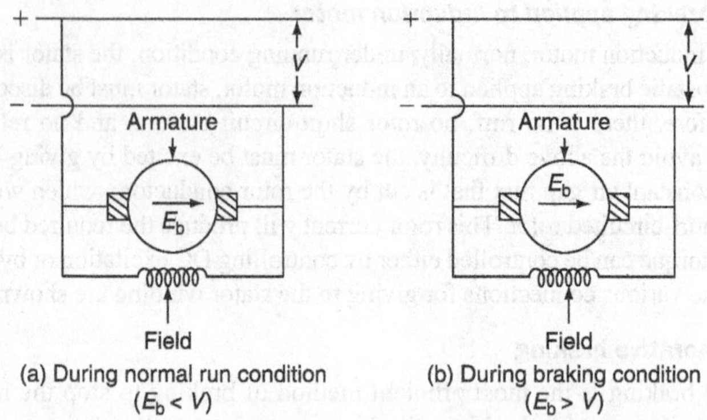

FIG. 9.14 *Regenerative braking of DC shunt motor*

This retarding torque helps to bring motor to rest. Connection diagram of DC shunt motor for regenerative braking is shown in Fig. 9.14.

DC series motor

In case of DC series motor, it is not easy to apply regenerative braking as of DC shunt motor. The main reasons of the difficulty of applying regenerative braking to DC series motor are:

(i) During the braking period, the motor acts as generator by reversing the direction of current flowing through the armature, but at the same time, the current flowing through the field winding is also reversed; hence, there is no retarding torque. And, a short-circuit condition will set up both back emf and supply voltage will be added together. So that, during the braking period, it is necessary to reverse the terminals of field winding.

(ii) Some sort of compensating equipment must be incorporated to take care of large change in supply voltage.

On doing some modifications during the braking period, the regenerative braking can be applied to DC series motor. Any one of the following methods is used.

Method-I (French method)

If one or more series motors are running in parallel, during the braking period, the field windings, of all series motors, are connected across the supply in series with suitable resistance. Thereby converting all series machines in shunt machines as shown in Fig. 9.15.

The main advantage of this method is, all armatures are connected in parallel and current supplied to one machine is sufficient to excite the field windings of all the machines, and the energy supplied by remaining all the machines is fed back to the supply system, during the braking period.

Method-II

In this method, the exciter is provided to excite the field windings of the series machine during the regenerative braking period. This is necessary to avoid the dissipation of energy or the loss of power in the external resistance.

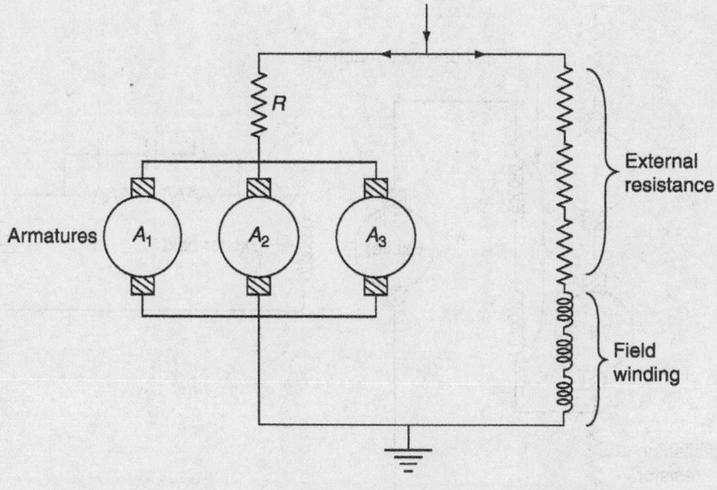

FIG. 9.15 *Regenerative braking of DC series motor*

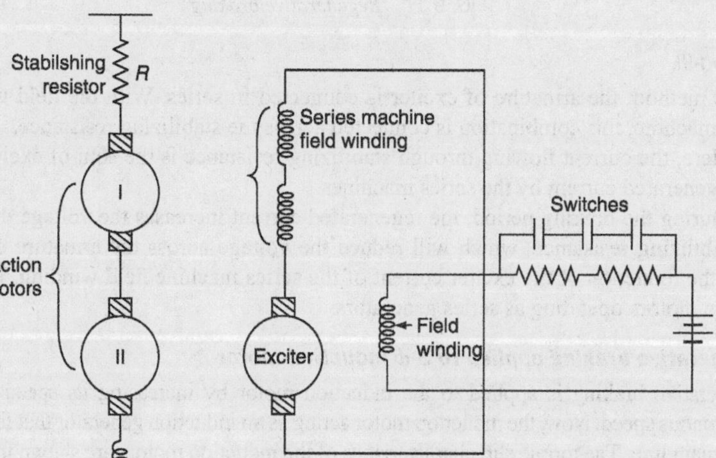

FIG. 9.16 *Regenerative braking*

Whenever the excitation of field winding is adjusted to increase the rotational emf more than the supply voltage, then the energy is supplied to the supply system. At that time, the field winding of the series machine is connected across an excited being driven by motor operated from an auxiliary supply. Now, during the braking period, the series machine acts as separately excited DC generator which supplies energy to the main lines. A stabilizing resistance is used to control the braking torque (Figs. 9.16 and 9.17).

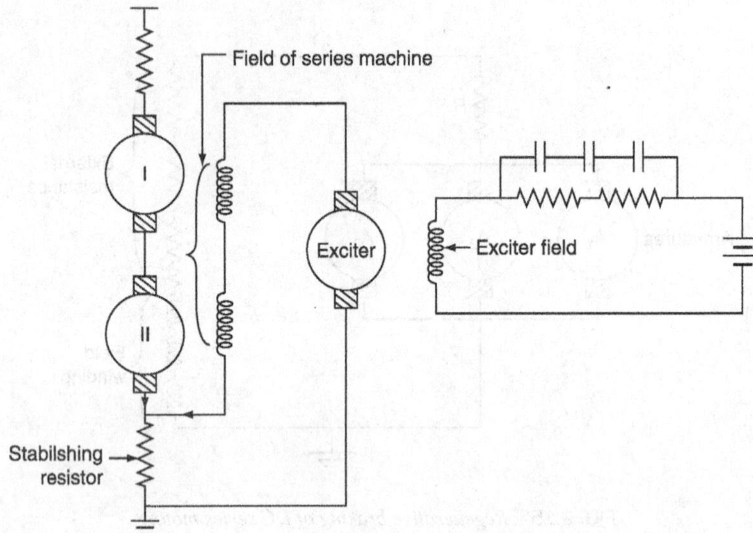

FIG. 9.17 *Regenerative braking*

Method-III

In this method, the armature of exciter is connected in series. With the field winding of series machine, this combination is connected across the stabilizing resistance.

Here, the current flowing through stabilizing resistance is the sum of exciter current and regenerated current by the series machines.

During the braking period, the regenerated current increases the voltage drop across the stabilizing resistance, which will reduce the voltage across the armature circuit and cause the reduction of the exciter current of the series machine field winding. Hence, the traction motors operating as series generators.

Regenerative braking applied to 3-ϕ induction motor

Regenerative braking is applied to the induction motor by increasing its speed above the synchronous speed. Now, the induction motor acting as an induction generator that feeds power to the main line. The torque slip characteristics of the induction motors are shown in Fig. 9.18.

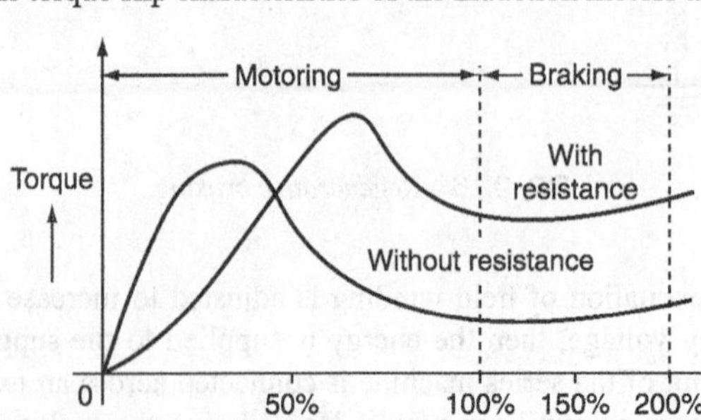

FIG. 9.18 *Torque vs slip characteristics*

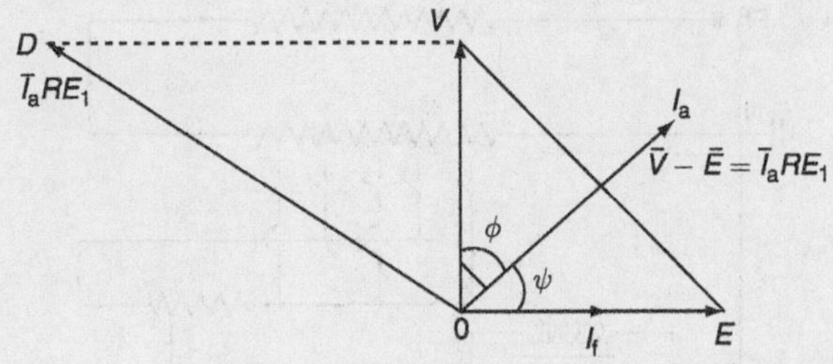

FIG. 9.19 *Phasor diagram*

The main advantage of the induction motor is during the braking period; no need of placing external resistance in the rotor circuit. The speed during the braking remains almost constant and independent of the gradient and the weight of the train.

This regenerative braking applied to an induction motor can save 20% of the total energy leads the reduction of operating cost.

Regenerative braking applied to AC series motors

It is not simple way to apply regenerative braking to an AC series motor. In this method, the armature of traction motor is connected to the top changing transformer through iron cored reactors RE_1 and RE_2 and commutating pole winding '*C*'.

An auxiliary transformer is provided to excite the field winding of the traction motor. Let us assume '*V*' be the voltage of tap-changing transformer and I_f is the field current of traction motor. Due to the presence of reactor, I_f lags V by an angle 90° of traction motor is phase with exciting current as shown in Fig. 9.19.

From the phasor diagram, the vector difference of $\bar{V}$ and $\bar{E}$ gives voltage across iron-cored reactor RE_1. Now, the armature current I_a lags $\left(\bar{I}_a\ RE_1\right)$ by 90°. And, the braking torque developed the series machine will be proportional to $I_a \cos\phi$. And, the power returned to the supply is also proportional $I_a \cos\phi$. So that, proper phase angle must be obtained for efficient braking effect arise in the regenerative braking applied to an AC series motor are:

- During the regenerative braking, the braking torque is proportional to the operating power factor. In order to operate the series motor at high power factor field, winding must be excited separately from other auxiliary devices.
- Proper phase-shifting device must be incorporated to ensure correct phase angle.

To overcome the difficulty stated above, a special arrangement is adopted that is known as Behn Eschenburg method of regenerative braking.

The circuit diagram for applying regenerative braking to an AC series motor is shown in Fig. 9.20.

9.10 TRACTION MOTOR CONTROL

Normally, at the time of starting, the excessive current drawn by the electric motor from the main supply causes to the effects. So that, it is necessary to reduce the current drawn by the traction motor for its smooth control such as:

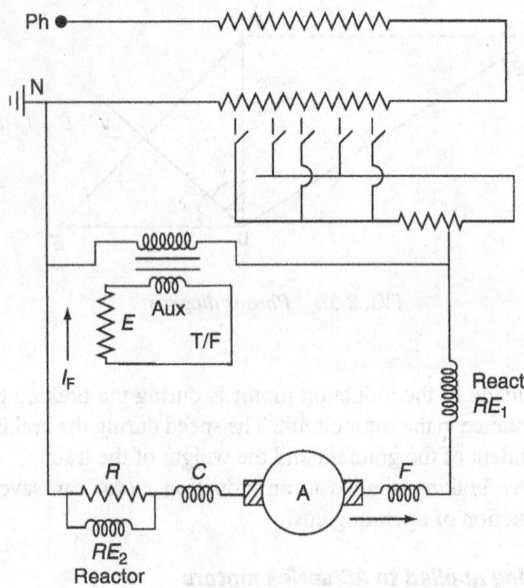

FIG. 9.20 *Regenerative braking of AC series motor*

(i) To achieve smooth acceleration without any jerking and sudden shocks.

(ii) To prevent damage to coupling.

(iii) To achieve various speed depending upon the type of services.

9.10.1 Control of DC motors

At the time of starting, excessive current is drawn by the traction motor when rated voltage is applied across its terminals. During the starting period, the current drawn by the motor is limited to its rated current. This can be achieved by placing a resistance in series with the armature winding. This is known as starting resistance; it will be cut off during the normal running period thereby applying rated voltage across its armature terminals. By the resistance of stating resistor, there is considerable loss of energy takes place in it.

∴ At the time of switching on, the back emf developed by the motor $E_b = 0$.

∴ Supply voltage, $V = I_a R_a + V_s$, **(9.11)**

where V_s is the voltage drop across starting resistance and $I_a R_a$ is the voltage drop in armature.

During the running condition:

$$V = I_a R_a + V_s + E_b. \quad \textbf{(9.12)}$$

At the end of accelerating period, the total starting resistance will be cut off from the armature then:

$$V = I_a R_a + E_b. \quad \textbf{(9.13)}$$

(a) Various drops during staring and running with armature resistance.

(b) Various drops during staring and running with negligible armature resistance.

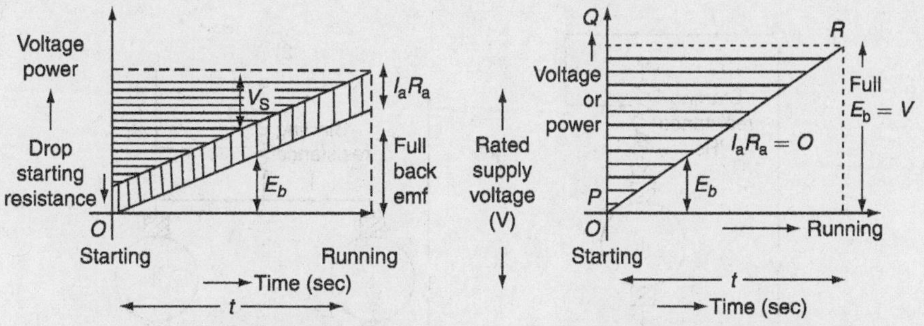

(a) Various drops during starting and running with armature resistance

(b) Various drops during starting and running with negligible armature resistance

FIG. 9.21 *Traction control of DC motor*

When armature resistance is neglected $R_a = 0$ and 't' is the time in seconds for starting, then total energy supplied is, VI_at watts-sec and the energy wasted in the starting resistance at the time of starting can be calculated from Fig. 9.21(b) as:

$= \text{Area of } \Delta^{le}PQR \times I_a$

$= \frac{1}{2}tVI_a$

$$= \frac{1}{2}VI_at\ W\text{-sec.} \tag{9.14}$$

That is whatever the electrical energy supplied to the motor, half of the energy is wasted during the starting resistor.

∴ The efficiency of the traction motor at time of starting, $\eta_{start} = 50\%$.

9.10.2 Series–parallel control

This method of traction motor control employs two motors. These motors are connected in series at the time of starting to achieve lower speeds and thereafter those are connected in parallel to achieve full speed during running condition. So, this method is known as the series–parallel control of the traction motor.

When traction motor control employing series–parallel control, at the time of starting, the motors are connected in series with starting resistance shown in Fig. 9.22 (a). Then, back emf developed by each motor is equal to half of the supply voltage minus IR drop. At this instant, traction motor control gives one running speed, due to the series connection of motors. Now, two motors are connected in parallel and in series with the starting resistance to get full running speed, as shown in Fig. 9.22 (b). Switch on the supply to the motor at the time of starting and gradually cut out the starting resistance. Then, back emf developed by each motor is equals to supply voltage minus IR drop. So that, the speed obtained during the running condition of the parallel connection is higher than the series connection (speed is proportional to the back emf developed by the motor).

Series operation

Let us consider two motors that are connected in series with the help of the starting resistance 'R_s' as shown in Fig. 9.22 (a). The current drawn by each electric motor is limited to its

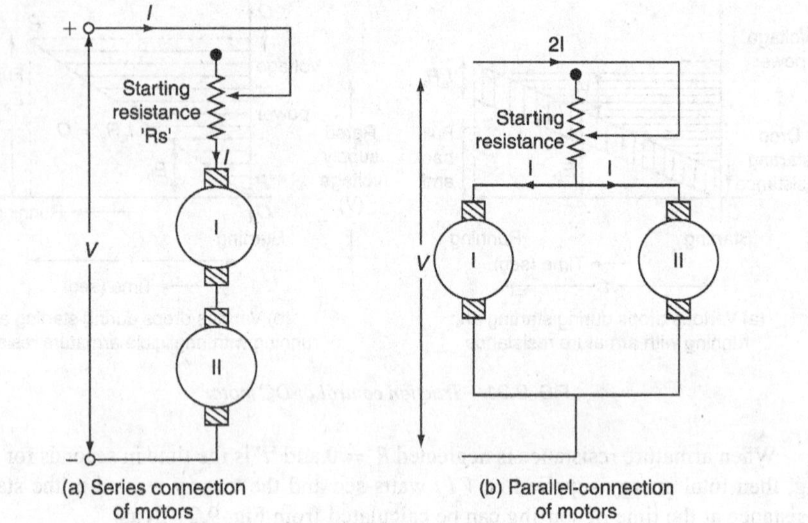

FIG. 9.22 *Series–parallel starting*

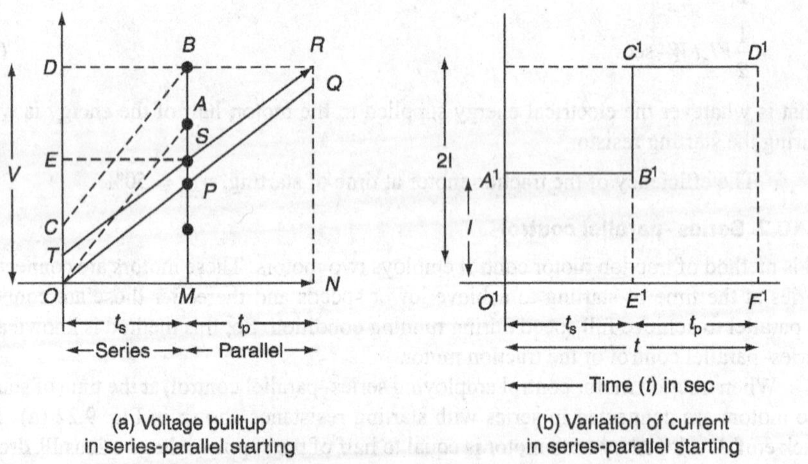

FIG. 9.23 *Series operation*

normal value by incorporating a starting resistance in series with the motors. Figure 9.23 (a) shows the voltage build-up by the motors both in series and parallel startings.

At the time of starting, $OT = TC = IR_a$ drop in each motor

OD = supply voltage.

Any point on the line 'BC' represents the sum of the back emf of two motors + IR_a drops of two motors + IR_s drop.

At any point 'M', i.e., at the end of series running period, the back emf developed by each motor:

$$E_b = \frac{V - 2IR_a}{2} = \frac{V}{2} - IR_a.$$

Now, the back emf developed by each motor is represented by the ordinate 'MS' in Fig. 9.23 (a).

$$\therefore MS - PS = E_b = \frac{V}{2} - IR_a.$$

Parallel operation

Now, at the instant 'M' motors are reconnected in parallel and this combination is in series with 'R_s'. Then, motors are switched onto the supply and current drawn by the parallel combination of motors from the supply is '$2I$'. The back emf developed by each motor in parallel connection at the time of starting is given by 'MP'.

And, the back emf developed by the motor is represented by 'PQ' during the running period. At the end of parallel running period, the back emf developed by the motor is equal to supply voltage minus IR_a drop of each motor.

$\therefore$ Supply voltage $V = NQ + QR$

= back emf developed by each motor + IR_a drop in each motor.

Determination of t_s, t_p, and η of starting

The duration of motors to remain in series connection (t_s) and in parallel connection (t_p) can be calculated from Fig. 9.23 (a) and (b).

Figure 9.22 (b) shows the variation of the current drawn by the motors during the series and the parallel connections.

From Fig. 9.23 (a), Δ^{lel} OPM and Δ^{lel} OQN are two similar triangles.

$$\therefore \frac{OM}{ON} = \frac{PM}{QN}$$

$$\therefore \frac{t_s}{t} = \frac{PM}{QN} = \frac{\frac{V}{2} - IR_a}{V - IR_a}$$

$$t_s = t\left[\frac{\frac{V}{2} - IR_a}{V - IR_a}\right]$$

$$= \frac{1}{2}\left[\frac{V - 2IR_a}{V - IR_a}\right] \times t. \tag{9.15}$$

$\therefore$ The duration of motors to remain in parallel connection:

$$t_p = t - t_s = t - \frac{1}{2}\left[\frac{V - 2IR_a}{V - IR_a}\right] \times t$$

$$= t\left[1 - \frac{1}{2}\left[\frac{V - 2IR_a}{V - IR_a}\right]\right], \tag{9.16}$$

where 't' is the total time of the motors to remain in both the series and the parallel connections.

Efficiency of starting(η)

If the armature resistance, R_a is neglected then drop $IR_a = 0$. And, from the Fig. 9.23 (a), at the end of series connection, the back emf developed by the motor is exactly equal to the supply voltage, and $DB = BR$, i.e., the duration of motors to remain in series combination is equals to the duration of motors to remain in parallel combination (Fig. 9.24).

$$\therefore t_s = t_p = \frac{t}{2}.$$

Energy dissipated in the starting resistance (R_s) = area under Δ^{lel} ODB +area under Δ^{lel} BSR

$$= \frac{1}{2} VI_a \times \frac{t}{2} + \frac{1}{2} \times \frac{V}{2} \times \frac{t}{2} \times 2I_a$$

$$= \frac{VI_a t}{2}.$$

$$\therefore \text{Total energy supplied} = V\,I_a \times \frac{t}{2} + V \times 2I_a \times \frac{t}{2}$$

$$= \frac{3VI_a t}{2}.$$

$\therefore$ The efficiency of starting:

$$\eta = \frac{\text{energy supplied} - \text{energy dissipated}}{\text{energy supplied}} \times 100$$

$$= \frac{\frac{3VI_a t}{2} - \frac{VI_a t}{2}}{\frac{3VI_a t}{2}} \times 100$$

$$= 2/3 \times 100 = 66.66\%.$$

If two motors are employed in series–parallel control, the efficiency of the starting is increased from 50% to 66.66%, i.e., the efficiency is increased by nearly by 17% thereby saving 15–20% in the energy. If four motors are employed for series–parallel control, the efficiency of starting can be increased to nearly 73%.

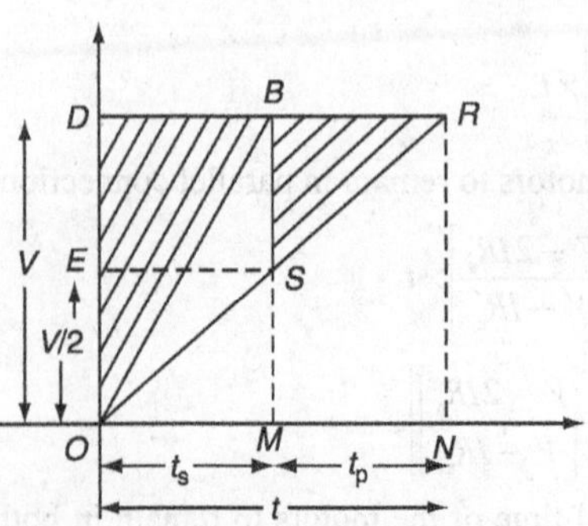

FIG. 9.24 *Parallel operation*

The series–parallel control change over connections can be carried out by the following methods:

a. Shunt transition.

b. Bridge transition.

Shunt transition

The connection diagram for the shunt transition of the series–parallel starting of the motor is shown in Fig. 9.25.

In series connection, two motors are in series by gradually cutting out the starting resistance in first four steps. Now, the series connection has to transform to shunt connection in Step '5' by reinserting the starting resistance.

Step 6: One motor is bypassed from the series connection.

Step 7: Bypassed motor is disconnected so that noticeable jerk is experienced by the vehicle.

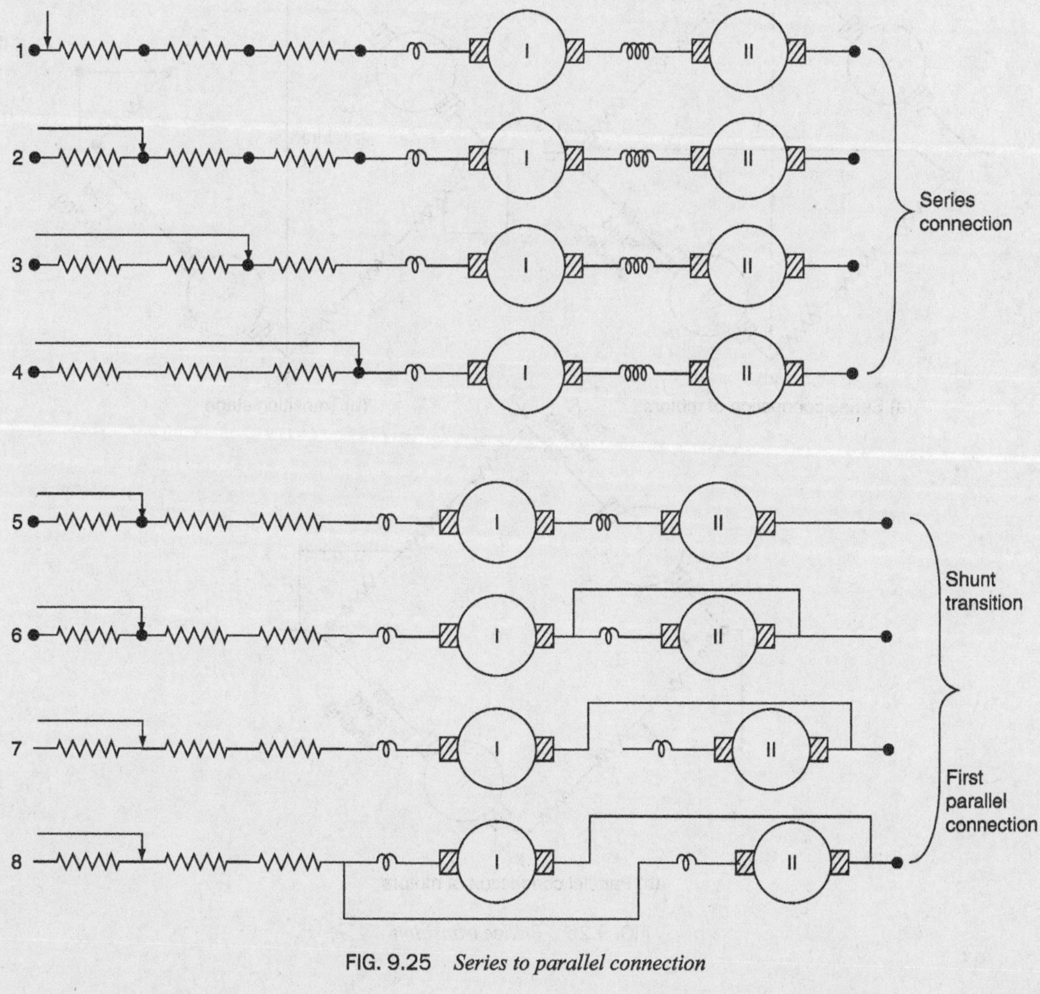

FIG. 9.25 *Series to parallel connection*

Step 8: In this step, the armature and field windings are directly connected as shown in Fig. 9.25, which gives the first parallel connection. Thereby changing the series connection of motors into parallel.

Bridge transition

In this method, normally, the motors and the starting resistances are connected in the form of a wheat stone bridge.

In changing the bridge connections, the starting resistances are connected to positive instead of joining the total starting resistance.

Now, the link 'L' is removed then the two motors connected in the parallel.

The bridge transition of the series connection of the motors into the parallel is shown in Fig. 9.26.

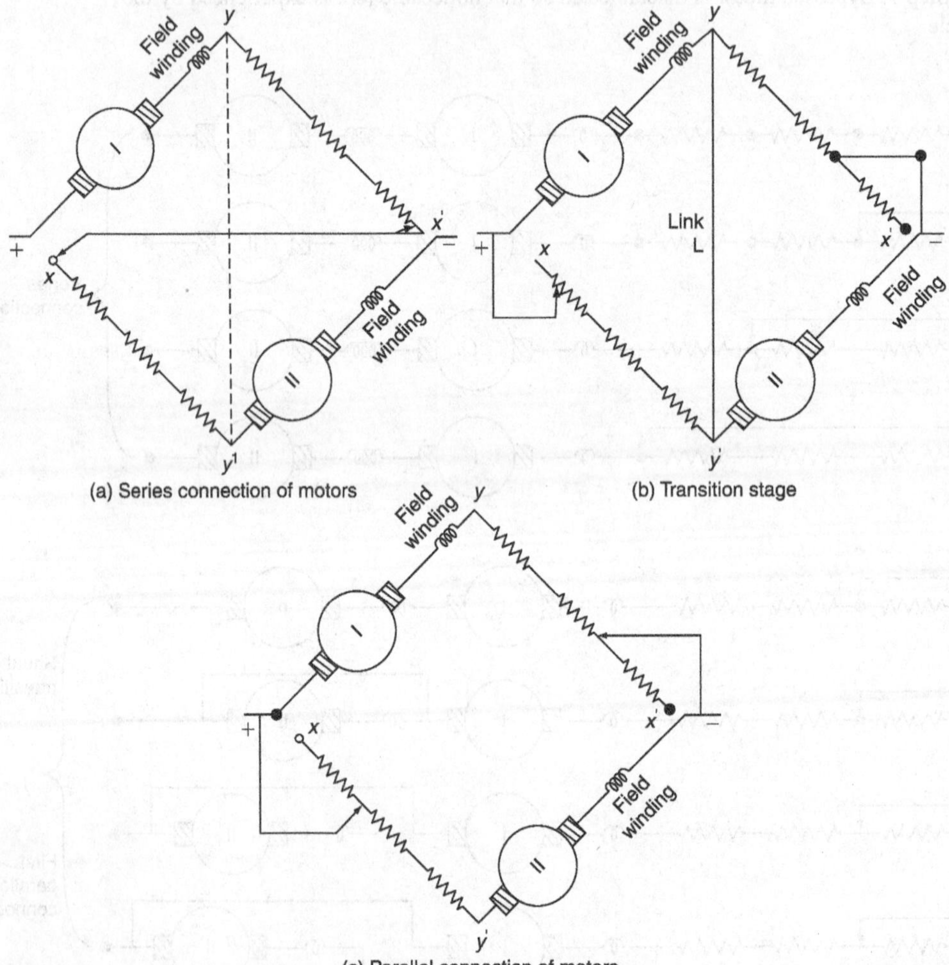

FIG. 9.26 *Bridge transition*

The main advantage of this method is, during the transition period, all the motors are connected to the supply unlike shunt transition; so that, the resistances can be adjusted to maintain constant current; the torque developed by the motors is constant. Hence, uniform acceleration can be obtained without any jerking and inconvenience to passengers.

9.11 OVER HEAD EQUIPMENT

Normally, the overhead equipment for tramways, trolley buses, and railways of 1,500 V and above is somewhat difficult to design than conductor-rail equipment. The material preferred for contact wires are silico-bronze, cadmium-copper, harddrawn copper, etc. Among all, silico-bronze is preferred due to its high electrical resistance, tensile strength, and wear-resisting properties.

9.11.1 Current collectors

Electric traction system uses the following current collecting systems for locomotives such as tramways and trolley buses.

Conductor–rail collectors

This conductor–rail system of supply has been used in many countries, for electric-traction. In this system, the current is supplied to the electrically operated vehicle through one-rail conductor or through two-rail conductors.

In case of one-rail conductor, the track rail is employed as the return conductor. The rails are mounted on insulators parallel with the track rails at a distance of 0.3–0.4 m from the running rail, whose upper most surfaces acting as contact surface and are fed at suitable points from the substations. When current drawn by the motor at starting is very large, this causes the wearing of rail conductor due to the traction of the collector shoes.

These rails are designed based on the electrical properties rather than mechanical properties. The main character based on which the rails are designed are (i) cost, (ii) shape and size of conductor rail, (iii) wearing qualities, (iv) electrical conductivity, and (v) contact surface available for the collector shoes.

The current is fed from the conductor rail to the train equipment through collector shoe, which is flat in shape with 20 cm in length and 7.6 cm in width. This system is suitable for heavy current collection, top contact system for voltages up to 750 V, and side contact system up to 1,200 V. The main advantage of this system is cheap and easy to repair and inspection (Fig. 9.27).

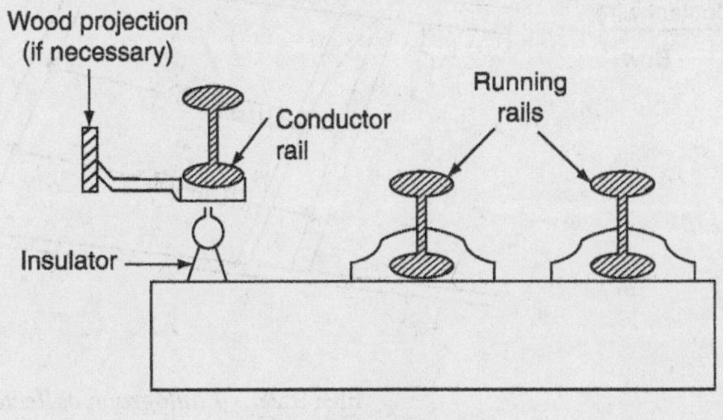

FIG. 9.27 *Conductor–rail system*

Trolley collectors

These trolley collectors are usually employed in tramways and trolley buses. This collector consists of grooved gun metal wheel trolley collector or grooved slide shoe with carbon attached to the long pole provided on the top of the vehicle. But, for trolley buses, two contact wires are to be required so that a separate trolley collector is provided for each wire.

The main drawback of the trolley collectors is, in order to reverse the direction of vehicle, collector has to be rotated through 180°. The trolley collector system is suitable for low speed of 22–30 kmph.

Bow collector

Bow collector consists of two trolley collector poles. At the end of these poles, a light metal strip of 1 m is placed for current collection. This bow collector has smaller inertia but it is not adoptable for the collection of large currents. The current collection metal strip is made up of soft materials such as copper, aluminum, or carbon. The main advantage of bow collector is that it can be used for high speeds (Fig. 9.28).

Pantograph collector

Pantograph is employed in electric traction system for the collection of currents. This collector is employed for the system whose operating speed is 100 or 130 kmph, and current to be collected are as large as 2,000 or 3,000 A.

Pantograph collectors are mounted on the roof of the vehicles similar to the bow collector. This collector carries a sliding shoe for contact with the overt head trolley wire. The contact shoes are usually about 1.2-m long. The material used for the pantograph is often steel, wearing plates of copper, or bronze inserted. There are basically two forms of pantograph collector namely 'diamond' type and single-ended 'faiveley' (Fig. 9.29).

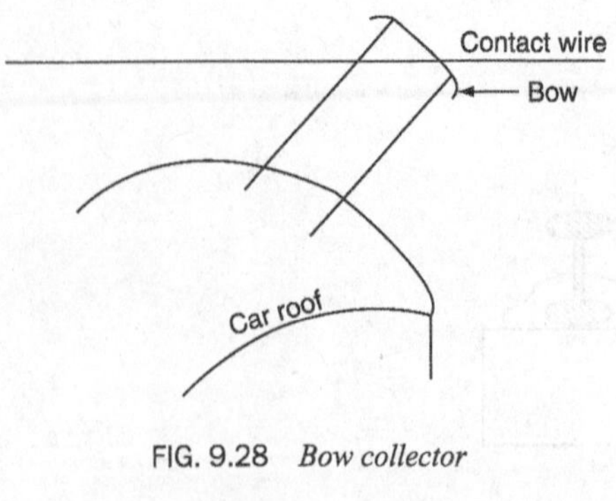

FIG. 9.28 *Bow collector*

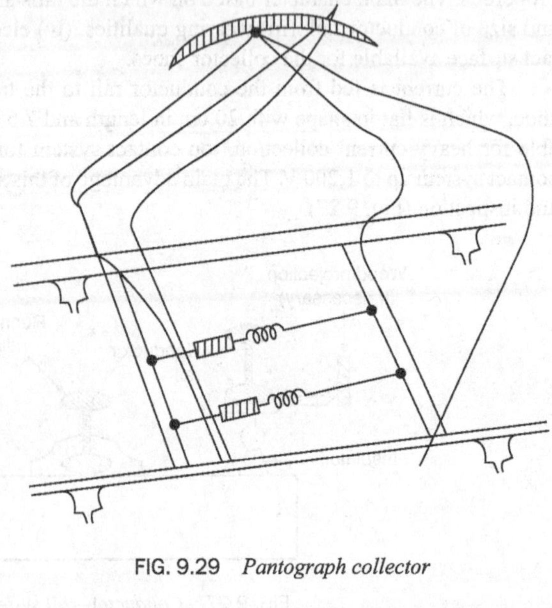

FIG. 9.29 *Pantograph collector*

9.11.2 Single catenary and compound catenary construction of railways

Railways for 1,500 V and above employ bow and pantograph collectors, depending upon their speeds. For such collectors, the contact wire is to be suspended with small sag to maintain high speeds between contact wire and the collector. This small sag of contact wire can be achieved without any excessive tension in the wire by employing the spans of comparatively short length (3–4.5 m).

For such short span suspension, a trolley wire is supported by another wire is known as '*catenary*' or '*messenger*' catenary, which is made up of more than seven strands of steel.

Normally, the catenary construction used for railways is of either single catenary construction or compound catenary construction.

Single catenary construction

In this construction, the catenary is made up of more than seven strands of still, which supports the trolley wire through dropper clipped to catenary. For a straight track, the span of catenary is 40–100 m with sag of 1–2 m, respectively, and the distance between droppers is nearly 2–5 m. Single catenary construction is shown in Fig. 9.30.

This type of construction with small sag provides sufficient fusibility for high speeds of 120 kmph without any interruption for the current collection. Single catenary construction is preferred to be suitable for most lightweight AC system whose operating speeds are low.

Compound catenary construction

In this construction, two messengers and one contact wire are arranged as shown in Fig. 9.31.

Main catenary, auxiliary catenary, and contact wires are arranged in the same vertical plane. Here, the auxiliary catenary is also known as '*intermediate catenary*'. The intermediate catenary not only increases the current carrying capacity but also provides more uniformity in elasticity.

9.12 AUXILIARY EQUIPMENT

A traction system comprises of the following auxiliary equipment in addition to the main traction motors required to be arranged in the locomotive are discussed below.

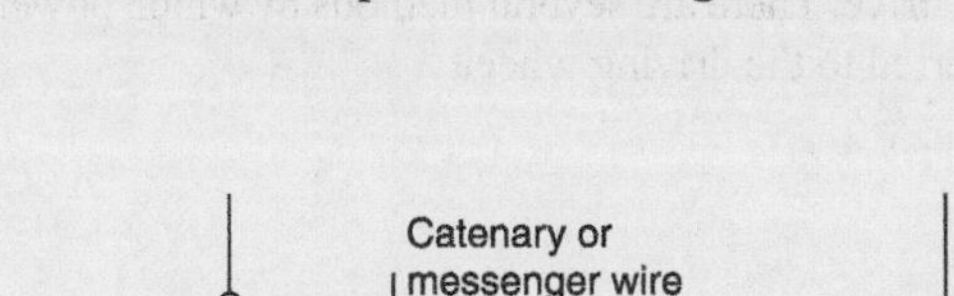

FIG. 9.30 *Single catenary construction*

FIG. 9.31 *Compound catenary construction*

9.12.1 Motor–generator set

Motor–generator set consists of a series motor and shunt generator. It is mainly used for lighting, control system, and the other power circuits of low voltages in the range 10–100 V. The voltage of generator is effectively controlled by automatic voltage regulator.

9.12.2 Battery

It is very important to use the battery as a source of energy for pantograph, to run auxiliary compressor, to operate air blast circuit breaker, etc. The capacity of battery used in the locomotive is depending on the vehicle. Normally, the battery may be charged by a separate rectifier.

9.12.3 Rectifier unit

If the track electrification system is AC motors and available traction motors are DC motors, then rectifiers are to be equipped with the traction motors to convert AC supply to DC to feed the DC traction motors.

9.12.4 Transformer or autotransformer

Depending on the track electrification system employed, the locomotive should be equipped with tap-changing transformers to step-down high voltages from the distribution network to the feed low-voltage traction motors.

9.12.5 Driving axles and gear arrangements

All the driving motors are connected to the driving axle through a gear arrangement, with ratios of 4:1 or 6:1.

9.13 TRANSMISSION OF DRIVE

Drive is a system used to create the movement of electric train. The electric locomotives are specially designed to have springs between the driving axles and the main body. This arrangement of springs reduces the damage not only to the track wings but also to the hammer blows.

The power developed by the armature of the traction motors must be transferred to the driving axels through pinion and gear drive. There are several methods by which power developed by the armature can be transferred to the driving wheel.

9.13.1 Gearless drive

Gearless drives are of two types.

Direct drive

It is a simple drive. The armatures of the electric motors are mounted directly on the driving axle with the field attached to the frame of locomotive. In this system, the poles of electric motors should be flat so that the armature can be able to move freely without affecting of the operation. Here, the size of the armatures of the traction motor is limited by the diameter of the driving wheels. The arrangement of direct drive is shown in Fig. 9.32.

Direct quill drive

Quill is nothing but a hallow shaft. Driving axle is surrounded by the hollow shaft attached by springs. The armature of the motor is mounted on a quill. The speed and the size of the armature are limited by the diameter of the driving wheels.

Geared drive

In this drive, the armature of the traction motor is attached to the driving wheel through the gear wheel system. Now, the power developed by the armature is transferred to the driving wheel through the gear system. Here, gear drive is necessary to reduce the size of the motor for given output at high speeds (Fig. 9.33). The gear ratio of the system is usually 3–5:1.

Brown–Boveri individual drive

In this drive, a special link is provided between the gear wheel and driving wheel, which provides more flexibility of the system.

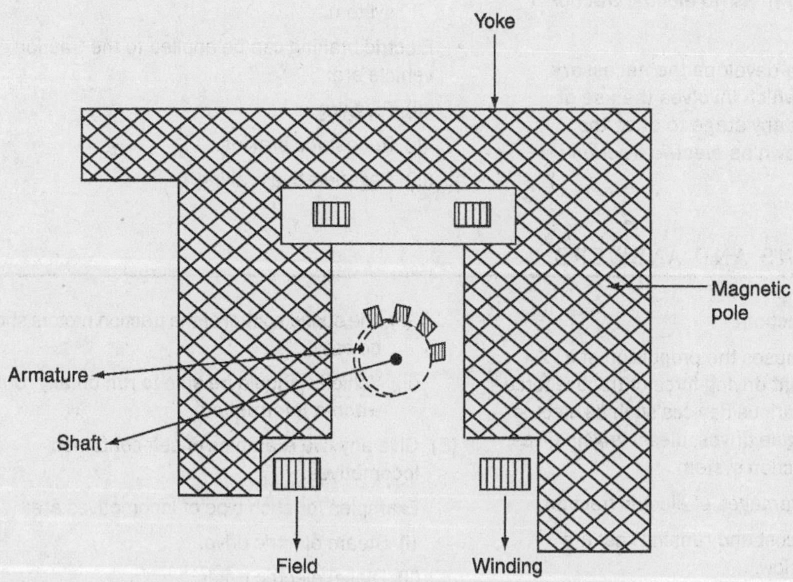

FIG. 9.32 *Direct drive*

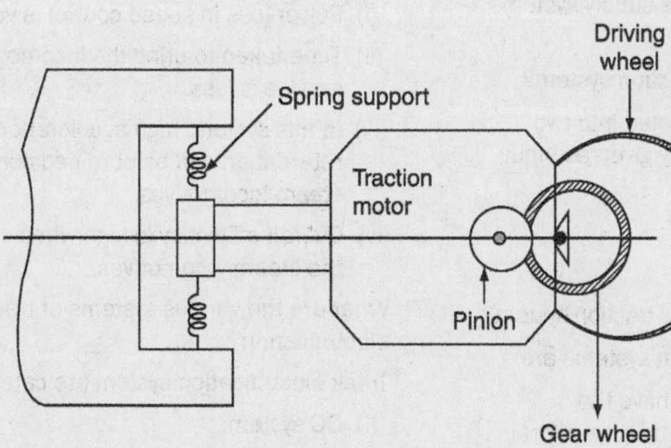

FIG. 9.33 *Geared drive*

KEY NOTES

- The system that causes the propulsion of a vehicle is known as *traction system*.
- *Traction system* is normally classified into two type:
  (i) Non-electric traction system.
  (ii) Electric traction system.
- The traction system develops the necessary propelling torque, which does not involve the use of electrical energy at any stage to drive the traction vehicle known as *no electric traction system*.
- The traction system develops the necessary propelling torque, which involves the use of electrical energy at any stage to drive the traction vehicle known as *electric traction system*.
- Based on the available supply, *track electrification system* are DC system:
  (i) Single-phase AC system.
  (ii) Three-phase AC system.
  (iii) Composite system.
- *Composite systems* are of two types:
  (i) Single-phase to DC system.
  (ii) Single-phase to three-phase system or kando system.
- Electric braking can be applied to the traction vehicle are:
  (i) Plugging.
  (ii) Rehostatic braking.
  (iii) Regenerative braking.

SHORT QUESTIONS AND ANSWERS

(1) What is electric traction?

The system that causes the propulsion of a vehicle in which that driving-force or tractive force is obtained from various devices such as electric motors, steam engine drives, diesel engine dives, etc. is know as traction system.

(2) Mention a few advantages of electric traction.

(i) Maintenance cost and running cost are comparatively low.
(ii) The speed control of the electric motor is easy.
(iii) Regenerative braking is possible so that the energy can be fed back to the supply system during the braking period.

(3) What are the types of electric traction system?

Traction system is normally classified into two types based on the type of energy given as input to drive the system and they are:

(i) Non-electric traction system.
(ii) Electric traction system.

(4) What are the requirements of ideal traction system?

The requirements of ideal traction systems are:

(i) Ideal traction system should have the capability of developing high tractive effort in order to have rapid acceleration.
(ii) The speed control of the traction motors should be easy.
(iii) Vehicles should be able to run on any route, without interruption.

(5) Give any two examples of self-contained locomotives.

Examples for such type of locomotives are:

(i) Steam electric drive.
(ii) Diesel electric trains.

(6) What are the advantages of self-contained locomotives?

(i) Power loss in speed control is very less.
(ii) Time taken to bring the locomotive into service is less.
(iii) In this system, high acceleration and braking retardation can be obtained compared to steam locomotives.
(iv) Overall efficiency is high when compared to the steam locomotives.

(7) What are the various systems of track electrification?

Track electrification system are categorized into:

(i) DC system.
(ii) Single-phase AC system.

(iii) Three-phase AC system.

(iv) Composite system.

(8) What are the causes lead to composite system?

1-ϕ AC system is preferable in view of distribution cost and in DC system. DC series motors have most desirable features and for 3-ϕ system, 3-ϕ induction motor has the advantage of automatic regenerative braking. So, it is necessary to combine the advantages of the DC/AC and 3-ϕ/1-ϕ systems. The above cause leads to the evolution of composite system.

(9) What is the need of electric braking?

The necessity of electric braking is, if at any time, it is required to stop an electric motor, then electric supply must be disconnected from its terminals to bring the motor to rest, to avoid accidents.

(10) What are the essential features of good braking system?

A good braking system must have the following features:

(i) Braking should be fast and reliable.

(ii) Equipment to stop the motor should be in such a way that the kinetic energy of rotating parts of motor should be dissipated as soon as the brakes are applied.

(11) What are the advantages of electric braking over mechanical braking?

The advantages of electric braking over mechanical braking are:

(i) Electric braking is smooth, fast, and reliable.

(ii) Heat produced in the electric braking is less and not harmful but heat produced in the mechanical braking will cause the failure of brakes.

(12) What are the disadvantages of electric braking?

Electric braking suffers from the following drawback. During the braking period, the traction motor acts as generator and the electric brakes can almost stop the motor but it cannot hold stationary. Hence, it is necessary to employ mechanical braking in addition to electric braking.

(13) What are the types of electric braking?

Electric braking can be applied to the traction vehicle by any one of the following methods:

(i) Plugging.

(ii) Rehostatic braking.

(iii) Regenerative braking.

(14) What is meant by plugging?

An electric motor is reconnected to the supply in such a way that it has to develop a torque in opposite direction to the movement of rotor is known as plugging.

(15) What is meant by rheostat or dynamic braking?

An electric motor is disconnected from the supply during the braking period and is reconnected across the same electrical resistance. But the field winding is continuously excited from the supply in the same direction to bring the motor to rest, which is known as rheostat braking.

(16) What is meant by regenerative braking?

The method of braking in which the back emf developed by the motor is adjusted more than the supply voltage is known as regenerative braking.

(17) What is the advantage of regenerative braking?

In this method of braking, no energy is drawn from the supply and some of the energy is fed back to the supply system.

MULTIPLE-CHOICE QUESTIONS

(1) Main traction systems used in India are those using:

(i) Steam engine locomotives.

(ii) Diesel engine locomotives.

(iii) Diesel electric locomotives.

(iv) Electric locomotives.

Of these, the correct statement(s) is/are:

(a) (i) and (ii).

(b) (i) and (iii).

(c) All.

(d) (iv).

(2) An ideal traction system should have:

(a) High-starting tractive effort and self-contained and compact locomotive of train unit.

(b) Equipment and capable of withstanding large temporary overloads of high efficiency and low-initial as well as maintenance cost.

(c) Easy speed control.

(d) All of the above.

(3) The steam engine provided on steam locomotive is:

(a) Double acting condensing type.

(b) Double acting non-condensing type.

(c) Single acting condensing type.

(d) Single acting non-condensing type.

(4) In case of steam locomotives, the tractive effort is provided by:

(a) Double cylinder, double acting steam engine.

(b) Double cylinder, single acting steam engine.

(c) Single cylinder, double acting stem engine.

(d) Single cylinder, single acting steam engine.

(5) In a steam locomotive, the electric power is provided through a/an:

(a) Small turbo-generator.

(b) Overhead wire.

(c) Diesel engine generator.

(d) Battery system.

(6) In case of a steam locomotive, an average coal consumption per km of run is around:

(a) 5–10 kg.

(b) 25–30 kg.

(c) 60–80 kg.

(d) 100–150 kg.

(7) The pressure of steam used in steam locomotives is about:

(a) 1–5 kgf/cm^2.

(b) 5–10 kgf/cm^2.

(c) 10–15 kgf/cm^2.

(d) 20–30 kgf/cm^2.

(8) The maximum horse power of steam locomotive is:

(a) Up to 1,500.

(b) 1,500–2,000.

(c) 2,000–3,000.

(d) 3,000–4,000.

(9) The speed of a steam locomotive is controlled by:

(a) Applying brakes.

(b) Gear box.

(c) Regulating steam flow to engine.

(d) Flywheel.

(10) The pulsating torque exerted by steam locomotives causes:

(a) Jolting and skidding.

(b) Hammer blow.

(c) Pitching.

(d) All of the above.

(11) Direct steam engine drive:

(a) Causes no interference to the communication lines running along the track.

(b) Needs low-initial investment in comparison to that for electric drive.

(c) Is cheap for low-density traffic areas in the initial stages of communication by rail.

(d) All of the above.

(12) Steam locomotives:

(a) Cannot be put into service at any moment.

(b) Cause considerable wear on the track.

(c) Need more repair and maintenance.

(d) All of the above.

(13) Steam engine drive is not suitable for urban and suburban services as:

(a) It is not clean drive.

(b) The coefficient of the adhesion of the steam locomotives is quite low.

(c) The steam locomotive causes considerable wear on the track.

(d) Steam locomotive has limited speed.

(14) Diesel electric traction has comparatively limited overload capacity because:

(a) Diesel engine is a constant output prime mover.

(b) Diesel engine has shorter life span.

(c) Regenerative braking cannot be employed.

(d) Diesel-electric locomotive is heavier than an ordinary electric locomotive.

(15) The range of the horse power of diesel locomotives is:

(a) 100–500.

(b) 1,500–2,500.

(c) 3,000–4,500.

(d) 4,500–5,000.

(16) The efficiency of the diesel locomotives is about:

(a) 75%.

(b) 50%.

(c) 25%.

(d) 10%.

(17) In India, diesel locomotives are manufactured at:

(a) Varanasi.

(b) Kolkata.

(c) Bangaluru.

(d) Ajmer.

(18) Electric traction in comparison to other traction systems has the advantage(s) of:

(a) Higher acceleration and braking retardation.

(b) Cleanest system and so ideally suitable for the underground and tube railways.

(c) Better speed control.

(d) All of the above.

(19) Electric railway can handle the traffic up to double the amount possible with steam railway. It is because of:

(a) Better speed control.

(b) Larger passenger carrying capacity.

(c) Higher schedule speed.

(d) Both (b) and (c).

(20) The most vital factor against electric traction is:

(a) Its high maintenance cost.

(b) The possibility of power failure.

(c) High initial cost in laying out overheat electric supply system.

(d) The necessity of providing negative booster.

(21) In India, the electric locomotives are manufactured at:

(a) Chittranjan.

(b) Varanasi.

(c) Bangaluru.

(d) Jabalpur.

(22) The first city to adopt electric traction in India was:

(a) Kolkata.

(b) Mumbai.

(c) New Delhi.

(d) Chennai.

(23) The latest traction system used in the world is:

(a) 3 phase, 3.7 kV.

(b) Single phase, 25 kV.

(c) DC 3 kV.

(d) DC 600 V.

(24) The voltage used for suburban railways in DC system is usually:

(a) 220 V.

(b) 600–750 V.

(c) 1,500–3,000 V.

(d) 15 kV.

(25) The voltage used for the main railways is:

(a) 600–750 V.

(b) 750–1,500 V.

(c) 1,500–3,000 V.

(d) 15 kV.

(26) Long-distance railways operate on:

(a) 600 V DC.

(b) 25 kV single-phase AC.

(c) 25 kV three-phase AC.

(d) 15 kV three-phase AC.

(27) The power supply frequency for 25-kV and single-phase AC system in India is:

(a) 50 Hz.

(b) 60 Hz.

(c) 25 Hz.

(d) 16 2/3 Hz.

(28) The traction motor used in tramways is:

(a) Three-phase induction motor.

(b) Single-phase AC series motor.

(c) DC series motor.

(d) DC shunt motor.

(29) The traction motor used in composite system employed in India is:

(a) An AC single-phase motor.

(b) A DC series motor.

(c) A DC shunt motor.

(d) A three-phase induction motor.

(30) Series motor has the main drawback of, when used in electric traction:

(a) Current surges after temporary interruption of supply.

(b) Self-relieving property.

(c) Low current drain on the heavy load torque.

(d) Poor commutation at heavy loads.

(31) The speed of DC series motors can be controlled by:

(a) Rheostatic control.

(b) Series–parallel control.

(c) Field control.

(d) Either of (a), (b), or (c).

(32) When two or more motors are used for traction service, the method of speed control used will be:

(a) Rheostatic control.

(b) Series–parallel control.

(c) Field control.

(d) Motor generator control.

(33) In motor–generator locomotive control:

(a) Rheostatic control is used.

(b) Series parallel control is used.

(c) The output voltage of generator is regulated by means of field control from exciter.

(d) Any one of the above method is used.

(34) The method of speed control adopted in 25-kv, single-phase, and 50-Hz traction is:

(a) Tap changing control of transformer.

(b) Reduced current method.

(c) Series–parallel control.

(d) Rheostatic control.

(35) The least efficient method of the speed control of the three-phase induction motors is:

(a) Pole changing.

(b) Rheostatic control.

(c) Cascade control.

(d) The combination of cascade and pole changing.

(36) Which of the following braking systems of the locomotives is costly?

(a) Vacuum braking on steam locomotives.

(b) Vacuum braking on diesel locomotives.

(c) Regenerative braking on electric locomotives.

(d) All braking systems are equally costly.

(37) The braking system employed with steam locomotives is:

(a) Vacuum system.

(b) Pneumatic system.

(c) Hydraulic system.

(d) Any of the above.

(38) The wheels of a train, engine, as well as bogies are slightly tapered so as to:

(a) Reduce friction.

(b) Increase friction.

(c) Facilitate in taking turns.

(d) Facilitate braking.

(39) In electric traction if contact voltage exceeds 1,500 V, the current collection is invariably by:

(a) A contact rail.

(b) A conductor rail.

(c) Overhead wire.

(d) Third rail.

(40) Conductor rail system of supply is:

(a) Cheap and easy to repair and inspection.

(b) Suitable for heavy current collection at voltages up to 1,200 V.

(c) Universally used for all AC railways.

(d) Both (a) and (b).

(41) Automatic signaling is used for:

(a) Urban/suburban electric trains.

(b) Mail/express trains.

(c) Superfast trains.

(d) All trains.

REVIEW QUESTIONS

(1) What are the various types of electric braking used?

(2) Explain how rheostat braking is done in DC shunt motors and series motors.

(3) Describe how plugging, rheostat braking, and regenerative braking are employed with DC series motor.

(4) What are the requirements of good electric braking?

(5) What are the various types of traction motors?

(6) What are the advantages of series–parallel control of DC motors?

(7) Why DC series motor is ideally suited for traction services?

(8) Briefly explain the AC motors used in traction.

(9) Review the existing electric traction systems in India.

(10) Explain the requirements for ideal traction system.

(11) Explain the different methods of the electric braking of the three-phase induction motor.

(12) What are the various electric traction systems in India? Compare them.

(13) Give the features of the various motors used in electric traction.

(14) What are the advantages of electric braking over mechanical braking?

(15) Explain the methods of plugging when induction motors are employed for electric traction.

(16) Briefly explain the AC motors used in traction.

ANSWERS

1. c
2. d
3. b
4. a
5. a
6. b
7. c
8. a
9. c
10. a
11. d
12. d
13. d
14. b
15. a
16. b
17. c
18. a
19. d
20. c
21. a
22. b
23. a
24. b
25. c
26. b
27. a
28. c
29. b
30. a
31. d
32. b
33. c
34. a
35. b
36. c
37. a
38. c
39. c
40. d
41. a

Electric Traction-II

OBJECTIVES

After reading this chapter, you should be able to:

- know the various types of services
- analyze the different speed–time curves
- understand the various factors affecting the propulsion of a traction vehicle

10.1 INTRODUCTION

The movement of trains and their energy consumption can be most conveniently studied by means of the speed–distance and the speed–time curves. The motion of any vehicle may be at constant speed or it may consist of periodic acceleration and retardation. The speed–time curves have significant importance in traction. If the frictional resistance to the motion is known value, the energy required for motion of the vehicle can be determined from it. Moreover, this curve gives the speed at various time instants after the start of run directly.

10.2 TYPES OF SERVICES

There are mainly three types of passenger services, by which the type of traction system has to be selected, namely:

(i) Main line service.

(ii) Urban or city service.

(iii) Suburban service.

10.2.1 Main line services

In the main line service, the distance between two stops is usually more than 10 km. High balancing speeds should be required. Acceleration and retardation are not so important.

10.2.2 Urban service

In the urban service, the distance between two stops is very less and it is less than 1 km. It requires high average speed for frequent starting and stopping.

10.2.3 Suburban service

In the suburban service, the distance between two stations is between 1 and 8 km. This service requires rapid acceleration and retardation as frequent starting and stopping is required.

10.3 SPEED–TIME AND SPEED–DISTANCE CURVES FOR DIFFERENT SERVICES

The curve that shows the instantaneous speed of train in kmph along the ordinate and time in seconds along the abscissa is known as '*speed–time*' curve.

The curve that shows the distance between two stations in km along the ordinate and time in seconds along the abscissa is known as '*speed–distance*' curve.

The area under the speed–time curve gives the distance travelled during, given time internal and slope at any point on the curve toward abscissa gives the acceleration and retardation at the instance, out of the two speed–time curve is more important.

10.3.1 Speed–time curve for main line service

Typical speed–time curve of a train running on main line service is shown in Fig. 10.1. It mainly consists of the following time periods:

(i) Constant accelerating period.

(ii) Acceleration on speed curve.

(iii) Free-running period.

(iv) Coasting period.

(v) Braking period.

Constant acceleration

During this period, the traction motor accelerate from rest. The curve 'OA' represents the constant accelerating period. During the instant 0 to T_1, the current is maintained approximately constant and the voltage across the motor is gradually increased by cutting out the starting resistance slowly moving from one notch to the other. Thus, current taken by the motor and the tractive efforts are practically constant and therefore acceleration

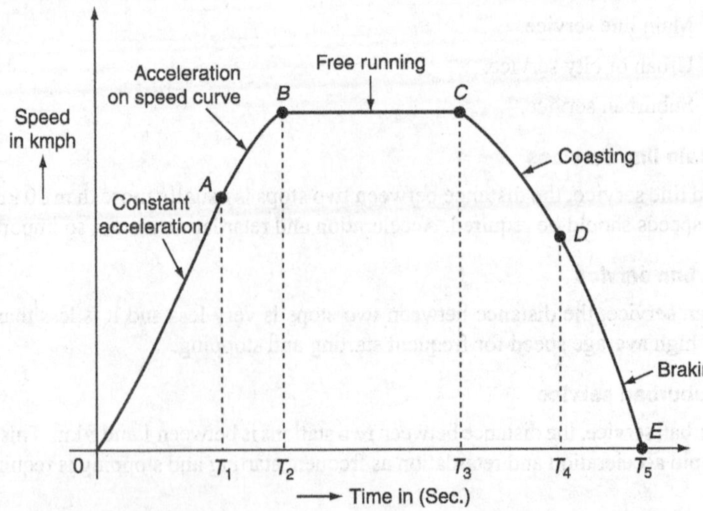

FIG. 10.1 *Speed–time curve for mainline service*

remains constant during this period. Hence, this period is also called as notch up accelerating period or rehostatic accelerating period. Typical value of acceleration lies between 0.5 and 1 kmph. Acceleration is denoted with the symbol 'α'.

Acceleration on speed-curve

During the running period from T_1 to T_2, the voltage across the motor remains constant and the current starts decreasing, this is because cut out at the instant 'T_1'.

According to the characteristics of motor, its speed increases with the decrease in the current and finally the current taken by the motor remains constant. But, at the same time, even though train accelerates, the acceleration decreases with the increase in speed. Finally, the acceleration reaches to zero for certain speed, at which the tractive effort excreted by the motor is exactly equals to the train resistance. This is also known as decreasing accelerating period. This period is shown by the curve '*AB*'.

Free-running or constant-speed period

The train runs freely during the period T_2 to T_3 at the speed attained by the train at the instant 'T_2'. During this speed, the motor draws constant power from the supply lines. This period is shown by the curve *BC*.

Coasting period

This period is from T_3 to T_4, i.e., from C to D. At the instant 'T_3' power supply to the traction, the motor will be cut off and the speed falls on account of friction, windage resistance, etc. During this period, the train runs due to the momentum attained at that particular instant. The rate of the decrease of the speed during coasting period is known as coasting retardation. Usually, it is denoted with the symbol 'β_c'.

Braking period

Braking period is from T_4 to T_5, i.e., from *D* to *E*. At the end of the coasting period, i.e., at 'T_4' brakes are applied to bring the train to rest. During this period, the speed of the train decreases rapidly and finally reduces to zero.

In main line service, the free-running period will be more, the starting and braking periods are very negligible, since the distance between the stops for the main line service is more than 10 km.

10.3.2 Speed–time curve for suburban service

In suburban service, the distance between two adjacent stops for electric train is lying between 1 and 8 km. In this service, the distance between stops is more than the urban service and smaller than the main line service. The typical speed–time curve for suburban service is shown in Fig. 10.2.

The speed–time curve for urban service consists of three distinct periods. They are:

(i) Acceleration.

(ii) Coasting.

(iii) Retardation.

For this service, there is no free-running period. The coasting period is comparatively longer since the distance between two stops is more. Braking or retardation period is comparatively small. It requires relatively high values of acceleration and retardation. Typical acceleration and retardation values are lying between 1.5 and 4 kmphp and 3 and 4 kmphp, respectively.

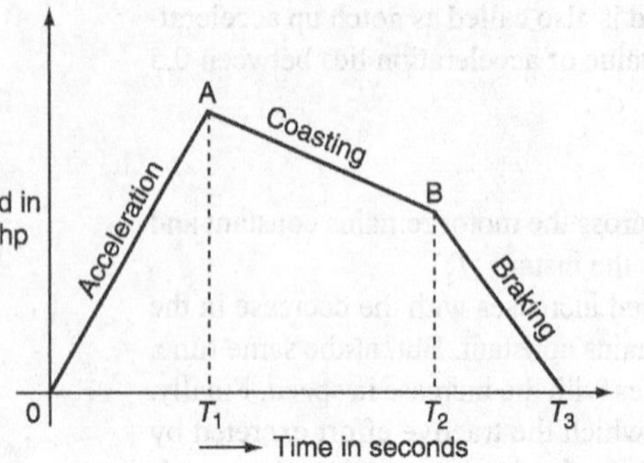

FIG. 10.2 *Typical speed–time curve for suburban service*

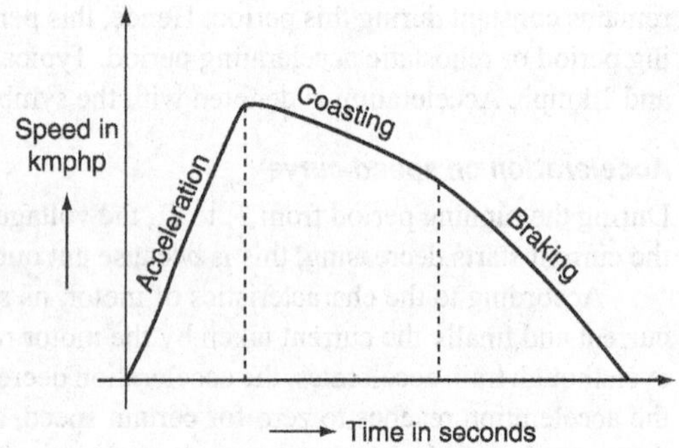

FIG. 10.3 *Typical speed–time curve for urban service*

10.3.3 Speed–time curve for urban or city service

The speed–time curve urban or city service is almost similar to suburban service and is shown in Fig. 10.3.

In this service also, there is no free-running period. The distance between two stop is less about 1 km. Hence, relatively short coasting and longer braking period is required. The relative values of acceleration and retardation are high to achieve moderately high average between the stops. Here, the small coasting period is included to save the energy consumption. The acceleration for the urban service lies between 1.6 and 4 kmphp. The coasting retardation is about 0.15 kmphp and the braking retardation is lying between 3 and 5 kmphp. Some typical values of various services are shown in Table. 10.1.

10.4 SOME DEFINITIONS

10.4.1 Crest speed

The maximum speed attained by the train during run is known as crest speed. It is denoted with 'V_m'.

10.4.2 Average speed

It is the mean of the speeds attained by the train from start to stop, i.e., it is defined as the ratio of the distance covered by the train between two stops to the total time of rum. It is denoted with 'V_a'.

TABLE 10.1 *Types of services*

	Mainline service	*Suburban service*	*Urban service*
Distance between stops in km	More than 10	1–8	1
Maximum speed in kmph	160	120	120
Acceleration in kmphp	0.5–0.9	1.5–4	1.5–4
Retardation in kmphp	1.5	3–4	3–4
Features	Long free-run period, coasting and acceleration braking periods are small	No free-running period, coasting period is long	No free-running period, coasting period is small

$$\therefore \text{Average speed} = \frac{\text{distance between stops}}{\text{actual time of run}}$$

$$V_a = \frac{D}{T},$$

where V_a is the average speed of train in kmph, D is the distance between stops in km, and T is the actual time of run in hours.

10.4.3 Schedule speed

The ratio of the distance covered between two stops to the total time of the run including the time for stop is known as schedule speed. It is denoted with the symbol 'V_s'.

$$\therefore \text{Schedule speed} = \frac{\text{distance between stops}}{\text{total time of run} + \text{time for stop}}$$

$$= \frac{\text{distance between stops}}{\text{shedule time}}$$

$$V_s = \frac{D}{T_s},$$

where T_s is the schedule time in hours.

10.4.4 Schedule time

It is defined as the sum of time required for actual run and the time required for stop.

i.e., $T_s = T_{run} + T_{stop}$.

10.5 FACTORS AFFECTING THE SCHEDULE SPEED OF A TRAIN

The factors that affect the schedule speed of a train are:

(i) Crest speed.

(ii) The duration of stops.

(iii) The distance between the stops.

(iv) Acceleration.

(v) Braking retardation.

10.5.1 Crest speed

It is the maximum speed of train, which affects the schedule speed as for fixed acceleration, retardation, and constant distance between the stops. If the crest speed increases, the actual running time of train decreases. For the low crest speed of train it running so, the high crest speed of train will increases its schedule speed.

10.5.2 Duration of stops

If the duration of stops is more, then the running time of train will be less; so that, this leads to the low schedule speed.

Thus, for high schedule speed, its duration of stops must be low.

10.5.3 Distance between the stops

If the distance between the stops is more, then the running time of the train is less; hence, the schedule speed of train will be more.

10.5.4 Acceleration

If the acceleration of train increases, then the running time of the train decreases provided the distance between stops and crest speed is maintained as constant. Thus, the increase in acceleration will increase the schedule speed.

10.5.5 Breaking retardation

High breaking retardation leads to the reduction of running time of train. These will cause high schedule speed provided the distance between the stops is small.

10.6 SIMPLIFIED TRAPEZOIDAL AND QUADRILATERAL SPEED TIME CURVES

Simplified speed–time curves gives the relationship between acceleration, retardation average speed, and the distance between the stop, which are needed to estimate the performance of a service at different schedule speeds. So that, the actual speed–time curves for the main line, urban, and suburban services are approximated to some from of the simplified curves. These curves may be of either trapezoidal or quadrilateral shape.

10.6.1 Analysis of trapezoidal speed–time curve

Trapezoidal speed–time curve can be approximated from the actual speed–time curves of different services by assuming that:

- The acceleration and retardation periods of the simplified curve is kept same as to that of the actual curve.
- The running and coasting periods of the actual speed–time curve are replaced by the constant periods.

This known as trapezoidal approximation, a simplified trapezoidal speed–time curve is shown in Fig. 10.4.

Calculations from the trapezoidal speed–time curve

Let D be the distance between the stops in km, T be the actual running time of train in second, α be the acceleration in km/h/sec, β be the retardation in km/h/sec, V_m be the

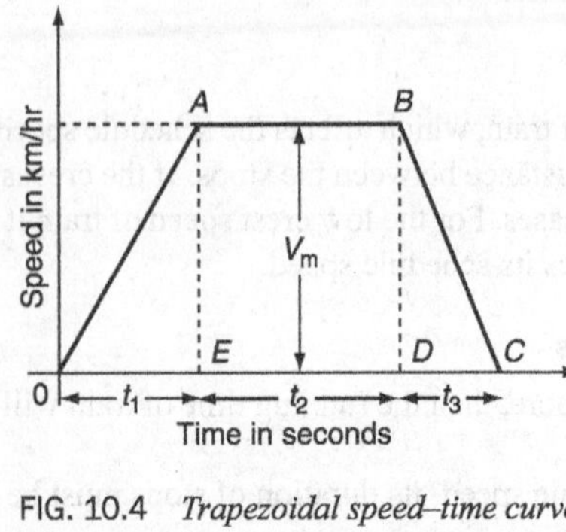

FIG. 10.4 *Trapezoidal speed–time curve*

maximum or the crest speed of train in km/h, and V_a be the average speed of train in km/h. From the Fig. 10.4:

Actual running time of train, $T = t_1 + t_2 + t_3$. **(10.1)**

Time for acceleration, $t_1 = \frac{V_m - 0}{\alpha} = \frac{V_m}{\alpha}$. **(10.2)**

Time for retardation, $t_3 = \frac{V_m - 0}{\beta} = \frac{V_m}{\beta}$. **(10.3)**

∴ Time for free-running period, $t_2 = T - (t_1 + t_3)$

$$= T - \left[\frac{V_m}{\alpha} + \frac{V_m}{\beta}\right]. \quad \textbf{(10.4)}$$

Area under the trapezoidal speed–time curve gives the total distance between the two stops (D).

∴ The distance between the stops (D) = area under triangle OAE + area of rectangle $ABDE$ + area of triangle DBC

= The distance travelled during acceleration + distance travelled during free-running period + distance travelled during retardation.

Now:

The distance travelled during acceleration = average speed during accelerating period × time for acceleration

$$= \frac{0 + V_m}{2} \times t_1 \text{ km/h} \times \text{sec}$$

$$= \frac{0 + V_m}{2} \times \frac{t_1}{3{,}600} \text{ km.}$$

The distance travelled during free-running period = average speed × time of free running

$$= V_m \times t_2 \text{ km/h} \times \text{sec}$$

$$= V_m \times \frac{t_2}{3{,}600} \text{ km.}$$

The distance travelled during retardation period = average speed × time for retardation

$$= \frac{V_m + 0}{2} \times t_3 \text{ km/h} \times \text{sec}$$

$$= \frac{0 + Vm}{2} \times \frac{t_3}{3{,}600} \text{ km.}$$

The distance between the two stops is:

$$D = \frac{V_m}{2} \times \frac{t_1}{3{,}600} + V_m \times \frac{t_2}{3{,}600} + \frac{V_m}{2} \times \frac{t_3}{3{,}600}$$

$$D = \frac{V_m t_1}{7,200} + \frac{V_m}{3,600}\left[T - V_m(t_1 + t_2)\right] + \frac{V_m t_3}{7,200}$$

$$D = \frac{V_m^2}{7,200\alpha} + \frac{V_m}{3,600}\left[T - V_m\left(\frac{1}{\alpha} + \frac{1}{\beta}\right)\right] + \frac{V_m^2}{7,200\beta}$$

$$3,600 \times D = \frac{V_m^2}{2\alpha} + \frac{V_m^2}{\beta} - V_m^2\left(\frac{1}{\alpha} + \frac{1}{\beta}\right) + V_m T$$

$$3,600\ D = V_m^2\left(\frac{1}{2\alpha} - \frac{1}{\alpha}\right) + V_m^2\left(\frac{1}{2\beta} - \frac{1}{\beta}\right) + V_m T$$

$$3,600\ D = \frac{-V_m^2}{2\alpha} - \frac{V_m^2}{2\beta} + V_m T$$

$$\therefore V_m^2\left[\frac{1}{2\alpha} + \frac{1}{2\beta}\right] - V_m T + 3,600 D = 0.$$

Let $\frac{1}{2\alpha} + \frac{1}{2\beta} = X = \frac{\alpha + \beta}{2\alpha\beta}$

$$\therefore V_m^2 X - V_m T + 3,600 D = 0. \quad \textbf{(10.5)}$$

Solving quadratic Equation (10.5), we get:

$$V_m = \frac{T + \sqrt{T^2 - 4 \times X \times 3,600D}}{2 \times X.}$$

$$= \frac{T}{2X} \pm \sqrt{\frac{T^2}{4X^2} - \frac{3,600D}{X}}.$$

By considering positive sign, we will get high values of crest speed, which is practically not possible, so negative sign should be considered:

$$V_m = \frac{T}{2X} - \sqrt{\frac{T^2}{4X^2} - \frac{3,600D}{X}} \quad \textbf{(10.6)}$$

$$\text{Or,} \quad V_m = \frac{\alpha\beta}{\alpha + \beta} T - \sqrt{\left(\frac{\alpha\beta}{\alpha + \beta}\right)^2 T^2 - 7,200\left(\frac{\alpha\beta}{\alpha + \beta}\right) D}.$$

10.6.2 Analysis of quadrilateral speed–time curve

Quadrilateral speed–time curve for urban and suburban services for which the distance between two stops is less. The assumption for simplified quadrilateral speed–time curve is the initial acceleration and coasting retardation periods are extended, and there is no free-running period. Simplified quadrilateral speed–time curve is shown in Fig. 10.5.

Let V_1 be the speed at the end of accelerating period in km/h, V_2 be the speed at the end of coasting retardation period in km/h, and β_c be the coasting retardation in km/h/sec.

Time for acceleration, $t_1 = \frac{V_1 - 0}{\alpha} = \frac{V_1}{\alpha}$.

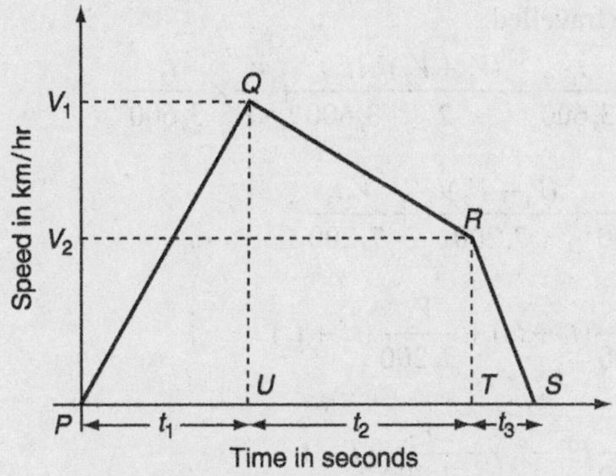

FIG. 10.5 *Quadrilateral speed–time curve*

Time for coasting period, $t_2 = \dfrac{V_2 - V_1}{\beta}$.

Time period for braking retardation period, $t_3 = \dfrac{V_2 - 0}{\beta} = \dfrac{V_2}{\beta}$.

Total distance travelled during the running period D:

= the area of triangle PQU + the area of rectangle $UQRS$ + the area of triangle TRS.

= the distance travelled during acceleration + the distance travelled during coasting retardation + the distance travelled during breaking retardation.

But, the distance travelled during acceleration = average speed × time for acceleration

$$= \frac{0 + V_1}{2} \times t_1 \text{ km/h} \times \text{sec}$$

$$= \frac{V_1}{2} \times \frac{t_1}{3{,}600} \text{ km.}$$

The distance travelled during coasting retardation $= \dfrac{V_2 + V_1}{2} \times t_2$ km/h × sec

$$= \frac{V_2 + V_1}{2} \times \frac{t_2}{3{,}600} \text{ km.}$$

The distance travelled during breaking retardation = average speed × time for breaking retardation

$$= \frac{0 + V_2}{2} \times t_3 \text{km/h} \times \text{sec}$$

$$= \frac{V_2}{2} \times \frac{t_3}{3{,}600} \text{ km.}$$

∴ Total distance travelled:

$$D = \frac{V_1}{2} \times \frac{t_1}{3,600} + \frac{(V_1 + V_2)}{2} \frac{(t_2)}{3,600} + \frac{V_2}{2} \times \frac{t_3}{3,600}$$

$$= \frac{V_1 t_1}{7,200} + \frac{(V_1 + V_2)t_2}{7,200} + \frac{V_2 t_3}{7,200}$$

$$= \frac{V_1}{7,200}(t_1 + t_2) + \frac{V_2}{7,200}(t_2 + t_3)$$

$$= \frac{V_1}{7,200}(T - t_3) + \frac{V_2}{7,200}(T - t_1)$$

$$= \frac{(V_1 + V_2)T}{7,200} - \frac{V_1 t_3}{7,200} - \frac{V_2 t_1}{7,200}$$

$$= \frac{(V_1 + V_2)T}{7,200} - \frac{V_1 V_2}{7,200\beta} - \frac{V_1 V_2}{7,200\alpha}$$

$$= \frac{\mathrm{T}}{7,200}(V_1 + V_2) - \frac{V_1 V_2}{7,200}\left(\frac{1}{\alpha} + \frac{1}{\beta}\right)$$

$$7,200D = (V_1 + V_2)\mathrm{T} - V_1 V_2 \left(\frac{1}{\alpha} + \frac{1}{\beta}\right). \tag{10.7}$$

Example 10.1: The distance between two stops is 1.2 km. A schedule speed of 40 kmph is required to cover that distance. The stop is of 18-s duration. The values of the acceleration and retardation are 2 kmphp and 3 kmphp, respectively. Then, determine the maximum speed over the run. Assume a simplified trapezoidal speed–time curve.

Solution:

Acceleration $\alpha = 2.0$ kmphp.

Retardation $\beta = 3$ kmphp.

Schedule speed $V_s = 40$ kmph.

Distance of run, D = 1.2 km.

$$\text{Schedule time, } T_s = \frac{D \times 3,600}{V_s}$$

$$= \frac{1.2 \times 3,600}{40}$$

$$= 108 \text{ s.}$$

Actual run time, $T = T_s -$ stop duration

$= 108 - 18$

$= 90$ s.

$$\text{Maximum speed } V_m = \frac{T}{2X} - \sqrt{\frac{T^2}{4X^2} - \frac{3,600D}{X}},$$

where

$$X = \frac{1}{2\alpha} + \frac{1}{2\beta}$$

$$= \frac{1}{2\times 2} + \frac{1}{2\times 3}$$

$$= 0.416.$$

$$\therefore V_{\text{m}} = \frac{90}{2\times 0.416} - \sqrt{\frac{(90)^2}{4\times(0.416)^2} - \frac{3,600\times 1.2}{0.416}}$$

$$= 108.173 - \sqrt{(1,1701.414) - (1,0384.61)}$$

$$= 71.88 \text{ kmph.}$$

Example 10.2: The speed–time curve of train carries of the following parameters:

(i) Free running for 12 min.

(ii) Uniform acceleration of 6.5 kmphp for 20 s.

(iii) Uniform deceleration of 6.5 kmphp to stop the train.

(iv) A stop of 7 min.

Then, determine the distance between two stations, the average, and the schedule speeds.

Solution:

Acceleration (α) = 6.5 kmphps.

Acceleration period $t_1 = 20$ s.

Maximum speed $V_{\text{m}} = \alpha t_1$

$= 6.5\times 20 = 130$ kmph.

Free-running time (t_2) = 12 × 60

= 720 s.

$$\text{Time for retardation, } (t_3) = \frac{V_{\text{m}}}{\beta}$$

$$= \frac{130}{6.5} = 20 \text{ s}.$$

The distance travelled during the acceleration period:

$$D_1 = \frac{1}{2}\frac{V_{\text{m}}t_1}{3,600}$$

$$= \frac{1}{2}\times\frac{130\times 20}{3,600}$$

$$= 0.36 \text{ km.}$$

The distance travelled during the free-running period:

$$D_2 = \frac{V_{\text{m}}t_2}{3,600}$$

$$= \frac{130 \times 720}{3,600}$$
$$= 26 \text{ km.}$$

The distance travelled during the braking period $D_3 = \frac{V_m t_3}{7,200}$

$$= \frac{130 \times 20}{7,200}$$
$$= 0.362 \text{ km.}$$

The distance between the two stations:

$$D = D_1 + D_2 + D_3$$
$$= 0.36 + 26 + 0.362$$
$$= 26.724 \text{ km.}$$

Average distance $(V_{avg}) = \frac{D \times 3600}{T}$

$$= \frac{26.724 \times 3600}{20 + 720 + 20}$$
$$= 126.58 \text{ kmph.}$$

Schedule speed $(V_s) = \frac{D \times 3600}{T + \text{stoptime}}$

$$= \frac{26.724 \times 3,600}{20 + 720 + 20 + 70 \times 60}$$
$$= 81.53 \text{ kmph.}$$

Example 10.3: An electric train is to have the acceleration and braking retardation of 0.6 km/hr/sec and 3 km/hr/sec, respectively. If the ratio of the maximum speed to the average speed is 1.3 and time for stop is 25 s. Then determine the schedule speed for a run of 1.6 km. Assume the simplified trapezoidal speed–time curve.

Solution:

Acceleration $\alpha = 0.6$ km/hr/s.
Retardation $\beta = 3$ km/hr/s.
Distance of run $D = 1.6$ km.
Let the cultural time of run be 'T' s.

Average speed $V_a = \frac{3,600D}{T}$

$$= \frac{3,600 \times 1.6}{T}$$
$$= \frac{5,760}{T} \text{ kmph.}$$

Maximum speed $= 1.3V_a$

$$= 1.3 \times \frac{5,760}{T}$$

$$= \frac{7,488}{T} \text{ km/hr}$$

$$V_m^2\left[\frac{1}{2\alpha}+\frac{1}{2\beta}\right]-V_m T+3,600=D$$

$$V_m^2=\frac{V_m T-3,600D}{\left(\frac{1}{2\alpha}+\frac{1}{2\beta}\right)}$$

$$=\frac{\frac{7,488}{T}\times T-3,600\times 1.6}{\left(\frac{1}{2\times 0.6}+\frac{1}{2\times 3}\right)}$$

$$=\frac{7,488-5,760}{0.833+0.166}$$

$$=1,729.729$$

$$\therefore V_m = 41.59 \text{ km/hr.}$$

Average speed, $(V_a)=\frac{V_m}{1.3}=\frac{41.59}{1.3}$

$(V_a) = 31.9923$ kmph.

Actual time of run $T=\frac{3,600D}{V_a}$

$$=\frac{3,600\times 1.6}{31.9923}$$

$T = 180.0433$ s.

Schedule time T_s = Actual time of run + time of stop

$= 180.0433 + 25$

$= 205.0433$ s.

Schedule speed $V_s=\frac{D\times 3,600}{T_s}$

$$=\frac{1.6\times 3,600}{205.0433}$$

$= 28.0916$ kmph.

Example 10.4: The distance between two stops is 5 km. A train has schedule speed of 50 kmph. The train accelerates at 2.5 kmphps and retards 3.5 kmphps and the duration of stop is 55 s. Determine the crest speed over the run assuming trapezoidal speed–time curve.

Solution:

Acceleration (α) = 2.5 kmphps.
Retardation (β) = 3.5 kmphps.

Distance of run (D) = 5 km.
Schedule speed (V_s) = 50 kmph.

$$\text{Schedule time, } T_s = \frac{D}{V_s} \times 3,600$$

$$= \frac{5}{50} \times 3,600$$

$$= 360 \text{ s.}$$

Actual time of run $T = T_s$ – Time of stop

$$= 360 - 55$$

$$= 305 \text{ s.}$$

By using the equation:

$$V_m = \frac{T}{2X} - \sqrt{\frac{T^2}{4X2} - \frac{3,600D}{X}}$$

$$X = \frac{1}{2\alpha} + \frac{1}{2\beta}$$

$$= \frac{1}{2\times 2.5} + \frac{1}{2\times 3.5}$$

$$= 0.2 + 0.1428$$

$$= 0.3428.$$

$$\therefore V_m = \frac{305}{2\times 0.3428} - \sqrt{\frac{(305)^2}{4\times(0.3428)^2} - \frac{3600\times 5}{0.3428}}$$

$$= 444.868 - \sqrt{197,905.5898 - 52,508.75146}$$

$$= 63.556 \text{ kmph.}$$

Example 10.5: A train is required to run between two stations 1.5 km apart at an average speed of 42 kmph. The run is to be made to a simplified quadrilateral speed–time curve. If the maximum speed is limited to 65 kmph, the acceleration to 2.5, kmphps, and the casting and braking retardation to 0.15 kmphs and 3 kmphs, respectively. Determine the duration of acceleration, costing, and braking periods.

Solution:

Distance between two stations D = 1.5 km.
Average speed V_a = 42 kmph.
Maximum speed V_m = 65 kmph.
Acceleration (α) = 2.5 kmphps.
Coasting retardation β_c = 0.15 kmphps.
Barking retardation β = 3 kmphps.

$$\text{The duration of acceleration } t_1 = \frac{V_m}{\alpha}$$

$$= \frac{65}{2.5}$$

$$= 26 \text{ s.}$$

The actual time of run, $T = \frac{3,600 \times D}{V_a} = \frac{3,600 \times 1.5}{42} = 128.57\ \text{s}.$

Before applying brakes; let the speed be V_2.

$$\text{The duration of coasting, } t_2 = \frac{V_m - V_2}{\beta_c}$$

$$= \frac{65 - V_2}{0.15}\ \text{s}.$$

$$\text{The duration of braking } t_3 = \frac{V_2}{\beta}$$

$$= \frac{V_2}{3}\ \text{s}.$$

The actual time of run, $T = t_1 + t_2 + t_3$

$$128.557 = 26 + \frac{65 - V_2}{0.15} + \frac{V_2}{3}$$

$$102.57 = 433.33 - 6.66V_2 + 0.33V_2$$

$$330.76 = 6.33V_2$$

$$V_2 = 52.252\ \text{km/hr}.$$

$$\text{The duration of coasting, } t_2 = \frac{V_m - V_2}{\beta_c}$$

$$= \frac{65 - 52.252}{0.15}$$

$$= 84.98\ \text{s}.$$

$$\text{Duration of braking } (t_3) = \frac{V_2}{\beta}$$

$$= \frac{52.252}{3}$$

$$= 17.4173\ \text{s}.$$

Example 10.6: A train has schedule speed of 32 kmph over a level track distance between two stations being 2 km. The duration of stop is 25 s. Assuming the braking retardation of 3.2 kmphps and the maximum speed is 20% grater than the average speed. Determine the acceleration required to run the service.

Solution:

Schedule speed $V_s = 32$ kmph.
Distance $D = 2$ km.
Duration of stop = 25 s.
Braking retardation = 3.2 kmphps.

$$\text{Schedule time} = \frac{D}{V_s}$$

$$= \frac{2}{32} \times 60 \times 60 = 225 \text{ s}.$$

Actual time of run $T = 225 - 25 = 200$ s.

$$\text{Average speed, } V_a = \frac{3,600 \times D}{T}$$

$$= \frac{3,600 \times 2}{200}$$

$$= 36 \text{ kmph}.$$

$$\text{Maximum speed, } (V_m) = 1.2\ V_a$$

$$= 1.2 \times 36$$

$$V_m = 43.2 \text{ kmph}$$

$$\therefore V_3^2\left(\frac{1}{2\alpha} + \frac{1}{2\beta}\right) - V_m\ T + 3,600$$

$$\frac{1}{2\alpha} + \frac{1}{2\beta} = \frac{V_m T - 3,600 \times D}{V_m^2}$$

$$= \frac{43.2 \times 200 - 3,600 \times 2}{(43.2)^2}$$

$$\frac{1}{2\alpha} + \frac{1}{2\beta} = 0.7716$$

$$\frac{1}{2\alpha} = 0.716 - \frac{1}{2 \times 3.2}$$

$$\frac{1}{2\alpha} = 0.61535$$

$$\alpha = 0.893 \text{ kmphps}.$$

Example 10.7: A suburban electric train has a maximum speed of 75 kmph. The schedule speed including a station stop of 25 s is 48 kmph. If the acceleration is 2 kmphps, the average distance between two stops is 4 km. Determine the value of retardation.

Solution:

Maximum speed $V_m = 75$ kmph.
The distance of run $(D) = 4$ km.
Schedule speed $(V_s) = 48$ kmph.
Acceleration $(\alpha) = 2$ kmphps.
The duration of stop = 25 s.

$$\text{Schedule time } (T_s) = \frac{D}{V_s}$$

$$= \frac{4}{48} \times 60 \times 60 = 300 \text{ s}.$$

$$\therefore V_m^2\left(\frac{1}{2\alpha}+\frac{1}{2\beta}\right)-V_m T+3,600\times D=0$$

$$\frac{1}{2\alpha}+\frac{1}{2\beta}=\frac{V_m T-3,600D}{V_m^2}$$

$$\frac{1}{2\times 2}+\frac{1}{2\beta}=\frac{75\times 275-3,600\times 4}{(75)^2}$$

$$0.25+\frac{1}{2\beta}=1.1066$$

$$\beta=0.5836 \text{ kmphps.}$$

Example 10.8: An electric train is accelerated at 2 kmphps and is braked at 3 kmphps. The train has an average speed of 50 kmph on a level track of 2,000 min between the two stations. Determine the following:

(i) Actual time of run.

(ii) Maximum speed.

(iii) The distance travelled before applying brakes

(iv) Schedule speed.

Assume time for stop as 12 s. And, run according to trapezoidal.

Solution:

Acceleration $(\alpha) = 2$ kmphps.
Retardation $(\beta) = 3$ kmphps.
Average speed $(V_a) = 50$ kmph.
Distance D = 2,000 min = 2 km.
The duration of stop = 12 s.

(i) Time of run $T=\frac{D}{V_a}$

$$=\frac{2}{50}\times 60\times 60=144 \text{ s.}$$

(ii) Maximum speed, $V_m=\frac{T}{2X}-\sqrt{\frac{T^2}{4X^2}-\frac{3,600D}{X}}$,

where

$$X=\frac{1}{2\alpha}+\frac{1}{2\beta}$$

$$=\frac{1}{2\times 2}+\frac{1}{2\times 3}$$

$$=0.4166.$$

$$\therefore V_m = \frac{144}{2\times 0.4166} - \sqrt{\frac{(144)^2}{4\times(0.4166)^2} - \frac{3,600\times 2}{0.4166}}$$

$$= 172.8276 - \sqrt{(29,869.397) - (17,282.765)}$$

$$= 60.63744 \text{ kmph.}$$

(iii) $t_3 = \frac{V_m}{\beta} = \frac{60.63744}{3}$

$$= 20.2148 \text{ s.}$$

$$D_3 = \frac{1}{2} V_m t_3$$

$$= \frac{1}{2} \times 60.63744 \times \frac{20.21248}{60\times 60} = 0.173 \text{km.}$$

The distance travelled before applying brakes

$$D_1 + D_2 = D - D_3$$

$$= 2 - 0.17 = 1.83 \text{ km.}$$

(iv) Schedule speed $V_s = \frac{D}{T + T_{stop}}$

$$= \frac{2}{\frac{144+12}{60\times 60}} = 46.153 \text{ kmph.}$$

Example 10.9: An electric train has an average speed of 40 kmph on a level track between stops 1,500 m apart. It is accelerated at 2 kmphps and is braked at 3 kmphps. Draw the speed–time curve for the run.

Solution:

Average speed V_a = 40 kmph.
The distance of run (D) = 1,500 m = 1.5 km.
Acceleration (α) = 2 kmphps.
Retroaction (β) = 3 kmphps.

The time of run $T = \frac{D}{V_a}$

$$= \frac{1.5}{40} \times 60 \times 60 = 135 \text{ s.}$$

Using the equation (Fig. P.10.1):

$$V_m = \frac{T}{2X} - \sqrt{\frac{T^2}{4X^2} - \frac{3,600D}{X}},$$

where

$$X = \frac{1}{2\alpha} + \frac{1}{2\beta}$$

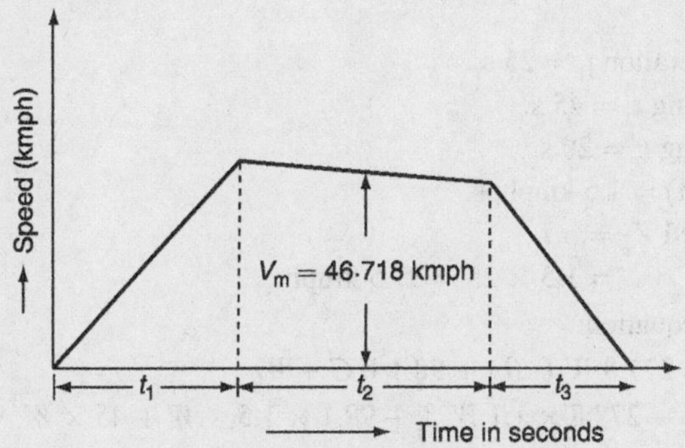

FIG. P.10.1

$$= \frac{1}{2\times 2} + \frac{1}{2\times 3} = 0.416.$$

$$\therefore V_m = \frac{135}{2\times 0.416} - \sqrt{\frac{(135)^2}{4\times(0.416)^2} - \frac{3600\times 1.5}{0.416}}$$

$$= 162.25 - \sqrt{(2,632 - 8.182) - (12,980.769)}$$

$$= 46.718 \text{ kmph.}$$

Acceleration period, $t_1 = \frac{V_m}{\alpha}$

$$= \frac{46.718}{2}$$

$$t_1 = 23.359 \text{ s.}$$

Braking period, $t_3 = \frac{V_m}{\beta}$

$$= \frac{46.718}{3} = 15.572.$$

Free-running period, $t_2 = T - (t_1 + t_3)$

$$= 135 - (23.359 + 15.572)$$

$$= 96.069.$$

Example 10.10: An electric train has quadrilateral speed–time curve as follows:

(1) Uniform acceleration from rest at 1.5 kmphps for 25 s.

(2) Coasting for 45 s.

(3) The duration of braking 20 s.

If the train is moving a uniform up gradient of 1.5%, the reactive resistance is 45 N/ton, the rotational inertia effect is 10% of dead weight, the duration of stop is 15 s, and the overall efficiency of transmission gear and motor is 80%. Find schedule speed.

Solution:

Time for acceleration $t_1 = 25$ s.
Time for coasting $t_2 = 45$ s.
Time for braking $t_3 = 20$ s.
Acceleration $(\alpha) = 1.5$ kmphps.
Maximum speed $V_m = \alpha\, t_1$
$= 1.5 \times 25 = 37.5$ kmph.

According to the equation:

$$F_t = 277.8\, W_e(-\beta_c) + 98.1\, WG + Wr$$

$$0 = -\,277.8 \times 1.1\, W\beta_c + 98.1 \times 1.5 \times W + 45 \times W$$

$$= -305.58\, W\beta_c + 147.15\, W + 45\, W$$

$$30.58\, W\beta_c = 192.15\, W$$

$$\beta_c = \frac{192.15\, W}{305.58\, W}$$

$\beta_c = 0.628$ kmphps.

The speed at the end of coating period $V_2 = V_m - \beta_c t_2$
$= 37.5 - 0.628 \times 45$
$= 9.24$ kmph.

The braking retardation $\beta = \dfrac{V_2}{t_3}$

$$= \frac{9.24}{20} = 0.462 \text{ kmphps.}$$

The distance travelled $D = \dfrac{V_m t_1}{7{,}200} + \dfrac{(V_m + V_2)t_2}{7{,}200} + \dfrac{V_2 t_3}{7{,}200}$

$$= \frac{37.5 \times 25}{7{,}200} + \frac{(37.5 + 9.24) \times 45}{7{,}200} + \frac{9.24 \times 20}{7{,}200}$$

$= 0.13 + 0.292 + 0.0256$
$= 0.4475$ km.

The schedule time $T_s = t_1 + t_2 + t_3 +$ duration of stop
$= 25 + 45 + 20 + 15$
$= 105$ s.

The schedule speed $V_s = \dfrac{3{,}600 \times D}{T_s}$

$$= \frac{3{,}600 \times 0.4476}{105}$$

$V_s = 15.346$ kmph.

10.7 TRACTIVE EFFORT (F_t)

It is the effective force acting on the wheel of locomotive, necessary to propel the train is known as '*tractive effort*'. It is denoted with the symbol F_t. The tractive effort is a vector quantity always acting tangential to the wheel of a locomotive. It is measured in newton.

The net effective force or the total tractive effort (F_t) on the wheel of a locomotive or a train to run on the track is equals to the sum of tractive effort:

(i) Required for linear and angular acceleration (F_a).

(ii) To overcome the effect of gravity (F_g).

(iii) To overcome the frictional resistance to the motion of the train (F_r).

$$\therefore F_t = F_a + F_g + F_r. \quad (10.8)$$

10.7.1 Mechanics of train movement

The essential driving mechanism of an electric locomotive is shown in Fig. 10.6. The electric locomotive consists of pinion and gear wheel meshed with the traction motor and the wheel of the locomotive. Here, the gear wheel transfers the tractive effort at the edge of the pinion to the driving wheel.

Let T is the torque exerted by the motor in N-m, F_p is tractive effort at the edge of the pinion in Newton, F_t is the tractive effort at the wheel, D is the diameter of the driving wheel, d_1 and d_2 are the diameter of pinion and gear wheel, respectively, and η is the efficiency of the power transmission for the motor to the driving axle.

Now, the torque developed by the motor $T = F_p \times \frac{d_1}{2}$ N-m.

$$\therefore F_p = \frac{2T}{d_1} \text{ N.} \quad (10.9)$$

The tractive effort at the edge of the pinion transferred to the wheel of locomotive is:

$$F_t = F_p \times \frac{d_2}{D} \text{ N.} \quad (10.10)$$

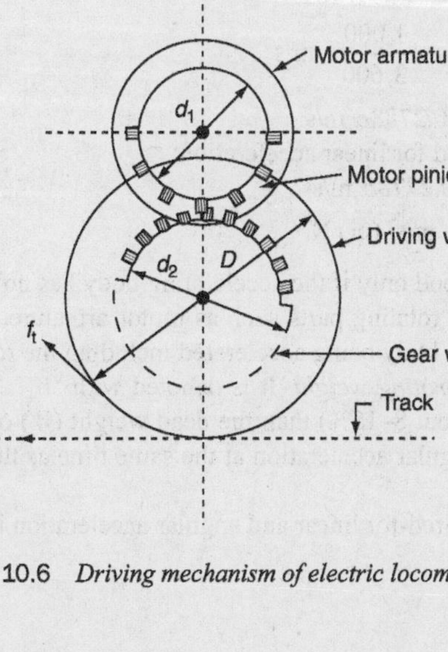

FIG. 10.6 *Driving mechanism of electric locomotives*

From Equations (10.9) and (10.10) $F_t = \eta \times \frac{2T}{d_1} \times \frac{d_2}{D}$

$$= \eta \cdot T \cdot \frac{2}{D}\left(\frac{d_2}{d_1}\right)$$

$$= \eta T \cdot \frac{2}{D} \cdot r,$$

where 'r'$= \left(\frac{d_2}{d_1}\right)$ is known as gear ratio.

$$\therefore F_t = 2\eta r \frac{T}{D} \text{ N.} \tag{10.11}$$

10.7.2 Tractive effort required for propulsion of train

From Equation (10.8), the tractive effort required for train propulsion is:

$$F_t = F_a + F_g + F_r,$$

where F_a is the force required for linear and angular acceleration, F_g is the force required to overcome the gravity, and F_r is the force required to overcome the resistance to the motion.

Force required for linear and angular acceleration (F_a)

According to the fundamental law of acceleration, the force required to accelerate the motion of the body is given by:

Force = Mass × acceleration

$F = ma.$

Let the weight of train be 'W' tons being accelerated at 'α' kmphps:

∴ The mass of train m = 1,000 W kg.

And, the acceleration = α kmphps

$$= \alpha \times \frac{1{,}000}{3{,}600} \text{ m/s}^2$$

$$= 0.2788\alpha \text{ m/s}^2.$$

The tractive effort required for linear acceleration:

F_a = 1,000 W kg × 0.2778α m/s²

$$= 27.88 \text{ W}\alpha \text{ kg} - \text{m/s}^2 \text{ (or) N.} \tag{10.12}$$

Equation (10.12) holds good only if the accelerating body has no rotating parts. Owing to the fact that the train has rotating parts such as motor armature, wheels, axels, and gear system. The weight of the body being accelerated including the rotating parts is known as *effective weight* or *accelerating weight.* It is denoted with 'W_e'. The accelerating weight '(W_e)' is much higher (about 8–15%) than the dead weight (W) of the train. Hence, these parts need to be given angular acceleration at the same time as the whole train is accelerated in linear direction.

∴ The tractive effort required-for linear and angular acceleration is:

$$F_a = 27.88\, W_e \alpha \text{ N.} \tag{10.13}$$

Tractive effort required to overcome the train resistance (F_r)

When the train is running at uniform speed on a level track, it has to overcome the opposing force due to the surface friction, i.e., the friction at various parts of the rolling stock, the fraction at the track, and also due to the wind resistance. The magnitude of the frictional resistance depends upon the shape, size, and condition of the track and the velocity of the train, etc.

Let 'r' is the specific train resistance in N/ton of the dead weight and 'W' is the dead weight in ton.

∴ The tractive effort required to overcome the train resistance $F_r = Wr$ N. **(10.14)**

Tractive effort required to overcome the effect of gravity (F_g)

When the train is moving on up gradient as shown in Fig. 10.7, the gravity component of the dead weight opposes the motion of the train in upward direction. In order to prevent this opposition, the tractive effort should be acting in upward direction.

∴ The tractive effort required to overcome the effect of gravity:

$$F_g = \pm \text{ mg } \sin\theta \text{ N}$$

$$= \pm 1{,}000 \text{ Wg } \sin\theta \quad [\because m = 1{,}000 \text{ Wkg}]. \qquad \textbf{(10.15)}$$

Now, from the Fig. 10.7:

$$\text{Gradient} = \sin\theta = \frac{BC}{AC} = \frac{\text{Elevation}}{\text{distance along the track}}$$

$$\% \text{ Gradient } G = \sin\theta \times 100. \qquad \textbf{(10.16)}$$

From Equations (10.15) and (10.16):

$$\therefore F_g = \pm 1{,}000 \text{ W g} \times \frac{G}{100}$$

$$= \pm 10 \times 9.81 \text{ W}G$$

$$= \pm 98.1 \text{ W}G \text{ N} \qquad [\text{since } g = 9.81 \text{ m/s}^2]. \qquad \textbf{(10.17)}$$

+ve sign for the train is moving on up gradient.

–ve sign for the train is moving on down gradient.

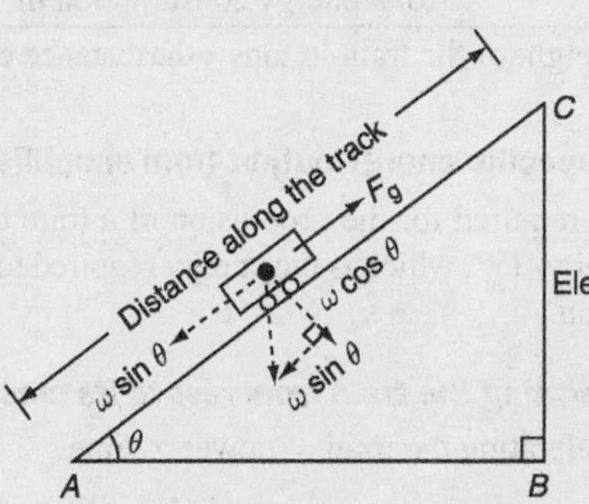

FIG. 10.7 *Train moving on up gradient*

This is due to when the train is moving on up a gradient, the tractive effort showing Equation (10.17) will be required to oppose the force due to gravitational force, but while going down the gradient, the same force will be added to the total tractive effort.

∴ The total tractive effort required for the propulsion of train $F_t = F_a + F_r \pm F_g$:

$$F_t = 277.8\, W_e\alpha + Wr \pm 98.1\, WG\ \text{N}. \quad \textbf{(10.18)}$$

10.7.3 Power output from the driving axle

Let F_t is the tractive effort in N and v is the speed of train in kmph.

∴ The power output (P) = rate of work done

$$= \text{Tractive effort} \times \frac{\text{distance}}{\text{time}}$$

$$= \text{Tractive effort} \times \text{speed}$$

$$= \frac{F_t \times v \times 1{,}000}{3{,}600}\ \text{W}$$

$$= \frac{F_t \times v}{3{,}600}\ \text{kW}. \quad \textbf{(10.19)}$$

If 'v' is in m/s, then $P = F_t \times v$ W.

If 'η' is the efficiency of the gear transmission, then the power output of motors,

$$P = \frac{F_t v}{\eta}\ \text{W}:$$

$$= \frac{F_t v}{3{,}600\eta}\ \text{kW}. \quad \textbf{(10.20)}$$

10.8 SPECIFIC ENERGY CONSUMPTION

The energy input to the motors is called the *energy consumption*. This is the energy consumed by various parts of the train for its propulsion. The energy drawn from the distribution system should be equals to the energy consumed by the various parts of the train and the quantity of the energy required for lighting, heating, control, and braking. This quantity of energy consumed by the various parts of train per ton per kilometer is known as specific energy consumption. It is expressed in watt hours per ton per km.

$$\therefore \left.\begin{array}{r}\text{Specific energy}\\ \text{consumption}\end{array}\right\} = \frac{\text{total energy consumption in W} - \text{h}}{\text{the weight of the train in tons} \times \text{the distance covered by train in km}}$$

10.8.1 Determination of specific energy output from simplified speed–time curve

Energy output is the energy required for the propulsion of a train or vehicle is mainly for accelerating the rest to velocity 'V_m', which is the energy required to overcome the gradient and track resistance to motion.

Energy required for accelerating the train from rest to its crest speed 'V_m'

The energy required for accelerating the train = power × time

$$= \frac{\text{work done}}{\text{time}} \times \text{time}$$

$= \text{tractive effort} \times \text{velocity} \times \text{time}$

$$= F_t \times \frac{V_m}{3,600} \times t_1 \text{ N-km/h-sec}$$

$$= F_t \times \frac{1}{2} \times \frac{V_m}{3,200} \times \frac{t_1}{3,600} \text{ N-km (or) kW-hr}$$

$$= \frac{1}{2} \times \frac{V_m^{\ 2}}{(3,600)^2 \alpha} F_t \text{ kw-hr} \left[\because t_1 = \frac{V_m}{\alpha}\right]$$

$$= \frac{1}{2} \times \frac{V_m^{\ 2}}{(3,600)^2 \alpha} [277.8 W_e \alpha + 98.1\, WG + Wr] \text{kW-hr.}$$

$[\because F_t = 277.8 W_e\, \alpha + 98.1\ WG + Wr]$.

Energy required for overcoming the gradient and tracking resistance to motion

Energy required for overcoming the gradient and tracking resistance:

$= \text{tractive effort} \times \text{velocity} \times \text{time}$

$$= F_t' \times \frac{V_m}{3,600} \times \frac{t_2}{3,600} \text{ kW-hr}$$

$$= \frac{V_m t_2}{(3,600)^2} [Wr + 98.1\, WG] \text{kW-hr,}$$

where F_t' is the tractive effort required to overcome the gradient and track resistance, W is the dead weight of train, r is the track resistance, and G is the percentage gradient.

Total energy output = energy required for acceleration + energy required to overcome gradient and to resistance to motion.

$$= \frac{V_m^{\ 2}}{2(3,600)^2 \alpha} [277.8\, W_e \alpha + 98.1\, WG + Wr] + \frac{V_m t_2}{(3,600)^2} [Wr + 98.1\, WG] \text{kW-hr}$$

$$= \frac{V_m^{\ 2}(1,000)}{2(3,600)^2 \alpha} [277.8\, W_e \alpha + 98.1\, WG + Wr] + \frac{V_m t_2 \times 1,000}{(3,600)^r} [Wr + 98.1\, WG] \text{ W-hr}$$

$$= \frac{V_m^{\ 2}(1,000)}{2\alpha(3,600)^2} [27.8\, W_e \alpha] + \left[\frac{V_m^{\ 2}(1,000)}{2\alpha(3,600)^2} + \frac{V_m t_2 \times 1,000}{(3,600)^2}\right] [Wr + 98.1\, WG] \text{ W-hr}$$

$$= 0.01072\ \text{W}_e V_m^{\ 2} + \frac{1,000}{(3,600)} [Wr + 98.1\, WG] \left[\frac{V_m^{\ 2}}{2\alpha 3,600} + \frac{V_m t_2}{3,600}\right] \text{W-hr}$$

$$= 0.01072\ \text{W}_e V_m^{\ 2} + 0.2778 [Wr + 98.1\, WG][D_1 + D_2] \text{W-hr,}$$

where $D_1 = \dfrac{V_m^{\ 2}}{2\alpha 3,600} = \dfrac{V_m^{\ 2}}{7,200\alpha}$.

$D_2 = \dfrac{V_m t_2}{3,600}$.

$$\therefore \text{The specific energy output} = \frac{\text{energy output in Whr}}{\text{weight of train in tons} \times \text{distance of running}}$$

$$= \frac{0.001072V_m^{\ 2}W_e + 0.2778[98.1\,WG + Wr][D_1 + D_2]}{W \times D}$$

$$= \frac{0.001072V_m^{\ 2}}{D}\left[\frac{W_e}{W}\right] + \left[\frac{98.1\,G + r}{D}\right] \times 0.2778 \times D',$$

where $D' = D_1 + D_2$.

For uniform level track $G = 0$:

$$\therefore \text{ The specific energy output} = \frac{0.001072V_m^{\ 2}}{D}\frac{W_e}{W} + 0.2778r \times \frac{D'}{D} \text{ W-hr/ton-km.}$$

$$\therefore \text{ The specific energy consumption} = \frac{\text{specific energy output}}{\text{efficiency of motors}}$$

$$= \frac{0.001072V_m^{\ 2}}{\eta D}\frac{W_e}{W} + 0.2778\frac{D'}{D}\frac{r}{\eta} \text{ W-hr/ton-km.} \qquad \textbf{(10.21)}$$

10.8.2 Factors affecting the specific energy consumption

Factors that affect the specific energy consumption are given as follows.

Distance between stations

From equation specific energy consumption is inversely proportional to the distance between stations. Greater the distance between stops is, the lesser will be the specific energy consumption. The typical values of the specific energy consumption is less for the main line service of 20–30 W-hr/ton-km and high for the urban and suburban services of 50–60 W-hr/ton-km.

Acceleration and retardation

For a given schedule speed, the specific energy consumption will accordingly be less for more acceleration and retardation.

Maximum speed

For a given distance between the stops, the specific energy consumption increases with the increase in the speed of train.

Gradient and train resistance

From the specific energy consumption, it is clear that both gradient and train resistance are proportional to the specific energy consumption. Normally, the coefficient of adhesion will be affected by the running of train, parentage gradient, condition of track, etc. for the wet and greasy track conditions. The value of the coefficient of adhesion is much higher compared to dry and sandy conditions.

10.9 IMPORTANT DEFINITIONS

10.9.1 Dead weight

It is the total weight of train to be propelled by the locomotive. It is denoted by 'W'.

10.9.2 Accelerating weight

It is the effective weight of train that has angular acceleration due to the rotational inertia including the dead weight of the train. It is denoted by 'W_e'.

This effective train is also known as accelerating weight. The effective weight of the train will be more than the dead weight. Normally, it is taken as 5–10% of more than the dead weight.

10.9.3 Adhesive weight

The total weight to be carried out on the drive in wheels of a locomotive is known as adhesive weight.

10.9.4 Coefficient of adhesion

It is defined as the ratio of the tractive effort required to propel the wheel of a locomotive to its adhesive weight.

$$F_t \propto W$$

$$= \mu W,$$

where F_t is the tractive effort and W is the adhesive weight.

$$\therefore \mu = \frac{F_t}{W}. \qquad (10.22)$$

Example 10.11: A 250-ton motor coach having four motors each developing 6,000 N-m torque during acceleration, starts from rest. If the gradient is 40 in 1,000, gear ration is 4, gear transmission efficiency is 87%, wheel radius is 40 cm, train resistance is 50 N/ton, the addition of rotational inertia is 12%. Calculate the time taken to attain a speed of 50 kmph. If the line voltage is 3,000-V DC and the efficiency of motors is 85%. Find the current during notching period.

Solution:

The weight of train $W = 250$ ton.

Parentage gradient $G = \frac{40}{1,000} \times 100 = 4\%$.

Gear ratio $r = 4$.

Wheel diameter $D = 2 \times 40 = 80$ cm.

Or, $D = 0.8$ m.

Train resistance $r = 50$ N/ton.

Rotational inertia = 12%.

Accelerating weight of the train $W_e = 1.10 \times 250 = 275$ ton.

Total torque developed $T = 4 \times 6,000 = 24,000$ Nm.

$$\text{Tractive effort } F_t = \frac{\eta T 2r}{D}$$

$$= \frac{0.87 \times 24,000 \times 2 \times 4}{0.8} = 208,800\text{ N}.$$

But,

$$F_t = 277.8\, W_e\, \alpha + 98.1\, WG + Wr$$

$$208,800 = 277.8 \times 275\, \alpha + 98.1 \times 250 \times 4 + 250 \times 50$$

$$\therefore \alpha = 1.285 \text{ kmphps}.$$

The time taken for the train to attain the speed of 50 kmph:

$$t = \frac{V_m}{\alpha}$$

$$= \frac{50}{1.285} = 38.89\,\text{s}.$$

Power output from the driving axles:

$$= \frac{F_t \times V_m}{3,600} = \frac{208,800 \times 50}{3,600}$$

$$= 2,900 \text{ kW}.$$

$$\text{Power input} = \frac{\text{power output}}{\eta_m}$$

$$= \frac{2,900}{0.85} = 3,411.76\,\text{kW}.$$

$$\text{Total current drawn} = \frac{\text{power input}}{V}$$

$$= \frac{3,411.76 \times 10^3}{3,000} = 1,137.25\,\text{A}.$$

$$\text{Current drawn by the each motor} = \frac{1,137.25}{4} = 284.31\,\text{A}.$$

Example 10.12: An electric train of weight 250 ton has eight motors geared to driving wheels, each is 85 cm diameter. The tractive resistance is of 50/ton. The effect of rotational inertia is 8% of the train weight, the gear ratio is 4–1, and the gearing efficiency is 85% determine. The torque developed by each motor to accelerate the train to a speed of 50 kmph in 30 s up a gradient of 1 in 200.

Solution:

The weight of train $W = 250$ ton.
The diameter of driving wheel $D = 0.85$ m.
Tractive resistance, $r = 50$N/ton.
Gear ratio $r = 4$.
Gearing efficiency $\eta = 0.85$.

Accelerating weight of the train:

$$W_e = 1.10 \times W$$

$$= 1.10 \times 250 = 275 \text{ ton}.$$

Maximum speed $V_m = 50$ kmph.

$$\text{Acceleration } \alpha = \frac{V_m}{t_1} = \frac{50}{30} = 1.66\,\text{kmpmph}.$$

$$\text{Tractive effort } F_t = 277.8\, W_e \alpha + 98.1\, WG + Wr$$

$$= 126,815.7 + 12,262.5 + 12,500$$

$$= 151,578.2 \text{ N}.$$

$$\text{Total torque developed } T = \frac{F_t \times D}{\eta \times 2\gamma}$$

$$= \frac{151,578.2 \times 0.85}{0.85 \times 2 \times 4}$$
$$= 18.947.25 \text{ N-m.}$$

Torque developed by each motor $= \dfrac{18,947.25}{8}$
$$= 2,368.409 \text{ N-m.}$$

Example 10.13: A tram car is equipped with two motors that are operating in parallel, the resistance in parallel. The resistance of each motor is 0.5 Ω. Calculate the current drawn from the supply mains at 450 V when the car is running at a steady-state speed of 45 kmph and each motor is developing a tractive effort of 1,600 N. The friction, windage, and other losses may be assumed as 3,000 W per motor.

Solution:

The resistance of each motor = 0.5 Ω.
Voltage across each motor $V = 450$ V.
Tractive effort $F_t = 1,600$ N.
Maximum speed $V_m = 45$ kmph.
Losses per motor = 3,000 W.

The power output of each motor $= \dfrac{F_t \times V_m}{3,600}$
$$= \frac{1,600 \times 45 \times 10^3}{3,600}$$
$$= 20,000 \text{ W.}$$

Copper losses $= I^2 R_m = I^2 \times 0.5$
Motor input = motor output + constant loss + copper losses
$$450 \times I = 20,000 + 3,000 + 0.5I^2$$
$$0.5 I^2 - 450I + 23,000 = 0.$$

After solving, we get $I = 54.39$ A.
Total current drawn from supply mains = 2 × 54.39
= 108.78 A.

Example 10.14: A locomotive exerts a tractive effort of 35,000 N in halting a train at 50 kmph on the level track. If the motor is to haul the same train on a gradient of 1 in 50 and the tractive effort required is 55,000 N, determine the power delivered by the locomotive if it is driven by (i) DC series motors and (ii) induction motors.

Solution:

Tractive effort $F_t = 35,000$ N.
Maximum speed $V_m = 50$ kmph.

Power output $= \dfrac{F_t \times V_m}{3,600}$
$$= \frac{35,000 \times 50 \times 10^3}{3,600}$$

$= 486{,}111.11$ W

$= 486.11$ kW.

The power delivered by the locomotive on up gradient track with the DC series motors:

$$= 486.11\sqrt{\frac{55{,}000}{35{,}000}}$$

$= 609.37$ kW.

Since the power output $\propto \sqrt{T} \propto \sqrt{F_t}$, the power delivered by the locomotive on up gradient with the induction motors is:

$$= 486.11 \times \frac{55{,}000}{35{,}000}$$

$= 763.8875$ W $(\because$ power output $\propto T \propto F_t)$.

Example 10.15: A train weighting 450 ton has speed reduced by the regenerative braking from 50 to 30 kmph over a distance of 2 km on down gradient of 1.5%. Calculate the electrical energy and the overage power returned to the line tractive resistance is 50 N/ton. And, allow the rotational inertia of 10% and the efficiency conversion 80%.

Solution:

The accelerating weight of the train $W_e = 1.1$ W

$= 1.1 \times 450 = 495$ ton.

The distance travelled $D = 2$ km.

Gradient $G = 1.5\%$

Track resistance $r = 50$ N/ton.

Efficiency $\eta = 0.8$.

The energy available due to the reduction in the speed is:

$= 0.01072\ W_e V_1^2 - V_2^2$

$= 0.1072 \times 495\ (50^2 - 30^2)$

$= 8{,}490.24$ W-hr

$= 8.49$ kW-hr.

The tractive effort required while going down the gradient:

$F_t = Wr - 98.1\ WG$

$= 450 \times 50 - 98.1 \times 450 \times 1.5$

$= -43{,}717.5$ N.

The energy available while moving down the gradient a distance of 2 km is:

$$\frac{F_t \times D \times 1{,}000}{1{,}000 \times 3{,}600}\ \text{kW-hr}$$

$$= \frac{43{,}717.5 \times 2 \times 1{,}000}{1{,}000 \times 3{,}600}$$

$= 24.2875$ kW-hr.

The total energy available $= 8.49 + 24.2875$

$= 32.7775$ kW-hr.

The energy returned to the line = 0.8 × 32.7775
= 26.222 kW-hr.

The average speed $= \frac{50+30}{2}$
= 40 kmph.

The time taken to cover $2\,\text{km} = \frac{2}{40} = \frac{1}{20}\,\text{h.}$

The average power $= \frac{\text{Energy returned to the line}}{\text{time}} = \frac{26.222}{1/20} = 524.44\ \text{kW.}$

Example 10.16: A train weighing 450 ton is going down a gradient of 20 in 1,000, it is desired to maintain train speed at 50 kmph by regenerative braking. Calculate the power fed into the line and allow rotational inertia of 12% and the efficiency of conversion is 80%. Traction resistance is 50 N/ton.

Solution:

The dead weight of train $W = 450$ ton.
The maximum speed $V_m = 50$ kmph.

Gradient $G = \frac{20 \times 100}{1,000} = 2\%.$

Tractive resistance $r = 50$ N/ton.
Rotational inertia = 12%.
The efficiency of conversion = 0.8
The tractive effort required while going down the gradient:
$= Wr - 98.1\ WG$
$= 450 \times 50 - 98.1 \times 450 \times 2$
$= -65,790$ N.

The power available $P = \frac{F_t \times V_m}{3,600}$

$= \frac{65,790 \times 50}{3,600}$

= 913.75 kW.

The power fed into the line = power available × efficiency of conversion
= 913.75 × 0.8
= 731 kW.

Example 10.17: The speed–time curve of an electric train on a uniform raising gradient of 10 in 1,000 comprise of:

(i) Uniform acceleration from rest at 2.2 kmphps for 30 s.

(ii) Wasting with power off for 30 s.

(iii) Braking at 3.2 kmphps to standstill the weight of the train is 200 ton. The tractive resistance of level track being 4 kg/ton and the allowance for rotary inertia 10%. Calculate the maximum power developed by traction motors and the total distance travelled by the train. Assume the transmission efficiency as 85%.

Solution:

$$\text{Gradient} = \frac{10}{1{,}000} \times 100 = 1\%.$$

Acceleration (α) = 2.2 kmphps.
Braking (β) = 3.2 kmph.
The dead weight of train W = 200 ton.
Track resistance r = 4 kg/ton = 4 × 9.81 = 39.24 N/ton.
Maximum velocity $V_m = \alpha t_1 = 2.2 \times 30 = 66$ kmph.
Tractive effort required:

$$F_t = 277.8\, W_e \alpha + 98.1\, WG + Wr$$
$$= 277.8 \times 8 \times 1.1 \times 200 \times 2.2 + 98.1 \times 200 \times 1 + 200 \times 39.24$$
$$= 161{,}923.2 \text{ N}.$$

$$\text{The maximum power output} = \frac{F_t V_m}{3{,}600}$$
$$= \frac{161{,}923.2 \times 66}{3{,}600}$$
$$= 2{,}968.592 \text{ kW}.$$

$$\text{The maximum power developed by the traction motor} = \frac{2{,}968.592}{0.85} = 3492.46 \text{ kW}.$$

Let, the coasting retardation be β_c:

$$F_t = 277.8\, W_e(-\beta_c) + 98.1\, WG + Wr$$

$$0 = -277.8 \times (1.1 \times 200) \times \beta_c + 98.1 \times 200 \times 1 + 200 \times 39.24$$

$$\beta_c = 0.449 \text{ kmphps}$$
$$V_2 = V_m - \beta_c V_2$$
$$= 66 - 0.449 \times 65$$
$$= 36.815 \text{ kmph}.$$

$$\text{Braking period } t_3 = \frac{V_2}{\beta} = \frac{36.815}{3.2} = 11.504 \text{ s}.$$

The total distance travelled by the train:

$$D = \frac{V_m t_1}{7{,}200} + \frac{(V_1 + V_2)t_2}{7{,}200} + \frac{V_2 t_3}{7{,}200}$$

$$= \frac{66 \times 30}{7{,}200} + \frac{(66 + 36.815) \times 65}{7{,}200} + \frac{36.815 \times 11.504}{7{,}200}$$

$$= 0.275 + 0.928 + 0.0588$$
$$= 1.26 \text{ km}.$$

Example 10.18: A 2,300-ton train proceeds down a gradient of 1 in 100 for 5 min, during which period, its speed gets reduced from 40 to 20 kmph by the application of the regenerative braking. Find the energy returned to the lines if the tractive resistance is 5 kg/ton, the rotational inertia 10%, and the overall efficiency of the motors during regeneration is 80%.

Solution:

The dead weight of the train $W = 2{,}300$ ton.

The accelerating weight of the train $W_e = 1.1 \times 2{,}300$ s

$= 2{,}530$ ton.

$$\text{Gradient} = \frac{1}{100} \times 100 = 1\%.$$

Tractive resistance $r = 5 \times 9.81 = 49.05$ N/ton.

Regenerative period $t = 5 \times 60$

$= 300$ s.

Overall efficiency $\eta = 0.8$.

The energy available due to the reduction in speed:

$= 0.01072\ W_e\ (V_1^2 - V_2^2)$

$= 0.01072 \times 2{,}530 \times (40^2 – 20^2)$

$= 32{,}545.92$

$= 32.54$ kW-hr.

The tractive effort required while going down the gradient:

$= Wr - 98.1\ WG$

$= 2{,}300 \times 49.05 – 98.1 \times 2{,}300 \times 1$

$= -112{,}815.$

The distance moved during regeneration:

$$= \frac{V_1 + V_2}{2} \times \frac{1{,}000}{3{,}600} \times t$$

$$= \frac{40 + 20}{2} \times \frac{1{,}000}{3{,}600} \times 300$$

$= 2{,}500$ m.

The energy available on the account of moving down the gradient over a distance of 2,500 m:

$$= \frac{112{,}815 \times 2{,}500}{2{,}600 \times 1{,}000}$$

$= 78.34$ kW-hr.

The total energy available $= 32.54 + 78.34$

$= 88.707$ kW-hr.

The energy returned to the line $= 0.8 \times 11.08$

$= 88.707$ kW-hr.

Example 10.19: An electric train has an average speed of 50 kmph on a level track betweenstops 1,500 m a part. It is accelerated at 2 kmphs and is braked at 3 kmphs. Estimate the energy consumption at the axle of the train per ton-km. Take the reactive resistance constant at 50 N/ton and allow 10% for rotational inertia.

Solution:

Acceleration (α) = 2 kmphs.

Retardation (β) = 3 kmphs.

The distance of run (D) = 1.5 km.

Average speed $V_a = 50$ kmph.

The time of run $T = \frac{D}{V_a} \times 3,600$

$$= \frac{1.5}{50} \times 3,600$$

$$= 108 \times 10^3 \text{ s.}$$

Using the equation:

$$V_m = \frac{T}{2X} - \sqrt{\frac{T^2}{4X^2} - \frac{3,600}{X}}$$

$$X = \frac{1}{2\alpha} + \frac{1}{2\beta}$$

$$= \frac{1}{2\times 2} + \frac{1}{2\times 3} = 0.416.$$

$$V_m = \frac{108}{2\times 0.416} - \sqrt{\frac{(108)^2}{4\times(0.416)} - \frac{3600\times 3.5}{0.416}}$$

$$= 129.807 - \sqrt{16850.036 - 1298.769}$$

$$= 67.603 \text{ kmph.}$$

Accelerating period, $t_1 = \frac{V_m}{\alpha}$

$$= \frac{67.603}{2}$$

$$= 33.8015 \text{ s.}$$

Braking period, $t_3 = \frac{V_m}{\beta}$

$$= \frac{67.603}{3}$$

$$= 22.534 \text{ s.}$$

The distance travelled during braking:

$$= 1/2 \times V_m \times \frac{t_3}{3,600}$$

$$= \frac{1}{2} \times 67.603 \times \frac{22.534}{3,600}$$

$$= 0.2115 \text{ km.}$$

$$D_1 = D - 0.2115$$

$$= 1.5 - 0.2115$$

$$= 1.288 \text{ km.}$$

Tractive resistance $r = 50$ N/ton

$$\frac{W_e}{W} = 1.1.$$

The energy consumption at the axle of the train per ton-km:

$$= \frac{0.01072V_m^{\ 2}}{D} \times \frac{W_e}{W} + 0.2778r\frac{D_1}{D}$$

$$= \frac{0.01072 \times (67.603)^2}{1.5} \times 1.1 + 2{,}778 \times 50 \times \frac{1.288}{1.5}$$

$= 35.927 + 11.926$
$= 47.853$ W-hr.

Example 10.20: An electric train has quadrilateral speed–time curve as follows.

(i) The uniform acceleration for rest at 2.2 kmphs for 30 s.
(ii) Coasting for 45 s.
(iii) The braking period of 20 s.

The train is moving in a uniform up gradient of 1%, the tractive resistance is 50 N/ton, the rotational inertia effect 10% of the dead weight the duration of the station stop 20 s and overall efficiency of transmission gear and motor as 80%. Determine the value of is schedule speed and specific energy consumption of run.

Solution:

Time of acceleration $t_1 = 30$ s.
Time of coasting $t_2 = 45$ s.
Time of braking $t_3 = 20$ s.
Acceleration $(\alpha) = 2.2$ kmphps.
Maximum speed $V_m = \alpha\, t_1 = 2.2 \times 30 = 66$ kmph.
Gradient $G = 1\%$.

Let the coasting retardation be β_c:

$F_t = 277.8\, W_e(-\beta_c) + 98.1\, WG + Wr.$
$0 = 277.8 \times 1.1\, W\beta_c + 98.1 \times W \times 1 + 50\, W$
$= -305.58\, W\beta_c + 98.1\, W + 50\, W.$
$\beta_c = 0.4846$ kmphps.
$V_2 = V_m - \beta_c t_2$
$= 66 - 0.4846 \times 45$
$= 44.193$ kmph.

Braking retardation, $\beta = \frac{V_2}{t_3} = \frac{44.193}{20}$
$= 2.207$ kmphps.

The distance travelled $D = \frac{V_m t_1}{7{,}200} + \frac{(V_m + V_2)t_2}{7{,}200} + \frac{V_2 t_3}{7{,}200}$

$$= \frac{66 \times 30}{7{,}200} + \frac{(66 + 44.193)}{7{,}200} \times 45 + \frac{44.193 \times 20}{7{,}200}$$

$= 0.275 + 0.688 + 0.122$
$= 1.085$ km.

Schedule time, $T_s = t_1 + t_2 + t_3 +$ stop duration
$= 30 + 45 + 20 + 20$
$= 115$ s.

Schedule speed, $V_s = \dfrac{3,600 \times D}{T_s}$

$= \dfrac{3,600 \times 1.085}{115}$

$= 33.965$ kmph.

When power is on, the distance travelled is:

$D_1 =$ distance travelled during acceleration period

$= \dfrac{V_m t_1}{7,200}$

$= \dfrac{66 \times 30}{7,200} = 0.275$ km.

The specific energy output:

$$= \frac{0.01072 V_m^2}{D} \times \frac{W_e}{W} + 0.2778(98.1G + r)\frac{D_1}{D}$$

$$= \frac{0.01072 \times 66^2}{1.085} \times 1.1 + 0.2778(98.1 \times 1 + 50) \times \frac{0.275}{1.085}$$

$= 47.341 + 10.427$
$= 57.768$ W-hr/ton-km.

The specific energy consumption $= \dfrac{57.768}{0.8}$

$= 72.21$ W-hr/ton-km.

Example 10.21: A train weighing 200-ton accelerates uniformly from rest to a speed of 40 kmph up a gradient of 1 in 100, the time taken being 30 s. The power is then cut off and train coasts down a uniform gradient of 1 in 1,000 for period of 40 s. When brakes are applied for period of 20 s so as to bring the train uniformly to rest on this gradient determine:

(i) The maximum power output from the driving axles.

(ii) The energy taken from the conductor rails in kW-hr assuming an efficiency of 70% assume tractive resistance to be 45 N/ton at all speeds and allow 10% for rotational inertia.

Solution:

Accelerating weight, $W_e = 1.1 \times 200$
$= 220$ ton.

Tractive resistance, $r = 45$ N/ton

Gradient $= \dfrac{1}{100} \times 100$

$= 1\%$.

Maximum speed $V_m = 40$ kmph.
Accelerating period $t_1 = 30$ s.

Acceleration $\alpha = \frac{V_m}{t_1}$

$= \frac{40}{30}$

$= 1.33$ kmphps.

Tractive effort required:

$F_t = 27.88\ W_e\ \alpha + 98.1\ WG + Wr$

$= 277.8 \times 220 \times 1.33 + 98.1 \times 200 \times 1 + 200 \times 45$

$= 109{,}904.28$ N.

(i) The maximum power output from driving axle:

$= \frac{F_t \times V_m}{3{,}600}$

$= \frac{109{,}904 \times 40}{3{,}600}$

$= 1{,}221.155$ kW.

Total energy required for the run:

$= \frac{1}{2} \times \frac{F_t V_m}{3{,}600} \times \frac{t_1}{3{,}600}$

$= \frac{1}{2} \times \frac{109{,}904 \times 40}{3{,}600} \times \frac{30}{3{,}600}$

$= 5.088$ kW-hr.

(ii) The energy taken for conductor rails $= \frac{5.088}{0.7}$

$= 7.268$ kW-hr.

Example 10.22: Calculate the energy consumption if a maximum speed of 12 m/sec and for a given run of 1,500 m, an acceleration of 0.36 m/s² desired. The tractive resistance during acceleration is 0.052 N/kg and during the coasting is 6.12 N/1,000 kg. Allow a 10% of rotational inertia, the efficiency of the equipment during the acceleration period is 60%. Assume quadrilateral speed–time curve.

Solution:

Accelerating weight of the train $W_e = 1.1$ W.
Maximum speed $V_m = 12$ m/s.
The distance of run $D = 1{,}500$ m.
Acceleration $\alpha = 0.36$ m/s².

Accelerating period $t_1 = \frac{V_m}{\alpha}$

$= \frac{12}{0.36}$

$= 33.33$ s.

The tractive resistance during acceleration $r = 0.52$ N/kg.
The tractive effort required during acceleration $F_t = W_e \alpha + Wr$

$= 1.1\ W \times 0.36 + W \times 0.052$

$= 0.448\ W$ N.

The total energy required for the run = average power during acceleration × accelerating period

$$= \frac{1}{2} F_t V_m \times t_1$$

$$= \frac{1}{2} \times 0.448 \times 12 \times 33.3$$

$$= 89.51 \text{ J}$$

$$= \frac{89.51}{3,600}$$

$$= 0.024 \text{ W-hr.}$$

$$\text{Specific energy output} = \frac{\text{energy required for the run}}{W \times D}$$

$$= \frac{0.024\ W}{W \times 1,500}$$

$$= 1.712 \times 10^{-5} \text{ W-hr/kg-m.}$$

$$\text{Specific energy consumption} = \frac{\text{specific energy output}}{\eta}$$

$$= \frac{1.713 \times 10^{-5}}{0.6}$$

$$= 2.85 \times 10^{-5} \text{ W-hr/kg-m.}$$

Example 10.23: A 100-ton weight train has a rotational inertia of 10%. This train has to be run between two stations that are 3 km a part and has an average speed of 50 km/hr. The acceleration and the retardation during braking are 2 kmphps and 3 kmphps, respectively. The percentage gradient between these two stations is 1% and the train is to move up the incline the track resistance is 50 N/ton, then determine:

(i) Maximum power at the driving axle.

(ii) Total energy consumption.

(iii) Specific energy consumption.

The combined efficiency of the alembic train is 70%. Assume simplified trapezoidal speed–time curve.

Solution:

The dead weight of the train, $W = 100$ ton.
The accelerating weight of the train, $W_e = 1.1 \times W = 1.1 \times 100 = 110$ ton.
The distance of run $(D) = 3$ km.
Average speed $V_a = 50$ kmph.
Acceleration $(\alpha) = 2$ kmphps.

Retardation (β) = 3 kmphps.
Gradient (G) = 1%.
Tractive resistance r = 50 N/ton.

$$\text{Duration of run} = \frac{3,600 \times D}{V_a}$$

$$= \frac{3,600 \times 3}{50} = 216\,\text{km},$$

$$\text{where } X = \frac{1}{2\alpha} + \frac{1}{2\beta} = \frac{1}{2 \times 2} + \frac{1}{2 \times 3} = 0.416.$$

Using the equation, the maximum speed:

$$V_m = \frac{T}{2X} - \sqrt{\frac{T^2}{4X^2} - \frac{3,600D}{X}}$$

$$= \frac{216}{2 \times 0.416} - \sqrt{\frac{(216)^2}{4 \times (0.416)^2} - \frac{3,600 \times 3}{0.416}}$$

$$= 259.615 - \sqrt{(67,400.147) - (25,961.538)}$$

$$= 56.05 \text{ kmph.}$$

$$\text{Accelerating period, } t_1 = \frac{V_m}{\alpha} = \frac{56.05}{2} = 28.025 \text{ s.}$$

$$\text{Braking period } t_3 = \frac{V_m}{\beta} = \frac{56.05}{3} = 18.683 \text{ s.}$$

Free-running period $t_2 = 216 - (28.025 + 18.683)$
$= 169.292$ s.

Tractive effort required $F_t = 278\, W_e\, \alpha + 98.1\, WG + Wr$
$= 277.8 \times 110 \times 2 + 98.1 \times 100 \times 1 + 100 \times 50$
$= 75,926$ N.

(i) Maximum power at the driving axle:

$$= \frac{F_t \times V_m}{3,600}$$

$$= \frac{75,926 \times 56.05}{3,600}$$

$$= 1,182.125 \text{ kW.}$$

Tractive effort required during free running is F_t^1:

$$F_t^1 = 98.1\,WG + Wr$$
$$= 98.1 \times 100 \times 1 + 100 \times 50$$
$$= 14{,}810 \text{ N}.$$

$$\text{Total energy output} = \frac{1}{2} F_t \times \frac{V_m}{3{,}600} \times \frac{t_1}{3{,}600} + \frac{F_t \times V_m}{3{,}600} \times \frac{t_2}{3{,}600}$$
$$= \frac{1}{2} \times 75{,}926 \times \frac{56.05}{3{,}600} \times \frac{28.025}{3{,}600} + \frac{14{,}810 \times 56.05}{3{,}600} \times \frac{169.292}{3{,}600}$$
$$= 4.6 + 10.843$$
$$= 15.44 \text{ kW-hr.}$$

(ii) Total energy consumption $= \dfrac{15.44}{0.70}$

$$= 22.06 \text{ kW-hr.}$$

(iii) $\left.\begin{array}{r}\text{Specific energy}\\ \text{consumption}\end{array}\right\} = \dfrac{\text{the total energy consumption in watt} - \text{hours}}{\text{the weight of the train in tons} \times \text{the distance of the run in km}}$

$$= \frac{22.06 \times 1000}{100 \times 3}$$
$$= 73.53 \text{ W-h/ton-km.}$$

Example 10.24: An electric train has quadrilateral speed–time curve as follows:

(i) Uniform acceleration from rest 2 kmphps for 30 s.

(ii) Coasting for 40 s.

(iii) Braking period of 25 s.

The train is moving a uniform down gradient of 1% and the tractive resistance of 50 N/ton. The rotational resistance is 10% of the dead weight, the duration of the stop is 20 s and the overall efficiency of the transmission the gear and the motor as 80%. Calculate its schedule speed and specific energy consumption.

Solution:

Acceleration (α) = 2 kmphps.

Acceleration period (t_1) = 30 s.

Gradient (G) = 1%.

The tractive of resistance (r) = 50 N/ton.

The duration of stop = 20 s.

Overall efficiency (η) = 80%.

Maximum speed $V_m = \alpha t_1$

$$= 2 \times 30 = 60 \text{ kmph.}$$

Let the coasting retardation be β_c:

Tractive effort:

$$F_t = 277.8\,W_e(-\beta_c) - 98.1\,WG + Wr$$
$$0 = -277.8 \times 1.1\,W\beta_c - 98.1 \times W \times 1 + 50\,W$$

$$\beta_c = \frac{-48.1W}{305.58}$$

$\beta_c = -0.157$ kmphps

$V_2 = V_m - \beta_c t_2$

$= 60 - (-0.517 \times 40)$

$= 66.28$ kmph.

The distance travelled, $D = \dfrac{V_1 t_1}{7,200} + \dfrac{(V_m + V_2)t_2}{7,200} + \dfrac{V_2 t_3}{7,200}$

$$= \frac{60 \times 30}{7,200} + \frac{(60 + 66.28)}{7,200} \times 40 + \frac{66.28 \times 25}{7,200}$$

$= 0.25 + 0.7 + 0.23$

$= 1.18$ km.

Schedule time, $T_s = t_1 + t_2 + t_3 +$ stop duration

$= 30 + 40 + 25 + 20$

$= 115$ s.

Schedule speed, $V_s = \dfrac{3,600 \times D}{T_s}$

$$= \frac{3,600 \times 1.18}{115}$$

$= 36.939$ kmph.

The specific energy output:

$$= \frac{0.01072 V_m^2}{D} \times \frac{W_e}{W} + 0.2778(98.1\,G + r)\frac{D_1}{D}$$

$$= \frac{0.01072 \times (60)^2}{1.18} \times 1.1 + 0.2778(98.1 \times 1 + 50) \times \frac{0.25}{1.18}$$

$= 35.975 + 8.716$

$= 44.69$ W-hr/ton-km.

The specific energy consumption $= \dfrac{44.69}{0.8} = 55.86$ W-hr/ton-km.

Example 10.25: The schedule speed of a electric train is 40 kmph. The distance between two stations is 3 km with each stop is of 30 s duration. Assuming the acceleration and the retardation to be 2 and 3 kmphps, respectively. The dead weight of the train is 20 ton. Assume the rotational inertia is 10% to the dead weight and the track resistance is 40 N/ton. Calculate:

(i) The maximum speed.

(ii) The maximum power output from driving axles.

(iii) The specific energy consumption is watt-hours per ton-km. The overall efficiency is 80%, assume simplified speed–time curve.

Solution:

Schedule speed $V_s = 40$ kmph.

The distance between the two stations $(D) = 3$ km.

The duration of stop $= 30$ s.

Acceleration $(\alpha) = 2$ kmphps.

Retardation $(\beta) = 3$ kmphps.

The dead weight of the train (w) = 20 ton.
The track resistance (r) = 40 N/ton.
The overall efficiency (η) = 80%.

The schedule time of run $T_s = \frac{3,600 \times D}{V_s}$

$= \frac{3,600 \times 3}{40} = 270\text{s}.$

The actual time of run, $T = 270 - 30$
$= 240$ s.

(i) The maximum speed, $V_m = \frac{T}{2X} - \sqrt{\frac{T^2}{4X^2} - \frac{3,600D}{X}}$,

where:

$$X = \frac{1}{2\alpha} + \frac{1}{2\beta}$$

$$= \frac{1}{2 \times 2} + \frac{1}{2 \times 3} = 0.416.$$

$$V_m = \frac{240}{2 \times 0.416} - \sqrt{\frac{(240)^2}{4 \times (0.416)^2} - \frac{3,600 \times 3}{0.416}}$$

$$= 288.46 - \sqrt{(83,210.059) - (25,961.538)}$$

$$= 49.193 \text{ kmph.}$$

(ii) The acceleration time, $t_1 = \frac{V_m}{\alpha}$

$= \frac{49.193}{2}$

$= 24.59$ s.

The duration of braking, $t_3 = \frac{V_m}{\beta} = \frac{49.193}{3} = 16.397\text{s}.$

The free-running time $t_2 = \text{T} - (t_1 + t_3)$
$= 240 - (24.59 + 16.397)$
$= 199.012$ s.

The tractive effort during acceleration:
$F_t = 277.8W_e \times \alpha + Wr$
$= 277.8 \times 1.1 \times 20 \times 2 + 20 \times 40$
$= 13,023.2$ N.

The maximum power output $= \frac{F_t V_m}{3,600}$

$= \frac{13,023.2 \times 49.193}{3,600}.$

The maximum power output = 177.958 kW.

(iii) The distance travelled during braking:

$$= \frac{1}{2}\frac{V_m \times t_3}{3,600}$$

$$= \frac{1}{2}\times\frac{49.193\times 16.397}{3,600}$$

$$= 0.112 \text{ km.}$$

The distance travelled with power is on:

$$D_1 = 3 - 0.112$$

$$= 2.88 \text{ km.}$$

The specific energy output:

$$= \frac{0.01072V^2{}_m}{D}\times\frac{W_e}{W} + 0.2778r\frac{D_1}{D}$$

$$= \left[\frac{0.01072\times(49.193)^2}{3}\times 1.1\right] + \left[0.2778\times 40\times\frac{2.88}{3}\right]$$

$$= 9.512 + 10.667$$

$$= 20.179 \text{ W-hr/ton-km.}$$

$$\text{The specific energy consumption} = \frac{20.179}{\text{efficiency}} = \frac{20.179}{0.8}$$

$$= 25.244 \text{ W-hr/ton-km.}$$

10.10 CALCULATION OF ENERGY RETURNED TO THE SUPPLY DURING REGENERATIVE BRAKING

When the train is accelerating, it acquires kinetic energy corresponding to that speed. During the coasting period, some of the kinetic energy is wasted, to propel the train against the friction and windage resistance.

While the train is moving on the down gradients or level track, the KE acquired by the rotating parts is converted into the electrical energy, which is fed back to the supply system. The amount of energy fed back to the system is depending on the following factors.

1. The initial and final speeds during the regenerative braking.
2. The train resistance and the gradient of the track.
3. The efficiency of the system.

Consider the initial and final speeds of the train during regenerative braking are V_1 and V_2 in KMPH, and the effective weight of the train is W_e tons.

$$\text{Thus, the mass of the train, } m = \frac{W_e}{g} \text{ tons/(m/s}^2)$$

$$= \frac{1000W_e}{9.81} \text{ kg/(m/s}^2).$$

The speed of the train = V_1 kmph

$$= \frac{V_1 \times 1{,}000}{3{,}600} \text{ m/s.}$$

The kinetic energy stored by the train at a speed of V_1 kmph:

$$= \frac{1}{2} \times mv^2 \text{J}$$

$$= \frac{1}{2} \times \frac{1{,}000W_e}{9.81} \times \left[\frac{v_1 \times 1{,}000}{3{,}600}\right]^2 \frac{\text{kg}}{\text{m/s}^2} \times \frac{\text{m}}{\text{s}}$$

$$= \frac{1}{2} \times \frac{1{,}000W_e}{9.81} \times \left[\frac{v_1 \times 1{,}000}{3{,}600}\right]^2 \times 9.81 \text{ N-m (or) } W\text{-sec}$$

[1 kg-m = 9.81 N-m]

$$= \frac{1}{2} \times \frac{1{,}000W_e}{9.81} \times \left[\frac{v_1 \times 1{,}000}{3{,}600}\right]^2 \times 9.81 \times \frac{1}{3{,}600} \text{ W-h}$$

[1 W-Sec = 1/3,600 W-h]

$$= 0.01072 V_1^2 W_e \text{ W-h}$$

$$= 0.01072 V_1^2 \left(\frac{W_e}{w}\right) \text{ W-h/ton.}$$

Thus, the kinetic energy at speed V_2 kmph:

$$= 0.01072 V_2^2 \left(\frac{W_e}{w}\right) \text{ W-h/ton.}$$

Therefore, the energy available during the regeneration:

$$= 0.01072 \left(\frac{W_e}{w}\right) \times \left(V_2^2 - V_1^2\right) \text{ W-h/ton.}$$

If D is the distance in km covered during the regenerative braking, then the energy fed back to the supply during the braking while the train is moving on down gradient:

$$= WD \times \frac{G}{100} \text{ ton-km}$$

$$= (1{,}000 \times w) \times (1{,}000 \times D) \times \frac{G}{100}$$

$= WDG \times 10^4$ kg-m (or) N-sec^2

$= 9.81 \times WDG \times 10^4$ N-m (or) W-sec

$$= \frac{9.81 \times WDG \times 10^4}{3{,}600} \text{ W-h}$$

$= 27.25\ WDG$ W-h

$= 27.25\ DG$ W-h/ton.

If r is the train resistance in N/ton, then the energy lost to overcome the resistance to the motion and friction, windage losses:

$= WrD$ N-km

$= WrD \times 1{,}000$ N-m (or) w-sec

$$= \frac{WrD \times 1{,}000}{3{,}600} \text{ W-h}$$

$$= 0.2778\,WrD \text{ W-h (or) } 0.2778\,rD\frac{\text{W-h}}{\text{ton}}.$$

Hence, the total energy available during regeneration:

$$= \left[0.01072\left(\frac{W_e}{W}\right)(V_1^2 - V_2^2) + 27.25\,DG - 0.2778rD\right] \text{W-h/ton}.$$

The energy returned to the supply system:

$$= \left[0.01072\left(\frac{W_e}{W}\right)(V_1^2 - V_2^2) + 27.25\,DG - 0.2778rD\right] \times \eta\, W\text{-h/ton},$$

where η is the efficiency of the system.

10.10.1 Advantages of regenerative breaking

1. In regenerative breaking, a part of the energy stored by the rotating parts is converted into the electrical energy and is fed back to the supply. This will lead to the minimum consumption of energy, thereby saving the operating cost.
2. High breaking retardation can be obtained during regenerative breaking.
3. Time taken to bring the vehicle to rest is less compared to the other breakings; so that, the running time of the vehicle is considerably reduced.
4. The wear on the brake shoes and tyre is reduced, which increases the life of brake shoe and tyre.

10.10.2 Disadvantages

In addition to the above advantages, this method suffers from the following disadvantages.

1. In addition to the regenerative breaking, to bring the vehicle to standstill, mechanical breaking is to be employed.
2. In case of DC traction, additional equipment is to be employed for regenerative breaking, which increases the cost and sometimes, substation are equipped with mercury arc rectifiers to convert AC to DC supply.
3. The electrical energy returned to the supply will cause the operation of substations complicated.

Example 10.26: A 450-ton train travels down gradient of 1 in 75 for 110 s during which its speed is reduced from 70 to 55 kmph. By the regenerative braking, determine the energy returned to the lines if the reactive resistance is 4.5 kg/ton and the allowance for the rotational inertia is 7% and the overall efficiency of the motor is 80%.

Solution:

Accelerating weight $W_a = 1.075 \times 450$

$= 483.75$ ton.

Track resistance $r = 9.81 \times 4.5$

$= 44.145$ N-m/ton

$= 9{,}904.782$ W-hr

$= 9.904$ kW-hr.

Factiva effort required during retardation:

$= Wr - 98.1\ WG$

$= 450 \times 44.45 - 98.1 \times 450 \times 4/3$

$= 19,865.25 - 58,860$

$= -38,994.75$ N.

The distance travelled during the retardation period:

$$= \frac{V_1 + V_2}{2} \times \frac{1,000}{3,600} \times T$$

$$= \frac{70+55}{2} \times \frac{1,000}{3,600} \times 110$$

$= 1,909.73$ m.

As the train is moving in downward gradient, so that the tractive effort will provide additional energy to the system. The energy available when the train moves over a gradient is given as:

$$= \frac{38,994.75 \times 1,909.73}{1,000 \times 3,600}.$$

$$\text{Gradient } G = \frac{1}{75} \times 100$$

$$= \frac{4}{3} \times 100.$$

The period of regeneration = 110 s.

Overall efficiency (η) = 80%.

The kinetic energy of the train at a speed of 70 kmph is:

$= 0.01092 V_1^2 \; Wa$

$= 0.01092 \times (70)^2 \times (483.75)$

$= 25,884.495$ W-hr.

The kinetic energy of the train at the speed of 55 kmph is:

$= 0.01092 V_2^2 \; Wa$

$= 0.01092 \times (55)^2 \times (483.75)$

$= 15,979.713$ W-hr.

The energy available due to the retardation by the regenerative braking:

$= 2,588.495 - 15,979.713$

$= 20.68$ kW-hr.

The energy returned to the supply system:

$= 0.8 \times$ total energy available

$= 0.8 \times (20.68 + 9.904)$

$= 24.467$ kW-hr.

KEY NOTES

- *Types of passenger services*, by which the type of traction system has to be selected are:
  - Main line service.
  - Urban or city service.
  - Suburban service.

- In *main line service,* the distance between the two stops is usually more than 10 km.
- In *urban service,* the distance between the two stops is very less and it is less than 1 km.
- In *suburban service,* the distance between the two stations in between 1 and 8 km.
- The curve that shows the instantaneous speed of train in kmph along the ordinate and time in seconds along the abscissa is known as *speed–time curve.*
- The curve that shows the distance between the two stations in km along the ordinate and time in seconds along the abscissa is known as *speed–distance curve.*
- The maximum speed attained by the train during run is known as *crest speed.*
- The ratio of the distance covered by the train between the two stops to the total time of run is known as *average speed.*
- The ratio of the distance covered between the two stops to the total time of run including time for the stop is known as *schedule speed.*
- The sum of the time required for actual run and the time required for stop is known as *schedule time.*
- The effective force acting on the wheel of the locomotive necessary to propel the train is known as *tractive effort.*
- The total weight of the train to be propelled by the locomotive is called as *dead weight.*
- The effective weight of the train that has angular acceleration due to the rotational inertia including the dead weight of the train is known as *accelerating weight.*
- The total weight to be carried out on the drive in wheels of a locomotive is known as *adhesive weight.*
- The coefficient of adhesion defined as the ratio of tractive effort required to propel the wheel of a locomotive to its adhesive weight.

SHORT QUESTIONS AND ANSWERS

(1) What are the various types of services?

There are mainly three types of passenger services, by which the type of traction system has to be selected, namely:

(i) Main line service.

(ii) Urban or city service.

(iii) Suburban service.

(2) What is meant by main line services?

Main line service in which the distance between the two stops is usually more than 10 km. High balancing speeds should be required. Acceleration and retardation are not so important.

(3) What is meant by urban services?

In this service, the distance between the two stops is very less and it is less than 1 km. It requires high average speed for frequent starting and stopping.

(4) What is meant by suburban services?

In this service, the distance between the two stations in between 1 and 8 km. This service requires rapid acceleration and retardation as frequent starting and stopping is required.

(5) Define crest speed.

The maximum speed attained by the train during run is known as crest speed. It is denoted with 'V_m'.

(6) Define average speed.

It is the mean of the speeds attained by the train from start to stop, i.e., it is defined as the ratio of the distance covered by the train between the two stops to the total time of run. It is denoted with 'V_a'.

(7) Define schedule speed.

The ratio of the distance covered between the two stops to the total time of run including time for stop is known as the schedule speed. It is denoted with the symbol 'V_s'.

(8) Define schedule time.

It is defined as the sum of time required for actual run and the time required for stop.

i.e., $T_s = T_{run} + T_{stop}$.

(9) What are the various factors affecting the schedule speed?

The factors that affect the schedule speed of a train are:

(i) Crest speed.

(ii) The duration of stops.

(iii) The distance between the stops.

(iv) Acceleration.

(v) Braking retardation.

(10) Define tractive effort.

It is the effective force acting on the wheel of the locomotive, which is necessary to propel the train is known as '*tractive effort*'.

(11) Define specific energy consumption.

This quantity of energy consumed by various parts of the train per ton per kilometer is known as specific energy consumption. It is expressed in watt hrs per ton per km.

(12) Define dead weight.

It is the total weight of the train to be propelled by the locomotive. It is denoted by 'W'.

(13) Define accelerating weight.

It is the effective weight of the train, which has angular acceleration due to the rotational inertia including the dead weight of the train. It is denoted by 'W_e'.

(14) Define adhesive weight.

The total weight to be carried out on the drive in the wheels of a locomotive is known as adhesive weight.

(15) Define coefficient of adhesion.

It is defined as the ratio of tractive effort required to propel the wheel of a locomotive to its adhesive weight.

(16) What is the relation between tractive effort and adhesive weight?

$F_t \propto W$

$F_t = \mu W.$

MULTIPLE-CHOICE QUESTIONS

(1) The acceleration rate for the urban or suburban services is:

(a) 1.5–4 kmphps.

(b) 3–4 kmphps.

(c) 5–10 kmphps.

(d) 0.5–1.5 kmphps.

(2) The braking retardation for the urban or suburban services is:

(a) 1.5–2.5 kmphps.

(b) 3–4 kmphps.

(c) 5–10 kmphps.

(d) 0.5–1.5 kmphps.

(3) The coasting retardation for the line railway services is about

(a) 10 kmphps.

(b) 3 kmphps.

(c) 0.16 kmphps.

(d) 0.01 kmphps.

(4) The maximum speed at which the trains run on the main line railway service is:

(a) 160 kmph.

(b) 120 kmph.

(c) 100 kmph.

(d) 200 kmph.

(5) The speed–time curve for the urban service has no:

(a) Coasting period.

(b) Free-running period.

(c) Braking period.

(d) Acceleration period.

(6) Free-running and coasting periods are generally long in case of:

(a) City service.

(b) Suburban service.

(c) Main line service.

(d) Outer suburban service.

(7) Trapezoidal speed–time curve pertains to:
(a) Main line service.
(b) Urban service.
(c) Suburban service.
(d) Urban/suburban service.

(8) Quadrilateral speed–time curve is the close approximation for:
(a) Urban service.
(b) Suburban service.
(c) Urban/suburban service.
(d) Main line service.

(9) Area under the speed–time curve represents:
(a) Total distance travelled.
(b) Average speed.
(c) Average acceleration.
(d) None of the above.

(10) The speed of a train estimated taking into account the stoppage time at a station in addition to the actual running time between stops is called the:
(a) Average speed.
(b) Schedule speed.
(c) Free-running speed.
(d) Notching speed.

(11) The average speed of a train is independent of:
(a) The duration of the stops.
(b) The acceleration and braking retardation.
(c) The distance between stops.
(d) The running time.

(12) The schedule speed of a given train when running on a given service (with given distance between stations) is affected by:
(a) Acceleration and braking retardation.
(b) Maximum or crest speed.
(c) Duration stop.
(d) All of the above.

(13) For a given value of an average speed, decrease in duration of stops causes:
(a) Decrease in schedule speed.
(b) Increase in schedule speed.
(c) Increase in crest speed.
(d) Decrease in crest speed.

(14) Skidding of a vehicle always occurs when:
(a) Braking effort exceeds its adhesive weight.
(b) Brake is applied suddenly.
(c) It negotiates a curve.
(d) It passes over points and crossings.

(15) The adhesive weight is the:
(a) Total weight of the locomotive and the train.
(b) Weight coming over the driving wheels.
(c) Same as the accelerating weight.
(d) None of the above.

(16) The coefficient of the adhesion is the ratio of tractive effort to slip the wheels and:
(a) Dead weight.
(b) Accelerating weight.
(c) Adhesive weight.
(d) None of the above.

(17) The normal value of the coefficient of adhesion is:
(a) 0.25.
(b) 0.35.
(c) 0.50.
(d) 0.65.

(18) The coefficient of the adhesion reduces due to the presence of:
(a) Dew on rails.
(b) Oil and grease on rails.
(c) Dry sand on rails.
(d) Both (a) and (b).

(19) The coefficient of the adhesion improves due to the presence of:
(a) Dry sand on rails.
(b) Rust on rails.
(c) Dust on rails.
(d) All of the above.

(20) The value of the coefficient of the adhesion will be high when rails are:
(a) Wet.
(b) Cleaned with sand.
(c) Greased.
(d) Sprayed with oil.

(21) The coefficient of the adhesion for wet or greasy rails is:

(a) 0.35.

(b) 0.25.

(c) 0.08.

(d) Zero.

(22) Tractive effort is required to:

(a) Accelerate the train mass.

(b) Overcome the train resistance (friction windage and curve resistance).

(c) Overcome the effect of gravity.

(d) All of the above.

(23) The resistance encountered by a train in motion is on account of:

(a) The resistance offered.

(b) The friction at the track.

(c) The friction at various parts of the rolling stock.

(d) All of the above.

(24) The air resistance to the movement is proportional to:

(a) 1/speed.

(b) (speed).

(c) $(\text{speed})^2$.

(d) $(\text{speed})^3$.

(25) The friction at the track is proportional to:

(a) Speed.

(b) $(\text{speed})^2$.

(c) $(\text{speed})^3$.

(d) $(1/\text{speed})^2$.

(26) If the resistance to electric train movement is given by $F_r = a + bv + cv^2$.

In the above expression b is likely to cover:

(a) Air resistance.

(b) Track resistance.

(c) Frictional resistance.

(d) None of the above.

(27) Tractive effort of an electric locomotive can be increased by:

(a) Using high output motors.

(b) Increasing the supply voltage.

(c) Increasing the dead weight over the driving axles.

(d) Both (a) and (c).

(28) In a train, the energy output of the driving axles is used in:

(a) Accelerating the train mass.

(b) Overcoming the train resistance.

(c) Overcoming the gradient.

(d) All of the above.

(29) Energy consumption in propelling the train is required for:

(a) Accelerating the train mass.

(b) Overcoming the gradient while moving up the gradient.

(c) Overcoming the train resistance.

(d) All of the above.

(30) Longer coasting period for a train results in:

(a) Higher schedule speed.

(b) Lower specific energy consumption.

(c) Higher retardation.

(d) Higher acceleration.

(31) Specific energy consumption is affected by:

(a) Acceleration and retardation values.

(b) The crest speed and nature of route.

(c) Distance between stops.

(d) All of the above.

(32) Specific energy consumption becomes:

(a) More when the distance between the stops is more.

(b) More with the higher values of acceleration (or retardation).

(c) More with high train resistance.

(d) Less with the increase in crest speed.

(33) Specific energy consumption is highest in:

(a) Urban service.

(b) Suburban service.

(c) Main line service.

(d) Equal for all types of services.

REVIEW QUESTIONS

(1) Derive an expression for specific energy output on level track using a simplified speed–time curve.

(2) Derive the expression for a trapezoidal speed–time curve of an electric train.

(3) Discuss the merits and demerits of DC and single-phase AC systems for the main and suburban line electrifications of the railways.

(4) Derive the expression for the tractive effort for a train on a level track.

(5) Derive the expression for the distance between the stops and the speed at the end of the coasting period for a quadrilateral speed–time curve.

(6) Explain the following terms.

  (i) Tractive effort.

  (ii) Coefficient of adhesion.

  (iii) Specific energy consumption.

  (iv) Tractive resistance.

(7) Explain briefly the tractive effort required while the train is moving up the gradient and down the gradient.

(8) Deduce from the first principles the relation between acceleration, retardation, maximum speed, running time, and the distance between the stops for simplified speed–time curves run for a electric train.

EXERCISE PROBLEMS

(1) An electric train is to have acceleration and braking retardation of 0.8 km/h/s and 2.5 km/h/s, respectively. If the ratio of the maximum speed to the average speed is 1.3 and the time for stop is 30 s. Then, determine the schedule speed for a run of 1.8 km. Assume simplified trapezoidal speed–time curve.

(2) The distance between the two stops is 5 km apart. A train has schedule speed of 70 kmph. The train accelerates at 2 kmphps and retards 4.5 kmphps the duration of stop is 45 s. Determine the crest speed over the run assuming trapezoidal speed–time curve.

(3) A train has schedule speed of 45 kmph over a level track distance between the two stations being 4 km. Duration of the stop is 53 s. Assuming the braking retardation of 2.8 kmphps and the maximum speed is 35%, which is greater than the average speed. Determine the acceleration required to run the service.

(4) A suburban electric train has a maximum speed of 80 kmph. The schedule speed including a station stop of 40 s is 80 kmph. If the acceleration is 5 kmphps, the average distance between the two stops is 8 km. Determine the value of retardation.

(5) An electric train has an average speed of 60 kmph on a level track between the stops 1,200 m apart. It is accelerated at 6 kmphps and is braked at 8 kmphps. Draw the speed–time curve for the run.

(6) A locomotive exerts a tractive effort of 20,000 N in halting a train at 60 kmph on the level track. If the motor is to haul the same train on a gradient of 1 in 75 and the tractive effort required is 45,000 N, determine the power delivered by the locomotive if it is driven by (i) DC series motors and (ii) induction motors.

(7) A train weighing 300 ton is going down a gradient of 40 in 1,000; it is desired to maintain train speed at 60 kmph by regenerative braking. Calculate the power fed into the line and allow rotational inertia of 15% and the efficiency of conversion is 88%. Traction resistance is 30 N/ton.

ANSWERS

1. a
2. b
3. c
4. a
5. b
6. c
7. a
8. c
9. a
10. b
11. a
12. d
13. b
14. a
15. b
16. c
17. a
18. d
19. d
20. b
21. c
22. d
23. d
24. c
25. a
26. b
27. d
28. d
29. d
30. b
31. d
32. c
33. a

Electrolysis

OBJECTIVES

After reading this chapter, you should be able to:

- understand the principle of electrolysis
- know the laws of electrolysis
- know the various applications of electrolytic process

11.1 INTRODUCTION

Electrolysis is nothing but the process by which electrical energy produce chemical changes. This process can be normally used for the extraction of pure metal from their ores, the refining of metals, the building up of worn parts in metallurgical, chemical, and in other industries.

11.2 PRINCIPLE OF ELECTROLYSIS

The basic principle of electrolysis is, whenever a DC electric current is made to pass through the solution of salt, some metals can be separated from them. These separated metals can be coated on any object to form a pure thin layer.

For example, a crucible filled with water in which two electrodes (anode and cathode) are immersed and those are supplied from a DC source as shown in Fig. 11.1. When sodium chloride (NaCl) salt is dissolved in water, it decomposes into positively charged Na^+ ion and negatively charge Cl^- ion, moving freely in the solution. The positively charged ion (Na^+) travels toward the cathode and the negatively charged ion (Cl^-) travels toward the anode. On reaching the cathode, each positively charged sodium ion takes one electron from it and forms a sodium metal. Similarly, each of the negatively charged chloride ion will give one electron to anode and cease to be anion. Now, as the sodium metal deposited at the cathode, the ions collected at the cathode react with water giving out oxygen and hydrogen chloride

$$4Cl + 2H_2O \rightarrow 4HCl + O_2. \qquad \textbf{(11.1)}$$

In case if the cathode is made up of sodium, again the hydrogen chloride reacts with sodium forming sodium chloride liberating hydrogen gas.

$$2HCl + 2Na \rightarrow 2NaCl + H_2. \qquad \textbf{(11.2)}$$

Thus, sodium metal from the sodium chloride in the water is deposited at the cathode. The above process is known as 'electrolysis'.

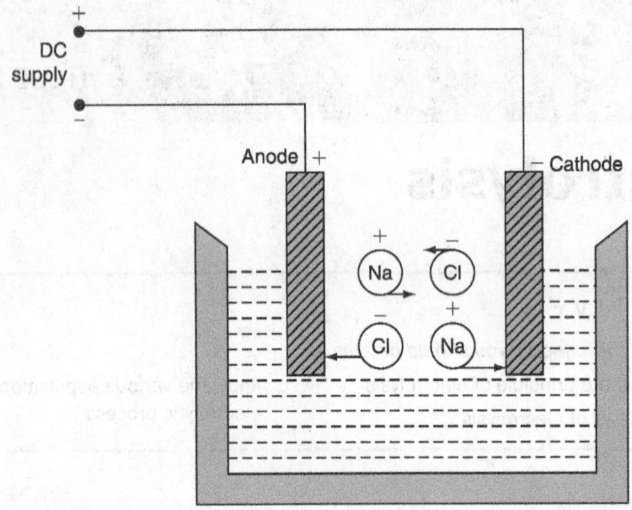

FIG. 11.1 *Electrolysis*

11.3 LAWS OF ELECTROLYSIS

The laws governing the electrolytic process were proposed by Michael Faraday. These laws are stated below.

11.3.1 Faraday's first law

This law states that 'the mass of substance deposited from an electrolyte is proportional to the quantity of electricity passing through the electrolyte in a given time'.

i.e., $m \propto It$

$$= ZIt, \tag{11.3}$$

where I is the steady current flowing through an electrolyte in amperes, t is the duration of current flowing through an electrolyte, Z is the constant of proportionality, and m is the mass of substance deposited.

An electrochemical equivalent 'Z' equals to the mass of substance deposited, when a steady electric current of $1A$ is passing through an electrolyte in 1 s:

i.e., if $I = 1\text{A}$ and $t = 1$ s, then, $Z = m$.

Usually, Z is expressed in terms of kilogram per coulomb (kg/c).

11.3.2 Faraday's second law

This law states that 'when the same quantity of electric current is passed through different electrolytes, the masses of the substances deposited are proportional to their respective chemical equivalents or equivalent weights'.

$$\left[\text{Chemical equivalent} = \frac{\text{atomic weight}}{\text{valency}}\right].$$

11.4 VARIOUS TERMS RELATED TO ELECTROLYTE PROCESS

Some of the important terms related to electrolytic process are discussed below.

(i) Voltage: It is necessary term for the electrolysis. The voltage required for the passage of current through the electrolysis must be equal to the voltage drop across the electrolyte and electrodes.

If $V_1$and V_2 are the voltage drops across the electrolytes and electrodes and 'V' is the total voltage applied across the electrodes, then:

$$V = V_1 + V_2. \quad \textbf{(11.4)}$$

(ii) Current efficiency: In the electrolytic process, secondary reactions are caused due to impurities. Therefore, the quantity of substance liberated from the electrodes is slightly less than that of the quantity actually calculated by Faraday's laws.

$$\text{Current efficiency} = \frac{\text{the actual quantity of substance liberated or deposited}}{\text{the theoritical quantity of substance liberated or deposited}}.$$

The practical value of the current efficiency lies between 90% and 98%.

(iii) Energy Efficiency: Due to the secondary reactions caused by the impurities present in the electrolyte, the actual value of the voltage required for the deposition of metal from the electrode is higher than the theoretical value of voltage.

$$\text{i.e., Energy efficiency} = \frac{\text{theoritical energy (voltage) required}}{\text{actual energy (voltage) required}}.$$

(iv) Equivalent weight: The equivalent weight of a substance is defined as the ratio of formula weight to its valency.

$$\text{i.e., Equivalent weight} = \frac{\text{formula weight}}{\text{valency}}.$$

11.5 APPLICATIONS OF ELECTROLYTIC PROCESS

Some of the applications of electrolysis used in the chemical industry and metal extraction are given below.

11.5.1 Manufacturers of chemicals

Various industrial applications of electrolysis such as the manufacturing of chemicals such as caustic soda, ammonium sulfate, hydrogen, oxygen, and chlorine. Here, the production of caustic soda and the production of hydrogen and oxygen by electrolysis process are explained below.

Production of caustic soda

The production of caustic soda can be done by two processes.

Diaphragm process

In this process, both anode and cathode compartments are separated by a diaphragm, to prevent the mechanical mixing of two solutions. During this process, chlorine is formed at anode; some of it is evolved as gas and the remaining goes into solution. And, sodium is discharged at the cathode reacts with the hydroxyl ions forming sodium hydroxide (NaOH) liberating hydrogen gas at the cathode. At this stage, brain solution is fed into the anode, which opposes the flow of hydroxyl ions toward the anode.

Mercury cathode process

It is another process of producing caustic soda by electrolytic process. In this process, mercury cells are built in various sizes from 1,000 A to very large 50,000 A per unit. Brain solution is fed into the cells for the reduction of salt concentration. Each mercury cell unit consists of two reactions. One of which has chlorine outlet, graphite anode, and mercury cathode and the other section consists of mercury amalgam anode and iron cathode. Throughout the process, the flow of mercury between two reactions closes the electrical connection.

The voltage required for each cell is about 4 V and the energy efficiency is of the order of 50–60%.

Production of hydrogen and oxygen by electrolysis

In this process, the gases such as hydrogen and oxygen obtained are of high purity at a cheap cost. This is mainly due to the low consumption of electrical energy for the production of gases by electrolysis.

In this process, the electrodes are made up of iron and nearly 15–20% of caustic soda is mixed with water. The chemical reactions take place at the electrodes as given below:

At cathode: $2Na + 2H_2O \rightarrow 2NaOH + H_2$.

At anode: $2OH \rightarrow H_2O + \frac{1}{2}O_2$.

Hydrogen gas is liberated at the cathode and oxygen is liberated at the anode. The voltage required for this process during the starting is 2–2.2 V and during the operation is 2.3–2.5 V, and energy required for this process is 6 kW-hr/m^3.

11.5.2 Electro metallurgy

Electro metallurgy indicates the extraction of metals from their ores and its refining.

Electroextraction

Electroextraction is the process by which metals can be extracted or separated from their ores. Depending upon the physical status of the ore, the metal can be extracted by the two ways: (i) when the ore is in solid state, it is to be treated with strong acid to obtain its salt, again this salt is to be electrolyzed to liberate the metal and (ii) when the metal is in liquid state or in molten state, it is directly electrolyzed in a furnace to liberate the metal.

Some of the materials used for the extraction of various metals from their ores are given below.

Extraction of zinc

Zinc can be extracted from the zinc ore. The zinc ore is in solid state that has high content of zinc oxide and it is to be treated with concentrated sulfuric acid, then it undergoes through various chemical process to get impurities such as cadmium, copper, and zinc sulfate solution. This solution is electrolyzed in a wooden box lined with lead having two electrodes. Both anodes and cathodes are lined with lead and aluminum. During the electrolysis, zinc is deposited at the cathode. For this process, the current density at the cathode is 1,000 A/m^2 and the voltage drop is nearly 3–3.5 V.

Extraction of aluminum

Aluminum can be extracted from its ores such as bauxite, cryolite, and barite. The ore of aluminum is treated chemically and reduced to aluminum oxide, then it is dissolved in fused cryolite. The mixture thus obtained is electrolyzed, in a furnace, lined with carbon and aluminum metal, and gets deposited at the cathode. The operating temperature of the furnace is about 1,000°C and the voltage required is about 8 V and the current of about 45,000 A.

Electro refining

Refining is a process by which the purity of the metal extracted from their ores can be improved (Table 11.1). Usually, the pure metal obtained from the metal extraction is nearly 90–95% only. The further improvement of the purity of such metal up to 99.5% can be done by electrolysis in refining. In the electrorefining, it is possible to get a metal of almost 100% purity at the cathode.

11.5.3 Electrodeposition

Electrodeposition is the process by which one metal is deposited over other metal or non-metal, by electrolysis.

Usually, electrodeposition is used for the decorative, protective, and functional purposes and it includes electroplating, electroforming, electrotyping, electrofacing, electrometallization, etc.

Factors affecting the quality of electrodeposition

The quality of the deposition of the metal is governed by the following factors.

Nature of electrolyte

The smoothness of the deposited metal over any other metal depends upon the nature of the electrolyte. For example, cyanide solution is one of the electrolytes that provide smooth deposit over any metal.

Current density

The deposited metal is very much strong and porous. If the rate of current is high, then the nuclei are formed and at low current density, the deposits are coarse and crystalline.

TABLE 11.1 *Refining of metals*

S. No	*Metal*	*Solution to be treated with metal*	*Energy consumed (kW-hr/ton)*
1	Nickel	Nickel chloride and sulfate	2,500–4,000
2	Iron	Iron ammonium sulfate	1,000–1,500
3	Silver	Nitric acid and silver nitrate	350–450
4	Gold	Chloride of gold	250–350
5	Copper	Copper sulfate	150–300

Temperature

The temperature of the electrolyte is different for different metals to have better deposit. At low temperatures, the electrolyte forms small crystals and at high temperatures, the electrolyte forms large crystals. In some cases, small variation of temperature will causes the reduction (50%) of the strength of metal deposited and on the other hand, high temperatures are beneficial due to increased conductivity, which also permits higher current densities and reduces the tendency to form trees.

Conductivity

The high conductivity of the electrolyte leads to the reduction of power consumption, and also prevents the formation of rough deposits.

Electrolytic concentration

If the concentration of the electrolyte is more, the higher will be the current density, which leads to the deposition of thin and uniform metal.

Addition of agents

Addition of agents such as glue, gum, and dextrin to the electrolyte influences the nature of the deposited metal. These additional agents are absorbed by the crystal nuclei that prevent the large growth and thus deposition will be fine-grained.

Throwing power

It is the ability of an electrolyte to produce uniform deposit even on irregular surface. Figure 11.2 shows uniform anode and cathode having irregular shape PQ and RS. The distance between the anode and the irregular surfaces of cathode PQ and RS is different. Therefore, the resistance to current path from the anode to PQ is more than that of RS. Hence, the thickness of the deposit on the surfaces PQ and RS are unequal.

Throwing power can be improved by the following two ways:

(i) By increasing the distance between the anode and the cathode.
(ii) By reducing the voltage drop at the cathode surface.

11.5.4 Electroplating

Electroplating is defined as the deposition of a metal over any metallic or non-metallic surfaces. Electroplating is usually employed to protect the metals from corrosion by

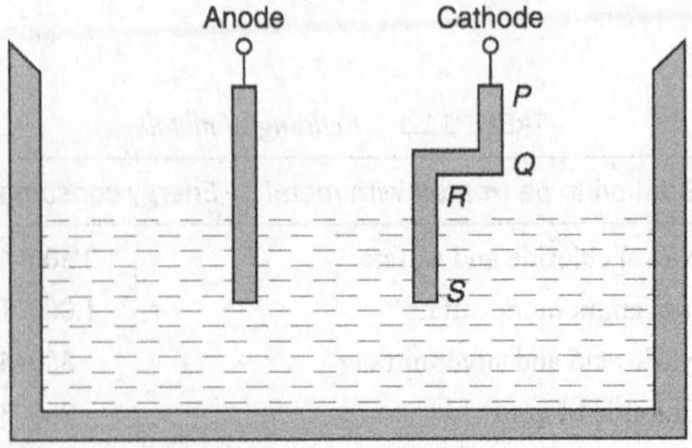

FIG. 11.2 *Description of throwing power*

atmospheric air, moisture, and CO_2, to give the reflecting properties to reflectors, to replace worn out metals, to give a shiny appearance to articles, etc.

Preparation for plating: Electroplating involves two functions. They are:

(i) cleaning operation and
(ii) plating operation.

Cleaning operation

In case if a metal is to be electroplated, it should be cleaned, i.e., metal should be polished, degreased oil, and any organic material, rust, scale, oxides, etc. is to be removed from the metal.

Plating operation

In plating process, the metal or article to be electroplated is arranged as the cathode and the anode is made up of the material that is to be deposited on the metal. And, salt is taken as solution in which the electrodes are immersed. The characteristic features of the plating of various metals are given below.

Copper plating

Copper plating baths used for the preparation of plating are of two types.

Acid bath

It is the bath in which solution is taken in a mixture of copper sulfate (15–200 gm) and H_2SO_4 (25–37 gm) per 1,000 cc of solution.

Current density maintained for copper plating is 200–400 A/m^2 and temperature is maintained at 25–50°. In this plating, the deposit obtained is thick and rough, so that polishing is required.

Cyanide bath

It consists of a solution with a mixture of 25 gm of copper cyanide, 25 gm of sodium cyanide, 5 gm of sodium carbonate, and 6 gm of sodium biphosphate per 100 cc of solution. The current density employed for this bath is 4–150 A/m^2 and the temperature is maintained at 35–50°. In both the methods, the anode is made up of copper.

If this type of bath is employed, the deposit obtained is so thin and smooth. But in both of the above baths, pure copper will be deposited at the anode.

The copper plating is usually employed to prevent the iron articles from rusting and the inner line coating for silver and nickel plating.

Nickel plating

In this plating, nickel bath is employed for steel and brass articles. This bath consists of solution; it is a mixture of 100 gm of nickel sulfate, 12 gm of ammonium chloride, and 12 gm of boric acid per 100 cc of solution. The temperature is maintained at 20–30° and the current density of 10–20 A/m^2 is employed.

In this plating, the anode is made up of nickel. For copper, zinc, and nickel platings, bath consists of solution with a mixture of solution nickel sulfate 150–240 gm, nickel chloride 36 gm, and boric acid 24 gm for 1,000 cc. Bath temperature is maintained at 40–65°. With a current density of 250–500 A/m^2.

In the above processes, pure nickel will be deposited at the anode.

Chromium plating

In this plating, bath consists of solution with a mixture of 180–300 gm of chromic acid and 2–3 gm of sulfuric acid per 1,000 cc. The working temperature is maintained at 40–70°C with the current density of 600–5,000 A/m² is employed. In this plating, the vats are made up of steel that is coated with lead chromium. The plating produces highly polished and extremely hard coating and it is proffered for the surface where it is to be protected from atmospheric condition.

11.5.5 Electrometallization

It is the process by which the metal can be deposited on a conduction base for decoration and for protective purposes. Any non-conductive base is made as conductive by depositing graphite layer over it.

11.5.6 Electropolishing

Electropolishing is mainly done for making the work as anode in a suitable position. This process makes the surface smoother.

11.5.7 Electrotyping

It is used to reproduce printing, set-up, engraving and metals, etc.

11.5.8 Electroparting or electrostripping

The process of separation of two or more metals electrolytically is known as electroparting or stripping. Usually, to stripe off copper from steel, the cathode is made up of iron, the work piece itself acts as anode and are immersed in a solution with a mixture of 75 gm of sodium cyanide and 25 gm of caustic soda in 1,000 cc of water. Here, during the electrolysis process, copper will be separated from the anode.

11.5.9 Anodizing

The process of deposition of oxide film on a metal surface is known as anodizing and oxidation. The formation of oxide film on a metal surface can be carried out in three steps.

(i) Initially, the metal surface on which oxide layer is to be deposited should be cleaned by various processes.

(ii) Oxide film is deposited on the cleaned metal surface by electrolysis.

(iii) Providing stability of oxide film and its desired color affect.

Oxide film on the metal surface can be made with different color affects. The desired color affects can be obtained by mixing proper acidic solutions. Sulfuric acid, chromic acid, and oxalic acids are used for the anodizing of various aluminum and aluminum alloy products.

The anodizing of the metal surface is required:

- For the attractive appearance and the shining of surface.
- To get the various desired color affects of metal base.
- To provide the protective coating on surface and in order to get smooth and bright surface prior to plating.

11.6 POWER SUPPLY FOR ELECTROLYTIC PROCESS

The power supply usually employed for electrolytic process is DC at very low voltage. Normally, the power supply for the electrodeposition process is of low voltage (10–12 V) and high current (100–200 A). This power supply can be obtained from a motor-generator set. These sets consist of an induction motor and a low voltage–high current DC generator. This method is not economical and less efficient. The same can be obtained from copper oxide rectifier, which has high operating efficiency and less space.

Solid state rectifiers are also used to get the low voltage–high current supply due to high rating of thyristor. This method is usually employed because of high efficiency, fast and reliable control, low maintenance cost, and occupies less space compared to the metal oxide rectifier.

The large amount of power supply is required for the extraction and refining of metals, the manufacturing of chemicals, the same can be obtained from the plants which are located nearer to the hydro-electric power stations.

Example 11.1: A plate of 2.35 cm^2 during electrolysis if a current of 1 A is passed for 90 min (the density of copper is 8.9 gm/cc and the ECE of copper is 0.0003295 gm/coulomb). Find the thickness of copper deposited on the plates?

Solution:

The electrochemical equivalent (ECE) of copper, $z = 0.0003295$ gm/coulomb.

Current strength $I = 1$ A

The time for which the current is passed through the solution, $T = 90$ min.

$$= 90 \times 60 = 5{,}400 \text{ s.}$$

The weight of copper deposited $m = ZIT$

$$= 0.0003295\ 1\times 5{,}400$$

$$= 1.7793 \text{ gm.}$$

The density of copper $d = 8.9$ gm/cc.

$$\therefore \text{The volume of copper deposited} = \frac{\text{weight } (m)}{\text{density } (d)}$$

$$= \frac{1.773}{8.9}$$

$$= 0.1999 \text{ cc.}$$

As we know, the volume, $v =$ area $\times$ thickness.

$$\text{Thickness} = \frac{v}{\text{Area}} = \frac{0.1999}{2.35} = 0.0850 \text{ cm.}$$

Example 11.2: Calculate the ampere hours required to deposit a coating of silver 0.08-mm thick on a sphere of 6-cm radius. Assume the electrochemical equivalent of silver = 0.001118 and the density of silver to be 10.5.

Solution:

The surface area of sphere, $S = 4\pi r^2 = 4\pi 6^2 = 452.389$ cm².

The thickness of coating, $t = 0.08$ mm $= 0.008$ cm.

The mass of silver to be deposited, $m = s \times t \times$ density of metal

$$= 452.389 \times 0.008 \times 10.5$$

$$= 38\text{gm}$$

$$= 0.038 \text{ kg.}$$

The ECE of silver, $z = 0.001118$ gm/coulomb $= \dfrac{0.001118 \times 3{,}600}{1{,}000}$

$$= 0.0040248 \text{ kg/A-h.}$$

Ampere-hours required $\dfrac{m}{z} = \dfrac{0.038}{0.004024} = 9.44.$

Example 11.3: If 17.5 gm of nickel is deposited by 90-A current flowing for 9 min. How much copper would be deposited by 45-A current in 5 min? The atomic weight of nickel and copper are 58.6 and 63.18, respectively, and valency of both is 2.

Solution:

The mass of nickel $m = 17.5$ gm.

The time of current flow $t = 9$ min.

The atomic weights of nickel = 58.6.

The atomic weights of copper = 63.18.

The ECE of nickel $z = \dfrac{m}{I\,t} = \dfrac{17.5 \times 10^{-3}}{90 \times 9 \times 60} = 36 \times 10^{-8}$ kg/c.

The ECE of copper $z^1 = z \times \dfrac{\text{the chemical equivalent of copper}}{\text{the chemical equivalent of nickel}}$.

$$= 36 \times 10^{-8} \times \frac{63.18/2}{58.6/2}$$

$$= 38.813 \times 10^{-8} \text{ kg/c.}$$

The mass of copper deposited $= z^1 \times I^1 \times t^1$

$$= 38.813 \times 10^{-8} \times 45 \times 5 \times 60$$

$$= 0.005239 \text{ kg.}$$

Example 11.4: Find the thickness of copper deposited on a plate area of 0.0003 m² during electrolysis. If a current of one ampere is passed for 100 min. The density of copper is 8,900 kg/m³ and the ECE of copper is 32.95×10^{-8} kg/coulomb.

Solution:

The ECE of copper, $z = 32.95 \times 10^{-8}$ kg/coulomb

The current strength $I = 1$ A

The copper density, $D = 8{,}900$ kg/m^3.
The time for which current is passed, $t = 100 \times 60 = 6{,}000$ s.
The mass of copper deposited, $m = z\,I\,T$

$$= 32.92 \times 10^{-8} \times 1 \times 6{,}000$$
$$= 0.001977 \text{ kg.}$$

The volume of copper deposited $v = \dfrac{m}{D} = \dfrac{0.001977}{8,900} = 0.222\times10^{-6}\,\text{m}^3.$

The thickness of copper deposited $t = \dfrac{v}{A} = \dfrac{0.222\times10^{-6}}{0.0003}$

$$= 0.74 \times 10^{-3}\text{ m} = 0.74 \text{ mm.}$$

Example 11.5: In a copper-sulfate voltmeter, the copper cathodes is increased in weight by 60 gm in 2 h, when the current maintained was constant. Calculate the value of the current.

The atomic weight of copper = 63.5.
The atomic weight of hydrogen = 1.
The atomic weight of silver = 108.
The electrochemical equivalent of silver = 0.001118 gm.

Solution:

The mass of copper deposited = 60 gm.
The time of flow of current $t = 2\text{ h} = 2 \times 60 \times 60 = 7{,}200$ s.
The ECE of silver = 0.001118 gm = 111.8×10^{-8} kg/c.

The chemical equivalent of silver $= \dfrac{\text{atomic weight}}{\text{valency}} = \dfrac{108}{1} = 108.$

The chemical equivalent of copper $= \dfrac{63.5}{2} = 31.75.$

The ECE of copper Z = ECE of sliver $\times \dfrac{\text{chemical equivalent of copper}}{\text{chemical equivalent of silver}}$

$$= 111.8\times10^{-8} \times \frac{31.75}{108} = 32.867\times10^{-8}\,\text{kgc}^{-1}.$$

The strength of current, $I = \dfrac{m}{Z\times t} = \dfrac{60\times10^{-3}}{32.867\times10^{-8}\times7,200} = 25.35A.$

Example 11.6: A copper-refining plant using 600 electrolytic cells carries a current of 6,000 A, voltage per cell being 0.3 V. If the plant was to work 10 h/week, calculate the energy consumption per tones, assuming the ECE of copper as 0.3281 mg/coulomb of electricity.

Solution:

The ECE of copper $Z = 0.3281$ mg/c $= 32.81 \times 10^{-8}$ kg/c
$= 32.81 \times 10^{-8} \times 3{,}600$ kg/A-h
$= 0.00118116$ kg/A-h.

The total number of ampere – hours per annum $= 600 \times 6{,}000 \times 40 \times 52$
$= 748.8 \times 10^7$ Ah.

The annual output of plant $= 0.00118116 \times 748.8 \times 10^7$
$= 8844.52$ ton.

The energy consumer per annum = A-hr per annum × voltage per-less
$= 748.8 \times 10^7 \times 0.3$
$= 224.64 \times 10^7$ W-h
$= 224.64 \times 10^4$ kw-hr.

$$\text{The energy consumption per tonne} = \frac{\text{annual consumption in kWh}}{\text{annual output in tons}}$$

$$= \frac{224.64 \times 10^4}{8{,}844.52}$$

$$= 253.98 \text{ kW-hr/ton}$$

Example 11.7: Determine the minimum voltage required for the electrolysis of water if one kg of hydrogen on oxidation to water liberation 13.3×10^7 J and the electrochemical equivalent of hydrogen is 1.0384×10^8 kg c^{-1}.

Solution:
During the electrolysis of water, the energy is required to decompose water into hydrogen and oxygen and this is equal to the energy expanded in the circuit in forcing the quantity of electricity through the electrolyte.

$$\text{The energy expanded during electrolysis} = \frac{1}{Z} \times V \text{ w-s/kg}$$

$$= \frac{V}{Z} \text{ J/kg.}$$

The energy liberated by 1 kg of hydrogen when it combines with oxygen $= 13.3 \times 10^7$.

$$\therefore \frac{V}{Z} = 13.3 \times 10^7.$$

But the electrochemical equivalent of hydrogen $Z = 1.0384 \times 10^8$ kg c^{-1}.

or, $V = 13.3 \times 10^7 \times 1.0384 \times 10^{-8}$

$= 1.381$ V.

Example 11.8: How much aluminum will be produced from aluminum oxide in 24 h if the average current is 3,000 A and the current efficiency in 90%. Aluminum in trivalent and atomic weight is 27. The chemical equivalent weight of silver is 107.98 and 0.00111 gm of silver is deposited by one coulomb.

Solution:
The ECE of silver = 0.00111 gm/c (or) 111×10^{-8} kg c^{-1}.
The chemical equivalent weight of silver = 107.98.

$$\text{The chemical equivalent weight of aluminum} = \frac{27}{3} = 9.$$

, $\therefore$ The ECE of aluminum $Z = \dfrac{\text{ECE of silver} \times \text{chemical equivalentwieght of aluminum}}{\text{chemical equivalent weight of silver}}$.

$$= \frac{111 \times 10^{-8} \times 9}{107.98} = 9.252 \times 10^{-8} \text{kg c}^{-1}.$$

Current efficiency = 90.1 = 0.9.
Average current I = 3,000 A.
The duration of flow of current $t = 24 \times 60 \times 60 = 86{,}400$ s.
The mass of aluminum produced $m = Z\,It \times$ current efficiency

$$= 9.252 \times 10^{-8} \times 3{,}000 \times 86{,}400 \times 0.9$$

$$= 21.583 \text{ kg}.$$

Example 11.9: A 18-cm long portion of a circular shaft having 8 cm diameter is to be water with a layer of 1.5-mm nickel. Determine the quantity of electricity in Ah and the time taken to the process. Assume a current density of 195 A/m² and a current efficiency of 90%. The specific gravity of nickel is 8.9 and its ECE is 1.0954 kg/1,000 Ah.

Solution:
The surface area of the circular shaft portion to be coats:

$$S = \pi dl = \pi \times 0.08 \times 0.18 = 0.0452 \text{ m}^2.$$

The mass of nickel to be deposits = surface area × thickness of coat × density of metal

$$= 0.0452 \times 1.5 \times 10^{-3} \times 8.9 \times 10^3$$

$$= 0.6034 \text{ kg}.$$

The theoretical value of the quantity of electricity required:

$$A = \frac{m}{z} = \frac{0.60.4}{1.0954/1{,}000} = 550.84 \text{ Ah}.$$

The actual value of the quantity of electricity required:

$$Q^1 = \frac{Q}{\text{current efficienty}} = \frac{550.84}{0.9} = 612.044 \text{ Ah}.$$

Since the current density used is 195 A/m² and the surface area of the shunt is 0.0452 m² therefore current used is:

$$I = 195 \times 0.0452 = 8.814 \text{ A}.$$

$$\text{Time required } t = \frac{Q^1}{I} = \frac{612.044}{8.814} = 69.43 \text{ h}.$$

Example 11.10: It is required to repair a wornout circular shaft 12 m in diameter and 30 m long by coating it with a layer of 1.5 mm of nickel. Determine the theoretical value of the quantity of electricity required and the time taken if the current density used in 210 A/m². The electrochemical equivalent of nickel is 30.4 × 10⁻⁸ kg/c of electricity and the density of nickel is 8.9 × 10³ kg/m³.

Solution:

$d = 12$ m = 0.12 m.
$L = 30$ m = 0.3 m.
The thickness of coating = 1.5 mm = 0.0015 m.
The current density $D = 210$ A/m^2.
The chemical equivalent of nickel $Z = 30.4 \times 10^{-8}$ kg/c.
The density of nickel is = 8.9×10^3 kg/m^3.

The quantity of electricity required Q:

The surface area of the shaft to be repaired $A_s = \pi d \times l = \pi \times 0.12 \times 0.3 = 0.113$ m^2.

The mass of nickel to be deposited is:

$$m = \text{the surface area} \times \text{the thickness of coating} \times \text{the density of nickel}$$
$$= 0.113 \times 0.0015 \times 8.9 \times 10^3$$
$$= 1.508 \text{ kg}.$$

The theoretical value of the quantity of electricity required $Q = \frac{m}{z} = \frac{1.508}{30.4 \times 10^{-8}} A-s$

$$= \frac{1.508}{30.4 \times 10^{-8} \times 3{,}600} \text{A-hr}$$

$$= 1{,}377.92 \text{ A-hr}.$$

Time taken t:

$$\text{Current density, } D = \frac{\text{current } (I)}{\text{surface area } (A_s)}$$

$$210 = \frac{I}{0.113}$$

$$\therefore I = 23.73 \text{ A}.$$

And, $Q = It$

$$t = \frac{Q}{I} = \frac{1{,}377.92}{23.73} = 58.06 \text{ h}.$$

Example 11.11: If a current of 9 A deposits 12.5 gm of silver from a silver nitrate solution in 20 min. Calculate the electrochemical equivalent of silver. Find the quantity of electricity and the steady current required to deposit 8 gm, and the quantity of copper from copper sulfate solution in 1 h. The electrochemical equivalent of copper is 0.3294 mgm/c.

Solution:

The strength of current $I = 9$ A.
The time of flow of current $T = 20 \times 60 = 1{,}200$ s.
The weight of silver deposited $m = 12.5$ gm.

$$\text{The now ECE of silver } Z = \frac{m}{It} = \frac{12.5}{9 \times 1{,}200} = 0.001157 \text{ gm/c}.$$

Example 11.12: Find the quantity of electricity and the steady current required to deposit 8 gm of copper from the copper sulfate solution in 1 hr. The electrochemical equivalent of copper is 0.3294 m gm/c.

Solution:

The weight of copper to be deposited $m = 8$ gm.
The ECE of copper $Z = 0.3294 \times 10^{-3}$ gm/c.

$$Q = It = \frac{m}{Z} \quad [\because m = ZIt]$$

$$= \frac{8}{0.3294 \times 10^{-3}} = 24,286.58 \text{ coulomb.}$$

$$\text{Steady current required } I = \frac{Q}{\text{time of deposite in sec}}$$

$$= \frac{24286.58}{1 \times 60 \times 60}$$

$$= 6.746 \text{ A.}$$

Example 11.13: A rectangular metal plate having $6 \times 5 \times 2$ cm as its dimensions is to be electroplates with nickel. How long will it take to deposit a layer of 0.1-mm thickness, when a current of 4.5 A flows through the circuit? The ECE of nickel = 0.000304 gm/c, and the density of nickel = 8.6 gm/cc.

Solution:

The volume of the metal plate before electroplating, $v = 6 \times 5 \times 2 = 60$ cc.
The final volume of the metal plate after electroplating $= 6.02 \times 5.02 \times 2.02 = 61.045$ cc.
$\therefore$ The volume of nickel deposited $= 61.045 - 60$

$= 1.045$ cc.

But, as we know:

Mass = volume × density.

So, the weight of the mass of the nickel deposited, m = $1.045 \times 8.6 = 8.987$ gm.
According to the law, if I be the current in the circuit flowing for time t seconds, then:

$$m = ZIt$$

$$8.987 = 0.000304 \times 4.5 \times t$$

$$t = \frac{8.987}{0.000304 \times 4.5 \times 60 \times 60}$$

$$= 1 \text{ h, } 49 \text{ min, } 29 \text{ s.}$$

Example 11.14: A rectangular plate of 15×10 cm is to be coated with nickel with a layer of 0.2-mm thickness. Determine the quantity of electricity in ampere hour and the time taken for the process. Current density is 190 A/m^2 and current efficiency is 92% and the specific gravity of nickel is 8.9.

Solution:

The area of the plate = 15 × 10 = 150 cm².
The volume of material to be plated v = 150 × 0.02 = 3 cc.
The weight of material to be deposited = 3 × 8.9.
= 26.7 gm.

Take Z for nickel = 0.0003043 gm/c.

$$It = \frac{m}{Z\eta}(Qm = ZIt \times \eta)$$

$$= \frac{26.7}{0.0003043 \times 0.92} \text{ A sec}$$

$$= \frac{26.7}{0.0003043 \times 0.92 \times 3{,}600} \text{ Ah}$$

$$= 26.492 \text{ A h.}$$

Current density = 190 A/m² (given)
∴ current I = current density × area

$$= 190 \times \frac{150}{10^4} = 2.85 \text{ A.}$$

$$\text{Time taken} = \frac{\text{A h}}{\text{A}} = \frac{26.492}{2.85} = \text{9h and 29 min.}$$

Example 11.15: If 95,600 coulombs of electricity liberates 1 gm equivalent of any substance. How long it will take for a current of 0.15 A to deposit 20 mg of copper from a solution of copper sulfate? The chemical equivalent of copper is to be taken as 32.

Solution:

Current through deposit = 0.15 A.
Chemical equivalent of copper = 32.

$$m = ZQ.$$

And, 32 g of copper is liberates by 95,600 coulomb of electricity:

$$\therefore Z = \frac{m}{Q} = \frac{32}{95{,}600} = 3.347 \times 10^{-4} \text{ gm/coulomb.}$$

∴ The amount of electricity Q in required to deposit 20 m gm of copper is:

$$Q_1 = \frac{m_1}{Z} = \frac{20 \times 10^{-3}}{3.347 \times 10^{-4}} = 59.755 \text{ coulombs.}$$

∴ The time required to deposit 20 mg when the current I = 0.15A is:

$$t = \frac{Q_1}{I} = \frac{59.755}{0.15} = 398.36\,\text{s}.$$

Example 11.16: A coating of nickel 1-mm thick is to be built on a cylinder 23 cm in diameter and 32 cm in length in 2 h. Calculate the electrical energy used in the process if the voltage is 10 V, the ECE of nickel is 0.000304, and the specific gravity of nickel is 8.9.

Solution:

The total surface area of the cylinder $= \pi\, dl = \pi \times 0.23 \times 0.32 = 0.2312\ \text{m}^2$.

So, the volume of nickel deposits = the surface area of cylinder × thickness of Ni

$$= 0.2312 \times 0.001$$

$$= 2.312 \times 10^{-4}\ \text{m}^3.$$

So, the mass of Ni deposited $= 2.312 \times 10^{-4} = 2.057$ gm.

Now, the mass deposited, m = ZIt.

Given:

The chemical equivalent of nickel, $Z = 0.000304$.

The time of current flow, $t = 2 \times 60 \times 60 = 7{,}200$ s.

$$\therefore m = ZIt$$

$$2.057 = 0.000304 \times I \times 7{,}200$$

$$I = 0.939\ \text{A}.$$

$\therefore$ The energy consumed in 2 hr $= VIt$

$$= 10 \times 0.939 \times 7{,}200$$

$$= 67{,}608\ \text{w-s}$$

$$= \frac{67{,}608}{1{,}000 \times 60 \times 60} = 0.01878\ \text{kg}.$$

KEY NOTES

- Electrolysis means producing chemical changes in an electrolyte by passing current through it.
- The principle of electrolysis is, whenever the DC electric current is made to pass through the solution of salt, some metals can be separated from it.
- The laws governing the electrolytic process were proposed by Michael Faraday.
- Faraday's first law states that 'The mass of substance deposited from an electrolyte is proportional to the quantity of electricity passing through the electrolyte in a given time'.
- Faraday's second law states that 'When the same quantity of electric current is passed through different electrolytes, the masses of the substances deposited are proportional to their respective chemical equivalents or equivalent weights'.
- Electroextraction is the process by which metals can be extracted or separated from their ores.
- The equivalent weight of a substance is defined as the ratio of formula weight to its valency.

SHORT QUESTIONS AND ANSWERS

(1) What is meant by electrolysis?

Electrolysis is nothing but the process by which electrical energy produces chemical changes.

(2) What are the uses of electrolysis?

This process can be normally used for the extraction of pure metal from their ores, the refining of metals, the building up of worn parts in metallurgical, chemical, and in other industries.

(3) Give the principle of electrolysis.

The basic principle of electrolysis is, whenever DC electric current is made to pass through the solution of salt, some metals can be separated from them. These separated metals can be coated on any object to form a pure thin layer.

(4) What are the laws of electrolysis?

Faraday's first law.

Faraday's second law

(5) State Faraday's first law of electrolytic process.

This law states that 'the mass of substance deposited from an electrolyte is proportional to the quantity of electricity passing through the electrolyte in a given time'.

(6) State Faraday's second law of electrolysis.

This law states that 'when the same quantity of electric current is passed through different electrolytes, the masses of the substances deposited are proportional to their respective chemical equivalents or equivalent weights'.

(7) List out various applications of electrolytic process.

The manufacturers of chemicals.

The production of caustic soda.

Electrometallurgy.

Electro refining.

Electro deposition.

(8) Define current efficiency.

$$\text{Current efficiency} = \frac{\text{the actual quantity of substance liberated or deposited}}{\text{the theoritical quantity of substance liberated or deposited}}$$

(9) Define energy efficiency.

$$\text{Energy efficiency} = \frac{\text{the theoritical energy (voltage) required}}{\text{the actual energy (voltage) required}}$$

(10) What is meant by electroextraction?

Electroextraction is the process by which metals can be extracted or separated from their ores.

(11) What is meant by electrodeposition?

Electrodeposition is the process by which one metal is deposited over other metal or non-metal by electrolysis.

(12) What is the use of electroplating?

Electroplating is usually employed to protect the metals from corrosion by atmospheric air, moisture, and CO_2.

(13) What is meant by anodizing?

The process of the deposition of oxide film on a metal surface is known as anodizing.

MULTIPLE-CHOICE QUESTIONS

(1) The deposition due to the flow of current through an electrolyte is directly proportional to the:

(a) The magnitude of steady current flow.

(b) The duration of current flow.

(c) The equivalent weight of the substance.

(d) All of the above.

(2) The voltage required to pass the necessary current through an electrolytic cell is of the order of:

(a) 1–2 V.

(b) 10–20 V.

(c) 100–200 V.

(d) 150–200 V.

(3) The energy consumption for the production of ammonium sulfate is of the order of:

(a) 2,000–25,000 kW-hr/ton.

(b) 3,000–4,000 kW-hr/ton.

(c) 4,000–7,000 kW-hr/ton.

(d) 70–80 kW-hr/ton.

(4) The materials used for copper plating are:

(a) Copper sulfate and sulfuric acid.

(b) Copper sulfate and nitric acid.

(c) Copper carbonate and ammonium carbonate.

(d) Copper nitrate and sulfuric acid.

(5) The materials used for chromium plating are:

(a) Chromium carbonate and sulfuric acid.

(b) Chromic acid and sulfuric acid.

(c) Chromium chloride and hydrochloride acid.

(d) None of the above.

(6) The gold plating is carried out:

(a) With a current density of 150–250 A/m^2 at a voltage of 1–2 V.

(b) With a current density of 50–150 A/m^2 at a voltage of 5–15 V.

(c) With a current density of 100–150 A/m^2 at a voltage of 1–4 V.

(d) With a current density of 50–150 A/m^2 at a voltage of 1 V.

(7) The power supply required for the electrolytic processes is:

(a) Alternating current (100–200 A) at very low voltage (10 or 12 V).

(b) Direct current (100–200 A) at very high voltage.

(c) Current (100–200 A) at very low voltage (10 or 12 V).

(d) Alternating current at very high voltage.

(8) The plants for the extraction and the direct refining of metals of large-scale manufacturing are located near the:

(a) Atomic power station.

(b) Hydro-electric power station.

(c) Steam power station.

(d) Either atomic power station or hydro-electric power station.

(9) The process of the coating of a metallic surface with a harder metal by electrodeposition is called:

(a) Electrofacing.

(b) Electroforming.

(c) Electrometallization.

(d) Either (a) or (b).

(10) Basically electroplating means:

(a) The formation of ions by two metallic plates in the acidic liquid.

(b) The electrodeposition of metal on electrodes.

(c) The electrodeposition of metal upon metallic surfaces.

(d) None of the above.

(11) Electroplating is done for:

(a) The replacement of wornout material.

(b) The protection of metals against corrosion.

(c) Giving a shining appearance to articles.

(d) All of the above.

(12) The preparation of an object for electroplating involves the:

(a) Removal of oil, grease, or other organic material.

(b) Removal of rust, scale, oxides, or other inorganic coatings adhering to the metal.

(c) Mechanical preparation of the metal surface by polishing, buffing, etc.

(d) Any or all of the above operations.

(13) The six-phase rectifier circuit meant for electroplating needs:

(a) Special AC generator.

(b) Normal three-phase mains.

(c) The system as (a) and (b) is just a theoretical possibility.

(d) None.

(14) In the process of electroplating, the circuitry involved is:

(a) Polarized.

(b) Non-polarized.

(c) Depends upon nature of plating.

(d) None out of above.

(15) The existence of a counter electrode is observed somewhere is the:

(a) Plating vats.

(b) Electrochemical cleaning baths.

(c) DC supply sources.

(d) Nothing as above is connected with the plating system.

(16) The capacitor bank installed in the rectifier system of any electroplating plant is meant for:

(a) Smoothing the effects of loads variation.

(b) Minimizing the ripple content of the DC supply.

(c) To improve the power factor and the line regulation of the mains feeding the rectifier system.

(d) All the above

(17) The object undergoing surface plating works as:

(a) Cathode.

(b) Anode.

(c) Depends upon the nature of supply source.

(d) None.

(18) The compound generator sets used for the purpose are:

(a) Differentially excited.

(b) Cumulatively excited.

(c) Depends upon plating load.

(d) None.

(19) The preferred vat polarity is:

(a) Positive.

(b) Negative.

(c) Zero potential without any polarity.

(d) An arbitrary choice.

(20) The current efficiency in some electrolytic process is the ratio of:

(a) The actual current density to the calculated current density for a given mass of coating.

(b) The mass of metal actually liberated to the calculated mass for a given current density.

(c) The actual current density to mass of metal actually liberated.

(d) None out of above.

(21) Mopping is an other name of:

(a) Grinding.

(b) Polishing.

(c) Abrasion.

(d) None out of above.

(22) The filtration of electrolyte is necessary:

(a) To remove the impurities going into the electrolyte along with the main salt.

(b) To remove the suspended salt particles and the other suspended impurities from the electrolyte.

(c) To make the agitation process more effective.

(d) Only some of the plating salts need filtration.

(23) The process as above is:

(a) Continuous.

(b) Intermittent.

(c) Done only once before plating commencement.

(d) Varies from electrolyte to electrolyte.

(24) The spongy coating of electroplating speaks of:

(a) Under current density.

(b) Over current density.

(c) Excessive electrolyte density.

(d) Poorer electrolyte density.

(25) A process known as hall bloating is done:

(a) Prior to subjecting a surface to electroplating.

(b) After plating a surface.

(c) Done in between primary and secondary plating layers.

(26) Ripple factor is being minimized:

(a) By incorporating the filter circuits along with the rectifier plant.

(b) By using the single-phase bridge rectifiers.

(c) With the help of multiphase rectifier unit without using additional filter network.

(d) Power capacitors are doing the needful.

(27) The metal being deposited is available in form of:

(a) Constituent of electrolyte.

(b) One of the electrodes.

(c) Both as above.

(d) None of the above.

(28) The shunt fields in such arrangement are:

(a) Connected in parallel to each other.

(b) Connected in antiparallel to each other.

(c) Connected in series across the outers.

(d) Field of generator (1) excited by armature (2) output and vice versa.

(29) Chrome plating is done as:

(a) Primary layer.

(b) Secondary layer.

(c) Tertiary layer.

(d) None of the above.

(30) Polarization on cathode surface can be checked through:

(a) Limiting current magnitude.

(b) The agitation of electrolyte.

(c) Periodical reverses plating.

(d) All of the above.

(31) The process by which electrical energy produces chemical changes is known as:

(a) Electrolysis.

(b) Electrofacing.

(c) Anodizing.

(d) Electroplating.

(32) Which bond is responsible for the formation of inorganic compound?

(a) Ionic.

(b) Covalent.

(c) Electrovalent.

(d) Both a and b.

(33) Which law states that the mass of substance liberated from an electrolyte is proportional to the quantity of electricity passing through it?

(a) Lenz law.

(b) Faradays first law.

(c) Faradays second law.

(d) Faradays laws of electromagnetic induction.

(34) Current density lies in between:

(a) 70% and 85%.

(b) 80% and 92%.

(c) 90% and 96%.

(d) 90% and 98%.

(35) The process of depositing one metal over the other metal is known as:

(a) Electrodeposition.

(b) Electrometallization.

(c) Electrofacing.

(d) Anodizing.

(36) The process of depositing metal on a conducting base for decoration purpose is known as:

(a) Electrodeposition.

(b) Electrometallization.

(c) Electrofacing.

(d) Anodizing.

(37) The process of coating a metal surface with a harder metal by electrodeposition is known as:

(a) Electrodeposition.

(b) Electrometallization.

(c) Electrofacing.

(d) Anodizing.

(38) The process of providing an oxide film is known as:

(a) Electrodeposition.

(b) Electrometallization.

(c) Electrofacing.

(d) Anodizing.

(39) Which is a process by which the purity of metal extracted from their ores can be improved?

(a) Electrodeposition.

(b) Refining.

(c) Electroplating.

(d) Anodizing.

(40) By electrorefining, it is possible to get metal of ——— purity.

(a) 60%.

(b) 80%.

(c) 90%.

(d) 100%.

REVIEW QUESTIONS

(1) What is meant by electrolysis? Explain the principle of electrolysis.

(2) State and explain Faraday's laws of electrolysis.

(3) What are the various applications of electrolysis?

(4) What is meant by anodizing?

(5) What do you mean by electrodeposition?

(6) What are the various factors effecting the quality of electrodeposition?

EXERCISE PROBLEMS

(1) A plate of 2 cm^2 during electrolysis if a current of 0.5 ampere is passed for 60 min (the density of copper is 8.9 gm/cc and the ECE of copper is 0.0003295 gm/coulomb). Find the thickness of copper deposited on the plates.

(2) A copper refining plant using 400-electrolytic cells carries a current of 1,200 A and voltage per cell being 0.175 V. If the plant was to work 10 h/week, calculate the energy consumption per tons, assuming the ECE of copper as 0.3281 mg/coulomb of electricities.

(3) Determine the minimum voltage required for the electrolysis of water if one kg of hydrogen on oxidation to water liberation 13.3×10^7 J and the electroequivalent of hydrogen is 1.0384×10^8 kg c^{-1}.

(4) Find the thickness of copper deposited on a plate area of 0.0056 m^2 during electrolysis. If a current of one ampere is passed for 300 min. The density of copper is 8,900 kg/m^3 and the ECE of copper is 32.95×10^{-8} kg/coulomb.

(5) In a copper-sulfate voltmeter, the copper cathodes is increased in weight by 60 gm in 2 h, when the current maintained was constant. Calculate the value of this current.

The atomic weight of copper = 63.5.

The atomic weight of hydrogen = 1.

The atomic weight of silver = 108.

The electrochemical equivalent of silver = 0.001118 gm.

(6) Calculate the ampere hours required to deposit a coating of silver 0.048-mm thick on a sphere of 8-cm radius. Assume the electrochemical equivalent of silver = 0.001118 and the density of silver to be 12.

ANSWERS

1. d	11. d	21. b	31. a
2. a	12. d	22. b	32. c
3. a	13. b	23. a	33. b
4. a	14. a	24. b	34. d
5. b	15. c	25. a	35. a
6. b	16. c	26. c	36. b
7. c	17. a	27. c	37. c
8. d	18. b	28. d	38. d
9. a	19. a	29. c	39. b
10. c	20. b	30. d	40. d

Refrigeration and Air-conditioning

OBJECTIVES

After reading this chapter, you should be able to:

- know refrigeration concept
- understand the principle of refrigeration
- know the air-conditioning process

12.1 INTRODUCTION

Energy can be neither created nor destroyed; it can be transferred from one form of energy to another form. The electrical energy is efficiently utilized by converting it into desire form for getting suitable output. Refrigeration and air-conditioning is mainly concerning with mechanical engines but the electrical energy used for the running of machines and for the process of refrigeration and air-conditioning, so that, all electrical engineers must have the knowledge of working of the refrigeration and air-conditioning machines.

12.2 REFRIGERATION

It is the process of reducing the temperature of an object from the normal surrounding temperature in a controlled way in order to cool them. The extraction of heat from the object to be cooled can be achieved by evaporation of a liquid refrigerant. The process of vapor compression refrigeration consists of the absorption of heat by the refrigerant that changes the refrigerant from liquid to gas. This gas is compressed and pumped into condenser where the heat is absorbed by circulation air, thereby bringing the refrigerant back to liquid state.

12.3 REFRIGERATOR

Refrigerator is a machine that reduces the temperature of substances by storing the heat in it. This machine has refrigerant nothing but gas or fluid, which is circulated through different stages to extract heat from the substances. Refrigerator is thermally insulated from the surrounding atmosphere.

12.4 PRINCIPLE OF REFRIGERATOR

Refrigeration is the process of reducing the temperature of substance. A refrigerator works on the principle of refrigeration, i.e., cooling caused by evaporation. The working principle of refrigerator is explained below.

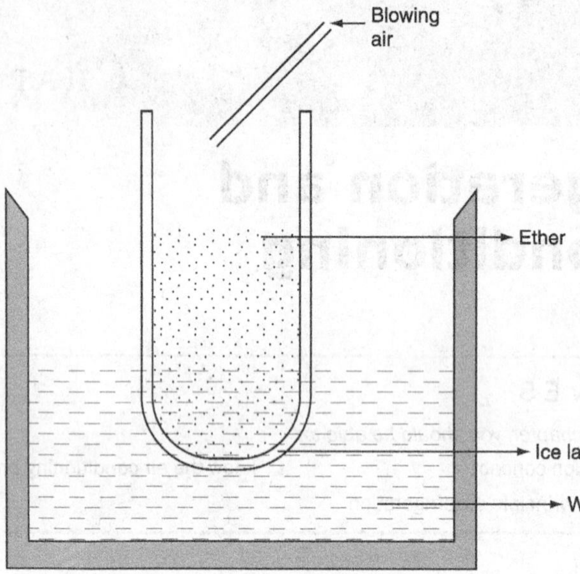

FIG. 12.1 *Principle of refrigerator*

A vessel is completely filled with water. Small amount of ether or spirit is taken in a test tube and dipped into the vessel, as shown in Fig. 12.1. Then, air is blown into the test tube through a small pipe; thus, ether or spirit present in the tube volatizes. Hence, it is observed the deposition of ice near the base of the test tube. The formation of ice at the base of the test tube is due to the evaporation of ether or spirit. When air is blown into the tube either gets evaporated by absorbing the temperature of water in the vessel that decreased the surrounding temperature, which leads to the deposition of ice.

12.5 REFRIGERANT

Refrigerant is nothing but the liquid that can be liquefied and vaporized to reduce the temperature of the substance to be cooled. CO_2 is used as refrigerant in ships, but while handling piping, problems arises with such refrigerant. Now, freon (CCl_2F_2 or $CHClF_2$) is used as refrigerant due to its special characteristics and some of the gases commonly used as refrigerants are ammonia (toxic) and SO_2 (again toxic). Due to the destructive nature of the chemical such as chloroflourocarbons and freons, these are extensively used in refrigerators and air-conditioners.

12.6 VAPOR COMPRESSION REFRIGERATION CYCLE

In this process of refrigeration, the liquid refrigerant gets evaporated by absorbing the temperature of surrounding medium to be cooled. Normally used gas refrigerants are carbon dioxide (CO_2), sulfur dioxide (SO_2), ammonia (NH_3), freon (F-12) [CCl_2F_2 or $CCHIF_2$], etc. Refrigeration or vapor compression cycle involves four stages such as compression, condensation, expansion, and evaporation as shown in Fig. 12.2. The purpose

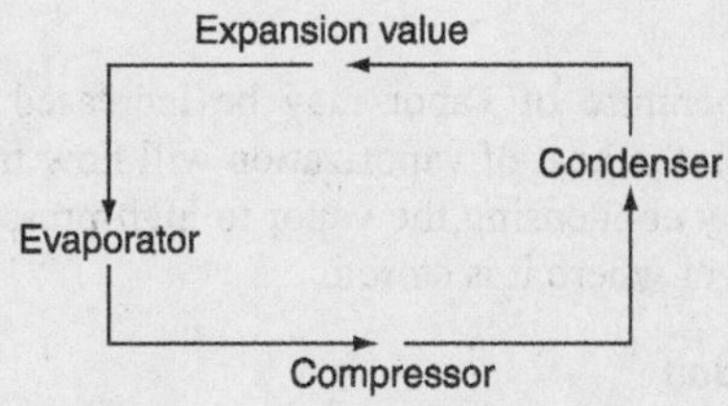

FIG. 12.2 *Vapor compression cycle*

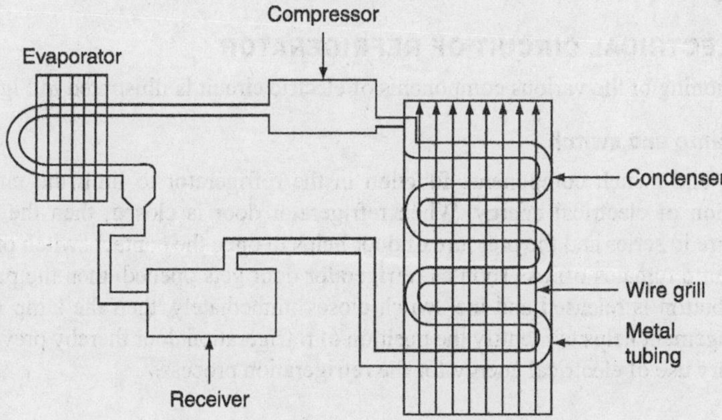

FIG. 12.3 *Vapor compression process*

of refrigeration cycle is to remove the heat absorbed by the refrigerant and bringing back into liquid form. Thus, the same refrigerant can be used for the extraction of heat from the substance to be cooled again.

The vapor compression system of various processes take place in the vapor compression systems and is illustrated in Fig. 12.3.

12.6.1 Evaporation

When refrigerant at low pressure and low temperature enters into the evaporate is in liquid states and is placed near or surrounding to the substance to be cooled. Thus, the refrigerant absorbs the heat from the substance transformed to a vapor state. This process of the transformation of liquid refrigeration in vapor is known as evaporation.

12.6.2 Compression

Vapors evolving out from the evaporator are then allowed to pass through the compressor through pipes. These vapors are compressed until the temperature of the vapor more than the condensing medium. Here, the compressor is driven by the mechanical energy supplied by an engine or electric motors.

12.6.3 Condensation

In compressor, the temperature of vapor may be increased above the condensating medium temperature, then the heat of vaporization will flow from the vapor to the condensating medium thereby condensing the vapor to high-pressurized liquid. This liquid is then flows to the receiver where it is stored.

12.6.4 Pressure reduction

The process of reducing the pressure of refrigerant is done by liquid control valve known as an expansion valve. This valve reduces the pressure of high-pressure liquid from the receiver to a low-pressure liquid capable of absorbing heat. This process is known as expansion.

12.7 ELECTRICAL CIRCUIT OF REFRIGERATOR

The functioning of the various components of electric circuit is illustrated in Fig. 12.4.

12.7.1 Lamp and switch

The lamp and switch components function in the refrigerator to limit the unnecessary consumption of electrical energy. When refrigerator door is closed, then the lamp and switches are in series and the pressure of door helps to open the contact switch of the lamp thus the lamp remains off. As soon as refrigerator door gets opened, then the pressure on the push button is released and the switch closes immediately, then the lamp will glow. This arrangement helps to identify the position of refrigeration door thereby preventing the unnecessary use of electrical energy for the refrigeration process.

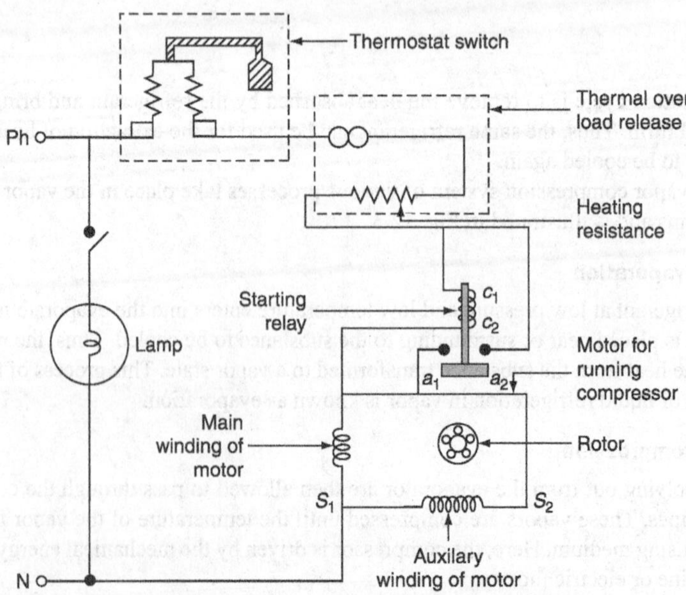

FIG. 12.4 *Electric circuit of refrigerator*

12.7.2 Thermostat switch

The main purpose of providing thermostat switch in the refrigerator is to maintain desired temperature in it. The function of the thermostat switch is given below.

When temperature of the domestic refrigerator is reached to 5–7°C, the thermostat switch automatically opens and cuts off the electric supply to the motor, thereby stopping the further cooling process. This switch gets opened for the duration as long as the above temperature is maintained.

When the temperature of the refrigerator increases from the normal value, the switch automatically closes and electrical energy is supplied to the motor temperature is maintained within the refrigerator.

12.7.3 Thermal overload release

The function of thermal overload release is to protect the motor compressor unit from excessive currents flowing through the motor winding when the temperature of compressor raises beyond the presetting value.

Here, the thermal overload release is nothing but a simple switch formed by a bimetallic switch in series with resistance. If excessive current is flowing through the motor winding due to overload, the resistance helps to heat the bimetallic strip and gets opened, thereby disconnecting the electric motor from the supply without any damage to it.

12.7.4 Starting relay

The function of starting relay provided in refrigerator is to start the split phase induction motor by connecting the auxiliary winding or starting winding across the main supply in addition to the main winding at the time of starting. This helps to make the split phase induction motor as self-induction motor is unable to start.

12.8 COMMON FAULTS IN REFRIGERATOR

Some of the faults usually occur on the refrigeration systems are discussed in the following sections.

12.8.1 Fault in the starting relay

If there is any fault in the starting relay, the motor is unable to start or it can start with humming sound.

In addition to the above, the motor starts with humming sound due to:

- The disconnection of starting winding at the time of starting.
- Low voltage applied across the refrigerator.

Starting relay function can be checked by measuring the resistance between the two ends of a relay coil. For a good working relay, the measured resistance should be zero. If there is any fault in the auxiliary winging, it must be replaced and the operating voltage of the refrigerator can be adjusted by incorporating a voltage stabilizer.

12.8.2 Fault in capacitor

In case of the refrigerator in which a single-phase split-phase induction motor is used to maintain a desired cooling in the system, it is not a self-starting one; to make it to self-start, a capacitor should be provided for phase splitting and to produce RMF. Thus, if there is any fault in the capacitor, the starting of the motor producing humming noise in such cases, it is necessary to replace the capacitor.

12.8.3 Fault in thermostat

Thermostat has to keep the motor circuit across the supply. If there is any change in cooling status of refrigerator, the fault is with thermostat, which is mainly due to the variation of resistance in series with the bimetallic strip due to environmental changes. Hence, the motor may or may not keep across the supply. So that, the thermostat contacts should be cleaned and the wiring must be redone.

12.9 APPLICATIONS OF REFRIGERATION

Some of the important applications of refrigeration are:

(i) The making of ice.

(ii) The air-conditioning of industries.

(iii) Metal manufacturing and their treatment.

(iv) The freezing purposes of the food products.

(v) The transportation of foods.

12.10 AIR-CONDITIONING

The process by which the temperature, humidity, purity, and circulation of air is controlled in an enclosed area is known as air-conditioning. The conditioning of air makes living conditions more comfortable and ensures the trouble-free operation of delicate equipment, industrial manufacturing, etc.

The various important factors that leads to efficient and complete air-conditioning are discussed in the following sections.

12.10.1 Temperature control

In an air-conditioning process, the temperature control of the working medium is required, i.e., the desired temperature should be maintained within the enclosed space even though the temperature of the surrounding may be either below or above the inside temperature. Sometimes, the atmospheric air must be fed into the chamber to raise or lower the inner temperature.

To raise the temperature of air in the enclosed space, the surrounding air must be heated and then fed into the chamber or to lower the temperature of the air in the space, the surrounding air must be cooled and then fed into the chamber.

Normally, the air is cooled in summer by the following methods:

- By circulating the cold water.
- By using the ice-activated systems.
- By using the water-evaporative systems using water.

And, the air is warmed up in winter by the following methods:

- By using electric heaters.
- Electromechanical heat pumps.
- By using heat exchangers using waste systems.

12.10.2 Humidity control

Humidity control is required to maintain sufficient water content of air for healthy and comfortable environment depending upon the seasonal effects of summer or winter. Humidity control helps to improve the quality of production and the cost of products such as drugs, automobiles, clay products, and electric goods. The control of humidity can be achieved by the process of humidification and dehumidification.

Humidification: The process of the addition of moisture or humidity is known as humidification.

Dehumidification: The process of the removal of moisture or humidity is known as dehumidification.

Humidification can be achieved by spraying steam, hot water, etc. Whereas dehumidification is accomplished by using the absorbent materials such as activated alumina, silica gel, and calcium chloride.

12.10.3 Air movement and circulation

Air movement and its circulation should be controlled for the equi-distribution of air throughout the enclosed space to be air-conditioned.

12.11 ELECTRIC CIRCUIT OF AN AIR-CONDITIONER

Air-conditioner is provided for the treatment of air in an enclosed space. The working principle of an air-conditioner is same as to that of a water cooler. An air-conditioner employs two fans; one fan is for the cooling of condenser and the other fan is blower, which circulates cooled air into the room.

For the small air-conditioners, two fans are arranged on both ends of the same shaft, but for the large air-conditioners, two fans are operated by two different motors.

The electric circuit for an air-conditioning system is shown in Fig. 12.5.

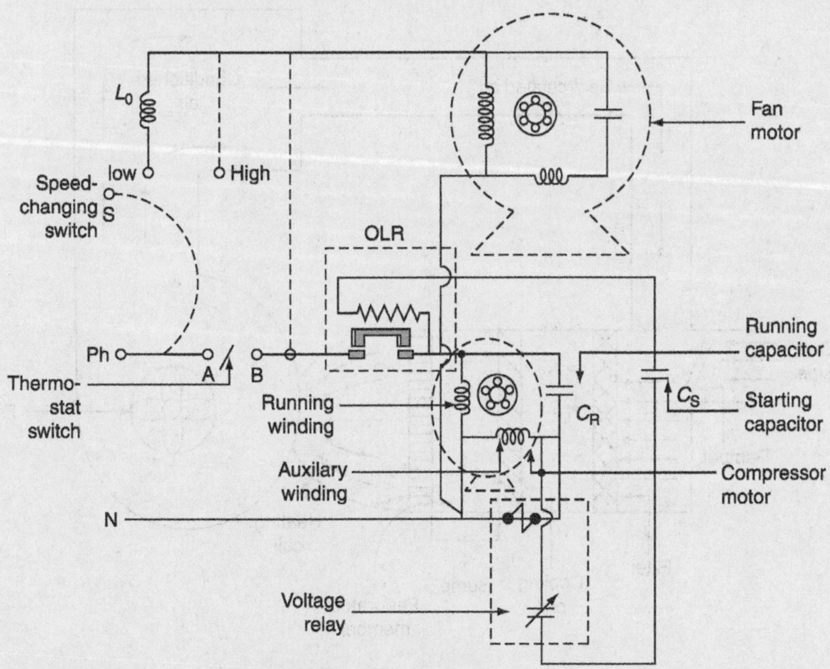

FIG. 12.5 *Electric circuit for air-conditioner*

The air-conditioning system that is employed with two fans is driven by two separate single-phase split-ring induction motors. When the thermostat switch is in open and the fan motor is directly connected across the supply through point 'A' and fan motor starts working. Here, the speed of the fan motor can be controlled by changing the speed changer switch position. The speed of the fan motor is reduced by closing the switch 'S' to low position by adding inductance 'Lo' in series with the motor circuit, and increased by closing the switch 'S' to high position without connecting any inductance in series with the circuit.

When thermostat switch closes, the compressor motor starts running with the help of both auxiliary and main windings. The compressor motor is a start capacitor that is run by a single-phase induction motor to develop 80–85% of synchronous speed, voltage relay automatically disconnects the auxiliary winding. Now the motor is able to run continuously with the running or the main winding alone. Overload release (OLR) function is to protect the compressor motor from overloads. In case if there is any overload occurring on the motor, overload release coil energize, which automatically disconnects the motor from the supply and causes the safe operation of compressor motor.

12.12 SUMMER AIR-CONDITIONING SYSTEM

Figure 12.6 shows the schematic arrangement of a summer air-conditioning system. It is one of the important applications of the air-conditioning system, in which the air is dehumidified.

In summer air-conditioning system, the outside atmospheric air is passed into the chamber through dampers and is mix up with the recirculated air from the conditioned space. The mixed air is passed through the filter to remove the dust, dirt, and other impurities. Now, the

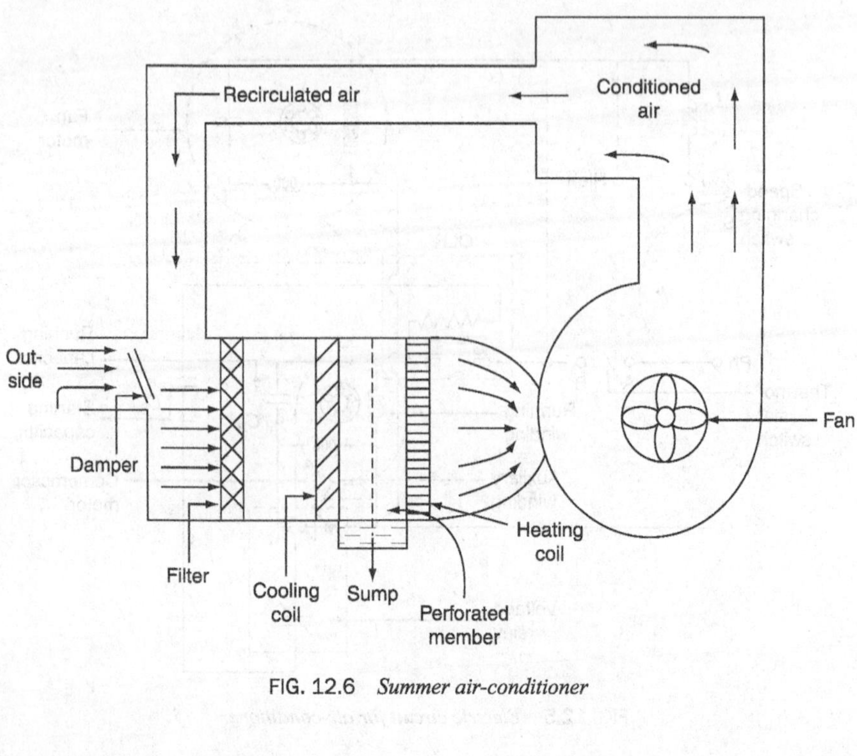

FIG. 12.6 *Summer air-conditioner*

pure air is allowed to pass through the cooling coil, which reduces the temperature of the air below and the temperature of the air in the conditioned area. This cooled air is passed over the perforated membrane to remove the moisture and is collected in the sump. After the removal of the moisture content, the air is then allowed to pass through heating coil to raise the temperature slightly equals to the temperature of the air in the conditioned space. Now, the fresh and pure air is passed into the conditioned space by a blowing fan to cool in. From the conditioned space, some of the air at the fan is exhausted to the atmosphere by the exhaust fans or ventilators. And the remaining part of the air is recirculated into the treatment chamber used again is known as recirculated air.

12.13 ROOM AIR-CONDITIONERS

Figure 12.7 shows the window-type room air-conditioning for conditioning the specified areas, i.e., hall, office room, residential room, etc. This system will control the temperature and humidity content of air automatically.

Window-type air-conditioner is subdivided into two parts; outdoor part and indoor part.

The outdoor part consists of completely sealed motor compressor unit, condenser, motor-driven evaporator, motor-driven fan, remote bulb, refrigerant control, control panel, an air filter, power connector, and a tray. There is a pipe line that connects the two trays of inner and outer parts.

A capillary tube through a refrigerant filter connects the outer condenser to inner part of evaporator. Evaporator is connected to a compressor motor through a suction pipeline. All parts of cooler, i.e., front and back of inner and outer portions, are fitted with the help of shutters.

In this system, the working substances are the vapors of ammonia, freon, carbon dioxide, etc., which can evaporate and condenses.

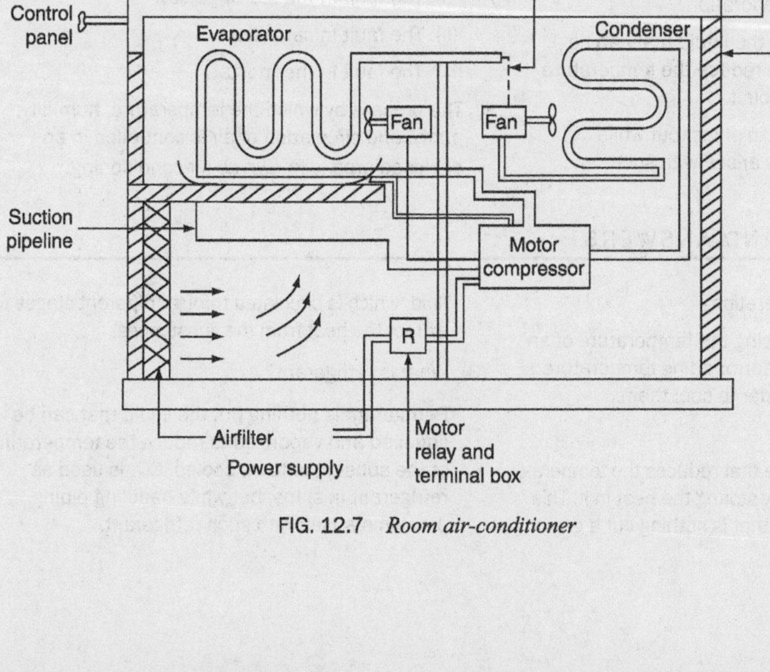

FIG. 12.7 *Room air-conditioner*

When low-pressure vapor from the evaporator is passed into the compressor unit through the suction pipe line, the compressor is able to deliver the vapor to the condenser at high pressure. The high-pressurized vapor is condensed and the heat is removed from the refrigerant vapor. Before the vapor flowing back into the evaporator, it is passed into the capillary control through filters, thus the cooled liquid refrigerant is circulating through the evaporator coils. Now, the air is drawn in the room of the condensed space through the motor-driven fan and then, the same is allowed pass over the evaporator coils. The cooled air from the evaporator coils is then supplied back into the room or condensed space. In this process, the dehumidification of the moisture from the air and the moisture from the evaporator coils is collected at the bottom of the evaporator into a pan.

This moisture in the indoor pan or the tray is pumped into the outdoor tray due to the gravity, which helps to cool the compressor and the condenser.

In outdoor part, the fan draws the atmospheric air and is circulated over the compressor and the condenser unit. The heat of the vapor refrigerant is absorbed by the air during its passage through the condenser and is cooled into the liquid form. And, the operation of the unit is automatically on and off by a control valve panel when the desired temperature is reached in the room.

KEY NOTES

- The process of reducing the temperature of an object from the normal surrounding temperature in a controlled way in order to cool them is known as refrigeration.
- Refrigerator is a machine that reduces the temperature of the substances by storing the heat in it.
- A refrigerator works on the principle of refrigeration, i.e., cooling caused by evaporation.
- Refrigerant is nothing but the liquid that can be liquefied and vaporized to reduce the temperature of the substance to be cooled.
- CO_2 is used as refrigerant in ships, but while handling piping, problems arises with such refrigerant. Now, freon (CCl_2F_2 or $CHClF_2$) is used as refrigerant due to its special characteristics.
- The commonly use as refrigerants are ammonia (toxic) and SO_2 (again toxic).
- The process of the transformation of liquid refrigeration in vapor is known as evaporation.
- The common faults in refrigerator are:
  (i) The fault in the starting relay.
  (ii) The fault in capacitor.
  (iii) The fault in thermostat.
- The process by which the temperature, humidity, purity, and circulation of air is controlled in an enclosed area is known as air-conditioning.

SHORT QUESTIONS AND ANSWERS

(1) What is meant by refrigeration?

It is the process of reducing the temperature of an object from the normal surrounding temperature in a controlled way in order to cool them.

(2) What is refrigerator?

Refrigerator is a machine that reduces the temperature of the substances by storing the heat in it. This machine has refrigerant that is nothing but a gas or fluid, which is circulated through different stages to extract the heat from the substances.

(3) What is refrigerant?

Refrigerant is nothing but the liquid that can be liquefied and vaporized to reduce the temperature of the substance to be cooled. CO_2 is used as refrigerant in ships, but while handling piping, problems arises with such refrigerant.

(4) Give some of the applications of refrigeration.

(i) The making of ice.

(ii) The Air-conditioning of industries.

(iii) The metal manufacturing and their treatment.

(5) What is meant by air-conditioning?

The process by which the temperature, humidity, purity, and circulation of air is controlled in an enclosed area is known as air-conditioning.

(6) What are the factors that lead to the efficient air-conditioning?

(i) Temperature control.

(ii) Humidity control.

(iii) Air movement and circulation.

(7) What are the cooling methods of air?

Normally, the air is cooled in summer by the following methods:

- By circulating the cold water.
- By using the ice-activated systems.
- By using the water-evaporative systems.

(8) What are the heating methods of air?

- By using electric heaters.
- By using electromechanical heat pumps.
- By using heat exchangers using waste systems.

MULTIPLE-CHOICE QUESTIONS

(1) The process of reducing the temperature of an object from the normal surrounding temperature in a controlled way in order to cool them is known as:

(a) Condensation.

(b) Refrigeration.

(c) Evaporation.

(d) Humidification.

(2) The extraction of heat from the object to be cooled can be achieved by ——— the liquid refrigerant.

(a) Cooling.

(b) Condensating.

(c) Evaporating.

(d) Humidification.

(3) Which is the liquid that can be liquefied and vaporized to reduce the temperature of substance to be cooled?

(a) Condensate.

(b) Refrigerant.

(c) Vapor.

(d) Moisture.

(4) The process of the transformation of liquid refrigeration in vapor is known as:

(a) Condensation.

(b) Refrigeration.

(c) Evaporation.

(d) Humidification.

(5) The process of the reducing pressure of refrigerant is done by liquid control valve is known as:

(a) Expansion valve.

(b) Open valve.

(c) Closed valve.

(d) Control valve.

(6) The art of measuring the moisture content is known as:

(a) Photometry.

(b) Psychometry.

(c) Optometry.

(d) None.

(7) Water coolers are used to produce cold water at about:

(a) 5–10%.

(b) 6–12%.

(c) 7–13%.

(d) 8–14%.

(8) The main characteristics of air refrigerant are that throughout the cycle the refrigerant remains in ——— state.

(a) Solid.

(b) Gaseous.

(c) Liquid.

(d) Vapor.

(9) The process by which the temperature, humidity, purity, and circulation of air is controlled in an enclosed area is known as:

(a) Condensation.

(b) Refrigeration.

(c) Evaporation.

(d) Air-conditioning.

(10) The process of the addition of moisture or humidity is known as:

(a) Condensation.

(b) Refrigeration.

(c) Evaporation.

(d) Humidification.

REVIEW QUESTIONS

(1) What is meant by refrigeration? Explain the working principle of the refrigerator.

(2) Explain in detailed the function of the refrigerator and also draw its electric circuit.

(3) What are the various faults commonly occur in a refrigerator? And also mention their remedies.

(4) Mention the various fields of applications of refrigeration.

(5) What is air-conditioning? Enumerate the factors that lead to efficient and complete air-conditioning.

(6) Draw and explain the electric circuit of air-conditioner.

(7) Write short notes on the following:

(i) Summer air-conditioner.

(ii) Room air-conditioners.

ANSWERS

1. b	4. d	7. c	10. d
2. c	5. a	8. b	
3. b	6. b	9. d	

Index